W0246192

The Reticuloendothelial System

A COMPREHENSIVE TREATISE

Volume 7A
Physiology

The Reticuloendothelial System
A COMPREHENSIVE TREATISE

General Editors:
Herman Friedman, *University of South Florida, Tampa, Florida*
Mario Escobar, *Medical College of Virginia, Richmond, Virginia*
and
Sherwood M. Reichard, *Medical College of Georgia, Augusta, Georgia*

MORPHOLOGY
Edited by Ian Carr and W. T. Daems

BIOCHEMISTRY AND METABOLISM
Edited by Anthony J. Sbarra and Robert R. Strauss

PHYLOGENY AND ONTOGENY
Edited by Nicholas Cohen and M. Michael Sigel

IMMUNOPATHOLOGY
Edited by Noel R. Rose and Benjamin V. Siegel

CANCER
Edited by Herman Friedman and Ronald B. Herberman

IMMUNOLOGY
Edited by Joseph A. Bellanti and Herbert B. Herscowitz

PHYSIOLOGY (In two parts)
Edited by Sherwood M. Reichard and James P. Filkins

PHARMACOLOGY
Edited by John Hadden, Jack R. Battisto, and Andor Szentivanyi

HYPERSENSITIVITY
Edited by S. Michael Phillips and Peter Abramoff

INFECTION
Edited by John P. Utz and Mario R. Escobar

The Reticuloendothelial System

A COMPREHENSIVE TREATISE

Volume 7A
Physiology

Edited by

SHERWOOD M. REICHARD
Medical College of Georgia
Augusta, Georgia

and

JAMES P. FILKINS
Loyola University Medical Center
Maywood, Illinois

PLENUM PRESS • NEW YORK AND LONDON

Library of Congress Cataloging in Publication Data

Main entry under title:

The Reticuloendothelial system.

Includes bibliographies and indexes.
CONTENTS: v. 1. Carr, I., Daems, W. T., and Lobo, A. Morphology.—v. 2. Biochemistry and metabolism—[etc.]—v. 7A Physiology.
1. Reticulo-endothelial system. 2. Macrophages. I. Friedman, Herman, 1931–
II. Escobar, Mario E. III. Reichard, Sherwood M. [DNLM: 1. Reticuloendothelial system. WH650 R437]
QP115.R47 591.2′95 79-25933

ISBN 978-1-4684-4576-3 ISBN 978-1-4684-4574-9 (eBook)
DOI 10.1007/978-1-4684-4574-9

Softcover reprint of the hardcover 1st edition 1984
A Division of Plenum Publishing Corporation
233 Spring Street, New York, N.Y. 10013

Contributors

DARRYL R. ABSOLOM • Immunochemistry Laboratory, Department of Microbiology, State University of New York, Buffalo, New York, and Research Institute, The Hospital for Sick Children, Toronto, and Department of Mechanical Engineering, University of Toronto, Toronto, Ontario, Canada

SAMIR K. BALLAS • Cardeza Foundation for Hematologic Research, Department of Medicine, Thomas Jefferson University, Philadelphia, Pennsylvania

JAMES A. COOK • Department of Physiology, Medical University of South Carolina, Charleston, South Carolina

W. J. DOUGHERTY • Department of Anatomy, Medical University of South Carolina, Charleston, South Carolina

MARC FELDMANN • ICRF Tumour Immunology Unit, Department of Zoology, University College London, London, England

JAMES P. FILKINS • Department of Physiology, Stritch School of Medicine, Loyola University of Chicago, Maywood, Illinois

HENRY GANS • Surgical, Research, and Pathology Services of the Danville Veterans Administration Medical Center, and University of Illinois School of Basic Medical Science and Clinical Medicine, and Division of Nutritional Sciences, College of Agriculture, University of Illinois at Urbana, Urbana, Illinois

ALBERT S. GORDON • Department of Biology, New York University, New York, New York

P. H. E. GROOT • Department of Biochemistry I, Medical Faculty, Erasmus University Rotterdam, The Netherlands

PAUL W. GUDEWICZ • Department of Physiology, Albany Medical College of Union University, Albany, New York

P.V. HALUSHKA • Departments of Pharmacology and Medicine, Medical University of South Carolina, Charleston, South Carolina

FARID I. HAURANI • Cardeza Foundation for Hematologic Research, Department of Medicine, Thomas Jefferson University, Philadelphia, Pennsylvania

NADER G. IBRAHAM • Department of Medicine, New York Medical College, Valhalla, New York

JOHN E. KAPLAN • Department of Physiology, Albany Medical College of Union University, Albany, New York

DAVID R. KATZ • Department of Pathology, The Middlesex Hospital Medical School, London, England

RICHARD D. LEVERE • Department of Medicine, New York Medical College, Valhalla, New York

DUDLEY G. MOON • Department of Physiology, Albany Medical College of Union University, Albany, New York

R. J. MOON • Division of Basic Sciences, School of Medicine, Mercer University, Macon, Georgia

ELIZABETH D. MOYER • Departments of Surgery and Biochemistry, State University of New York, Buffalo, New York. *Present address:* Department of Intravenous Nutrition, Cutter Group of Miles Laboratories, Berkeley, California

BRIAN A. NAUGHTON • Department of Biology, New York University, New York, New York

JOHN C. NELSON • Department of Medicine, New York Medical College, Valhalla, New York

A. WILHELM NEUMANN • Research Institute, The Hospital for Sick Children, Toronto, and Department of Mechanical Engineering, University of Toronto, Toronto, Ontario, Canada

SIGURD J. NORMANN • Department of Pathology, College of Medicine, University of Florida, Gainesville, Florida

MICHAEL C. POWANDA • Biochemistry Branch, U.S. Army Institute of Surgical Research, Fort Sam Houston, Texas. *Present address:* Division of Cutaneous Hazards, Letterman Army Institute of Research, Presidio of San Francisco, California

LAURENCE A. SHERMAN • Missouri/Illinois Regional Red Cross Blood Services, St. Louis, Missouri

CARLETON C. STEWART • Experimental Pathology Group, Los Alamos National Laboratory, Los Alamos, New Mexico

GEOFFREY H. SUNSHINE • ICRF Tumour Immunology Unit, Department of Zoology, University College London, London, England. *Present address:* Department of Surgery, Tufts University Veterinary School, Boston, Massachusetts

T. J. C. VAN BERKEL • Department of Biochemistry I, Medical Faculty, Erasmus University Rotterdam, Rotterdam, The Netherlands

CAREL J. VAN OSS • Immunochemistry Laboratory, Departments of Microbiology and Chemical Engineering, State University of New York, Buffalo, New York

A. VAN TOL • Department of Biochemistry I, Medical Faculty, Erasmus University Rotterdam, The Netherlands

W. C. WISE • Department of Physiology, Medical University of South Carolina, Charleston, South Carolina

Foreword

This comprehensive treatise on the reticuloendothelial system is a project jointly shared by individual members of the Reticuloendothelial (RE) Society and biomedical scientists in general who are interested in the intricate system of cells and molecular moieties derived from those cells which constitute the RES. It may now be more fashionable in some quarters to consider these cells as part of what is called the mononuclear phagocytic system or the lymphoreticular system. Nevertheless, because of historical developments and current interest in the subject by investigators from many diverse areas, it seems advantageous to present in one comprehensive treatise current information and knowledge concerning basic aspects of the RES, such as morphology, biochemistry, phylogeny and ontogeny, physiology, and pharmacology as well as clinical areas including immunopathology, cancer, infectious diseases, allergy, and hypersensitivity. It is anticipated that by presenting information concerning these apparently heterogeneous topics under the unifying umbrella of the RES attention will be focused on the similarities as well as interactions among the cell types constituting the RES from the viewpoint of various disciplines. The treatise editors and their editorial board, consisting predominantly of the editors of individual volumes, are extremely grateful for the enthusiastic cooperation and enormous task undertaken by members of the biomedical community in general and especially by members of the American as well as European and Japanese Reticuloendothelial Societies. The assistance, cooperation, and great support from the editorial staff of Plenum Press are also valued greatly. It is hoped that this unique treatise, the first to offer a fully comprehensive treatment of our knowedge concerning the RES, will provide a unified framework for evaluating what is known and what still has to be investigated in this actively growing field. The various volumes of this treatise provide extensive in-depth and integrated information on classical as well as experimental aspects of the RES. It is expected that these volumes will serve as a major reference for day-to-day examination of various subjects dealing with the RES from many different viewpoints.

Herman Friedman
Mario R. Escobar
Sherwood M. Reichard

Introduction

The physiology of any body system—be it the cardiovascular, endocrine, body fluid, pulmonary, etc.—embraces three categories of functional analysis:

1. The fundamental *mechanisms* of the system.
2. The *regulation* of the mechanisms by either intrinsic or extrinsic influences.
3. The coordinated interactions of the system in the *integration* of total organismic functions.

Thus, this treatise on the physiology of the reticuloendothelial system by definition will focus on the mechanisms, regulation, and integrative role of the body macrophage system.

While the phagocytic or more properly endocytic functions—which for many years almost exclusively defined the RES—are given predominant attention, the current state of the growing body of knowledge on RES exocytosis, i.e., secretory functions of the macrophages, is thoroughly reviewed.

The clinical physiology of the RES has also been included, since the application of basic knowledge to the prevention of human disease is the goal of all biomedical investigation.

The vast knowledge of the functions of the RES as influencing immune functions, host defense in infection, and neoplasia has been omitted, since these areas will be developed in other volumes of this treatise.

Herman Friedman
Mario R. Escobar
Sherwood M. Reichard

Preface

This volume is divided into three sections: I, Fundamental Mechanisms and Regulation of Phagocytosis; II, Regulatory Interactions with the Blood Elements; and III, Regulatory Interactions with Blood Metabolites and Constituents.

The first three chapters of Section I develop the cellular physiology of phagocytosis—with special attention to surface forces (van Oss *et al.*), the factors governing phagocytic proliferation in cultures of macrophages (Stewart), and the available means of quantitation of macrophage phagocytosis *in vitro* (Gudewicz). The second set of three chapters in Section I is devoted to the mechanisms of phagocytosis in the vascular macrophages and deals with the fundamentals of the kinetics of vascular clearance (Normann), the intravascular phagocytosis of microorganisms (Moon), and the macrophage's role in endotoxemia control (Gans).

Section II deals with the interaction of the RES with the cellular blood elements and emphasizes macrophage functions in erythropoiesis (Naughton and Gordon), erythroclasia and bilirubin metabolism (Nelson *et al.*), hemostasis (Sherman and Kaplan), platelet activities (Kaplan and Moon), and leukocyte interaction (Feldmann *et al.*).

Section III summarizes the current state of knowledge regarding RES involvement in the metabolic physiology and pathophysiology of glucose regulation (Filkins), lipid and lipoprotein metabolism (van Berkel *et al.*), protein metabolism (Powanda and Moyer), iron metabolism (Haurani and Ballas), and lead and cadmium interactions (Cook *et al.*). Volume A should provide the fundamental physiology and pathophysiology of the RES necessary to progress into the contents of Volume B: macrophage secretory functions and regulation, the clinical physiology of the RES, and integrative function of the RES.

We are grateful to the authors who accepted the difficult task of summarizing the state of knowledge in rapidly evolving areas of current investigation.

Sherwood M. Reichard
James P. Filkins

Contents

I. Fundamental Mechanisms and Regulation of Phagocytosis

1. Surface Forces in Phagocytosis

CAREL J. VAN OSS, DARRYL R. ABSOLOM, and A. WILHELM NEUMANN

1. Introduction 3
2. Thermodynamics of Particle Adhesion and Particle Engulfment 4
   2.1. Particle Adhesion 4
   2.2. Particle Engulfment 6
   2.3. Adhesion and Engulfment in Liquid Media of Lower Surface Tension Than Saline Water 7
3. Thermodynamics of Protein (Especially IgG) Adsorption 10
   3.1. Protein Adsorption and Opsonization 10
   3.2. Thermodynamics 10
4. Opsonization 14
   4.1. Role of Specific and Aspecific IgG 14
   4.2. Smallest Size of Immune Complexes Likely to Be Ingested 15
   4.3. Role of Complement and IgM 17
   4.4. Other Opsonizing Agents 17
   4.5. Role of IgA 17
   4.6. Opsonization *in Vivo* 18
5. Pathological Phagocytes 18
6. "Activation" of Macrophages and Other Phagocytes 21
7. Effect of Various Agents on the Surfaces of Phagocytes and/or Bacteria *in Vitro* 22
   7.1. Agents Mainly Affecting Phagocytes 22
   7.2. Agents Mainly Affecting Bacteria 22
   7.3. Agents Affecting Both Phagocytes and Bacteria 23
8. Influence of Electrokinetic Surface Potential and of Cell Shape 23
9. Methods for Measuring Surface Tensions of Cells, Particles, and Proteins 24
   9.1. Partition Methods 24
   9.2. Contact Angle Methods 25
   9.3. Adhesion and Adsorption Methods 26
   9.4. Solidification Front Method 27

9.5. Droplet Sedimentation Method 28
9.6. Phagocytic Ingestion in Liquids of Different Surface Tensions 28
10. Phagocytosis and Recognition *in Vitro* and *in Vivo* 28
10.1. *In Vitro* Phagocytosis in Aqueous Media 28
10.2. Phagocytosis in Biological Fluids 29
10.3. Recognition 30
References 31

2. Regulation of Mononuclear Phagocyte Proliferation

Carleton C. Stewart

1. Introduction 37
2. Colony Formation by Murine MNP 38
3. Proliferative Capacity of Peritoneal Exudate Cells 41
4. Proliferative Capacity of Bone Marrow Cells 43
5. Surface Markers on Differentiating MNP 45
6. Consumption of MGF by Macrophages 47
7. Serum Dialyzable Activity 48
8. Growth-Promoting Activity of Erythrocytes 49
9. Inhibition of MNP Proliferation 50
10. Genetics of Colony Formation 51
11. Local Proliferation Revisited 52
References 54

3. Quantitation of Macrophage Phagocytosis *in Vitro*

Paul W. Gudewicz

1. Introduction 57
2. Sources of Macrophages for *in Vitro* Phagocytic Studies 58
2.1. Sources of Macrophage Populations 58
2.2. *In Vitro* Assay Conditions for Phagocytosis 60
3. Test Particles Used for Phagocytosis *in Vitro* 61
3.1. Artificial Particles 62
3.2. Biological Particles 63
4. Methods to Quantify Phagocytosis *in Vitro* 64
4.1. Morphological and Bacterial Counting Methods 64
4.2. Radioactive and Chemical Methods 66
5. Summary 67
References 68

4. Kinetics of Vascular Clearance of Particles by Phagocytes

SIGURD J. NORMANN

1. Introduction 73
2. Kinetics of Vascular Clearance: A Zero- or First-Order Rate Equation? 74
3. Vascular Clearance, Surface-Saturation Kinetics, and Phagocytosis 77
4. Derivation of the Particle–Membrane Constant (K_p) and Maximum Phagocytic Velocity (V_{max}) 78
5. Affinity of Particles for the Phagocyte Surface: The Association Constant K_a 81
6. Particle Selection and the Clearance Behavior of Different Particles 84
   6.1. Particle Stability 84
   6.2. Particle Charge 86
   6.3. Monodispersity 87
   6.4. Summary 87
7. Host Factors Modulating Vascular Clearance 88
   7.1. Liver Blood Flow 88
   7.2. Species 89
   7.3. Kupffer Cell Number 90
   7.4. Opsonins 91
8. Clearance Inhibition by Particle Injection 91
   8.1. Immediate-Onset RES Paralysis 92
   8.2. Delayed-Onset RES Paralysis 93
   8.3. Opsonin Depletion 93
   8.4. Competitive Inhibition between Particles 94
9. Coda 96

References 97

5. Vascular Clearance of Microorganisms

R. J. MOON

1. Introduction 103
2. *In Vivo* Studies 103
   2.1. Viruses 103
   2.2. Bacteria 104
   2.3. Fungi 106
3. Liver Perfusion Studies 106
4. Future Directions 109
5. Conclusions 110

References 111

6. **RES Control of Endotoxemia**

HENRY GANS

1. Introduction 115
2. Detection of Endotoxin 116
3. Sources of Infection 118
4. Effect of Endotoxin at the Cellular Level 119
5. Endotoxin Transfer across the Normal Gut Wall 120
6. Endotoxin Absorption in Disease 123
7. Enteric Endotoxin's Potentiation of Various Hepatotoxins 125
8. The Role of the Kupffer Cell in Processing Enteric Endotoxins 126
9. Endotoxin-Induced Host Responses 128
10. Host Immune Responses to Endotoxin 132

References 136

II. Regulatory Interactions with the Blood Elements

7. **The Reticuloendothelial System and Erythropoiesis**

BRIAN A. NAUGHTON and ALBERT S. GORDON

1. Introduction: General Role of the RES in Erythropoiesis 147
2. The Role of the Macrophage in Erythropoiesis 148
   2.1. Reticular Cells of the Bone Marrow: Medullary Erythropoiesis 148
   2.2. Reticular Cells of the Liver and Spleen: Extramedullary Erythropoiesis 149
   2.3. Hormones and Chemical Agents: Relation of the RES to Hematopoiesis 160

References 178

8. **Erythroclasia and Bilirubin Metabolism**

JOHN C. NELSON, NADER G. IBRAHAM, and RICHARD D. LEVERE

1. Erythrocyte Destruction 189
   1.1. The Life Span of the Erythrocyte 189
   1.2. Changes in the Aging Erythrocyte 190

1.3. Sites of Erythrocyte Destruction 194
1.4. Mechanisms of Erythrocyte Destruction 195
2. Hemoglobin and Heme Metabolism 199
2.1. Regulation of Heme Metabolism 199
2.2. Mechanisms of Heme Degradation 203
2.3. Transport, Conjugation, and Excretion of Bilirubin 210
References 212

9. Hemostasis

Laurence A. Sherman and John E. Kaplan

1. Introduction 221
2. Clearance of Coagulation System Activators 222
3. Clearance of Procoagulants and Antiproteases 224
4. Clearance of Fibrinogen Derivatives 225
5. Production of Hemostatically Active Substances 228
5.1. Synthesis 228
5.2. Regulation of Synthesis 230
6. Influences of the Coagulation System on the RES 231
7. Conclusion 232
References 232

10. Platelets

John E. Kaplan and Dudley G. Moon

1. Introduction 237
2. Physiology and Anatomy of Platelets 238
3. Phagocytosis of Platelets 239
3.1. Hypersplenic and Thrombocytopenic States 239
3.2. Normal and Thrombotic States 243
4. Recognition of Platelets by the RES 246
4.1. Immune Recognition 246
4.2. Senescent Recognition—Role of Sialic Acid 247
4.3. Recognition during Thrombosis 248
5. Platelet–Particle Interaction 250
5.1. Phagocytosis by Platelets 250
5.2. Platelets and Colloid Clearance 252
6. RES Uptake of Platelet Activators 253
7. Platelet–Macrophage Interactions 254
7.1. Immune Interactions 254
7.2. Role of Prostaglandins 255

7.3. Role of Leukotrienes 256
7.4. Platelet-Activating Factor and Other Interactions 256
8. Summary and Concluding Statement 257
References 258

11. RES–Leukocyte Interactions

MARC FELDMANN, DAVID R. KATZ, and GEOFFREY H. SUNSHINE

1. Introduction 267
2. Types of Interaction between RES Cells and Lymphocytes 267
3. Interaction of RES Cells and Nonlymphoid Leukocytes 270
4. Interaction of RES Cells and T Lymphocytes 271
4.1. Introduction 271
4.2. Proliferative Responses 272
4.3. T Helper Cells 273
4.4. Dendritic Cell–Lymphocyte Interaction 273
4.5. Development of the T Repertoire 275
4.6. Mechanisms of RES–T Lymphocyte Interaction 276
5. Interaction of RES Cells and B Lymphocytes 279
6. Lymphocyte Activation of RES Cells 280
6.1. Introduction 280
6.2. *In Vivo* Phenomena 280
6.3. *In Vitro* Interactions between Lymphocytes and Macrophages 281
7. Conclusion 283
References 284

III. Regulatory Interactions with Blood Metabolites and Constituents

12. Glucose Regulation and the RES

JAMES P. FILKINS

1. Introduction 291
2. Glucoregulatory Alterations in Endotoxicosis 292
2.1. Phases of Blood Glucose Responses 292
2.2. Changes in Inputs and Outputs of the Glucose Pool 292
2.3. Insulin Changes in Endotoxicosis 293

3. Glucoregulatory Alterations after RES Perturbations 294
3.1. Changes in Inputs to the Glucose Pool 294
3.2. Changes in Outputs from the Glucose Pool 295
3.3. Insulin Changes 295
4. Glucoregulatory Monokines 296
4.1. Leukocytic Endogenous Mediator (LEM) 296
4.2. Glucocorticoid-Antagonizing Factor (GAF) 296
4.3. Macrophage Insulin-like Activity (MILA) 297
4.4. Macrophage Insulin-Releasing Activity (MIRA) 297
5. Glucose Regulation and RES Endocytic Functions 298
5.1. Intravascular Clearance Defects during Hypoglycemia 298
5.2. Unifying Schema 298
6. Concluding Remarks 300
References 300

13. Interaction of the Reticuloendothelial System with Blood Lipid and Lipoprotein Metabolism

T. J. C. VAN BERKEL, P. H. E. GROOT, and A. VAN TOL

1. Introduction 305
2. Interaction of Lipids with Free Macrophages 306
2.1. Artificial Lipid Substrates 306
2.2. Natural Lipid Substrates 307
3. Interaction of Lipids with Tissue Macrophages 309
3.1. Artificial Lipid Substrates 309
3.2. Natural Lipid Substrates 314
4. Perspectives in the Use of Macrophage Activity in the Prevention or Treatment of Atherosclerosis 325
References 326

14. Selected Aspects of Protein Metabolism in Relation to Reticuloendothelial System, Lymphocyte, and Fibroblast Function

MICHAEL C. POWANDA and ELIZABETH D. MOYER

1. Introduction 331
2. Interactions of Specific Proteins with RES, Lymphocytes, and Fibroblasts 332
2.1. α_1-Antitrypsin 332
2.2. α_2-Macroglobulin 334
2.3. C-Reactive Protein 336
2.4. Transferrin 338
2.5. Lipoproteins 339
2.6. Fibronectin 341

3. Summary and Conclusions 342
References 344

15. Iron Metabolism

Farid I. Haurani and Samir K. Ballas

1. Introduction 353
2. Uptake of Iron Carriers by the RES 353
   2.1. Haptoglobin and Hemopexin 353
   2.2. Erythrocyte 355
3. Processing and Release of Iron from the RES 360
   3.1. Concepts and Mechanisms of Iron Release 360
   3.2. Methods of Study of RES Activity in Relation to Iron 366
   3.3. Factors That Decrease Iron Release 367
   3.4. Factors That Increase Iron Release 371
References 372

16. Lead and Cadmium: Effect on Host Defense Mechanisms and Toxic Interactions with Bacterial Endotoxin

James A. Cook, W. J. Dougherty, W. C. Wise, and P. V. Halushka

1. Relative Toxicity of Lead and Cadmium on Phagocytes 380
2. Effect of Lead and Cadmium on Phagocytic Capacity 381
   2.1. *In Vitro* Studies 381
   2.2. Respiratory Exposure 381
   2.3. Parenteral Administration 382
   2.4. Oral Administration 385
3. Metabolic Effects of Lead and Cadmium on Phagocytes 385
4. Effect of Lead and Cadmium on Humoral and Cellular Immunity 387
   4.1. Humoral Immunity 387
   4.2. Lymphocyte Blast Cell Transformation 388
   4.3. Cellular Immunity 388
5. Effect of Lead and Cadmium on Susceptibility to Infections 389
   5.1. Bacterial Infections 389
   5.2. Viral Infections 390
   5.3. Parasitic Infections 390
6. Toxic Interactions of Lead and Cadmium with Bacterial Endotoxin 391
7. Conclusion 396
References 396

Index 401

Contents of Volume 7B

IV. Regulation and Macrophage Secretions

1. **Endocrinelike Activities of the RES: An Overview**
   ROBERT N. MOORE and L. JOE BERRY
2. **Regulation of Complement Synthesis in Mononuclear Phagocytes**
   ROBERT C. STRUNK and HARVEY R. COLTEN
3. **The Synthesis of Arachidonic Acid Oxygenation Products by Macrophages**
   P. DAVIES, R. J. BONNEY, J. L. HUMES, and F. A. KUEHL, JR.
4. **Lysosomal Hydrolases**
   MARCO BAGGIOLINI
5. **Macrophage Neutral Proteinases: Nature, Regulation, and Role**
   SIAMON GORDON and R. ALAN B. EZEKOWITZ
6. **Interferon and Macrophages**
   DAVID O. LUCAS and LOIS B. EPSTEIN

V. Clinical Physiology of the RES

7. **Leukocytic Endogenous Mediator in Nonspecific Host Defenses**
   ROGER H. MITCHELL and RALPH F. KAMPSCHMIDT
8. **Evaluation of RES Clearances in Man**
   JOHN W. B. BRADFIELD
9. **RES Function in Experimental and Human Liver Disease**
   HEINRICH LIEHR

10. **Regional Phagocytosis in Man**

JULIA W. BUCHANAN and HENRY N. WAGNER, JR.

11. **Inflammatory Cell Dynamics in Man**

JOHN W. REBUCK

12. **Fibronectin and Reticuloendothelial Clearance of Blood-Borne Particles: Clinical Studies in Septic Shock**

THOMAS M. SABA

VI. Integrative Functions of the RES

13. **Physiology and Pathophysiology of the Pulmonary Macrophages**

JOSEPH D. BRAIN

14. **Temperature Regulation and Fever**

HARRY A. BERNHEIM

15. **Microcirculatory Regulation and Dysfunction: Relationship to RES Function and Resistance to Shock and Trauma**

BURTON M. ALTURA

16. **Radiation Effects on Phagocytic Cells of the RES**

K. B. P. FLEMMING and SHERWOOD M. REICHARD

17. **Role of the Reticuloendothelial System in Shock**

SHERWOOD M. REICHARD and ANDY C. REESE

18. **Toxic Oxygen Products in Shock**

SHERWOOD M. REICHARD

19. **The RES and the Turnover of Circulating Lysosomal Enzymes in Shock**

G. HORPACSY

Index

I

Fundamental Mechanisms and Regulation of Phagocytosis

1

Surface Forces in Phagocytosis

CAREL J. VAN OSS, DARRYL R. ABSOLOM, and A. WILHELM NEUMANN

1. INTRODUCTION

In the past decade considerable advances have been made by our groups, in Buffalo and in Toronto, in the measurement of the physical surface forces involved in cell adhesion and phagocytic engulfment (by using the contact angle method), as well as in the interpretation of the relative roles played by various surface phenomena in *in vitro* and *in vivo* phagocytosis. In 1975 (van Oss *et al.*) and in 1978 (van Oss) we published reviews that may be regarded as interim reports as far as our theoretical understanding of the role of surface phenomena in phagocytosis is concerned. By 1979 the realization that van der Waals interactions could be repulsive and the discovery of the precise conditions under which these interactions become either attractive or repulsive (Neumann *et al.*, 1979a; van Oss *et al.*, 1979a) opened new vistas in the fields of cell–particle as well as cell–protein and even antigen–antibody interactions (van Oss *et al.*, 1980a). This research culminated in the development of new (non-contact angle) methods for measuring cellular surface tensions (Neumann *et al.*, 1979b; Absolom *et al.*, 1979), which also facilitated the theoretical treatment of cellular interactions in liquid media of surface tensions lower than that of water, i.e., in media more closely resembling various biological fluids. An extension of these new methods, involving cell adhesion to various polymer surfaces (Neumann *et al.*, 1979b), to protein adsorption to similar polymer surfaces (van Oss *et al.*, 1979b), made it

CAREL J. VAN OSS • Immunochemistry Laboratory, Departments of Microbiology and Chemical Engineering, State University of New York, Buffalo, New York 14214. DARRYL R. ABSOLOM • Immunochemistry Laboratory, Department of Microbiology, State University of New York, Buffalo, New York 14214, and Research Institute, The Hospital for Sick Children, Toronto, and Department of Mechanical Engineering, University of Toronto, Toronto, Ontario M5S 1A4, Canada. A. WILHELM NEUMANN • Research Institute, The Hospital for Sick Children, Toronto, and Department of Mechanical Engineering, University of Toronto, Toronto, Ontario M5S 1A4, Canada.

possible to estimate the surface thermodynamic role played by protein (especially IgG) adsorption onto various bacteria in their phagocytic ingestion, in aqueous media as well as in media of lower surface tension (Neumann *et al.*, 1982). These new developments for the first time open the way to attempts to arrive at a quantitative estimation of the principal physical surface forces involved in phagocytic engulfment in aqueous media *and* in biological fluids.

2. THERMODYNAMICS OF PARTICLE ADHESION AND PARTICLE ENGULFMENT

2.1. PARTICLE ADHESION

The first step, and a *conditio sine qua non* for the inception of phagocytosis, is the adhesion of a bacterium (B) to the surface of a phagocyte (P), while both are immersed in a liquid medium (L). The free energy change of this process (which has to be negative for adhesion to take place) is

$$\Delta F_{adh} = \gamma_{PB} - \gamma_{PL} - \gamma_{BL} \tag{1}$$

in which γ represents the interfacial tensions between the various materials indicated by the subscripts. Once the surface tensions with respect to vapor (γ_{XV}) of the various materials have been determined by one of the methods discussed below (Section 8) via the equation of state (Neumann *et al.*, 1974a) and Young's equation (Young, 1805), the various values for γ occurring in equation (1) can be obtained with the help of a computer program (Neumann *et al.*, 1980a), or found by means of published tables (Neumann *et al.*, 1980b).

As a general rule, in aqueous media such as Hank's balanced salt solution (HBSS) ($\gamma_{LV} \approx 72.8$ ergs/cm^2): the more hydrophobic the bacteria, the higher the negative value of ΔF_{adh} and thus the stronger the adhesion (van Oss *et al.*, 1979b). This is clear from Table 1, in which the values of ΔF_{adh} between normal human polymorphonuclear leukocytes (PMNL) and a number of bacteria are given. These values have been computed from the contact angle data (van Oss and Gillman, 1972a; van Oss *et al.*, 1975) for aqueous media ($\gamma_{LV} = 72.8$ ergs/cm^2), e.g., HBSS, and are compared with the degree to which the bacteria become ingested by normal human PMNL. The adhesion stage, as the first step toward phagocytic engulfment, is schematically depicted in Fig. 1. As seen from Table 1 all values for ΔF_{adh} are greater than $-$ 0.3 erg/cm^2 which is significantly more than the empirically established minimum requirement for adhesion to occur of $-$ 0.2 erg/cm^2 (Neumann *et al.*, 1979a; van Oss *et al.*, 1979a). Thus, adhesion of bacteria to phagocytes, while a minimum requirement for subsequent phagocytic ingestion, always should take place to a certain extent, at least in aqueous media, as there are to our knowledge few microorganisms even more hydrophilic than *S. aureus* s. Smith. Also, residual bacterial adhesion alone played no role in the quantitative *in vitro* phagocytosis tests, as various precau-

TABLE 1. FREE ENERGIES OF ADHESION (ΔF_{adh}; SEE FIG. 1) AND FREE ENERGIES OF ENGULFMENT ($\Delta F_{eng\text{-}1}$ AND $\Delta F_{eng\text{-}2}$; SEE FIG. 2) OF VARIOUS BACTERIA *VIS-À-VIS* NORMAL PMNL[a] IN AQUEOUS MEDIA[b], AND THE DEGREE TO WHICH THEY BECOME PHAGOCYTIZED *IN VITRO* IN SUCH MEDIA[c]

Bacteria	Contact angle with drops of saline water (degrees)[d]	Phagocytic activity ($N \pm$ S.E.)[e]	ΔF_{adh}[f]	$\Delta F_{eng\text{-}1}$[f,g]	$\Delta F_{eng\text{-}2}$[f,g]	$\Delta F_{eng\text{-}net}$ (= $\Delta F_{eng\text{-}1} + \Delta F_{eng\text{-}2}$)
Staphylococcus aureus s. Smith	16.5	0.2 ± 0.1	−0.31	−0.11	−0.01	−0.12
Escherichia coli t.0111	17.2	0.6 + 0.2	−0.35	−0.13	−0.03	−0.16
Staphylococcus aureus	18.7	1.6 ± 0.3	−0.41	−0.16	−0.06	−0.22
Staphylococcus epidermidis	24.5	2.5 ± 0.4	−0.69	−0.30	−0.20	−0.50
Listeria monocytogenes	26.5	2.1 ± 0.4	−0.84	−0.37	−0.28	−0.65

[a] The contact angle of normal human PMNL is 18.0 ± 0.5°.
[b] For example, Hank's balanced salt solution, γ_{LV} = 72.8 ergs/cm^2.
[c] From Absolom *et al.* (1982a).
[d] From van Oss and Gillman (1972a) and van Oss *et al.* (1975).
[e] Average number of bacteria ingested per PMNL (van Oss *et al.*, 1975).
[f] In ergs/cm^2 (1 erg/cm^2 = 1 mJ/m^2).
[g] See Fig. 2.

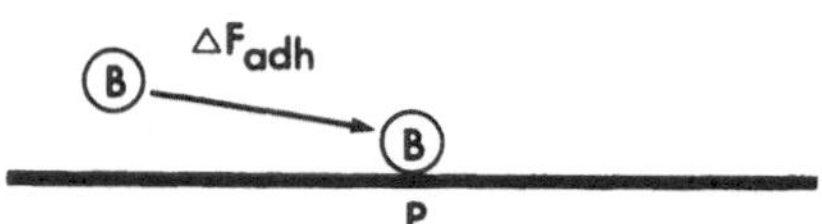

FIGURE 1. Schematic presentation of the first major step, bacterial adhesion, essential in the initiation of phagocytic engulfment of a bacterium (B) by a phagocyte, the interior side of which is indicated by P. The free energy of adhesion, $\Delta F_{adh} = \gamma_{PB} - \gamma_{PL} - \gamma_{BL}$ (see text).

tions were taken to obviate this (see van Oss *et al.*, 1975).* Residually adhering nonengulfed bacteria were removed by thorough washing, which was obviously sufficient to overcome adhesive energies of ≈ − 0.3 to − 0.8 erg/cm².

It can be seen from Table 1 that the more hydrophilic bacteria hardly become phagocytized at all, indicating that simple adherence, although a prerequisite, does not in itself suffice for completing phagocytosis. For that completion a further, ingestive, process is needed.

2.2. PARTICLE ENGULFMENT

The process of phagocytic engulfment is subdivided into two steps, which are schematically represented in Fig. 2. Theoretically a third ingestion step might be considered (Neumann *et al.*, 1972, 1974b; van Oss *et al.*, 1975), but that last step (consisting of the movement of an already completely ingested bacterium, away from the one remaining point of attachment to the phagocyte's membrane, further into its interior) is thermodynamically of only small quantitative consequence. Table 1 also shows that significant engulfment ensues only when the total ΔF_{eng}, or $\Delta F_{eng\text{-}net}$, is even more negative than the critical − 0.2 erg/cm².† In Section 5 (Pathological Phagocytes) it will be seen that the $\Delta F_{eng\text{-}2}$ step is as important as the earlier $\Delta F_{eng\text{-}1}$ step, in view of which it is useful to note that phagocytic ingestion only is achieved when $\Delta F_{eng\text{-}1}$ *as well as* $\Delta F_{eng\text{-}2} < -0.2$ erg/cm² (see Table 1; see also Table 6). The values presented in Table 1, and in further tables, are *typical* examples; similar values, obtained under identical circumstances, have been found many times over. The differences in various free energies are statistically significant and quite reproducible. Phagocytic activities measured are constant within a set of experiments done on any given day; results may vary somewhat from day to day, depending on changes in bacterial growth: phagocytic activity also strongly depends on the particle concentration of the bacterial suspension used. For commodity's sake the *in vitro* phagocytosis test first described by Newsome (1967) was adopted and adapted

*The efficacy of the washing procedure was monitored with the help of lysostaphin, a *Staphylococcus*-dissolving enzyme (van Schaik *et al.*, 1975; Tan *et al.*, 1971) in the case of phagocytosis of staphylococci, and by hemolysis tests (hemolysis at low ionic strength only lyses adhering, but not ingested erythrocytes; Gigli and Nelson, 1968). Recently a more generally applicable procedure has emerged, based on the quenching of the fluorescence of fluorescent antibodies by crystal violet extracellularly, but not intracellularly (Hed, 1977, 1979).

†A process is thermodynamically favored if the change in the Helmholtz free energy (ΔF) for the process is negative. The outcome of a process can be accurately predicted if ΔF is greater than the absolute value of 0.2 erg (Neumann *et al.*, 1979a).

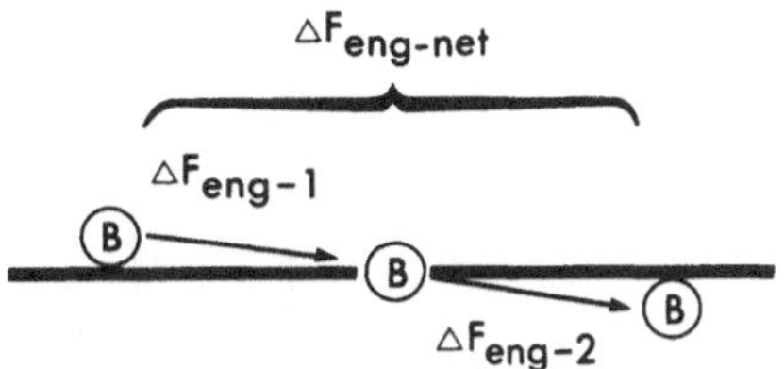

FIGURE 2. Schematic presentation of the two engulfment steps in the phagocytosis of a bacterium (B) by a phagocyte. The free energy of the first step of engulfment, $\Delta F_{\text{eng}-1} = \frac{1}{2}(\gamma_{PB} - \gamma_{BL}) - \frac{1}{2}\gamma_{PL}$; the free energy of the second step of engulfment, $\Delta F_{\text{eng}-2} = \frac{1}{2}(\gamma_{PB} - \gamma_{BL}) + \frac{1}{2}\gamma_{PL}$. The sum of these two steps have a free energy $\Delta F_{\text{eng-net}} = \gamma_{PB} - \gamma_{BL}$ (see text).

(van Oss and Gillman, 1972a; van Oss *et al.*, 1975); this test also has been done in parallel with *in vitro* tests of phagocytosis in whole blood in rotating tubes; the results were on the whole superimposable on one another (except in the case of nonopsonized bacteria) (see Table 2). There are two rather important differences in principle to be noted between these two methods:

1. Whole blood has a lower surface tension than saline water (or HBSS) (van Oss *et al.*, 1980c).
2. Whole blood contains sizeable amounts of IgG, which cannot fail to opsonize most bacteria (van Oss and Stinson, 1970; Stinson and van Oss, 1971).

The phenomena occurring in liquid media with surface tensions lower than that of water are discussed in the following subsection (2.3), and the mechanisms of opsonization are treated in Section 4. In anticipation it may be stated that the relatively low degree of phagocytosis of *S. epidermidis* (nonopsonized) (see Table 2) is indeed mainly due to the absence of IgG opsonization.

2.3. ADHESION AND ENGULFMENT IN LIQUID MEDIA OF LOWER SURFACE TENSION THAN SALINE WATER

Rather puzzling at first sight is the fact that the phagocytic engulfment of *S. epidermidis*, opsonized by IgG, is exactly the same in liquids of both higher (HBSS) and lower surface tension (whole blood) (see Table 2). This is in contrast with cell adhesion which under similar circumstances is known to vary considerably in liquids of lower surface tension, even when the liquids' surface tensions are lowered by means of the addition of proteins (Neumann *et al.*, 1979b; Absolom *et al.*, 1979). In order to study the influence of the surface tension of the liquid medium on phagocytic engulfment, phagocytosis of *nonopsonized bacteria* of various surface energies was measured in liquids of different surface tensions (see Table 3). From Table 3 it becomes clear that phagocytic ingestion largely parallels cell adhesion in that, as with cell adhesion, a minimum occurs when the surface tension of the particle equals that of the liquid medium ($\gamma_{BV} = \gamma_{LV}$) (Neumann *et al.*, 1979b; Absolom *et al.*, 1979). At first sight, protein solutions appear to have little influence on the phagocytic ingestion of a few typical bacteria (see Table 4).

However, in general, cell adhesion and thus also cell engulfment, in aqueous media of lower surface tension than that of the cells themselves, contrary to their behavior in HBSS, *favors the adhesion* (and thus the engulfment) *of the more*

TABLE 2. PHAGOCYTIC INGESTION OF OPSONIZED AND NONOPSONIZED *STAPHYLOCOCCUS EPIDERMIS* BY HUMAN PMNL[a] IN AN AQUEOUS MEDIUM[b], MEASURED IN MACKANNESS CHAMBERS, AND MEASURED IN WHOLE EDTA ANTICOAGULATED BLOOD[c] IN ROTATING TUBES[d]

Bacteria	Contact angle with drops of saline water (degrees)	Phagocytic activity ($N \pm$ S.E.)[e]: In Mackanness chambers[f]		Phagocytic activity ($N \pm$ S.E.)[e]: In rotating tubes of whole blood[g]
Staphylococcus epidermidis, opsonized with 1% normal human IgG	19.8	7.0 ± 0.4	$p < 10^{-5}$[h] (between 7.0 ± 0.4 and 4.7 ± 0.2)	7.5 ± 0.6
Staphylococcus epidermidis, as is, nonopsonized	19.2	4.7 ± 0.2		7.1 ± 0.6
		$p < 2 \times 10^{-4}$ (between 4.7 ± 0.2 and 7.1 ± 0.6)		

[a]The contact angle of normal human PMNL is $18.0 \pm 0.5°$.
[b]HBSS, $\gamma_{LV} = 72.8$ ergs/cm^2.
[c]γ_{LV} of whole blood ≈ 50 ergs/cm^2; γ_{LV} of blood ultrafiltrate ≈ 70.4 ergs/cm^2.
[d]From Absolom *et al.* (1982a).
[e]Average number of baceteria ingested per PMNL (van Oss *et al.*, 1975).
[f]Measured with monolayers of PMNL in Mackanness chambers, in HBSS.
[g]Measured with PMNL in whole EDTA anticoagulated blood in rotating tubes.
[h]Statistical significance of the difference between these two values of $p < 10^{-5}$ indicates that the likelihood that the difference occurred accidentally was less than 1 in 100,000 and less than 2 in 10,000, respectively.

TABLE 3. PHAGOCYTIC INGESTION OF NONOPSONIZED BACTERIA BY HUMAN PMNL[a] IN LIQUID MEDIA OF DIFFERENT SURFACE TENSIONS (γ_{LV})[b]

	Phagocytic activity (N + S.E.)[c]				Liquid medium	
	E. coli 055	*S. aureus* 49	*S. epidermidis* 47	*L. monocytogenes*		% DMSO
Contact angle θ:	16.7°	18.8°	23.4°	25.3°	γ_{LV}	(v/v)
γ_{BV} (ergs/cm^2):	69.5	68.8	67.0	66.2	(ergs/cm^2)	in HBSS
	0.9 ± 0.2	1.3 ± 0.1	3.8 ± 0.7	5.6 ± 0.2	72.8	0
	0.8 ± 0.2	1.4 ± 0.2	3.6 ± 0.1	5.8 ± 0.2	72.5	1
	0.9 ± 0.2	1.6 ± 0.1	3.7 ± 0.1	5.6 ± 0.2	71.0	2.5
	1.0 ± 0.1	1.2 ± 0.1	4.0 ± 0.1	5.1 ± 0.1	69.7	5
	0.8 ± 0.1	0.5 ± 0.1	2.3 ± 0.2	3.4 ± 0.1	68.5	7.5
	1.6 ± 0.1	0.9 ± 0.1	0.7 ± 0.1	1.5 ± 0.1	67.2	10
	2.8 ± 0.1	2.0 ± 0.1	1.4 ± 0.1	0.9 ± 0.1	64.0	15
	3.5 ± 0.2	2.7 ± 0.1	1.9 ± 0.1	1.3 ± 0.1	60.8	20
	3.6 ± 0.1	2.7 ± 0.1	2.0 ± 0.1	1.4 ± 0.2	58.2	25[d]

[a]The contact angle of these PMNL was 18.6°, corresponding to γ_{PV} = 68.9 ergs/cm^2.
[b]From Neumann *et al.* (1982).
[c]Average number of bacteria ingested per PMNL (van Oss *et al.*, 1975).
[d]At this concentration of DMSO, PMNL start manifesting changes in morphology.

TABLE 4. PHAGOCYTIC ACTIVITY ($N \pm$ S.E.)[a] OF NONOPSONIZED BACTERIA BY HUMAN PMNL IN HBSS AND IN HBSS + 3% BOVINE SERUM ALBUMIN

Bacteria		
Staphylococcus epidermidis	*Listeria monocytogenes*	Medium
5.4 ± 0.5	5.8 ± 0.3	HBSS[b]
5.7 ± 0.7	5.3 ± 0.4	HBSS + 3% BSA[c]

[a]Average number of bacteria ingested per PMNL (van Oss *et al.*, 1975).
[b]Surface tension of HBSS: γ_{LV} = 72.8 ergs/cm^2.
[c]Surface tension of 3% (w/v) BSA in HBSS: $\gamma_{LV} \approx$ low 50s ergs/cm^2 (Absolom *et al.*, 1981).

hydrophilic cells (Newmann *et al.*, 1979b; Absolom *et al.*, 1979, 1980a), i.e., in low-surface-tension media, the more hydrophilic bacteria should become *more* engulfed by phagocytic cells than hydrophobic bacteria, which is not only contrary to what happens *in vitro* in aqueous media (van Oss and Gillman, 1972a) but also the opposite of what occurs *in vivo* (van Oss *et al.*, 1975; Davis *et al.*, 1973). Protein adsorption onto bacterial surfaces (and its consequences with respect to opsonization), however, is not taken into account here, but will be considered in the following sections.

In conclusion, *in vitro* phagocytic ingestion of nonopsonized bacteria largely parallels *in vitro* cell adhesion; both phenomena are dependent on the surface tension γ_{LV} of the liquid medium as well as on the surface tensions γ_{PV} and γ_{BV} of the cells. The ΔF_{eng} steps (see Fig. 2) come under further scrutiny in Section 5, where the behavior of certain pathological phagocytes is treated.

3. THERMODYNAMICS OF PROTEIN (ESPECIALLY IgG) ADSORPTION

3.1. PROTEIN ADSORPTION AND OPSONIZATION

Although there are body fluids, serving as a liquid medium for phagocytosis, in which very little protein and extremely small amounts of IgG are present (e.g., in the urinary bladder), most body fluids contain sizeable amounts of protein, usually including IgG as well as albumin. Apart from a general tendency of lowering the surface tension of the aqueous medium, of which the relevance to phagocytic engulfment has been discussed in the preceding section (2.3), the degree to which various proteins become physically adsorbed onto bacterial surfaces must be taken into account in interpreting the interfacial interactions between bacteria and phagocytes under conditions approaching those prevailing *in vivo*. Especially important is the physical adsorption of IgG by bacteria, as that, in most (but not in all) cases, is tantamount to opsonization (van Oss and Stinson, 1970; Stinson and van Oss, 1971).

3.2. THERMODYNAMICS

Protein adsorption follows in many instances the same rules as cell adhesion, from a thermodynamic point of view, so that, in analogy with equation (1):

$$\Delta F_{ads} = \gamma_{PrB} - \gamma_{PrL} - \gamma_{BL} \tag{2}$$

in which the subscript Pr stands for protein and the other subscripts remain as defined for equation (1). As with cells, surface tensions of proteins (γ_{PrV}) can be determined by a few different procedures (see Section 8) so that here also, via the equation of state (Neumann *et al.*, 1974b) and appropriate tables (Neumann *et al.*, 1980b), the free energy differences of protein adsorption onto various bacterial and other surfaces can be calculated. The adsorption of two plasma proteins is studied here in detail: human serum albumin (the protein present in plasma in by far the highest concentration) and human IgG (the principal heat-stable opsonin; see van Oss and Stinson, 1970; Stinson and van Oss, 1971), in the form of pooled polyclonal IgG (Cohn's Fraction II), as well as purified human monoclonal IgG, in this case IgG3, which does not bind to staphylococci (see van Oss, 1979a, and Table 5). The adsorption of all these proteins (in 1% concentration) is measured through the residual radioactivity (^{125}I was used as a label) of the various bacteria, after extensive washing following the adsorption step (Absolom *et al.*, 1982a). The surface tensions (γ_{PrV}) of human serum albumin (HSA) and IgG are 70.2 and 67.5 ergs/cm^2, respectively (van Oss *et al.*, 1980b), corresponding to contact angles of $\approx 15°$ and $\approx 22°$, respectively (van Oss *et al.*, 1975, 1980b). Table 5 shows the adsorption of HSA and IgG onto a number of bacteria. Clearly, the more hydrophobic the bacteria, the more they adsorb of both HSA and IgG, and any given bacterium adsorbs more IgG than HSA, in accordance with the fact that IgG is more hydrophobic than HSA; there is a striking concordance between the values of ΔF_{ads} and the various adsorptivities. All four bacteria tested adsorbed more pooled polyclonal (FII) than monoclonal IgG, most likely on account of the fact that pooled polyclonal IgG still comprises various more or less specific antibodies against the different bacterial antigenic determinants, which is not the case with monoclonal IgG. Thus, IgG does, as it should for thermodynamic reasons, become physically adsorbed onto bacterial surfaces to a significantly greater extent than HSA; IgG becomes adsorbed most strongly to the more hydrophobic bacteria. Even in whole plasma or serum, monoclonal IgG still physically adsorbs to a significantly greater extent to all bacteria tested, again the most strongly to the most hydrophobic species (see Table 6). Although less IgG becomes adsorbed to the bacteria in the presence of whole plasma than in HBSS, even under these conditions 3000 to 30,000 IgG molecules become adsorbed per bacterium, which still is ample for opsonization (see Section 4).

The lesser IgG adsorption in whole serum or plasma (Table 6), as compared to its adsorption in HBSS (Table 5), is more likely to be due to competition with other proteins in serum or plasma, than to the lower liquid surface tension (γ_{LV}) of serum and plasma as compared to the high γ_{LV} of HBSS (van Oss *et al.*, 1980c). Table 7 shows that the surface tension of the liquid (γ_{LV}) has to be lowered (with DMSO) to ≈ 64 ergs/cm^2 to obtain a significant decrease in IgG adsorption (it is interesting to compare this with the ≈ 62 ergs/cm^2 needed to prevent a van der Waals-type antigen–antibody interaction; van Oss *et al.*, 1979c). Lowering the γ_{LV} with HSA, however, has no significant effect on IgG-adsorption, except perhaps ever so slightly at the higher HSA concentration, which is more likely due to protein competition (compare Tables 5 and 6). Apart from the difficulty of in-

TABLE 5. ADSORPTION OF 1% HUMAN SERUM ALBUMIN (HSA) AND 1% HUMAN IgG ONTO BACTERIA IN AN AQUEOUS MEDIUM (HBSS)[a,b]

Bacteria	Contact angle of bacteria with drops of saline water (degrees)	γ_{BV} of bacteria (ergs/cm^2)	Protein adsorbed (fg)/bacterial particle ± S.D.			ΔF_{ads} (ergs/cm^2)	
			HSA	IgG (FII)	IgG (monoclonal)[c]	HSA	IgG
Escherichia coli 055	16.7	69.6	11 ± 5	92 ± 23	56 ± 7	−0.30	−0.67
Staphylococcus aureus 49	18.8	68.8	16 ± 8	170 ± 23	92 ± 18	−0.40	−0.77
Staphylococcus epidermidis 47	23.4	67.0	28 ± 7	521 ± 63	201 ± 19	−0.58	−1.09
Listeria monocytogenes	25.3	66.2	58 ± 17	724 ± 81	259 ± 28	−0.66	−1.25

[a]The surface tensions of HBSS, HSA, and IgG are 72.8, 70.2, and 67.5 ergs/cm^2, respectively.
[b]Absolom *et al.* (1982a).
[c]Myeloma IgG #1024 (IgG3).

TABLE 6. ADSORPTION OF 1% HUMAN MONOCLONAL IgG[a] ONTO BACTERIA IN WHOLE SERUM AND IN WHOLE PLASMA[b]

Bacteria	Contact angle of bacteria with drops of saline water (degrees)	γ_{BV} of bacteria (ergs/cm^2)	IgG adsorbed (fg)/bacterium ± S.D.	
			In whole serum	In whole plasma
Escherichia coli 055	16.7	69.6	1.8 ± 0.7	0.6 ± 0.3
Staphylococcus aureus 49	18.8	68.8	3.8 ± 1.0	2.3 ± 1.1
Staphylococcus epidermidis 47	23.4	67.0	5.4 ± 1.2	3.6 ± 1.5
Listeria monocytogenes	25.3	66.2	8.3 ± 1.5	6.3 ± 1.1

[a]Myeloma #1024 (IgG3).
[b]Absolom *et al.* (1982a).

terpreting the meaning of γ_{LV} values caused by protein solutions and of dissociating these values from the methodology used in measuring them (Absolom *et al.*, 1981), there remains the fact that while protein–ligand interactions can be prevented, and even reversed, by lowering the surface tension of the medium through the admixture of appropriate low-molecular-weight solutes (van Oss *et al.*, 1979c,d), lowering the surface tension of the medium by means of dissolved proteins cannot (on account of the large size of the individual protein molecules, compared with the small dimensions of the protein–ligand determinants and distances involved) and does not directly affect such protein–ligand interactions (van Oss *et al.*, 1980c).

In conclusion, bacterial surfaces readily physically adsorb human IgG (from aqueous media as well as from whole plasma or serum), and the more hydrophobic the bacteria, the more IgG they adsorb.

TABLE 7. ADSORPTION OF 1% HUMAN MONOCLONAL IgG3 ONTO BACTERIA AS A FUNCTION OF THE SURFACE TENSION (γ_{LV}) OF THE LIQUID MEDIUM[a]

Liquid medium	γ_{LV} of liquid medium (ergs/cm^2)	IgG adsorbed (fg)/bacterium ± S.D.			
		E. coli	*S. aureus*	*S. epidermidis*	*L. monocytogenes*
HBSS	72.8	58 ± 8	107 ± 23	192 ± 18	264 ± 21
1% DMSO[b]	72.5	60 ± 9	115 ± 15	191 ± 22	268 ± 20
5% DMSO[b]	69.7	67 ± 13	139 ± 20	192 ± 15	268 ± 27
15% DMSO[b]	64.0	17 ± 5	14 ± 4	23 ± 6	28 ± 10
1% HSA[b]	high 50s[c]	54 ± 14	127 ± 15	218 ± 17	258 ± 24
2.5% HSA[b]	low 50s[c]	40 ± 6	121 ± 14	178 ± 17	239 ± 31

[a]Absolom *et al.* (1982a).
[b]In HBSS.
[c]Absolom *et al.* (1981); the surface tension of albumin (and other protein) solutions is strongly dependent on the method of measurement; it is not meaningful to attempt to give more precise values than the approximations given here.

4. OPSONIZATION

4.1. ROLE OF SPECIFIC AND ASPECIFIC IgG

The opsonizing role of IgG, specific, as well as aspecific (FII) and nonspecific (monoclonal), is well established (see, e.g., van Oss and Stinson, 1970; Stinson and van Oss, 1971). The part of the IgG molecule endowed with opsonizing power is the Fc moiety (of IgG subclasses IgG1 and IgG3) while the Fab and $F(ab')_2$ fractions are totally devoid of it (van Oss *et al.*, 1973, 1975). Studies with the strongly encapsulated *S. aureus* s. Smith indicated that IgG only is active as an opsonin when situated at the outermost surface of the bacterial cell, or, if present, of its capsule (van Oss and Stinson, 1970).

Opsonization with IgG is accompanied with an increase of hydrophobicity of the bacterial surface (van Oss and Gillman, 1972b; Cunningham *et al.*, 1975), which of course should (and does) give rise to increased phagocytic ingestion *in vitro*, in aqueous media. It is tempting to ascribe the increased phagocytic engulfment mainly to the greater surface hydrophobicity (van Oss and Gillman, 1972b; van Oss *et al.*, 1974), but in 1975 it was noted (van Oss *et al.*) that the normally already rather hydrophobic *Listeria monocytogenes* (contact angle $\theta \approx 25°$), which therefore cannot and does not become more hydrophobic upon opsonization with IgG, nevertheless after incubation with IgG becomes noticeably more phagocytized. This strengthens the suspicion that there is more to the function of IgG than simply to increase the hydrophobicity of bacteria (or to trigger the fixation of complement), and that interactions between IgG (via its Fc moiety) and IgG receptors on phagocytic cells play an important role in phagocytic ingestion.

Table 8 shows that in all cases opsonization with both kinds of IgG significantly enhances phagocytic ingestion. With *E. coli* and *S. aureus*, that enhancement parallels an increase in contact angle, but with *S. epidermidis* and *L. monocytogenes*, no such increase in contact angle is noticeable, nor given the contact angle with drops of saline water for IgG of $\theta \approx 23.5°$, could such an increase in contact angle be expected with the latter two bacterial species (van Oss *et al.*, 1975, 1980b). Treatment with HSA in all cases causes a slight decrease in contact angle as well as in phagocytic engulfment, which was to be expected, in view of the low contact angle of HSA of 15° (van Oss *et al.*, 1980b). The data in Table 8 pertaining to *S. epidermidis* and *L. monocytogenes* clearly show that IgG can play an important opsonic role independent of changes in bacterial hydrophobicity (see also Davies, 1975).

Lowering the surface tension (γ_{LV}) of the liquid medium with up to 20% DMSO, which causes pronounced minima in the phagocytic ingestion of nonopsonized bacteria (see Table 3), *has no influence whatever on the phagocytic ingestion by PMNL of bacteria opsonized with IgG* (Absolom *et al.*, 1982a). This indicates (1) that detachment of IgG from bacteria (contrary to the prevention of its attachment; see Table 7) takes a fairly large amount of energy (see also van Oss *et al.*, 1979c) and (2) that the detachment of IgG from the Fc receptors on PMNL requires an even larger amount of energy. Even if electrostatic forces play no important role in the IgG-Fc receptor linkage, the probable high specificity and

TABLE 8. COMPARISON BETWEEN THE PHAGOCYTIC INGESTION OF NONOPSONIZED BACTERIA, AND BACTERIA OPSONIZED WITH 1% IGG (FII, AS WELL AS MONOCLONAL), AND BACTERIA TREATED WITH HSA[a]

Treatment	Phagocytic activity ($N \pm$ S.E.)[b]			
	E. coli	*S. aureus*	*S. epidermidis*	*L. monocytogenes*
Nonopsonized	0.8 ± 0.2 (16.7°)[c]	1.5 ± 0.1 (18.8°)[c]	3.6 ± 0.1 (23.4°)[c]	5.4 ± 0.2 (25.3°)[c]
Opsonized with 1% IgG (FII)	2.4 ± 0.2 (19.4°)[c]	3.6 ± 0.2 (21.7°)[c]	5.6 ± 0.1 (23.6°)[c]	8.2 ± 0.1 (25.5°)[c]
Opsonized with 1% IgG (monoclonal)[d]	2.0 ± 0.1 (18.9°)[c]	3.8 ± 0.2 (21.5°)[c]	4.8 ± 0.2 (22.6°)[c]	7.8 ± 0.1 (24.8°)[c]
Treated with 1% HSA	0.6 ± 0.2 (16.5°)[c]	1.4 ± 0.1 (18.2°)[c]	3.2 ± 0.2 (21.6°)[c]	4.8 ± 0.2 (23.8°)[c]

[a] Absolom *et al.* (1982a).
[b] Average number of bacteria ingested per PMNL (van Oss *et al.*, 1975).
[c] The contact angles with drops of saline water, in degrees, of each of the bacteria after the treatment indicated in the left-hand column are given in parentheses.
[d] Myeloma #1024 (IgG).

fairly small surface area of attachment make it unlikely that DMSO concentrations of much less than 40–50% would favor its dissociation, if that linkage is at all comparable to the antigen–antibody bond (van Oss *et al.*, 1979c, 198Od). However, DMSO concentrations ≥ 25% (see Table 3) are detrimental to PMNL morphology, and probably also to their function.

IgG subclasses IgG1 and IgG3 are mainly implicated in opsonization, at least vis-à-vis monocytes and macrophages (Nisonoff *et al.*, 1975) and most likely also as far as PMNL are concerned. It is interesting to note that IgG1 and IgG3 also appear to be the IgG forms that are principally adsorbed onto (polystyrene) latex particles from whole human serum (Weening *et al.*, 1979).

In conclusion, human IgG (principally IgG1 and IgG3), specific as well as aspecific, is a powerful opsonin, favoring the phagocytic ingestion of bacteria by PMNL, in part through an increase in bacterial surface hydrophobicity (which mainly plays a role *in vitro*), and for a probably more important part through (strong and specific) interaction with Fc receptors on the surface of phagocytic cells. For opsonization, IgG is only active when situated at the outer periphery of the bacterial surface or of its capsule; the Fc moiety comprises the opsonizing determinant. Specific as well as aspecific IgG1 and IgG3 are the principal "heat-stable" opsonins.

4.2. SMALLEST SIZE OF IMMUNE COMPLEXES LIKELY TO BE INGESTED

Every freely mobile molecule or particle, regardless of its size, has a translational kinetic (or Brownian) energy equal to $1.5kT$, where k is Boltzmann's constant (= 1.38×10^{-16} erg/degree) and T is the absolute temperature in degrees Kelvin. Thus, an immune complex that in the course of its movement becomes attached to a phagocytic cell surface, with an energy of attachment smaller than $1.5kT$, will be able to escape again thanks to its own kinetic energy of $1.5kT$.

We do not know anything about the precise specific shape of the end of the Fc tail of the IgG molecule, nor of the complementary shape of the Fc receptor on the phagocyte, but we do know something about its most likely surface area (≈ 1250 Å²; see Valentine and Green, 1967; Labaw and Davies, 1971; van Oss *et al.*, 1974). We also may assume that the Fc receptor has a rather low electric surface potential, if it does not differ greatly from the average electric surface potential of phagocytic cells, so that the interaction between Fc and Fc receptor is likely to be mainly a van der Waals attraction. The surface tension of the Fc moiety (γ_{FcV} = 67.8 ergs/cm², from θ = 21.5°; see van Oss *et al.*, 1975), as well as the surface tension of PMNL (γ_{PV} = 69.1 ergs/cm², from θ = 18°) and of macrophages (γ_{MV} = 67.8 ergs/cm², from θ = 21.5°; see van Oss *et al.*, 1975) are known. Accordingly, it is then possible, through equation (2), to derive the values for $\Delta F_{adh}^{Fc\text{-}PMN}$ for Fc–Fc receptors on PMNL and also for $\Delta F_{adh}^{Fc\text{-}M\phi}$ for Fc–Fc receptors on macrophages, under conditions approaching those which exist *in vivo*. To that effect we postulate the surface tension of the biological fluids in which the reaction between Fc and Fc receptors takes place to be closest to that of a plasma ultrafiltrate, i.e., γ_{LV} = 70.5 ergs/cm², as the proteins in plasma cannot contribute to the liquid surface tensions at distances of the order ≈ 2 Å (van Oss *et al.*, 1980c). Thus, from equation (2):

$$\Delta F_{adh}^{Fc\text{-}PMN} \approx -0.14 \text{ erg/cm}^2$$

and

$$\Delta F_{adh}^{Fc\text{-}M\phi} \approx -0.24 \text{ erg/cm}^2$$

As $1.5kT$, at 37°C, corresponds to 641.7×10^{-16} erg, this, for the 1250 Å² of the postulated receptor area, amounts to a maximum escape energy of the immune complex due to Brownian motion after accidental attachment of the Fc tails to Fc receptors, of:

$$\Delta F_{escape} \approx 0.51 \text{ erg/cm}^2$$

Thus, immune complexes comprising up to three Fc tails can escape attachment to PMNL, and immune complexes comprising up to two Fc tails can escape attachment to macrophages (see also van Oss *et al.*, 1971). These conclusions conform closely to the observations that all immune complexes comprising three or more IgG molecules rapidly disappear from the circulation *in vivo* (Arend and Mannik, 1971; Mannik *et al.*, 1971; Mannik and Arend, 1971). More recent work further confirmed, in experimental immune complex disease in rabbits, elicited with bovine serum albumin (BSA) as the antigen, that only complexes of molecular weight (MW) ≈ 400,000 to 450,000 (IgG_2-BSA to IgG_2-BSA_2) and those of MW 600,000 to 670,000 (IgG_3-BSA_2 to IgG_3-BSA_3) remain in circulation, i.e., are not cleared by phagocytosis (Fagundus, 1980; see also Germuth *et al.*, 1972). These data from *in vivo* observations fit exactly into the thermodynamically imposed constraints.

4.3. ROLE OF COMPLEMENT AND IgM

The addition of complement to bacteria sensitized with specific immunoglobulins (IgG and IgM) strongly enhances their phagocytic ingestion (van Oss and Gillman, 1972b; van Oss *et al.*, 1975). Investigations using the consecutive addition of isolated complement (C) component to sensitized *E. coli* showed that opsonization took place at the *E. coli*–antibody–$C\overline{1423}$ stage (van Oss and Gillman, 1973). Enhanced phagocytic engulfment of that complex was accompanied by a marked increase in its hydrophobicity. *In vivo,* however, C receptors (or more precisely, C3b receptors) on phagocytic cells are also bound to play the preponderant role in this type of opsonization. Addition of C5, C6, C7, C8, C9 has no further effect on either hydrophobicity or phagocytic ingestion. Thus, $C\overline{1423}$ is the principal "heat-labile" opsonin. Experiments with rough forms of *salmonellae* and heat-inactivated C would seem to indicate that C3b-mediated opsonization of certain gram-negative bacteria via the alternate pathway is also possible (van Oss *et al.*, 1975). It is doubtful that IgM alone, under conditions where C is inactive, has any opsonic activity vis-à-vis PMNL (van Oss and Stinson, 1970). This agrees well with its rather pronounced hydrophilicity, as judged from its contact angle of 17.9° (van Oss *et al.*, 1975, 1980b), and its γ_{PrV} of 69.4 ergs/cm^2 (van Oss *et al.*, 1980b). It is interesting to note that alveolar macrophages have receptors for IgG and C3b, but not for IgM (Reynolds *et al.*, 1975).

4.4. OTHER OPSONIZING AGENTS

In principle a number of naturally occurring substances that can aspecifically adhere to microorganisms may have an opsonizing influence. Fibronectin, a known opsonin and general adhesion-inducing agent, is also known under a wide variety of other names (Yamada and Olden, 1978): cold-insoluble globulin, antigelatin factor, microfibrillar protein, opsonic protein, fibroblast surface antigen, galactoprotein-a, cell attachment factor, large external transformation-sensitive (LETS) protein, cell surface protein, zeta, cell-spreading factor, as well as opsonic α_2 surface-binding (α_2SB) glycoprotein (Blumenstock *et al.*, 1978; Saba, 1975). The aspecific and specific opsonic functions of IgG have been treated above.

4.5. ROLE OF IgA

IgA is quite hydrophilic; its contact angle is 15.6° (van Oss *et al.*, 1975), and particles and bacteria coated with it resist phagocytosis *in vitro* (van Oss and Stinson, 1970). IgA (as well as sIgA) may be regarded as a true dysopsonin, or a phagocytosis-inhibiting substance. It prevents the adherence of microorganisms to epithelial surfaces, as well as their ingestion by PMNL and macrophages (Reynolds *et al.*, 1978). The hydrophilicity of IgA also makes it uniquely suited for preventing viruses from penetrating cells (van Oss and Gillman, 1975). The dysopsonic function of IgA and sIgA is especially important in preventing infection (particularly virus infections) in the upper respiratory and lower digestive tracts, i.e., exactly in those parts of the mammalian anatomy where exclusion and subsequent removal of microorganisms (e.g., by coughing) is feasible without the help of phagocytosis (see also Heremans, 1975).

4.6. OPSONIZATION *IN VIVO*

As is already apparent from Table 2, whole blood contains a factor (lacking in HBSS) that causes an enhancement in the phagocytic uptake of *S. epidermidis*, to the same degree as can be achieved by prior opsonization with pooled human IgG. Table 9 illustrates that the more hydrophobic bacteria (in this case *S. epidermidis*) also become more phagocytized in whole blood, which correlates well with the higher degree of adsorption of IgG from whole plasma or serum by the more hydrophobic bacteria (see Table 6). Preopsonization with 1% pooled IgG caused a higher phagocytic uptake of both species and a smaller difference in uptake between the two species, which agrees well with the higher total adsorption of IgG in the absence of other blood proteins (see Table 5), compared with the adsorption of IgG in whole serum or plasma (Table 6).

5. PATHOLOGICAL PHAGOCYTES

A few children with recurrent upper respiratory infections (Bernstein and Gillman, 1976) or juvenile periodontal disease (Cianciola *et al.*, 1977; van Oss *et al.*, 1979d) have PMNL with significantly diminished phagocytic activity *in vitro*, and a significantly increased contact angle ($\theta \approx 20°$). The increase in hydrophobicity of these children's PMNL, manifesting itself in a decrease in γ_{PV} (as measured via the contact angle method), from a normal 69.0 ergs/cm^2 to $\approx$ 68.2 ergs/cm^2, has been confirmed with the leukocyte adhesion method (Absolom *et al.*, 1980a). Table 10 shows a comparison of the thermodynamic parameters of the adhesion and engulfment of normal with those of pathological PMNL of this type. Again (see also Table 1), the value of ΔF_{adh}, as long as it is below a certain critical value (most probably ≈ -0.2 erg/cm^2), does not appear to be decisive for the ultimate outcome of the ingestion. In this particular case, $\Delta F_{eng\text{-}1}$ was the same in both cases. Thus, the decreased phagocytic engulfment of the more hydrophobic (pathological) PMNL correlates only with the much decreased value of $\Delta F_{eng\text{-}2}$, corresponding to the final step of ingestion (see Fig. 2).

One of the interesting aspects of these pathological PMNL is that the abnormality in their contact angle and phagocytic activity resides in the patients' serum. Incubation of a given patient's PMNL in his or her own serum up to the

TABLE 9. PHAGOCYTIC UPTAKE OF NONOPSONIZED AND OPSONIZED (WITH 1% POOLED HUMAN IGG) *S. AUREUS* AND *S. EPIDERMIDIS* BY PMNL IN (NA_2 EDTA ANTICOAGULATED) WHOLE BLOOD[a]

Bacteria	Contact angle of bacteria (degrees)		Phagocytic activity ($N \pm$ S.E.)[b]			
	Nonopsonized	Opsonized	Nonopsonized		Opsonized	
Staphylococcus aureus 49	18.7	19.5	5.4 ± 0.4	$p < 0.05$[c]	6.4 ± 0.6	no significant difference
Staphylococcus epidermidis 47	19.1	19.8	7.1 ± 0.8		7.5 ± 0.6	

[a] Absolom *et al.* (1982a).
[b] Average number of bacteria ingested per PMNL (van Oss *et al.*, 1975).
[c] Statistical significance of the difference between these two values of $p < 0.05$ indicates that the likelihood that this difference arose by accident is less than 5%.

TABLE 10. COMPONENTS OF FREE ENERGIES OF ADHESION (ΔF_{adh}) AND ENGULFMENTS ($\Delta F_{eng\text{-}1}$ AND $\Delta F_{eng\text{-}2}$) OF *S. AUREUS* (OPSONIZED 1% WITH NORMAL HUMAN IGG) BY NORMAL AND PATHOLOGICAL HUMAN PMNL[a]

PMNL	Phagocytic activity ($N \pm$ S.E.)[b]	Contact angle of PMNL (degrees) ± S.E.	ΔF_{adh}[c]	$\Delta F_{eng\text{-}1}$[c]	$\Delta F_{eng\text{-}2}$[c]
Normal	6.9 ± 0.5	18.4 ± 0.2	−0.53	−0.18	−0.18
Patient G[a]	3.8 ± 0.4	20.3 ± 0.4	−0.64	−0.20	−0.05

[a]Absolom *et al.* (1980a).
[b]Average number of *S. aureus* particles ingested per PMNL (van Oss *et al.*, 1975).
[c]In ergs/cm^2 (1 erg/cm^2 = 1 mJ/m^2); ΔF values obtained by adhesion experiments agree with those obtained from contact angles, to the second decimal.

last moment before washing and determination of contact angle and phagocytic activity shows the abnormalities most clearly. Incubation of a patient's PMNL in *normal* serum, or in the patient's serum supplemented with an extra 0.5% HSA make both their contact angle and phagocytic activity revert to normal. With the patients with upper respiratory tract infections, treatment of normal PMNL with a patient's serum *in vitro* will make them manifest the same abnormalities of contact angle and phagocytosis (van Oss *et al.*, 1979d) (see also Hosking *et al.*, 1977, who reported on several hundred similar cases among children with chronic infections). It has not yet been possible to determine whether the patients' sera contain an excess of an unknown detrimental factor, or not enough of a factor necessary for normal phagocytosis. Although the HSA level of these patients' sera often is on the low-normal side, it rarely is below normal. Whatever it is, a modest increase in the HSA level (by 0.5% w/v) makes the pathological PMNL revert to normal (*in vitro*), although that treatment does not affect normal PMNL. The patients' PMNL also have chemotactic defects (Cianciola *et al.*, 1977). It may be speculated that an increased surface hydrophobicity possibly also reflects an increased surface viscosity, which would of course tend to impede engulfment. The addition of HSA to the pathological sera, which enhances the surface hydrophilicity of the pathological PMNL, may well also induce a decrease in their surface viscosity. It would not seem that differences in Fc receptors could play a role here, as added HSA should not influence either the quantity or the quality of such receptors.

Thus, while differences in hydrophobicity among bacteria showed the importance of $\Delta F_{eng\text{-}1} + \Delta F_{eng\text{-}2}$ ($= \Delta F_{eng\text{-}net}$) in phagocytic engulfment (see Table 1), the accident of nature, in the form of the pathological PMNL discussed above, singles out the importance of the later stages of ingestion ($\Delta F_{eng\text{-}2}$) in deciding engulfment (Table 10), at least under *in vitro* conditions.

In conclusion, among children with chronic upper respiratory or periodontal infections, a proportion of cases is found with PMNL that (after having been incubated in the patients' own serum) have a diminished phagocytic activity *in vitro*, as well as an increased surface hydrophobicity. Incubation of the patients' PMNL in normal serum, or in the patients' serum containing additional HSA makes the PMNL revert to normal. The *in vitro* changes are indicative of the importance of the last thermodynamic step of phagocytic ingestion (see Fig. 2). The meaning of the defect in the *in vivo* situation is not clear as yet, but it is speculated that the patients' PMNL, immersed in the patients' own serum, have an increased surface viscosity, which depresses phagocytosis *in vivo*.

6. "ACTIVATION" OF MACROPHAGES AND OTHER PHAGOCYTES

Macrophages, "activated" by, e.g., lymphokines such as migration inhibition factor (MIF), are characterized by enhanced: spreading ability, phagocytic activity, respiration rate, microbicidal activity, hydrolytic enzymes, proliferation (Holden and Englard, 1979; Vernon-Roberts, 1972). The salient characteristics of "activated" macrophages that are most suited to quantitative determination *in vitro* are macrophage migration inhibition (Yoshida *et al.*, 1972) and enhancement of macrophage adhesion (Dy *et al.*, 1974). Thrasher *et al.* (1973) found that *in vitro* treatment with MIF decreased the contact angle of guinea pig macrophages from 21.2° (control cells) to 17.8° ("activated" macrophages), while concomitantly the phagocytic activity (vis-à-vis sensitized erythrocytes) of the "activated" macrophages approximately doubled (as compared to the control cells). There also, the only component showing a significant ($\approx$ 40%) *increase* in free energy paralleling the decrease in contact angle and the increase in phagocytic activity accompanying "activation" is $\Delta F_{eng\text{-}2}$ (see Fig. 2) (compare also with Table 10). At the same time, an increased adhesion, and a decrease in migrating ability through glass capillaries (due to increased adhesion to the capillary wall) goes parallel with an *increase* in the negative value of ΔF_{adh}. One may assume that in the presence of albumin and/or fetal calf serum proteins ($\gamma_{LV} \approx 60$ ergs/cm^2), the glass surface (of the capillaries used in MIF tests) and the polystyrene surface (of the cell culture plates used in cell adhesion tests) become coated with albumin and acquire a surface tension $\gamma_{SV} \approx 70$ ergs/cm^2 (van Oss *et al.*, 1980b). In such circumstances, the more hydrophilic ("activated") macrophages should indeed adhere more strongly to these surfaces than the more hydrophobic control cells (Absolom *et al.*, 1980a).

Ingestion of hydrophobic particles, such as mycobacteria, polystyrene latex beads, or lipid droplets, also leads to a decrease in contact angle of guinea pig macrophages and human PMNL, concomitant with an increase in phagocytic activity (van Oss *et al.*, 1972b, 1975). Stimulation of macrophages with phospholipids also leads to "activation," *in vitro* as well as *in vivo* (Fauve, 1974). Cooper and West (1962) also noted stimulation of macrophages with suspensions of droplets of triolein, but Cooper and Houston later (1964) ascribed these effects to macrophage proliferation. It is unlikely, however, that proliferation is the only mechanism of the observed heightened activity, as essentially the same effect is obtained with PMNL, which do not proliferate. The increase in surface hydrophilicity of phagocytic cells, "activated" by hyperingestion of hydrophobic particles or droplets, has been explained in the following manner: "When particles are phagocytized, part of the phagocytes' cell membrane also disappears into the phagocytes' interior. This must give rise to stretching (and thus to dilution with water) of the material making up the phagocytes' cell membrane, which in turn renders the membrane more hydrophilic" (van Oss *et al.*, 1975). At the same time, however, a decrease in surface viscosity of the phagocytes' cell membranes probably ensues concomitantly with the increase in surface hydration. Such a decrease in surface viscosity also may well contribute to the observed increase in phagocytic activity, especially under *in vivo* conditions, as at $\gamma_{LV} \approx 50$ ergs/cm^2 the differences between "activated" and nonactivated macrophages with respect to ΔF_{adh} and $\Delta F_{eng\text{-}1}$ and $\Delta F_{eng\text{-}2}$, become negligible. Whatever the mechanism may be, it is likely that the "activation" of macrophages following the administration of hydrophobic particles such as mycobacteria or polystyrene latex particles is connected with the adjuvanticity of such hydrophobic particles (van Oss *et al.*, 1976).

In conclusion, the "activation" of phagocytic cells causes, *in vitro*, an increased surface hydrophilicity, an increased adhesiveness to various surfaces in the presence of proteins, and an increased phagocytic activity. The latter correlates with an enhancement of the last thermodynamic step of phagocytic ingestion (see Fig. 2). The mechanism of the "activation" *in vivo* may well be linked with a decrease in surface viscosity.

7. EFFECT OF VARIOUS AGENTS ON THE SURFACES OF PHAGOCYTES AND/OR BACTERIA *IN VITRO*

7.1. AGENTS MAINLY AFFECTING PHAGOCYTES

Hageman factor, or clotting factor XII, causes a decrease in the contact angle of PMNL ($\theta \approx 14.3°$) and an increase in their phagocytic activity (van Oss *et al.*, 1975); it is also known to enhance the adhesiveness of PMNL (Alexander and Good, 1970).

Heparin causes a decrease in the contact angle of macrophages ($\theta \approx 18.8°$) and an increase in their phagocytic activity (van Oss *et al.*, 1975; see also Kitchen and Megirian, 1971).

Levamisole (L-2,3,5,6-tetrahydro-6-phenylimidazo(2,1-*b*) thiazole HCP), an immunostimulating agent (Symoens, 1977), causes a decrease in the contact angle of macrophages ($\theta \approx 17.6°$) and an increase in their phagocytic activity (van Oss *et al.*, 1975).

Lectins, such as phytohemagglutinin and Con A, cause increases in the contact angles of macrophages (to respectively $\theta \approx 25.6°$ and $\theta \approx 24.3°$) and decreases in their phagocytic activity (van Oss *et al.*, 1975).

7.2. AGENTS MAINLY AFFECTING BACTERIA

A number of antibiotics may, in subinhibitory doses, cause a decrease in the contact angle of various bacteria and thus depress their phagocytic uptake, e.g., penicillin, chloromycetin, polymyxin, bacitracin (van Oss *et al.*, 1975) [see also Munoz and Geister (1950) with regard to aureomycin, and Louria (1974) concerning the effect of high doses of oral phenethicillin and injectable penicillin].

7.3. AGENTS AFFECTING BOTH PHAGOCYTES AND BACTERIA

The role of serum albumin has already been discussed in Section 5.

Surfactants (e.g., sodium desoxycholate, Tween, SDS) act on both phagocytic cells and bacteria, by decreasing their contact angles (to $\theta \approx 13°$), which depresses cell adhesion as well as phagocytic ingestion, and favors exocytosis (when the surfactants are administered after ingestion already has taken place) (van Oss *et al.*, 1975).

Ampicillin also causes a decrease in contact angle of phagocytic cells, and a strong decrease in θ of bacteria, resulting in a depression of phagocytic engulfment. Gentamycin on the other hand causes a *decrease* in the θ of phagocytes and a strong increase in the θ of bacteria, while results in a marked enhancement of phagocytic ingestion (van Oss *et al.*, 1975). Adam *et al.* (1974) reported an enhanced phagocytic uptake of L. monocytogenes by human macrophages in the presence of subinhibitory doses of ampicillin, chloramphenicol, and tetracycline.

8. INFLUENCE OF ELECTROKINETIC SURFACE POTENTIAL AND OF CELL SHAPE

The electrokinetic, or ζ-potential (as measured by means of cell microelectrophoresis; see van Oss, 1975, 1979b; Seaman and Brooks, 1979; van Oss and Fike, 1979) of human blood cells, under physiological conditions, is −11, −12, −16, and −18 mV for platelets, PMNL, lymphocytes, and erythrocytes, respectively (van Oss *et al.*, 1972c). That potential suffices to prevent any of these cells from approaching one another more closely than to within ≈ 50–80 Å, as long as they are reasonably smooth, and spherical or discoid shaped. However, as soon as cells have protruding pseudopodia, spicules, villi, or other projections, with a radius of curvature of ≤1000 Å (Bangham, 1964; van Oss *et al.*, 1972c), the diffuse electrostatic repulsion layer extending to the "secondary minimum" of 50–80 Å is easily pierced by such protrusions, upon which contact to within a few Ångströms can occur and van der Waals attractions prevail. PMNL as well as macrophages are characterized by their various protrusions of small radius of curvature, so that their contact with (and subsequent adhesion to and engulfment of) bacterial and other particles generally ensues without being hindered by the negative charge of virtually all bacteria. This accounts for the fact that attempts to correlate bacterial ζ-potentials with the degree to which they become phagocytized remain unrewarding (see, e.g., Joffe and Mudd, 1935; Daniels, 1967; Kozel *et al.*, 1980).

High glucose levels cause a depression in the phagocytic activity of PMNL (van Oss, 1971) and of macrophages, without, however, a change in contact angle of either cell sort (van Oss *et al.*, 1980). High glucose levels also result in diminished cell adhesiveness. The cause of the decrease in phagocytic activity and cell adhesiveness at high glucose concentrations (0.2 to 0.8%) appears to lie in the suppression of pseudopod formation and the increased tendency to assume a spherocytic shape by phagocytic cells under these conditions, favoring electrostatic repulsion between phagocytes and bacteria as well as between phagocytes and glass coverslips (van Oss *et al.*, 1972a, 1975). This phenomenon is most probably one of the major underlying causes of the strongly increased incidence of bacterial infections among uncontrolled diabetics (van Oss, 1971).

When platelets become "sticky" under the influence of ADP, their contact angle does not change (van Oss *et al.*, 1972a) but they become spiculated (White, 1968; van Oss *et al.*, 1972a), which allows them to overcome the electrostatic repulsion of other negatively charged cells or tissues. Unaided hemagglutination by blood-group antibodies of the IgG class is only feasible in those cases where these antibodies cause spiculation in erythrocytes, as with, e.g., anti-A and anti-B (but not with anti-D) antibodies (van Oss and Mohn, 1970; van Oss *et al.*, 1978). It is also interesting to note that viruses (e.g., adenoviruses) achieve contact with cells by means of spiky appendages with very small radii of curvature (van Oss *et al.*, 1972a).

In conclusion, spherical or otherwise smooth cells cannot normally approach one another closely enough to close-range interaction to occur, on account of the electrostatic repulsion between the negatively charged electrical double layers surrounding the cells. However, when pseudopodia or other processes of small radius of curvature (< 1000 Å) are protruded, such projections can penetrate the electrical double layer of other cells, and thus establish contact and initiate phagocytosis.

9. METHODS FOR MEASURING SURFACE TENSIONS OF CELLS, PARTICLES, AND PROTEINS

9.1. PARTITION METHODS

The partition of microorganisms between a hydrophobic liquid and water was first utilized for the (qualitative) classification of bacterial surfaces according to their hydrophobicity, by the pioneering efforts of Mudd and Mudd (1924, 1933). Albertsson (1971) introduced the use of immiscible aqueous dextran and polyethylene glycol solution in partition separations. Stendahl *et al.* (1973) used that method to demonstrate that the more hydrophobic (rough forms) of salmonellae are the ones that become phagocytized to the greatest extent; these results were later correlated with the more quantitative contact angle method (see below) by Cunningham *et al.* (1975). Stendahl *et al.* (1974, 1977) also showed via partition that salmonellae opsonized with IgG antibodies become more hydrophobic, and that this increased hydrophobicity appeared to correlate with an enhanced phagocytic uptake.

9.2. CONTACT ANGLE METHODS

Although the thermodynamic equilibrium condition relating surface tensions to contact angles has been known for at least 175 years (Young, 1805), coherent attempts to characterize solid surfaces through contact angles were made only in the second half of this century. The approaches taken may be divided into microscopic and macroscopic. In the main microscopic approach (Good, 1964), detailed knowledge of intermolecular forces between liquid and solid substrate is used, in conjunction with contact angle data, to predict the surface tension γ_{SV} of the substrate. In the case of biological systems, this knowledge will normally not be available so that we have to turn to macroscopic approaches. The oldest macroscopic approach is due to Zisman and his coworkers (Zisman, 1963, 1964). Zisman used contact angles obtained with many different liquids, of various surface tensions, plotting the cosines of the contact angles (θ) versus the surface tensions of the liquids used, and finding, by extrapolation to $\cos\theta = 1$, "the critical surface tension of wetting γ_c," of the solid surfaces.

However, apart from the uncertain thermodynamic status of γ_c, for osmotic and biochemical reasons, the use of drops of various organic liquids on monolayers of, e.g., leukocytes is liable to give aberrant results. In addition, the necessity of using liquids of even higher surface tension than the already quite hydrophilic (and thus high energy) cell surfaces severely limits the choice of liquids, and virtually reduces the choice to concentrated solutions of salts (e.g., $NaNO_3$) in water (van Oss *et al.*, 1975).

The interpretation of contact angles θ of cell monolayers became possible with the development of the equation of state by Neumann *et al.* (1974a), which made it possible to obtain the surface tensions of solid surfaces (γ_{SV}) from values of θ obtained with a single liquid only, which made it feasible to measure contact angles with drops of physiological saline water, the only clearly defined liquid (γ_{LV} = 72.8 ergs/cm^2) that is entirely compatible with all living cells under investigation (phagocytic cells as well as bacteria) (van Oss *et al.*, 1975). The angles θ are determined with the help of a telescope, with cross hairs attached to a goniometer (e.g., Gaertner Scientific Co., Chicago, Ill.; van Oss and Gillman, 1972a) (see Fig. 3). Before measuring θ on cell monolayers (of phagocytes as well as of bacteria), some preliminary air-drying is essential, to remove a surface layer of water that would otherwise yield $\theta = 0°$. To ascertain on the one hand that the water layer has disappeared, and to safeguard on the other hand against excessive desiccation, for each cell monolayer system a plateau value must be established for θ, by plotting θ values found versus drying time. The most dependable value for θ (see also Section 9.3) is found at the horizontal plateau, where, for a

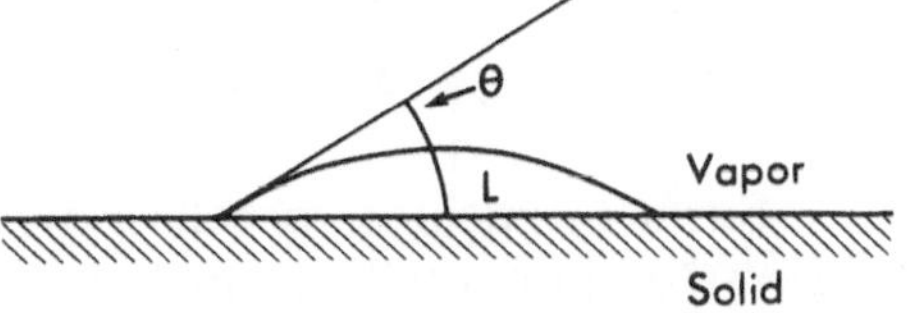

FIGURE 3. Contact angle θ of a liquid drop on a flat solid surface. The contact angle θ is the angle the tangent to the drop makes at the solid/liquid/air meeting point, measured through the drop.

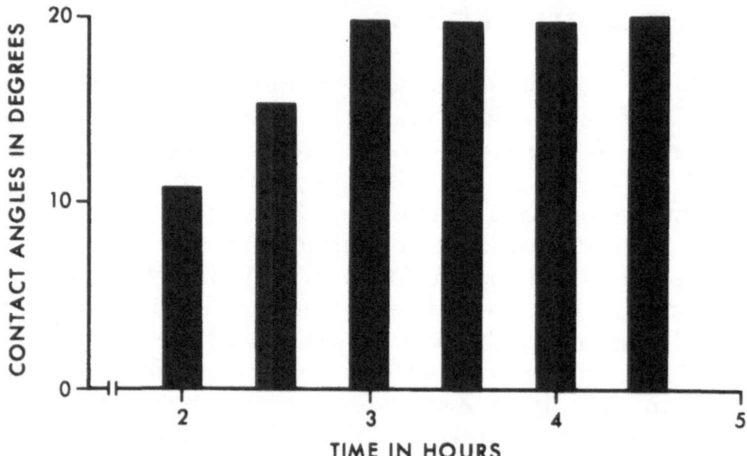

FIGURE 4. Contact angles of *Staphylococcus aureus* 49 vs. time of air-drying (on 2% agar, containing 10% glycerol). The plateau value clearly is ≈ 19.5°, persisting between 3 and 4 hr *post*-drying.

significant length of time (of the order of an hour or more), θ does not change (see Fig. 4).

Monolayers of phagocytic cells are obtained by depositing the cells from a suspension or isolate on siliconed glass surfaces. Human PMNL can be deposited directly from a few drops of finger-prick blood. After 20–30 min in a moist chamber at 37°C, the PMNL adhere firmly to the siliconed glass surface and the erythrocytes and other nonphagocytic cells may be washed off with saline (Newsome, 1967), resulting in a PMNL monolayer of 200–500 cells/mm (van Oss *et al.*, 1975; van Oss, 1978). Plateau values are usually reached after 1 hr air-drying.

Monolayers of bacteria from the sediment of a 17-hr broth culture (washed twice in saline) are best deposited on a flat layer of (2%) agar (containing saline and 10% glycerol); 2 to 3 hr air-drying is then generally needed before the plateau value is reached (van Oss *et al.*, 1975).

Contact angles of proteins are best measured by first obtaining a relatively thick and concentrated layer of protein, from a solution of the protein, by pressure ultrafiltration onto an anisotropic ultrafilter membrane of a pore size small enough to stop all of the protein in question (van Oss *et al.*, 1975). There also, plateau values have to be reached, after preliminary air-drying of the protein-coated membrane, after removal from the ultrafilter (van Oss *et al.*, 1980b).

With the help of a computer program (Neumann *et al.*, 1974a, 1980a) or by means of conversion tables (Neumann *et al.*, 1980b), contact angles θ may be converted to values for γ_{PV} or γ_{PL} (or γ_{BV} and γ_{BL}, etc.).

9.3. ADHESION AND ADSORPTION METHODS

Surface tensions of cells now can also be determined by a method that differs fundamentally from the contact angle method, i.e., by cell adhesion.

When cells are allowed to adhere to various different polymer surfaces (of high, medium, and low surface energy), while suspended in various aqueous media (adjusted to different surface tensions by the admixture of various proportions of DMSO), the surface tension γ_{LV} of the liquid medium in which the cells will adhere to all the different polymer surfaces to exactly the same degree is the same as the surface tension γ_{PV} of the cells: $\gamma_{LV} = \gamma_{PV}$. Using this method with, e.g., normal human PMNL, a value for $\gamma_{PV} = 69.0$ ergs/cm^2 was found (Absolom *et al.*, 1979), which correlates well with the value of $\gamma_{PV} = 69.1$ ergs/cm^2 found by the contact angle method, for $\theta = 18°$. The surface tensions of normal and pathological PMNL, found via the contact angle method (respectively 18.4° and 20.3°, corresponding to γ_{PV} values of 69.0 and 68.3 ergs/cm^2), also correlate well with the γ_{PV} values of respectively 69.0 and 68.2 ergs/cm^2 found by the cell adhesion method (Absolom *et al.*, 1980a) (see also Section 5).

In the same manner as with cell adhesion, surface tensions of proteins in their native hydrated state can be determined by protein adsorption. When proteins are allowed to adsorb onto various different polymer surfaces, while dissolved in various aqueous media (adjusted to different surface tensions by the admixture of various proportions of DMSO), the surface tension γ_{LV} of the liquid medium in which the cells will adsorb onto all the different polymer surfaces to exactly the same degree is the same as the surface tension γ_{PrV} of the proteins: $\gamma_{LV} = \gamma_{PrV}$. Using this method with four different human serum proteins: α_2-macroglobulin (α_2M), HSA, IgM, and IgG, surface tensions (γ_{PrV}) of respectively 71.0, 70.2, 69.4, and 67.3 ergs/cm^2 were found, correlating well with the surface tensions found for the same proteins via the contact angle method, with values for θ of 13.4, 15.2, 17.9, and 23.3°, corresponding to 70.9, 70.3, 69.4, and 67.3 ergs/cm^2, respectively (van Oss *et al.*, 1980b).

The cell adhesion and protein adsorption methods are considerably more tedious than the contact angle method, but as it is essentially a different and independent procedure, the close correlation between the results obtained with the method and those obtained by contact angle measurements, furnishes an important confirmation of the correctness and the accuracy of the plateau values obtained with the contact angle method.

9.4. SOLIDIFICATION FRONT METHOD

Particles or cells immersed in the liquid phase of a solidifying melt will be pushed by the advancing solidification front if the free energy for engulfment is positive. However, as the solidification front velocity is increased, a limiting or critical velocity V_c will be reached at which viscous drag becomes greater than the thermodynamic repulsion, so that engulfment then occurs (Neumann *et al.*, 1979a; Omenyi, 1978). From experimental data an empirical realtion could be established between ΔF_{adh} of the particles or cells, their diameter D, V_c, and other system properties (Smith *et al.*, 1980). This method was used for the determination of the surface tension (γ_{CV}) of glutaraldehyde-fixed human erythrocytes by Spelt *et al.* (1980), using both thymol and water as melt systems. A value for $\gamma_{CV} \approx 61.9$ ergs/cm^2 was found at 51°C (the melting point of thymol), corresponding to $\gamma_{CV} \approx 64.5$ ergs/cm^2 at 26°C. For human PMNL (in water/ice systems) a value for $\gamma_{PV} \approx 68$–69 ergs/cm^2 was obtained (Spelt, 1980).

9.5. DROPLET SEDIMENTATION METHOD

The maximum stability of suspensions prevails when the van der Waals attraction between particles or cells is reduced to zero, which occurs when the surface tension of the liquid suspending medium (γ_{LV}) equals that of the cells (γ_{CV}): $\gamma_{LV} = \gamma_{CV}$. Thus, by determining the γ_{LV} at which the highest concentration of cells can be supported by a density gradient (e.g., of D_2O) without giving rise to instability (as judged by droplet sedimentation), the value for γ_{CV} of the cells is automatically found (Omenyi *et al.*, 1980). For glutaraldehyde-fixed human erythrocytes, $\gamma_{CV} \approx 65$ ergs/cm^2 (at 26°C) was found, which conforms well with the value of 64.5 ergs/cm^2 found for the same cells with the solidification front method (see Section 9.4).

The last two methods are especially useful for γ_{CV} measurement of cells hardened by aldehyde fixation, as the contact angle and cell adhesion approaches have hitherto proved unsuitable with these rather rigid cell types. Glutaraldehyde treatment appears to decrease the rather high γ_{CV} of unaltered human erythrocytes from ≈ 70.1 ($\theta = 15°$) to ≈ 65 ergs/cm^2, which latter value would correspond to $\theta \approx 28°$.

9.6. PHAGOCYTIC INGESTION IN LIQUIDS OF DIFFERENT SURFACE TENSIONS

As already has been indicated above (see Section 2.2), and as appears clearly from Table 3, phagocytic ingestion of nonopsonized particles by PMNL closely parallels particle adhesion. In both phenomena a minimum exists when the surface tension of the particle is equal to that of the liquid medium ($\gamma_{BV} = \gamma_{LV}$) (Neumann *et al.*, 1979b; Absolom *et al.*, 1979, 1980a). Thus, *in vitro* phagocytic ingestion of nonopsonized bacteria and other cells and particles, in liquids of different surface tensions, also can be used as a method for the determination of the surface tension of such bacteria and other cells or particles. This not only applies to phagocytosis by PMNL, but also to ingestion by nonprofessional phagocytes, such as platelets (Zingg *et al.*, 1981; Absolom *et al.*, 1982b).

10. PHAGOCYTOSIS AND RECOGNITION *IN VITRO* AND *IN VIVO*

10.1. *IN VITRO* PHAGOCYTOSIS IN AQUEOUS MEDIA

In vitro, in aqueous media ($\gamma_{LV} \approx 72.8$ ergs/cm^2), the more hydrophobic the nonopsonized bacteria, the more they become engulfed by phagocytes. With increasing hydrophobicity of the bacteria, the negative values of ΔF_{adh} (see Fig.

1), as well as the $\Delta F_{eng\text{-}1}$ and $\Delta F_{eng\text{-}2}$ (see Fig. 2), become even more negative (see Table 1). The importance (*in vitro*) of the final step of engulfment in the guise of $\Delta F_{eng\text{-}2}$ (see Fig. 2) becomes apparent when the phagocytic activity of normal PMNL is compared with that of pathological (more hydrophobic) PMNL (see Table 10), which also agrees well with the increase in *in vitro* phagocytic activity concomitant with macrophage "activation," which is paralleled by a decrease in macrophage hydrophobicity, as well as by an increase in macrophage adhesiveness (in diluted protein solutions) (see Section 6).

The important role of ΔF_{adh} (see Fig. 1), particularly in the initiation of engulfment, is illustrated in Table 3, where it is shown that exactly as in cell adhesion, a minimum in phagocytic engulfment is found when the surface tension of the liquid medium (γ_{LV}) equals that of the bacterial surface (γ_{BV}): $\gamma_{LV} = \gamma_{BV}$.

Opsonization by IgG1 and IgG3 (specific as well as aspecific) causes an increase in surface hydrophobicity of the opsonized particles (to $\theta \simeq 23°$), while further opsonization with complement ($C\overline{1423}$, with C3b as active fraction) causes another increase in surface hydrophobicity (to $\theta = 25°$). This increased surface hydrophobicity is an important factor in the increased phagocytic uptake of the particles. However, as work with opsonized bacteria that already are rather hydrophobic (*S. epidermidis, L. monocytogenes;* see Table 8) has shown, specific receptors on the surface of phagocytic cells also play an important role in phagocytosis *in vitro*.

10.2. PHAGOCYTOSIS IN BIOLOGICAL FLUIDS

In biological fluids, with γ_{LV} values of 50–60 ergs/cm^2, phagocytic engulfment can be as pronounced as in aqueous media of $\gamma_{LV} \approx 72.8$ ergs/cm^2 (see Table 3), but, at low γ_{LV}, hydrophilic bacteria such as *Streptococcus pneumoniae* would be more readily engulfed than the more hydrophobic ones such as *Staphylococcus epidermidis* or *L. monocytogenes*. It is known that this is not the case *in vivo*. However, in biological fluids such as blood serum or plasma, significant amounts of IgG become aspecifically adsorbed to bacterial surfaces (Table 6), and the more so, the more hydrophobic the bacterial surface (Table 5 and 6). Decreases in the surface tension γ_{LV} of the liquid medium only depress the adsorption of IgG onto bacteria when γ_{LV} is lowered by means of low-molecular-weight solutes such as DMSO, but to a much lesser extent when γ_{LV} is lowered by proteins (Table 7). This is due to the fact that surface-tension-lowering agents that are of the same order of magnitude ($\approx$ 100 Å) as the proteins to be adsorbed, cannot easily influence surface–protein interactions that take place at distances as small as $\sim$ 2 Å (van Oss *et al.*, 1980c).

For the same reason one should regard the interaction between hydrophilic encapsulated bacteria and phagocytes *in vivo* (in biological media and blood) rather as interactions between capsular polysaccharide molecules and surface polysaccharide molecules of phagocytes, occurring in what for all practical purposes should be regarded as *ultrafiltrates* of blood or similar liquids (van Oss *et al.*, 1980c). As the surface tensions of granulocytes, capsules, as well as of ultrafiltrates of biological liquids all are close to ~ 70 ergs/cm^2 (within less than 1 erg/cm^2), their van der Waals attraction ($\Delta F \approx -0.02$ erg/cm^2) is negligible. On the other hand, the negative surface potential of both capsular polysaccharides and the sialic acid of PMNL causes a net Coulombic repulsion between phagocytes and encapsulated bacteria in biological fluids. This significant electrostatic repulsion, coupled to the lack of van der Waals attraction, forms the mechanism by which encapsulated microorganisms resist phagocytic ingestion *in vivo* (in the absence of specific antibodies). Aspecific IgG does not adsorb onto the outer periphery of capsules (Stinson and van Oss, 1971) and thus does not influence ingestion via Fc receptors. Only *specific* anticapsular antibodies of the IgG1 and IgG3 types, and *specific* anticapsular antibodies of the IgG1, IgG2, IgG3, and IgM types plus complement, can opsonize the outer surface of encapsulated bacteria; see the following section (10.3).

In conclusion, in biological fluids the more hydrophobic bacteria also become engulfed to the greatest extent, due to their capacity to adsorb more IgG (which amounts to enhanced opsonization) (see Table 9), directly through phagocytic Fc receptors, as well as indirectly, after fixation of complement, through C3 receptors on the phagocytes. Very hydrophilic (encapsulated) bacteria resist phagocytic ingestion (in the absence of specific antibody) through electrostatic repulsion, coupled with virtually zero van der Waals attraction.

10.3. RECOGNITION

Aspecific recognition *in vivo* of nonvirulent bacteria (which comprise the vast majority of bacterial species) is to an important extent based on their relative surface hydrophobicity, which causes them to adsorb IgG (mainly aspecifically), which amounts to opsonization through the adsorption of IgG1 and IgG3 alone (via phagocytic Fc receptors), and/or through the adsorption of IgG1, IgG2, and IgG3, through complement fixation (via C3-receptors on the phagocytic cells).

Specific recognition (*in vivo*) is effected through:

1. Specific antibodies of the IgG1 and IgG3 classes, which give rise to opsonization (via Fc receptors on the phagocytes).
2. Specific antibodies of the IgG1, IgG2, IgG3, and IgM classes, which, after fixation of complement, give rise to opsonization (via C3 receptors on the phagocytes). This C fixation also gives rise (through C5a) to chemotaxis of phagocytes, toward the scene of interaction.
3. Specific T lymphocytes, which give rise to lymphokines that (aspecifically) "activate" phagocytic cells (increasing their phagocytic activity, through an increase in their surface hydrophilicity and, possibly, a concomitant decrease in their surface viscosity), and chemotactic agents that cause movement of phagocytes to the scene of interaction.
4. Specific antibodies of the IgA class, which in more diluted body fluids (secretions), by coating microorganisms (especially viruses) prevent their attachment to and their subsequent internalization by cells.

REFERENCES

Absolom, D. R., Neumann, A. W., Zingg, W., and van Oss, C. J., 1979, Thermodynamic studies of cellular adhesion, *Trans. Am. Soc. Artif. Intern. Organs* **25**:152.

Absolom, D. R., van Oss, C. J., Genco, R. J., Francis, D. W., Zingg, W., and Neumann, A. W., 1980a, Surface thermodynamics of normal and pathological human granulocytes, *Cell Biophys.* **2**:113.

Absolom, D. R., van Oss, C. J., Neumann, A. W., and Zingg, W., 1980b, Determination of surface tensions of proteins. II. Surface tension of serum albumin altered at the protein–air interface, 2nd International Chemical Congress on the North American Continent, Abstracts ACS Meeting, Las Vegas.

Absolom, D. R., Zingg, W., van Oss, C. J., and Neumann, A. W., 1981, Determination of surface tensions of proteins. II. Surface tension of serum albumin altered at the protein–air interface, *Biochim. Biophys. Acta* **670**:74.

Absolom, D. R., van Oss, C. J., Zingg, W., and Neumann, A. W., 1982a, Phagocytosis as a surface phenomenon: Opsonization by aspecific adsorption of IgG as a function of bacterial hydrophobicity, *J. Reticuloendothelial Soc.* **31**:59.

Absolom, D. R., Francis, D. W., Zingg, W., van Oss, C. J., and Neumann, A. W., 1982b, Phagocytosis of bacteria by platelets: Surface thermodynamics, *J. Colloid Interface Sci.* **85**:168.

Adam, D., Schaffert, W., and Marget, W., 1974, Enhanced *in vitro* phagocytosis of *Listeria monocytogenes* by human monocytes in the presence of ampicillin, tetracycline and chloramphenicol, *Infect. Immun.* **9**:811.

Albertsson, P. A., 1971, *Partition of Cell Particles and Macromolecules*, Wiley–Interscience, New York.

Alexander, J. W., and Good, R. A., 1970, *Immunobiology for Surgeons*, p. 10, Saunders, Philadelphia.

Arend, W. P., and Mannik, M., 1971, Studies on antigen–antibody complexes. II. Quantification of tissue uptake of complement-depleted rabbits, *J. Immunol.* **107**:63.

Bangham, A. D., 1964, The adhesiveness of leukocytes with special reference to zeta potential, *Ann. N.Y. Acad. Sci.* **116**:945.

Bernstein, J. M., and Gillman, C. F., 1976, Phagocytic dysfunction as a cause of recurrent upper respiratory disease, *Trans. Am. Acad. Ophthalmol. Otolaryngol.* **82**:509.

Blumenstock, F. A., Saba, T. M., Weber, P., and Laffin, R., 1978, Biochemical and immunological characterization of human opsonic α_2 SB glycoprotein: Its identity with cold insoluble globulin, *J. Biol. Chem.* **253**:4287.

Cianciola, L. J., Genco, R. J., Patters, R. J., McKenna, M. R., and van Oss, C. J., 1977, Defective polymorphonuclear leukocyte function in a human periodontal disease, *Nature (London)* **265**:445.

Cooper, G. N., and Houston, B., 1964, Effects of simple lipids on the phagocytic properties of peritoneal macrophages. II. Studies on the phagocytic potential of cell populations, *Aust. J. Exp. Biol. Med. Sci.* **42**:429.

Cooper, G. N., and West, D., 1962, Effects of simple lipids on the phagocytic properties of peritoneal macrophages. I. Stimulatory effects of glyceryl trioleate, *Aus. J. Exp. Biol.* **40**:485.

Cunningham, R. K., Söderström, T. O., Gillman, C. F., and van Oss, C. J., 1975, Phagocytosis as a surface phenomenon. V. Contact angles and phagocytosis of rough and smooth strains of *Salmonella typhimurium*, and the influence of specific antiserum, *Immunol. Commun.* **4**:429.

Daniels, S. L., 1967, Separation of bacteria by adsorption onto ion-exchange resins, Ph.D. dissertation, University of Michigan, Ann Arbor.

Davies, W., 1975, Interactions between particles and cells, Ph.D. dissertation, University of Sydney, Australia.

Davis, B. D., Dulbecco, R., Eisen, H. N., Ginsberg, H. S., and Wood, W. B., 1973, *Microbiology*, pp. 633–634, Harper & Row (Hoeber), New York.

Dy, M., Dimitriu, A., Thomson, N., and Hamburger, J., 1974, A macrophage adherence test, *Ann. Immunol. Inst. Pasteur* **125c**:451.

Fagundus, A. M., 1980, Physico-chemical and immunochemical studies on circulating immune complexes in systemic chronic serum sickness of the rabbit, Ph.D. dissertation, SUNY, Buffalo, p. 117.

Fauve, R. M., 1974, Immunostimulation with phospholipids, in: *Activation of Macrophages* (W. H. Wagner, H. Hahn, and R. Evans, eds.), pp. 157–176. Excerpta Medica, Amsterdam.

Germuth, F. C., Sentrefit, L. B., and Dreesman, G. R., 1972, Immune complex disease. V. The nature of circulating immune complexes with glomeruli alterations in the chronic BSA-rabbit system, *Johns Hopkins Med. J.* **130**:344.

Gigli, I., and Nelson, R. A., 1968, Complement dependent immune phagocytosis, *Exp. Cell Res.* **51**:45.

Good, R. J., 1964, Theory for the estimation of surface and interfacial energies. VI. Surface energies of some fluorocarbon surfaces from contact angle measurements, *Adv. Chem. Ser.* **43**:74.

Hed, J., 1977, The extinction of crystal violet and its use to differentiate between attached and ingested microorganisms in phagocytosis, *FEMS Lett.* **1**:357.

Hed, J., 1979, Studies on phagocytosis by human polymorpho-nuclear leukocytes using a new assay which allows distinction between attachment and ingestion, M.D. dissertation, Linköping University, Medical Microfilms #74, Linköping.

Heremans, J. F., 1975, The secretory immune system, *Int. Convoc. Immunol.* **4**:376.

Holden, J. W., and Englard, A., 1979, Macrophage activation and proliferation, in: *Phagocytosis* (Y. Kokubun and N. Kobayashi, eds.), pp. 147–168, University Park Press, Baltimore.

Hosking, C. S., Fitzgerald, M. G., and Shelton, M. J., 1977, Results of immune function testing in children with recurrent infections, *Aus. Paediatr. J.* **13**:(Suppl.):61.

Joffe, E. W., and Mudd, S., 1935, A paradoxical relation between zeta potential and suspension stability on S and R variants of intestinal bacteria, *J. Gen. Physiol.* **18**:599.

Kitchen, A. G., and Megirian, R., 1971, Heparin enhancement of Kupffer cell phagocytosis *in vitro*, *J. Reticuloendothelial Soc.* **9**:13.

Kozel, T. R., Reiss, E., and Cherniak, R., 1980, Concomitant but not casual association between surface charge and inhibition of phagocytosis by cryptococcal polysaccharide, *Infect. Immun.* **29**:295.

Labaw, L. W., and Davies, D. R., 1971, An electron microscopic study of human gamma G_1 immunoglobulin crystals, *J. Biol. Chem.* **246**:3760.

Louria, D. B., 1974, Superinfection: A partial overview, in: *Opportunistic Pathogens* (J. E. Prier and H. Friedman, eds.), pp. 1–18, University Park Press, Baltimore.

Mannik, M., and Arend, W. P., 1971, Fate of preformed immune complexes in rabbits and rhesus monkeys, *J. Exp. Med.* **134**:19s.

Mannik, M., Arend, W. P., Hall, A. P., and Gilliland, B. C., 1971, Studies on antigen–antibody complexes. I. Elimination of soluble complexes from rabbit circulation, *J. Exp. Med.* **133**:713.

Mudd, E. B. H., and Mudd, S., 1933, Process of phagocytosis: Agreement between direct observation and deductions from theory, *J. Gen. Physiol.* **16**:625.

Mudd, S., and Mudd, E. B. H., 1924, Certain interfacial tension relations and the behavior of bacteria in films, *J. Exp. Med.* **40**:647.

Munoz, J., and Geister, R., 1950, Inhibition of phagocytosis by aureomycin, *Proc. Soc. Exp. Biol. Med.* **75**:367.

Neumann, A. W., van Oss, C. J., and Szekely, J., 1972, Thermodynamics of particle engulfment: Particle engulfment by solidifying melts and phagocytosis, *Kolloid Z. Polym.* **251**:415.

Neumann, A. W., Good, R. J., Hope, C. J., and Sejpal, M., 1974a, An equation-of-state approach to determine surface tensions of low-energy solids from contact angles, *J. Colloid Interface Sci.* **49**:291.

Neumann, A. W., Gillman, C. F., and van Oss, C. J., 1974b, Phagocytosis and surface free energies, *Electroanal. Chem. Interfacial Electrochem.* **49**:393.

Neumann, A. W., Omenyi, S. N., and van Oss, C. J., 1979a, Negative Hamaker coefficients. I. Particle engulfment or rejection at solidification fronts, *Colloid Polym. Sci.* **257**:413.

Neumann, A. W., Absolom, D. R., van Oss, C. J., and Zingg, W., 1979b, Surface thermodynamics of leukocyte and platelet adhesion to polymer surfaces, *Cell Biophys.* **1**:79.

Neumann, A. W., Hum, O. S., Francis, D. W., Zingg, W., and van Oss, C. J., 1980a, Kinetic and thermodynamic aspects of platelet adhesion, *J. Biomed. Mater. Res.* **14**:499.

Neumann, A. W., Absolom, D. R., Francis, D. W., and van Oss, C. J., 1980b, Conversion tables of contact angles to surface tensions, *Sep. Purif. Methods* **9**:69.

Neumann, A. W., Absolom, D. R., Francis, D. W., Zingg, W., and van Oss, C. J., 1982, Surface thermodynamics of phagocytic ingestion of non-opsonized bacteria by granulocytes in liquids of different surface tensions, *Cell Biophys.* **4**:285.

Newsome, J., 1967, Phagocytosis by human neutrophils, *Nature (London)* **214**:1092.

Nisonoff, A., Hopper, J. E., and Spring, S. B., 1975, *The Antibody Molecule*, pp. 96–97, Academic Press, New York.

Omenyi, S. N., 1978, Attraction and repulsion of particles by solidifying melts, Ph.D. dissertation, University of Toronto.

Omenyi, S. N., Snyder, R. S., van Oss, C. J., Absolom, D. R., and Neumann, A. W., 1980, Effects of zero van der Walls and zero electrostatic forces on droplet sedimentation, 2nd International Chemical Congress on the North American Continent, Abstracts ACS Meeting, Las Vegas.

Reynolds, H. Y., Atkinson, J. P., Newball, H. H., and Frank, M. M., 1975, Receptors for immunoglobulin and complement on human alveolar macrophages, *J. Immunol.* **114**:1813.

Reynolds, H. Y., Merrill, W. M., Amento, E. P., and Nagel, G. P., 1978, Immunoglobulin A in secretions in human from the lower respiratory tract, in: *Secretory Immunity and Infection* (J. R. McGhee, J. Mestecky, and J. L. Babb, eds.), pp. 533–564, Plenum Press, New York.

Saba, T. M., 1975, Aspecific opsonins, *Int. Convoc. Immunol.* **4**:489.

Seaman, G. V. F., and Brooks, D. E., 1979, Analytical cell electrophoresis, in: *Electrokinetic Separation Methods* (P. G. Rhigetti, C. J. van Oss, and J. W. Vanderhoff, eds.), pp. 95–110, Elsevier, Amsterdam.

Smith, R. P., Omenyi, S. N., and Neumann, A. W., 1980, Dimensional analysis of behaviour of small particles at solidification fronts, 54th Colloid Surf. Sci. Symp., Lehigh University, Bethlehem, Pa.

Spelt, J. K., 1980, Surface tension measurements of biological cells using the freezing front technique, M.Sc. thesis, University of Toronto.

Spelt, J. K., Absolom, D. R., Neumann, A. W., van Oss, C. J., and Zingg, W., 1980, Surface tension and wettability studies of human erythrocytes using the freezing front technique, 54th Colloid Surf. Sci. Symp., Lehigh, Pa.

Stendahl, O., Magnusson, K. E., Tagesson, C., Cunningham, R., and Edebo, L. B., 1973, Influence of hyperimmune immunoglobulin G on the physicochemical properties of the surface of *Salmonella typhimurium* 395 MS in relation to their interaction with phagocytic cells, *Infect. Immun.* **7**:573.

Stendahl, O., Tagesson, C., and Edebo, M., 1973, Partition of *Salmonella typhimurium* in a two-polymer aqueous phase system in relation to liability to phagocytosis, *Infect. Immun.* **8**:36.

Stendahl, O., Tagesson, C., Magnusson, K. E., and Edebo, L. B., 1977, Physicochemical consequences of opsonization of *Salmonella typhimurium* with hyperimmune IgG and complement, *Immunology* **32**:11.

Stinson, M. W., and van Oss, C. J., 1971, Immunoglobulins as specific opsonins. II. The influence of specific and aspecific immunoglobulins on the *in vitro* phagocytosis of noncapsulated, capsulated, and decapsulated bacteria by human neutrophils, *J. Reticuloendothelial Soc.* **9**:503.

Symoens, J., 1977, Levaucisole, an antianergic chemotherapeutic agent: an overview, in: *Control of Neoplasia by Modulation of the Immune System* (M. A. Chirigos, ed.), pp. 1–24, Raven Press, New York.

Tan, J. S., Watanakunakorn, C., and Phair, J. P., 1971, A modified assay of neutrophil function: Use of Lysostaphin to differentiate defective phagocytosis from impaired intracellular killing, *J. Lab. Clin. Med.* **78**:316.

Thrasher, S. G., Yoshida, T., van Oss, C. J., Cohen, S., and Rose, N. R., 1973, Alteration of macrophage interfacial tension by supernatants of antigen-activated lymphocyte cultures, *J. Immunol.* **110**:321.

Valentine, R. C., and Green, N. M., 1967, Electron microscopy of an antibody–hapten complex, *J. Mol. Biol.* **27**:615.

van Oss, C. J., 1971, Influence of glucose levels on the *in vitro* phagocytosis of bacteria by human neutrophils, *Infect. Immun.* **4**:54.

van Oss, C. J., 1975, Influence of the size and shape of molecules on their electrophoretic mobility, *Sep. Purif. Methods* **4**:167.

van Oss, C. J., 1978, Phagocytosis as a surface phenomenon, *Annu. Rev. Microbiol.* **32**:19.

van Oss, C. J., 1979a, The immunoglobulins, in: *Principles of Immunology* (N. R. Rose, F. Milgrom, and C. J. van Oss, eds.), Macmillan Co., New York.

van Oss, C. J., 1979b, Electrokinetic separation methods, *Sep. Purif. Methods* **8**:119.

van Oss, C. J., and Fike, R. M., 1979, Analytical cell electrophoresis, in: *Electrokinetic Separation*

Methods (P. G. Rhigetti, C. J. van Oss, and J. W. Vanderhoff, eds.), pp. 111–120, Elsevier, Amsterdam.

van Oss, C. J., and Gillman, C. F., 1972a, Phagocytosis as a surface phenomenon. I. Contact angles and phagocytosis of non-opsonized bacteria, *J. Reticuloendothelial Soc.* **12:**283.

van Oss, C. J., and Gillman, C. F., 1972b, Phagocytosis as a surface phenomenon. II. Contact angles and phagocytosis of encapsulated bacteria before and after opsonization by specific antiserum and complement, *J. Reticuloendothelial Soc.* **12:**497.

van Oss, C. J., and Gillman, C. F., 1973, Phagocytosis as a surface phenomenon. III. Influence of C1423 on the contact angle and on the phagocytosis of sensitized encapsulated bacteria, *Immunol. Commun.* **2:**415.

van Oss, C. J., and Gillman, C. F., 1975, Phagocytosis and immunity, *Int. Convoc. Immunol.* **4:**505.

van Oss, C. J., and Mohn, J. F., 1970, Scanning electron microscopy of red cell agglutination, *Vox Sang.* **19:**432.

van Oss, C. J., and Stinson, M. W., 1970, Immunoglobulins as aspecific opsonins. I. The influence of polyclonal and monoclonal immunoglobulins on the *in vitro* phagocytosis of latex particles and staphylococci by human neutrophils, *J. Reticuloendothelial Soc.* **8:**397.

van Oss, C. J., Good, R. J., and Neumann, A. W., 1971, The connection of interfacial free energies and surface potentials with phagocytosis and cellular adhesiveness, *Electroanal. Chem. Interfacial Electrochem.* **37:**387.

van Oss, C. J., Gillman, C. F., and Good, R. J., 1972a, The influence of the shape of phagocytes on their adhesiveness, *Immunol. Commun.* **1:**627.

van Oss, C. J., Rose, N. R., Cohen, S., Thrasher, S. G., and Gillman, C. F., 1972b, Alteration of the interfacial tension of phagocytes by the ingestion of mycobacteria and latex particles, Annu. Meet. Am. Soc. Microbiol. Abstr. p. 896.

van Oss, C. J., Good, R. J., and Neumann, A. W., 1972c, The connection of interfacial free energies and surface potentials with phagocytosis and cellular adhesiveness, *J. Electroanal. Chem.* **37:**387.

van Oss, C. J., Woeppel, M. S., and Marquart, S. E., 1973, Immunoglobulins as aspecific opsonins. III. The opsonizing power of fragments of polyclonal and monoclonal immunoglobulin G, *J. Reticuloendothelial Soc.* **13:**221.

van Oss, C. J., Gillman, C. F., and Neumann, A. W., 1974, Phagocytosis as a surface phenomenon. IV. The minimum size and composition of antigen–antibody complexes that can become phagocytized, *Immunol. Commun.* **3:**77.

van Oss, C. J., Gillman, C. F., and Neumann, A. W., 1975, *Phagocytic Engulfment and Cell Adhesiveness*, Dekker, New York.

van Oss, C. J., Gillman, C. F., and Singer, J. M., 1976, The influence of particulate carriers and of mineral oil in adjuvant on the antibody response in rabbits to human gamma globulin, *Immunol. Commun.* **5:**181.

van Oss, C. J., Mohn, J. F., and Cunningham, R. J., 1978, Influence of various physiocochemical factors on hemagglutination, *Vox Sang.* **34:**351.

van Oss, C. J., Omenyi, S. N., and Neumann, A. W., 1979a, Negative Hamaker coefficients. II. Phase separation of polymer solutions, *Colloid Polym. Sci.* **257:**737.

van Oss, C. J., Absolom, D. R., and Neumann, A. W., 1979b, Repulsive van der Waals forces. II. The mechanism of hydrophobic chromatography, *Sep. Sci. Technol.* **14:**305.

van Oss, C. J., Absolom, D. R., Grossberg, A. L., and Neumann, A. W., 1979c, Repulsive van der Waals forces. I. Complete dissociation of antigen–antibody complexes by means of negative van der Waals forces, *Immunol. Commun.* **8:**11.

van Oss, C. J., Bernstein, J. M., Park, B. H., Cianciola, L. J., and Genco, R. J., 1979d, Physicochemical aspects of phagocytosis and of some phagocytic disorders, *Int. Convoc. Immunol.* **6:**311.

van Oss, C. J., Absolom, D. R., and Neumann, A. W., 1980a, Applications of net repulsive van der Waals forces between different particles, macromolecules or cells in liquids, *Colloid Surf.* **1:**45.

van Oss, C. J., Absolom, D. R., Neumann, A. W., and Zingg, W., 1980b, Determination of surface tensions of proteins. I. Surface tensions of native serum proteins in aqueous media, 2nd International Chemical Congress on the North American Continent, Abstracts ACS Meeting, Las Vegas.

van Oss, C. J., Neumann, A. W., Good, R. J., and Absolom, D. R., 1980c, The influence of extremely small as well as repulsive van der Waals forces on cell interactions, in: *Bioelectrochemistry: Ions, Surfaces and Membranes* (M. Blank, ed.), *Adv. Chem. Ser.* **188**:107, Academic Press, New York.

van Oss, C. J., Beckers, D., Engelfriet, C. P., Helmerhorst, F. M., Bruyns, E. C., Absolom, D. R., and Neumann, A. W., 1980d, Elution of blood group antibodies from blood cells by means of repulsive van der Waals forces, 2nd International Chemical Congress on the North American Continent, Abstracts ACS Meeting, Las Vegas.

van Schaik, M. L. J., Weening, R. S., Voetman, A., and Roos, D., 1975, Phagocytosis and killing of *S. aureus* by human granulocytes, Ann. Rep. Karl Landsteiner Found., Cntrl. Lab. Netherlands Red Cross Blood Transfusion Service, Amsterdam, pp. 59–61.

Vernon-Roberts, B., 1972, *The Macrophage*, pp. 150–151, Cambridge University Press, London.

White, J. G., 1968, Fine structural alterations induced in platelets or adenosine diphosphate, *Blood* **31**:604–622.

Weening, R. S., Roos, D., and van Schaik, M. L. J., 1979, Defective initiation of the metabolic stimulation in phagocytizing granulocytes, in: *Inborn Errors of Immunity and Phagocytosis* (F. Gittler, J. W. T. Seakins, and R. A. Harkness, eds.), p. 291, University Park Press, Baltimore.

Yamada, K. M., and Olden, K., 1978, Fibronectins—Adhesive glycoproteins of cell surface and blood, *Nature (London)* **275**:179.

Yoshida, T., Janeway, C. A., and Paul, W. E., 1972, Activity of migration inhibitory factor in the absence of antigen, *J. Immunol.* **109**:201.

Zingg, W., Absolom, D. R., Neumann, A. W., Francis, D. W., and van Oss, C. J., 1981, Platelet phagocytosis as a model for the study of forces involved in platelet adhesion, Abstr. 2nd Eur. Conf. Biomaterials, Gothenburg, Sweden.

Zisman, W. A., 1963, Influence of constitution on adhesion, *Ind. Eng. Chem.* **55**:18.

Zisman, W. A., 1964, Relation of equilibrium contact angle to liquid and solid constitution, *Adv. Chem. Ser.* **43**:1.

2

Regulation of Mononuclear Phagocyte Proliferation

CARLETON C. STEWART

1. INTRODUCTION

The diversity of functions assigned to mononuclear phagocytes (MNP) leads to the view that either the system is composed of multiple functionally distinct subpopulations (Walker, 1976; Walker *et al.*, 1981) or that expression of a particular function represents differentiation states through which all cells pass as they mature. Superimposed on either developmental sequence, environmental stimuli may modulate the expression fo MNP function. For example, surface markers such as Ia or C3b receptors might be the result of clonal expansion of a subpopulation (Walker *et al.*, 1981). The expression of Ia could also be a modulatable activity. Several investigators have shown that incubation of macrophages with mitogen-stimulated lymphocyte products can induce Ia on the surface membrane of macrophages (Farr *et al.*, 1979; Beller and Unanue, 1981).

In contrast, such properties as adherence, phagocytosis, and expression of Fc receptors appear to be accommodated within the framework of a single-lineage differentiation scheme with MNP acquiring these properties as they mature (Rabellino and Metcalf, 1975; van Furth *et al.*, 1979). Other functions of MNP appear to be induced by the environment in which they reside. These include neutral protease secretion (Wahl *et al.*, 1975; Gordon and Werb, 1976; Vassalli and Reich, 1977; Vassalli *et al.*, 1977), lymphokine-induced tumoricidal (Ruco and Meltzer, 1977; Leonard *et al.*, 1978) and microbicidal activity (Cole, 1975; Nacy *et al.*, 1981), and, as mentioned, expression of Ia. Even though phagocytosis is an accepted property of all MNP, it too is an opsonin-stimulated function. Thus, it is possible to envision this host defense system as a series of subpopulations, some of which exhibit certain distinct functions while at the same time other functions are shared by all cells within the series.

CARLETON C. STEWART • Experimental Pathology Group, Los Alamos National Laboratory, Los Alamos, New Mexico 87545.

TABLE 1. COLONY-FORMING CELLS OBTAINED FROM VARIOUS MURINE TISSUES

Source	Colonies per 10^4 cells	
	Day 7	Day 14
Bone marrow	80	310
Spleen	5	145
Blood	1	121
Bronchial lavage	0	530

The present nomenclature of the mononuclear phagocyte system (MPS) was originally proposed by van Furth *et al.* (1972; van Furth, 1980). This orderly progression of differentiated states represents a good working hypothesis upon which to build the functional categorization of MNP. Accordingly, we will refer to the various cells of the MNP system according to the van Furth model. One of the major controversies concerns whether MNP proliferate locally in the tissue as suggested by Volkman (1976a,b) or whether they all come from the bone marrow as suggested by van Furth *et al.* (1972; van Furth, 1980). Our results clearly show that MNP within the tissue have the capacity to proliferate (Stewart, 1980; Stewart *et al.*, 1975). We will review evidence showing that MNP from all murine sources have the capacity to proliferate.

2. COLONY FORMATION BY MURINE MNP

MNP from all murine tissues that have been tested will form discrete colonies from single progenitor cells when grown in medium containing serum and macrophage growth factor (MGF) (Lin and Stewart, 1973; Stewart, 1980, 1981; van der Zeijst *et al.*, 1978). As summarized in Table 1, there is a variable frequency of colony-forming cells (CFC) which can be obtained from various murine tissues. When bone marrow cells (BMC) are cultured in liquid culture medium with L-cell conditioned medium, a potent source of MGF (Stewart and Lin, 1978), large colonies about 3 mm in diameter will form by day 7 (Fig. 1). The spleen is the only other source of MNP that contain progenitors capable of forming colonies by day 7. This is because the mouse spleen is also an active hematopoietic organ.

For bone marrow, about 0.8 ± 0.3% of cells from C3H mice will form colonies by day 7. As discussed later, these CFC are found in the nonadherent

FIGURE 1. Colonies formed by mononuclear phagocytes from various sources. Cells were isolated and appropriately diluted with growth medium and 3 ml was plated in 35-mm culture dishes. (A) 10^4 bone marrow cells after 7 days' incubation. (B) The edge of one colony from the plate shown in (A) at 100×. (C) 10^4 spleen cells after 14 days' incubation. (D) The edge of one colony shown in (C). (E) 10^4 peritoneal cells elicited with thioglycollate medium. (F) A colony shown in (E). (G) 10^3 alveolar cells after 14 days' incubation. (H) A colony shown in (G). (From Stewart, 1979.)

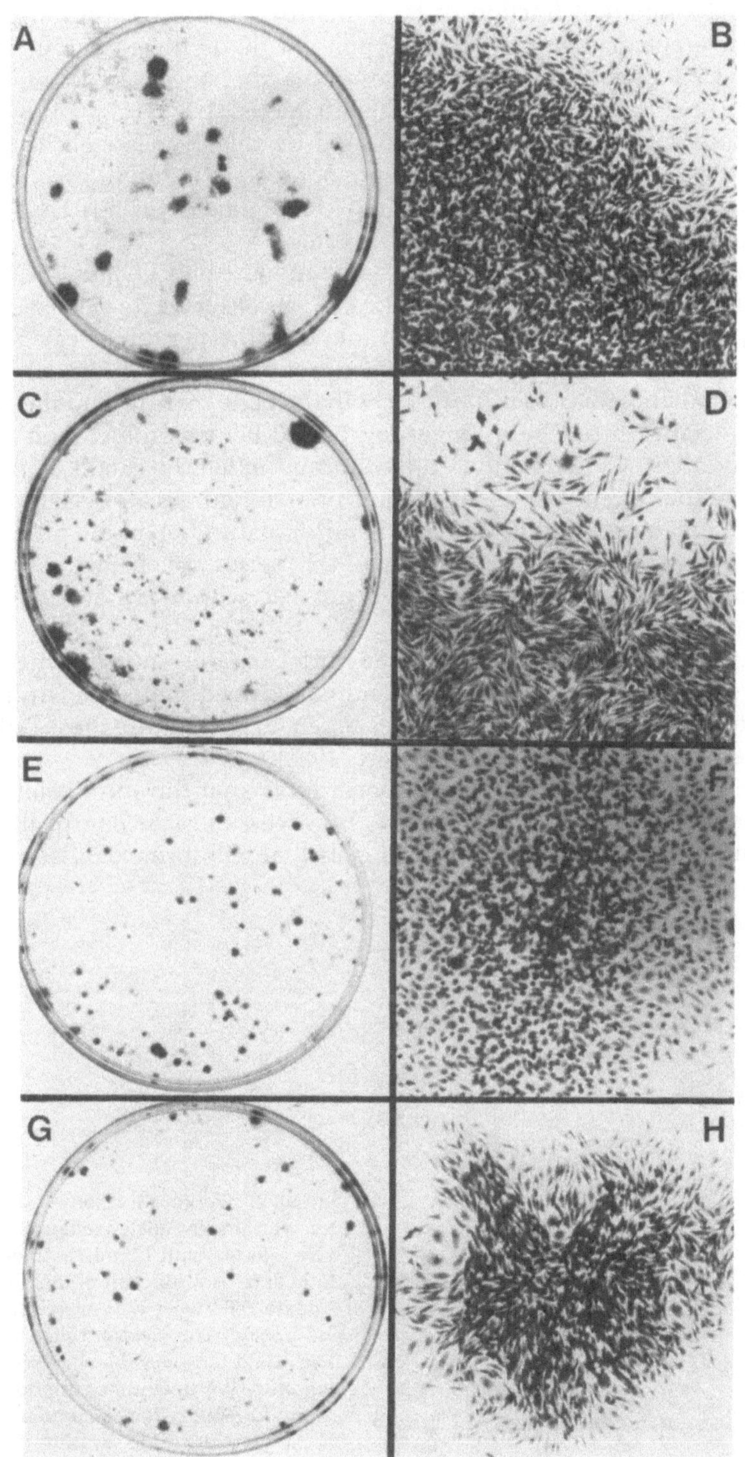
A
B
C
D
E
F
G
H

and loosely adherent fraction of bone marrow. These colonies are quite large with an average diameter of 2.7 ± 0.7 mm. By day 14 there is a fourfold increase to 3.1 ± 1.2% of cells forming colonies of macrophages. The latter colonies are nearly all small colonies whose average diameter (1.1 ± 0.1 mm) is not significantly different from the colonies formed by thioglycollate-elicited peritoneal exudate cells (PEC). Colonies of this phenotype are found in the adherent fraction which represents about 6% of the BMC originally plated. Thus, about half the adherent BMC form macrophage colonies.

Unlike bone marrow, no colonies are found by day 7 when PEC (thioglycollate-elicited) are cultured; by day 14, 15% of cells have formed colonies. Thus, one population of BMC which is not found in the peritoneal cavity, is characterized by a short lag period, extensive proliferation, and nonadherence. The second population of cells is found in both the bone marrow and the peritoneum and it is characterized by a longer lag period before proliferation begins (4–6 days), and these CFC are adherent. We propose that the extent of proliferation, as well as the length of the lag period prior to entrance of cells into cycle, is related to the degree of differentiation and/or maturation of the progenitor cells. The most immature progenitors in the MNP series are characterized by a high proliferative potential and they have not yet acquired the property of firm adherence.

The typical appearance of colonies from various sources when grown in liquid culture medium with L-cell conditioned medium is shown in Fig. 1. A complete description of the procedures to obtain and culture these cells has been published (Stewart *et al.*, 1981; Stewart, 1981).

The frequency of resident peritoneal MNP that can form colonies is quite low, about 0.1%. As shown in Fig. 2, however, the injection of thioglycollate medium produces a rapid increase of cells that will form colonies. The infiltra-

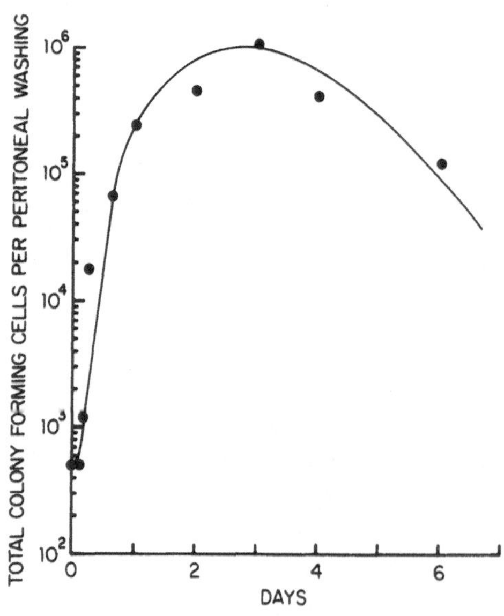

FIGURE 2. Peritoneal colony-forming cells elicited with thioglycollate medium. C3Hf/An mice were injected with 1.5 ml 3% thioglycollate medium and, as a function of time, the peritoneal exudate cells were harvested from three mice and pooled. The yield of cells and the fraction that would form macrophage colonies were determined. We determined the total number of colony-forming cells per peritoneal washing by multiplying the yield by the fraction of cells forming colonies. (From Stewart *et al.*, 1975.)

TABLE 2. COLONY FORMATION BY PERITONEAL EXUDATE CELLS

Agent	Yield per mouse	Colonies per 1000 cells
None (resident cells)	2.7 ± 1	1 ± 1
Needle puncture	2.5	4 ± 2
Saline	3.0	8 ± 3
Pyran copolymer		
Day 7	4.3 ± 1	—
Day 3	2.9 ± 1	20 ± 5
Starch	6.0	80 ± 7
BCG (10^6)		
Day 7	2.9 ± 1	82 ± 9
Day 3	2.3 ± 1	63 ± 10
5 mM sodium periodate	7.1 ± 1	69 ± 13
Thioglycollate		
3%	15.0 ± 1	153 ± 30
1%	7.4 ± 2	52 ± 13
0.3%	2.8 ± 1	27 ± 4

tion of CFC offers a convenient measure of the potency of a particular phlogogenic agent. As shown in Table 2, the number of CFC increases slightly with the injection of pyrogen-free saline. As more potent stimuli are used, there is both an increase in cells as well as an increase in CFC that can be obtained. Even with a small change in cell yield, there can be well over an order of magnitude increase in CFC, e.g., pyran copolymer and starch.

In conclusion, there can be no question that these cells have the capacity to proliferate locally within the tissue. The more important question would appear to concern the means by which their proliferation is regulated and with what frequency they proliferate locally.

3. PROLIFERATIVE CAPACITY OF PERITONEAL EXUDATE CELLS

The growth of PEC was determined using cells obtained from C3Hf/An mice 3 days after injection of 3% thioglycollate medium. Briefly, cells were plated in growth medium containing L-cell conditioned medium. Each day the number of cells was determined on duplicate cultures using the Pronase cetrimide procedure (Stewart *et al.*, 1981). Cetrimide is a detergent that lyses the cell's cytoplasm quantitatively liberating nuclei, thus providing a convenient means of removing the adherent macrophages. Other duplicate cultures were pulsed for 1 hr with 1 μCi/ml [^{3}H]-TdR. The cells were washed to remove unincorporated [^{3}H]-TdR and counted. As shown in Fig. 3, cells increase exponentially from day 1 to day 12 with a doubling time of about 42 hr. The incorporation of [^{3}H]-TdR into the cells parallels the cell count data after day 4. Through day 3, however, [^{3}H]-TdR incorporation increases rapidly as cells enter cycle for the first time.

There are two populations of cycling cells. One population representing

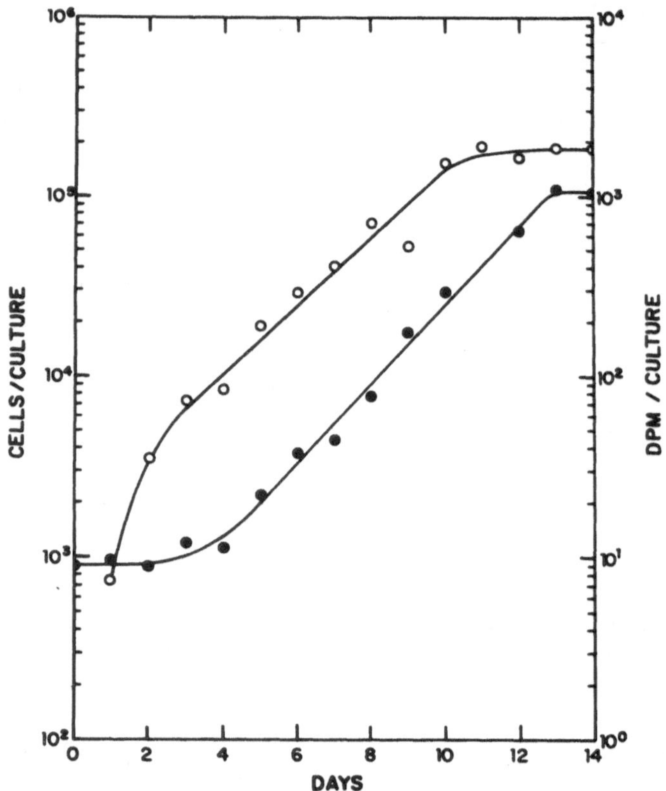

FIGURE 3. Proliferation of peritoneal exudate cells. C3Hf/An mice were injected with 1.5 ml 3% thioglycollate medium and harvested 3 days later. One thousand cells in 3 ml growth medium were plated in 35-mm culture dishes. The number of cells were determined daily by removing them with cetrimide. Other cultures were pulsed 1 hr with 1 Ci/ml [^{3}H]-TdR, rinsed, cells removed with 1.5 ml cetrimide, and the radioactivity incorporated determined using a liquid scintillation counter.

about 75% of CFC has a generation time of about 16 hr. The second population has a generation time of 24 hr (Stewart, 1980). We have not determined whether these two populations are functionally distinct.

To determine the proliferative capacity of MNP, PEC were subcultured. Since PEC are adherent, our procedures to remove them all in a viable condition have been imperfect. Using a 15-min incubation with 0.3% lidocaine in medium with 10% serum, we were able to recover 30–50% viable (after 24 hr) cells. Briefly, PEC elicited with 3% thioglycollate medium were initially plated at 10^4 cells in 3 ml growth medium. Cell counts were performed at frequent intervals and when there were 1–2 × 10^5 cells per culture, they were subcultured by diluting them to 10^4 cells in 3 ml growth medium and they were replated.

The results are summarized in Table 3. As shown in column 2, the number of cells obtained at the time of passage decreased with each subculture. This decrease in the proliferative ability is also reflected by the decrease in plating

TABLE 3. PROLIFERATIVE CAPACITY OF PERITONEAL EXUDATE MACROPHAGES

Elapsed time[a]	Number at passage ($\times 10^3$)	Increase[b]	Cumulative cells produced[c] ($\times 10^6$)	Plating efficiency[d]
7	160	16	0.16	20.1 ± 1.2
14	120	12	1.9	6.2 ± 0.5
23	82	8.2	16	2.6 ± 0.5
30	38	3.8	60	1.0 ± 0.5

[a]Time in days after plating the initial exudate cells that were subcultured.
[b]Fractional increase in cell number achieved during that subculture.
[c]Number of cells that would have been produced if all the progeny from the initial culture had been subcultured.
[d]Colonies per 100 cells ± S.E.

efficiency (colonies/100 cells) of the passaged cells shown in column 5. It is not possible, however, to determine from these data whether the decrease is due to an inherent trait of the precursor cell which limits the number of divisions possible or whether an artifact is introduced by the lidocaine treatment. Fibroblasts, however, have also been shown to have a limited proliferative capacity (Hayflick, 1973). We calculate, however, that each CFC is capable of giving rise to an average of 25,000 progeny representing 14.6 divisions and that had all the PEC obtained from a single mouse been cultured, 4×10^{10} macrophages (30 ml of packed cells) would have been obtained at the end of the last subculture. These results are of great practical significance since the progeny derived from a single precursor are sufficient in number to measure any macrophage membrane marker and some functions. Thus, by cloning individual cells it may be possible to determine if the heterogeneity of function is an innate property of macrophages or whether individual precursors give rise to discrete homogeneous populations (or clones) of macrophages expressing a single function all of which sum together to produce the heterogeneity observed in the whole population.

4. PROLIFERATIVE CAPACITY OF BONE MARROW CELLS

The growth of bone marrow MNP is shown in Fig. 4. The number of nonadherent cells recovered from culture was fairly constant between day 3 and 7 while the number of adherent cells increased with a doubling time of about 24 hr. Between day 7 and 10, there is a rapid increase in the number of nonadherent cells. This increase, however, is not due to an expansion of nonadherent progenitor cells but rather is an artifact caused by a deterioration of culture conditions resulting in the detachment of previously adherent cells. In collaboration with Dr. Noel Warner and Edwin Walker and using the flow cytometer at the Los Alamos National Laboratory, it was found that nonadherent cells exhibited low autofluorescence while adherent cells were highly autofluorescent. When cells were derived before day 6, the nonadherent cells had an eight-fold lower mean corpuscular volume and 16-fold lower mean fluorescence than did their ad-

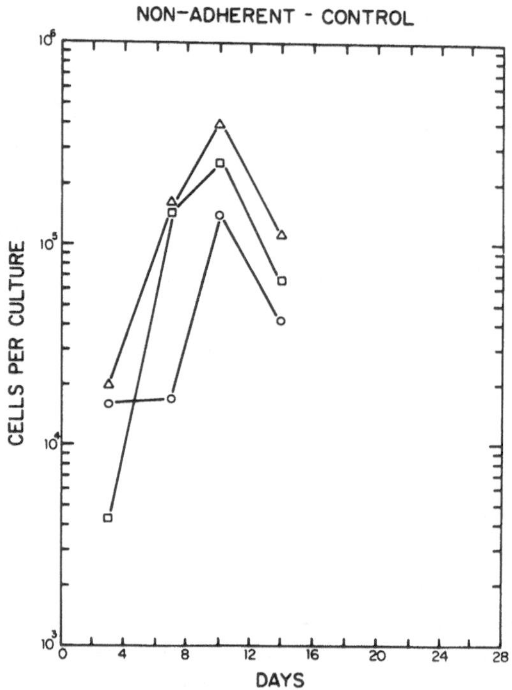

FIGURE 4. Growth of bone marrow-derived MNP. Bone marrow cells were derived from the femur of C3H mice (yield 10^7 cells/femur). Cells were adjusted to 3×10^4 in 3 ml growth medium (alpha MEM supplemented with 10% fetal (or newborn) bovine serum and 10% LCM) and plated in 35-mm culture dishes. As a function of time cells were counted (△ = total). Nonadherent cells (○) were removed and counted separately from the adherent cells (□). Adherent cells were removed with the detergent cetrimide as previously described (Stewart *et al.*, 1981).

herent counterparts. When cells were obtained after day 6, however, significant contamination of the nonadherent cells by cells with the adherent highly autofluorescent phenotype was found. Thus, the increase in "nonadherent" cells after day 7, as shown in Fig. 4, is due to the removal of loosely adherent cells.

When the nonadherent cells were subcultured, an estimate of the frequency of MNP progenitor cells could be obtained. There is a continuous increase in the frequency of CFC so that, by day 6, about 60% of the nonadherent cells will form macrophage colonies. We believe this enrichment in CFC is due to the death of BMC in culture combined with the proliferation of the nonadherent progenitor cell pool.

Taking advantage of the fact that bone marrow MNP progenitors are nonadherent, the proliferative capacity of these cells was determined by rinsing established cultures of BMC every 3 or 4 days and passing them to new cultures. Growth medium was added to the cells which remained attached. As shown in Fig. 5, when the nonadherent bone marrow progenitors are passed every 3 to 4 days, they give rise to both nonadherent and adherent cells. Over a 3-week period, the average yield of nonadherent cells per culture remained fairly constant while the adherent cells derived from each passage increased geometrically with a doubling time of 21 hr.

As shown in Fig. 5, nonadherent cells produce progeny which are adherent. Since nonadherent cells do not increase in number, one daughter of a nonadherent cell must, on the average, remain nonadherent while the other daughter becomes adherent. Each population continues to proliferate so that adherent

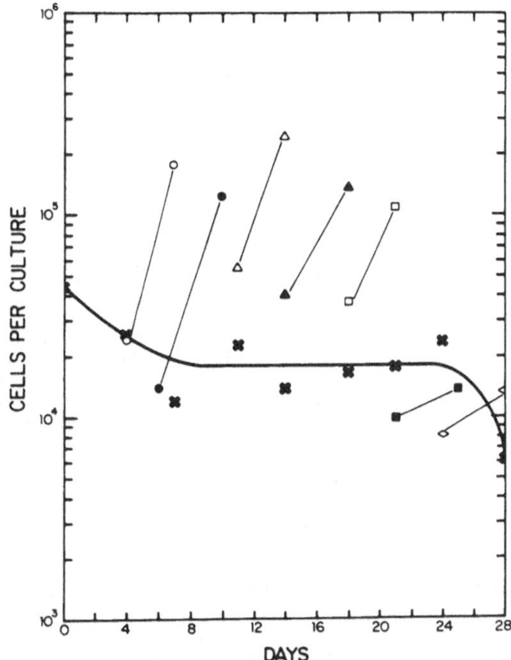

FIGURE 5. Proliferative capacity of bone marrow-derived MNP. 3×10^4 bone marrow cells were cultured in 3 ml growth medium containing 10% fetal bovine serum, 5% horse serum, and 10% LCM. Every 3 or 4 days the nonadherent cells were rinsed from the plates and transferred to new 35-mm dishes. Fresh growth medium was added to the adherent cells whose numbers were determined 7 days later. **X**, number of nonadherent cells recovered per plate; other symbols, adherent cells recovered per plate. (From Stewart, 1981.)

cells are produced from both the nonadherent and the adherent compartments. The nonadherent cells, therefore, have a limited capacity of self-renewal. We have, however, been unable to maintain this proliferative response and, after the third week, the yield of nonadherent progenitor cells begins to decrease and, by 30 days, none can be recovered for passage. Nevertheless, we calculate that a single nonadherent progenitor cell goes through about 27 divisions and can produce at least 10^8 progeny, a sufficient number of cells for analysis of function. As we will discuss below, MNP proliferation is absolutely dependent upon MGF and other factors which the cells consume as they proliferate.

5. SURFACE MARKERS ON DIFFERENTIATING MNP

In collaboration with Dr. William Walker *et al.* (1981), we have been using colonies to study differentiation of MNP using surface markers. Our preliminary results are shown in Table 4. Nearly all colonies formed by day 4 from bone marrow have a variable proportion of Fc receptor-positive cells. This is consistent with the notion that these colonies contain cells in various stages of maturation—some cells are mature enough to have acquired the receptor while others are not. This kind of data suggests a single-lineage differentiation wherein cells acquire (or lose) a particular property as they mature.

In contrast, Ia appears to be a clonally restricted marker. There are some colonies which have less than 25% positive cells in both the C3H (relevant) and

TABLE 4. PERCENTAGE OF COLONIES WITH MARKER-POSITIVE CELLS[a]

	Percent positive cells per colony				
Receptor	0	1–25	26–50	51–75	76–100
Fc	17	5	14	14	47
Ia.2					
C3H	57	18	0	9	17
DBA/2	78	22	0	0	0

[a]Bone marrow from C3H (H_2k) or DBA/2 (H_2d) mice was cultured in growth medium at 10^4 cells/ml in Lab-Tek slides for 4 days. (Plastic was found to increase background adherence of SRBC.) SRBC were coated with the appropriate monoclonal antibody reagent using the chromic chloride procedure (Parish *et al.*, 1974). In the body of the table the percentage of colonies having the percentage of positive cells within the range indicated along the top is shown. For example, 78% of DBA/2-derived colonies had no positive cells for Ia.2 and in 22% of colonies 1–25% of cells were positive (actually less than 10% of cells were positive).

the DBA/2 (irrelevant) haplotype for the anti-Ia.2 monoclonal antibody used; we interpret this as background. Only in the case of the relevant haplotype were colonies found containing 50–100% positive cells. We found that 15% of nonadherent progenitor cells obtained 4 days after culture of bone marrow were also Ia.2 positive. This is similar to the percentage of colonies containing greater than 50% Ia.2-positive cells. Thus, the progenitor cells acquire this marker even before they acquire the property of adherence.

If the cells are truly cloning for Ia.2, then why aren't all the cells positive in some colonies? We believe colonies which contain less than 90% positive cells are actually mixed colonies produced by an Ia.2-positive and an Ia.2-negative progenitor cell in close proximity or aggregated when they were originally plated. The evidence for this possibility is as follows. Ia.2-positive cells within the mixed colonies are always in a group: they are never found randomly dispersed throughout the colony. In addition, when the number of cells within a mixed colony are counted on day 4, more cells are found in it than could have been produced by a single progenitor cell proliferating with the observed mean 16-hr generation time.

These results support the contention that some macrophage functions may be ascribed to distinct clonal subpopulations of MNP. For expression of membrane Ia, this would seem to be at variance with the recent findings that Ia can be induced in macrophages by lymphokine-rich supernatants (Farr *et al.*, 1979; Steinman *et al.*, 1980; Beller and Unanue, 1981). Induction of Ia, however, is not mutually exclusive of our result that some macrophages have Ia on their membranes all the time because they are the differentiated progeny of Ia-positive progenitor cells. Furthermore, two other laboratories have also shown, in the absence of any inducing agents, that some macrophages derived from mass cultures of BMC have Ia markers on them (Lee and Wong, 1980; Stern *et al.*, 1979; Erb *et al.*, 1980).

TABLE 5. PROLIFERATION OF PERITONEAL EXUDATE CELLS IN PRIMARY CULTURE

Initial cultured[a]	Stationary time	Phase no.[b]	Fractional increase over initial	No. of divisions[c]
340,000	5	750,000	2	3
68,000	8	750,000	11	6
13,000	12	825,000	60	8
2,700	14	420,000	155	10

[a]Total cells in 6 ml of growth medium in 60-mm culture dishes.
[b]Day cells stopped doubling every 42 hr and number of cells at plateau.
[c]Number of divisions of the clonogenic cells.

6. CONSUMPTION OF MGF BY MACROPHAGES

We found that peritoneal CFC stop growing before they establish a confluent monolayer. Medium derived from these cultures will not support the growth of freshly isolated cells. The results are summarized in Table 5. Whether 3.4×10^5 cells or 2.7×10^3 cells were initially cultured, cells entered stationary phase at nearly the same total cell number (4 to 8×10^6 cells per culture, 0.8 to 1.6×10^5 cells/ml). As summarized in Table 5, when the highest number of cells was cultured, stationary phase was achieved at 5 days after a two-fold increase in total cells; for the lowest cell number cultured, cells proliferated for 14 days when there was a 155-fold increase in cell number. These results show that some cells within the population have the capacity of many divisions, and limited proliferative capacity is not the reason for their leaving cell cycle.

The requirement for MGF is shown in Fig. 6. When MGF is added to PEC at 3×10^3 cells per culture, they begin to proliferate. If MGF is not removed, they continue proliferating exponentially with a doubling time of 42 hr and by day 10 they enter stationary phase. If, during the exponential proliferative phase, MGF is removed (on day 6), cells leave cell cycle. In the absence of MGF, these cells tend to become less adherent. When MGF is readded to the cultures (day 9), the cells reenter cell cycle and proliferate with a doubling time similar to that found prior to MGF removal. These results illustrate the total dependency of these cells for MGF to proliferate. They further show that the cell cycle state can be manipulated for studies involving functional expression as related to cell cycle status.

One possible way to explain the observation that cells leave cell cycle when a concentration of $1–2 \times 10^5$ cells/ml is achieved is that the cells have depleted MGF to nonstimulatory levels. In collaboration with Dr. E. R. Stanley and L. Guilbert, the amount of MGF remaining in culture was measured. The results are shown in Table 6 for PEC. By day 6, cells are no longer proliferating exponentially and a nadir in MGF concentration was reached. Thereafter, it began to accumulate in the culture so that by day 11 there was twice the concentration of MGF in culture than was initially added. This MGF is produced by the fibroblasts, which contaminate the initial suspension and proliferate within the culture. We find a frequency of one clonogenic fibroblast per 1000 PEC. The results

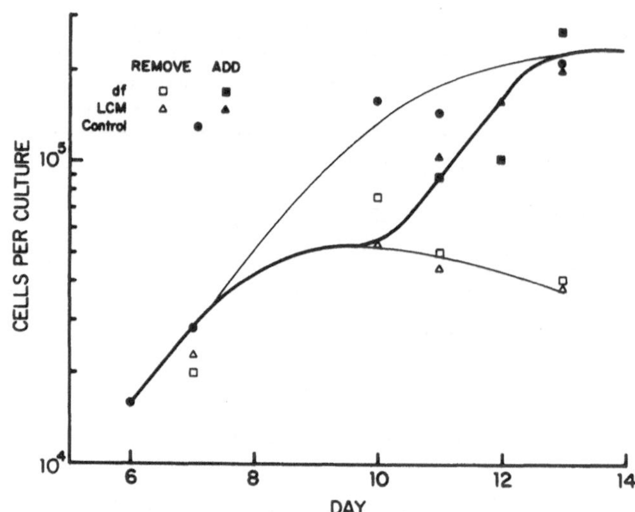

FIGURE 6. Regulation of PEC growth by MGF and SDA. PEC were established at 10^3 cells in 3 ml growth medium and allowed to incubate for 6 days. LCM refers to L-cell conditioned medium, the source of MGF, and df refers to the serum dialyzable activity (SDA). See text for details.

show that MGF, while consumed, is not the limiting substrate in these cultures. From this datum, for the various cell concentrations that were cultured, the consumption rate was found to be 0.017 ± 0.09 unit per cell per day.

7. SERUM DIALYZABLE ACTIVITY

Accordingly, we sought other factors which might be limiting. Virolainen and Defendi (1967) showed that PEC in medium containing MGF and fetal bovine serum (FBS) would proliferate but that if the serum was dialyzed, no proliferation occurred. No further studies were done concerning this observation until recently when we renewed the investigation (Stewart *et al.*, 1980). As

TABLE 6. CONSUMPTION OF MGF BY PERITONEAL EXUDATE CELLS[a]

Day	Cells/ml ($\times 10^3$)	Units remaining/ml
0	50	550
2	45	500
4	108	250
6	140	200
8	120	300
10	110	500
12	100	1000

[a]Medium from cultures established at 5×10^4 cells/ml was collected and analyzed for remaining MGF using an immunoassay (Stanley, 1979).

TABLE 7. EFFECT OF SERUM ON COLONY FORMATION

	Colonies per 10^4 cells	
	PEC	BMC
Neat fetal bovine serum	1320	163
Dialyzed serum		
Retentate only	0	69
Dialyzate only	0	90
Reconstituted	1210	180

shown in Table 7, when PEC or BMC are cultured in complete growth medium, they form colonies. If, however, the FBS is dialyzed, no colonies are formed in the cultures containing PEC and there is a reduction in bone marrow-derived colonies. An activity in the dialyzable portion of serum is required for proliferation of PEC since a complete restoration can be achieved when the serum is reconstituted with the dialyzate.

In order to further explore the growth regulation of PEC by both MGF and the serum dialyzable activity (SDA), cultures of PEC were established containing 10^3 cells in complete growth medium. On day 6 cultures were screened to eliminate those with obvious fibroblast contamination (which would produce MGF). At this time, the medium on cultures was replaced with medium containing MGF and dialyzed FBS while other cultures received medium containing complete FBS but no MGF. The results are shown in Fig. 6. When either SDA or MGF was removed, cells entered stationary phase while those with no medium change continued to proliferate exponentially. If, on day 9, SDA was added to the group containing the dialyzed FBS or MGF was added to the group depleted of it, proliferation resumed and by day 13 cells had achieved the same culture cellularity as found in those which had not been manipulated. These results show that, like MGF, SDA is absolutely required for PEC to proliferate.

We propose the following model. Early nonadherent bone marrow progenitors proliferate and some produce adherent progeny, neither of which requires SDA (thus, some colonies are formed by BMC). The adherent cells continue to divide as well as to differentiate. Some will reach a point in their differentiation state wherein they require SDA to continue proliferating. In its absence they stop proliferating; these cells are now like PEC.

8. GROWTH-PROMOTING ACTIVITY OF ERYTHROCYTES

It has been reported that hemolyzate will augment the colony-forming ability of BMC grown in agar culture (Morton and Isacs, 1967; Bradley *et al.*, 1971; Chen and Lin, 1981a). When mononuclear cells (MNC) from blood were contaminated with irradiated erythrocytes (5000 rads to prevent contaminating monocytes from proliferating) they too produced considerably larger colonies.

The growth rates of MNC with and without erythrocytes were found to be exactly the same; the doubling time was 42 hr. While the growth rate for the MNC without added erythrocytes was not different from when they were present, the lag period was shorter by about 2 days. Therefore, the total number of cells produced at any time was greater when erythrocytes were present.

In order to determine maximal augmentation, cultures of MNC were established containing different erythrocyte concentrations. The optimal number of murine erythrocytes for maximal stimulation was 1.7×10^8 erythrocytes in 6 ml growth medium in 60-mm culture dishes. We need to determine if concentration (3×10^7 RBC/ml) or density (6×10^6 RBC/cm^2) of erythrocytes is the most important variable.

9. INHIBITION OF MNP PROLIFERATION

It is well known that MGF is at stimulatory levels in most tissues (Stewart and Lin, 1978) and that it is absolutely required for MNP to proliferate (van der Zeijst *et al.*, 1978). Why are not all the CFC in the tissues in cycle? We reasoned there must also be inhibitors present in the tissues at concentration levels which inhibit most proliferation. One of these inhibitors is probably prostaglandin E_2 (Kurland *et al.*, 1978, 1979; Williams, 1979). As a result, we tested the cell-free lavaged fluid for inhibitory activity on the colony-forming ability of PEC and found that the peritoneal fluid does indeed contain a nondialyzable, reversible inhibitor capable of totally inhibiting colony formation (Yen and Stewart, 1981). The data are shown in Table 8. The most interesting finding, however, was that BMC were only partially affected. This partial effect could be caused by the acquisition of sensitivity of cells to the inhibitor as they mature, similar to the model described above for the acquisition of sensitivity to SDA. These data strongly suggest the inhibitor may be specific for MNP. Chen and Lin (1981b) have coined the term *colony inhibitory factor* (CIF) to describe an inhibitory activity found in LCM. There may be a relationship between the CIF found in LCM and that which we have found in the peritoneal lavage. They both inhibit peripheral

TABLE 8. INHIBITION OF COLONY FORMATION BY CELL-FREE PERITONEAL FLUID[a]

% peritoneal lavage	Percent inhibition		
	MNP source: Bone marrow	Blood	PEC
40	63 ± 7	100	100
20	58 ± 18	88 ± 48	92 ± 16
10	37 ± 11	65 ± 16	69 ± 19
5	16 ± 6.2	47 ± 7.8	38 ± 3.8
2.5	11 ± 4.1	24 ± 7.5	27 ± 4.8

[a]Cells were cultured in growth medium containing 10% fetal bovine serum, 5% horse serum, and 5% L-cell conditioned medium and the percentage of peritoneal lavage fluid indicated. Results are from two experiments. The mean ± S.D. are shown.

macrophage proliferation but have a lesser effect on immature bone marrow-derived progenitor cells.

In summary, there are at least three distinct activities which regulate MNP growth. MGF, the only one which has been characterized (Stewart and Lin, 1978; Stanley *et al.*, 1976; Stanley, 1981), is required by all MNP to proliferate. SDA is required for all sources of peripheral MNP to proliferate but apparently not by immature bone marrow MNP. Finally, there is an activity derived from the erythrocyte fraction which is totally uncharacterized and appears to promote proliferative activity by reducing the lag period; it is not essential for proliferation. There are also growth inhibitory activities. MNP will not proliferate when effective concentrations of these activities are reached no matter how much stimulating activity is present. The concentration of the growth factors, promoters, and inhibitors within the tissue may exquisitely regulate the local proliferation of MNP as they respond in the host defense against environmental pathogens.

10. GENETICS OF COLONY FORMATION

Lin *et al.* (1978) studied the ability of thioglycollate medium-elicited cells to form colonies as a function of the age of the mouse and as a function of mouse strain. They found that the frequency of CFC increased up to 3 weeks of age and, thereafter, no significant differences were found up to 40 weeks. While no difference in colony formation was found between males and females, there was a striking difference in the frequency of CFC found among the animal strains tested. Strains could be divided into three major groups. Animals having a high frequency of CFC (> 7% of PEC) were DBA/2, DBA/1, C3H/He, C3Hf/An, and A strain mice; animals showing moderate CFC (2–7% of PEC) included AKR, BALB/c, and Swiss mice; C57BL mice were found to be low-responder mice whose PEC contained less than 2% CFC.

Due to the extreme differences in CFC between high- and low-responder strains, we investigated whether the differences had a genetic basis. For these studies, performed in collaboration with Drs. E. Skamene and P. Kongshavn, we selected A mice as the high-responder strain and B10.A mice as the low-responder strain (Stewart *et al.*, 1980). We determined the frequency of colony formation by MNP from both bone marrow and thioglycollate medium-elicited peritoneal cells. Detailed procedures for culturing these cells have been described (Stewart, 1981; Stewart *et et al.*, 1981).

The results are shown in Table 9. For bone marrow, 5.5 colonies per 1000 cells were selected and, for PEC, 90 colonies per 1000 cells were selected to separate the high- and low-responder mice. For the A strain 83% of mice yielded BMC which produced more than 5.5 colonies per 1000 cells and 90% of mice yielded PEC which produced more than 90 colonies per 1000 cells. In contrast, B10.A mice yield cells whose ability to form colonies was less than the above. When the F_1 was tested, it behaved like the B10.A parent. The F_2 yielded CFC consistent with a 25–75% high- to low-responder phenotype. The backcross of

TABLE 9. GENETICS OF COLONY FORMATION

	Percentage of mice			
	Bone marrow		Peritoneal exudate cells	
Strain — Colonies[a]:	> 5.5	< 5.5	> 90	< 90
A	83	17	90	10
B10.A	0	100	10	90
F_1	6	94	10	90
F_2	28	72	23	77
$F_1 \times A$	52	48	51	49

[a]Number of colonies per 1000 cells.

F_1 with high-responder mice yielded progeny, half of which were of the high- and half of which were of the low-responder phenotype.

Taken together, these results strongly support the concept that expression of low macrophage CFC content formation is controlled by a single dominant gene. As mentioned above, inhibition of colony formation has been found in L-cell conditioned medium and in the peritoneal cavity. Chen and Lin (1981b) showed that C57BL/6 CFC were more sensitive to the LCM inhibitor than were those from C3H mice. If the inhibitor was removed from the conditioned medium, then both C57BL (low responder) and C3H mice (high responder) had the same high plating efficiency. Thus, the frequency of CFC in a population of exudate cells appears to be similar among the mouse strains tested. These results suggest that the gene which controls whether a mouse is a high or low responder is actually controlling the cell's sensitivity to the inhibitor found in the LCM.

11. LOCAL PROLIFERATION REVISITED

Taken together, the above discussion clearly shows that MNP from many murine tissues retain the ability to proliferate. The question is whether these cells proliferate in the tissues as suggested by Volkman (1976a,b), or whether all tissue macrophages are derived from the bone marrow as proposed by van Furth *et al.* (1972; van Furth, 1980), or whether both schools of thought can be accommodated. While there is no definitive data to answer the question, it is possible to determine what the local production rates might be based on published labeling indices for them. These data are summarized in Table 10. Utilizing these data, it is possible to calculate the number of cells produced locally when the tissue is in a steady state. A steady state is that state which exists when the labeling index and the net number of cells within the tissue are not appreciably changing with time.

TABLE 10. EXTENT OF LOCAL PROLIFERATION REVISITED

	Growth fraction[a] (%)	Total cells[b]	Turnover time (days)	
			Calculated	Reported
Promonocytes	80	5×10^5	0.8	0.7
Monocytes				
Bone marrow	1.8	26×10^5	42	3.3
Blood	2.6	11×10^5	29	1.3
Kupffer cells	2.0	91×10^5	37	21
Lung	2.9	19×10^5	26	27
Peritoneum	2.8	2×10^6	20	30

[a]Locally labeled cells after 2 hr after a single injection of [^{3}H]-TdR. The growth fraction = tcxLI/ts = 2 × LI.
[b]Total cells in the compartment during steady-state conditions.

To illustrate the basis for the calculations, consider the following model: the labeling index and number of peritoneal cells obtained, as shown in Table 1, are not appreciably different for the donor mice at any time. A simple model can be used to calculate the turnover time of MNP for the tissue. The turnover time is the length of time for all cells in the compartment to be replaced by new cells. We must determine if the time for replacement can be accomplished by the locally cycling cells or whether contributions from outside the tissues are necessary, for example by infiltrating bone marrow-derived blood monocytes.

Since the number of cells in the compartment is represented by a fixed number of cells, we can represent the tissue as a box of 32 compartments, so that one compartment represents the proliferating cells, i.e., 1/32 = 3.125% and 1.56% of cells would be labeled after a pulse of [^{3}H]-TdR. Since S phase represents only half the cycling cells when the pulse time is short, the others are in G1, G2, and mitosis and would not be labeled, the fraction of cells in cycle is really 0.03125 (3% are in cell cycle). The question is how long it will take for all 31 boxes to be replaced by the single box representing the cycling cells; only one box turns over with each generation time. Since this is a steady state and the labeling index does not change nor is there any increase in the number of cells, then, for each division, one daughter must leave cell cycle and one cell (not necessarily the same cell) must leave the compartment (either by death or by migration). Therefore, for each round of division one compartment is replaced. Therefore, after 31 rounds of cell division all the compartments would be replaced. Since the mean generation time for cycling macrophages is 15.5 hr (Stewart, 1980), it will take (32 compartments − 1) × 15.5 hr = 480.5 hr or 20 days to completely replace all the resident cells by local proliferation. The published turnover time is 30 days (van Furth *et al.*, 1973). Are we to assume, then, that the peritoneal cavity is a source of MNP? No. This is just an example to illustrate how a low labeling index may actually reflect a rather formidable proliferative activity. Perhaps what we ought to wonder is where the locally produced cells go and what they do.

REFERENCES

Beller, D. I., and Unanue, E. R., 1981, Regulation of macrophage populations. II. Synthesis and expression of Ia antigens by peritoneal exudate macrophage is a transient event, *J. Immunol.* **126**:263.

Bradley, P. A., Telfer, P. A., and Fry, P., 1971, The effect of erythrocytes on mouse bone marrow colony development *in vitro, Blood* **38**:353.

Chen, D., and Lin, H., 1981a, Differential enhancement of the clonal growth of various mononuclear phagocytes by hemolysates, *J. Reticuloendothelial Soc.* **29**:465.

Chen, D., and Lin, H., 1981b, Inhibitor of peritoneal mononuclear phagocyte colony forming cells, *Exp. Hematol.* **9**:240.

Cole, P., 1975, Activation of mouse peritoneal cells to kill *Listeria monocytogenes* by T-lymphocyte products, *Infect. Immun.* **12**:35.

Erb, P., Stern, A. C., Alkan, S. S., Studer, S., Zoombou, E., and Gissler, R. H., 1980, Characterization of accessory cells required for helper T-cell induction *in vitro:* Evidence for a phagocytic F - receptor and Ia-bearing cell type, *J. Immunol.* **125**:2504.

Farr, A. G., Kiely, J., and Unanue, E. R., 1979, Macrophage T-cell interactions involving *Listeria monocytogenes*—Role of the H-2 gene complex, *J. Immunol.* **122**:2395.

Gordon, S., and Werb, Z., 1976, Secretion of macrophage neutral proteinase is enhanced by colchicine, *Proc. Natl. Acad. Sci. USA* **73**:872.

Hayflick, L., 1973, The biology of human aging, *Am. J. Med. Sci.* **265**:433.

Kurland, J. I., Bockman, R. S., Broxmeyer, H., and Moore, M. A. S., 1978, Limitation of excessive myelopoiesis by the intrinsic modulation of macrophage-derived prostaglandin E, *Science* **199**:552.

Kurland, J. I., Pelus, L. M., Ralph, P., Bockman, R. S., and Moore, M. A. S., 1979, Induction of prostaglandin E synthesis in normal and neoplastic macrophages: Role for colony-stimulating factor(s) distinct from effects on myeloid progenitor cell proliferation, *Proc. Natl. Acad. Sci. USA* **76**:2326.

Lee, K.-C., and Wong, M., 1980, Functional heterogeneity of culture-grown bone marrow-derived macrophages. I. Antigen presenting function, *J. Immunol.* **125**:86.

Leonard, E. J., Ruco, L. P., and Meltzer, M. S., 1978, Characterization of macrophage activation factor: A lymphokine that causes macrophages to become cytotoxic for tumor cells, *Cell. Immunol.* **41**:347.

Lin, H., and Stewart, C., 1973, Colony formation by mouse peritoneal exudate cells *in vitro, Nature New Biol.* **243**:176.

Lin, H., Kuhn, C., and Stewart, C. C., 1978, Peritoneal exudate cells. V. Influence of age, sex, strain and species on the induction and the growth of macrophage colony forming cells, *J. Cell. Physiol.* **96**:133.

Morton, H. J., and Isacs, R., 1967, Cultivation of rat bone marrow. I. Preliminary studies and some effects of hemolyzed blood as nutrient, *J. Natl. Cancer Inst.* **39**:796.

Nacy, C. A., Leonard, E. J., and Meltzer, M. S., 1981, Macrophages in resistance to rickettsial infections: Characterization of lymphokines that induce rickettsiacidal activity in macrophages, *J. Immunol.* **126**:204.

Parish, C., Kirov, C. S., Bowern, N., and Blanden, R., 1974, A one step procedure for separating mouse T and B lymphocytes, *Eur. J. Immunol.* **4**:808.

Rabellino, E. M., and Metcalf, D., 1975, Receptors for C_3 and IgG on macrophage, neutrophil and eosinophil colony cells grown *in vitro, J. Immunol.* **115**:688.

Ruco, L. P., and Meltzer, M. S., 1977, Macrophage activation for tumor cytotoxicity: Induction of tumoricidal macrophages by supernatants of PPD-stimulated bacillus Calmette-Guerin-immune spleen cell culture, *J. Immunol.* **119**:889.

Stanley, E. R., 1979, Colony-stimulating factor (CSF) radioimmunoassay: Detection of a CSF subclass stimulating macrophage production, *Proc. Natl. Acad. Sci. USA* **76**:2969.

Stanley, E. R., 1981, Colony stimulating factors, in: *The Lymphokines* (R. Stewart and J. Haddon, eds.), Humana Press, Clifton, N.J.

Stanley, E. R., Cifone, M., Heard, P. M., and Defendi, V., 1976, Factors regulating macrophage production and growth: Identity of colony-stimulating factor and macrophage growth factor, *J. Exp. Med.* **143**:631.

Steinman, R. M., Nogueira, N., Witmer, M. D., Tydings, J. D., and Mellman, I. S., 1980, Lymphokine enhances the expression and synthesis of Ia antigens on cultured mouse peritoneal macrophages, *J. Exp. Med.* **152**:1248.

Stern, A. C., Erb, P., and Gisler, R. H., 1979, Ia-bearing bone marrow-cultured macrophages induce antigen-specific helper T cells for antibody synthesis, *J. Immunol.* **123**:612.

Stewart, C. C., 1979, The use of cloned mononuclear phagocytes to study immunoregulation, in: *Regulatory Roles of Mononuclear Phagocytes in Immunity* (A. S. Rosenthal and E. R. Unanue, eds.), p. 455, Academic Press, New York.

Stewart, C. C., 1980, Formation of colonies of mononuclear phagocytes outside the bone marrow, in: *Mononuclear Phagocytes: Functional Aspects* (R. van Furth, ed.), p. 377, Nijhoff, The Hague.

Stewart, C. C., 1981, Murine mononuclear phagocytes from bone marrow, in: *Methods for Studying Mononuclear Phagocytes* (D. O. Adams, H. Koren, and P. Edelson, eds.), p. 5, Academic Press, New York.

Stewart, C. C., and Lin, H., 1978, Macrophage growth factor and its relationship to colony stimulating factor, *J. Reticuloendothelial Soc.* **23**:269.

Stewart, C. C., Yen, S., and Senior, R. M., 1981, Colony-forming ability of mononuclear phagocytes, in: *Manual of Macrophage Methodology* (H. B. Herscowitz, H. T. Holden, J. A. Bellanti, and A. Ghaffer, eds.), p. 171, Dekkar, New York.

Stewart, C. C., Lin, H., and Adles, C., 1975, Proliferation and colony-forming ability of peritoneal exudate cells in liquid culture, *J. Exp. Med.* **141**:1114.

Stewart, C. C., Skamene, E., and Kongshaun, P. H. L., 1980, The genetic basis of macrophage colony formation, in: *Genetic Control of Natural Resistance to Infection and Malignancy* (E. Skamene and P. Kongshaun, eds.), Academic Press, New York.

van der Zeijst, B. A. M., Stewart, C. C., and Schlesinger, S., 1978, Proliferative capacity of mouse peritoneal macrophages *in vitro*, *J. Exp. Med.* **147**:1253.

van Furth, R. (ed.), 1980, Cells of the mononuclear phagocyte system: Nomenclature in terms of sites and conditions, in: *Mononuclear Phagocytes: Functional Aspects*, p. 1, Nijhoff, The Hague.

van Furth, R., Cohn, Z. A., Hirsch, J. G., Humphry, J. H., Spector, W. G., and Langewoort, H. L., 1972, The mononuclear phagocyte system: A new classification of macrophages, monocytes and their precursors, *Bull. W. H. O.* **46**:845.

van Furth, R., Diesselhoff-dendulk, and Mattie, H., 1973, Quantitative study on the production and kinetics of mononuclear phagocytes during an acute inflammatory reaction, *J. Exp. Med.* **138**:1314.

van Furth, R., Raeburn, J. A., and van Zwet, T. L., 1979, Characteristics of human mononuclear phagocytes, *Blood* **54**:485.

Vassalli, J., and Reich, E., 1977, Macrophage plasminogen activator: Induction by products of activated lymphoid cells, *J. Exp. Med.* **145**:429.

Vassalli, J., Hamilton, J., and Reich, E., 1977, Macrophage plasminogen activator: Induction by concanavalin A and phorbol myristate acetate, *Cell* **11**:695.

Virolainen, M., and Defendi, V., 1967, Dependence of macrophage growth *in vitro* upon interaction with other cell types, *Wistar Inst. Symp. Monogr.* **7**:67.

Volkman, A., 1976a, Monocyte kinetics and their changes in infection, in: *Immunobiology of the Macrophage* (D. S. Nelson, ed.), p. 291, Academic Press, New York.

Volkman, A., 1976b, Disparity in origin of mononuclear phagocyte populations, *J. Reticuloendothelial Soc.* **19**:249.

Wahl, L. M., Wahl, S. M., Mergenhagen, S. E., and Martin, G. R., 1975, Collagenase production by lymphokine-activated macrophages, *Science* **187**:261.

Walker, W. S., 1976, Functional heterogeneity of macrophages, in: *Immunobiology of the Macrophage* (D. S. Nelson, ed.), p. 91, Academic Press, New York.

Walker, W. S., Hester, R. B., Gandour, D. M., and Stewart, C. C., 1981, Evidence of a distinct progenitor for the Ia-bearing (Ia^+) murine bone-marrow-derived mononuclear phagocyte (MNP), in: *Heterogeneity of Mononuclear Phagocytes* (Z. Cohn, D. Foester, and M. Landy, eds.), p. 229 Academic Press, New York.

Williams, N., 1979, Preferential inhibition of murine macrophage colony formation by prostaglandin E, *Blood* **53:**1089.

Yen, S.-E., and Stewart, C. C., 1981, Macrophage growth inhibitors derived from the murine peritoneal cavity, *In Vitro* **17:**871.

3

Quantitation of Macrophage Phagocytosis *in Vitro*

PAUL W. GUDEWICZ

1. INTRODUCTION

Phagocytosis is the cellular process that regulates the uptake of exogenous particulate material by eukaryotic cells. As a result of nearly 100 years of intensive research, initiating with the classic works of Metchnikoff, the importance of phagocytosis with regard to homeostasis, defense mechanisms against invading infectious agents and particulate matter, bulk transport of macromolecules, and cellular nutrition is now apparent. In their attempts to uncover the mechanisms regulating phagocytic activity, biologists have developed and utilized a number of techniques and particles to measure this process. In recent years, it has become increasingly apparent that most investigators interested in unraveling the mechanisms as well as the intracellular elements involved in the ingestion process have favored an *in vitro* approach to the study of phagocytosis. Although the ability to phagocytize particulate matter is a functional property exhibited by most cells to some degree, by far the most active cell types, and those most studied in terms of their phagocytic activity, are the "professional" phagocytes of the blood and tissues, namely, the polymorphonuclear leukocyte (PMNL) and the mononuclear phagocyte or macrophage.

The purpose of this chapter will be to examine *in vitro* methodology and the types of particles now in use to quantify phagocytosis. This chapter will emphasize those methods that are being used to study phagocytosis in isolated cell preparations of macrophages and PMNL as opposed to the use of tissue slice procedures or isolated organ perfusion systems in the measurement of particle uptake. Whenever possible, this chapter will also focus on current methodology that is applicable to macrophage as opposed to PMNL preparations. The use of macrophages offers several distinct advantages to the investigator interested in

PAUL W. GUDEWICZ • Department of Physiology, Albany Medical College of Union University, Albany, New York 12208.

the study of phagocytosis by *in vitro* techniques. For example, macrophages can be obtained in large numbers from several tissue sites, e.g., peritoneal cavity, lung, spleen, and peripheral blood. Second, unlike PMNL, a stable population of macrophages can be maintained *in vitro*, by culture techniques, for periods of time ranging from hours to weeks. A clear advantage of using *in vitro* as opposed to *in vivo* techniques in assessing macrophage phagocytic activity is that the variables regulating phagocytosis can be examined independently in an attempt to reproduce both physiological and pathophysiological conditions. This chapter will also attempt to point out some of the advantages and disadvantages of each of the phagocytic assays described as well as the type of particle utilized to measure uptake. In any discussion of phagocytosis, one must keep in mind a problem inherent to all measurements of particle uptake, that is, the ability of each method to distinguish between true internalization of the particle and cell surface binding or attachment. Since the physiology and biochemistry of phagocytosis are being intensely studied in many laboratories, the reader is referred to a number of recent reviews on selected aspects of this topic (Silverstein *et al.*, 1977, 1978; Zuckerman and Douglas, 1979; Walters and Papadimitriou, 1978; Kavet and Brain, 1980; Stossel, 1974).

2. SOURCES OF MACROPHAGES FOR *IN VITRO* PHAGOCYTIC STUDIES

A major concern for investigators studying phagocytosis *in vitro* is the ability to obtain phagocytic cells in adequate numbers and in a high degree of viability. Fortunately, there exist many methods for obtaining macrophages from tissues with a significant phagocytic component (e.g., liver, lung, spleen, and lymphoid tissue) as well as from peripheral blood (monocytes) and from inflammatory sites (e.g., acute peritoneal exudates). This section will briefly summarize the major sources and methods commonly used for obtaining adequate numbers of macrophages for *in vitro* studies and also basic assay conditions necessary for effective phagocytosis.

2.1. SOURCES OF MACROPHAGE POPULATIONS

2.1.1. Monocytes

The importance of the monocyte in cell-mediated immune reactions as well as its role as a precursor of the tissue macrophage is now actively being investigated. However, until recently, the use of monocytes for *in vitro* functional and metabolic studies has been limited by the relatively small yield of monocytes from peripheral blood samples from either animal or human subjects. Although monocytes comprise only a few percent of the total circulating white blood cell population, recent advances in cell isolation methodology have increased monocyte yields as well as their viability. The most widely used research technique for monocyte separation and isolation employes a Ficoll–Hypaque gradient centrifugation of peripheral blood buffy coat cells followed by further purification by adherence to glass or plastic surfaces (Brodersen and Burns, 1973; Musson and Henson, 1979; Wardley *et al.*, 1980). One disadvantage in this method is the considerable loss of viable monocytes during the adherence procedures. Larger numbers of monocytes have been obtained by continuous-flow filtration leukophoresis (Hart and Fidler, 1979) and by elutriation techniques (Fogelman *et al.*, 1979); however, these methods are far more costly and usually require specialized equipment. Another problem that has plagued investigators utilizing monocytes in the past is the difficulty of maintaining monocytes in culture. Methods are now available to culture human and animal monocytes for long periods of time (Zuckerman *et al.*, 1979; Wardley *et al.*, 1980), thereby expanding the potential uses of monocytes for *in vitro* studies.

2.1.2. Peritoneal Macrophages

Large numbers of phagocytic cells are easily obtainable from the peritoneal cavity of laboratory animals, such as the mouse, rat, guinea pig, and rabbit. Resident peritoneal macrophages (PM) can be isolated by simple lavage of the peritoneal cavity with a physiological salt solution containing heparin (Cohn and Benson, 1965a). The yield of peritoneal cells from the mouse is approximately 5×10^6 cells, of which 50% of the cells are macrophages. The number of PM can be markedly enhanced by the intraperitoneal administration of an irritant 3 to 4 days prior to cell harvesting (Reed and Tepperman, 1969). Total peritoneal cell yields as high as 1 to 5×10^8 cells per animal are possible from rats, guinea pigs, and rabbits. As is the case with nonelicited PM, these cell populations are contaminated with PMNL (10–40%) and nonphagocytic mononuclear cells (2–10%) and adherence procedures are required to purify the macrophage population. For example, 1- to 2-hr adherence at 37°C of peritoneal cells in plastic culture dishes or any glass-bottomed flask will remove most of the nonadhering cells and further incubation of the resulting monolayers will eliminate most of the short-lived PMNL that were initially adhering to the vessel surface.

2.1.3. Alveolar Macrophages

Alveolar macrophages (AM) are readily available from the lungs of rabbit, cat, guinea pig, rat, and mouse following repeated lung lavage with physiological saline by a method originally described by Myrvik *et et al.* (1961). The average yield of AM from unstimulated rabbit lungs by lavage is about 20×10^6 cells and, from rat lungs, 10×10^6 cells. The number of AM can be greatly increased (greater than 100-fold) when lungs are lavaged 3 to 4 weeks after an intravenous challenge of complete Freund's adjuvant (Myrvik *et al.*, 1962). As in the case with monocytes and PM, techniques are now available to maintain AM in culture for several weeks in order to study their phagocytic and metabolic characteristics (Cohen and Cline, 1971).

2.1.4. Kupffer Cells

The central importance of the liver Kupffer cell in the clearance and detoxification of foreign material from the circulation has been well recognized for many years. However, our knowledge of the physiological controls of Kupffer cell function has been hampered by the difficulty in isolating viable Kupffer cells in high yields. Until recently, most of the kinetic information regarding Kupffer cell phagocytosis has been dependent upon studies utilizing either a perfused whole liver system (Filkins and Smith, 1965; Bonventre and Oxman, 1965; Koenig *et al.*, 1965) or *in vitro* liver slice preparations (Saba *et al.*, 1966; Molnar *et al.*, 1977). A method for isolating rat (Pisano *et al.*, 1968) and rabbit (Melly *et al.*, 1972) Kupffer cells has been described. More recently, Munthe-Kaas *et al.* (1975) have described a procedure for isolating Kupffer cells in viable condition and in adequate numbers for *in vitro* functional studies. In this procedure, the liver is dispersed by collagenase perfusion *in situ* followed by enzymatic digestion of contaminating parenchymal cells by Pronase. This method resulted in high yields of viable Kupffer cells, relatively free of contaminating parenchymal cells, that were avidly phagocytic and could also be maintained in culture (Munthe-Kaas, 1976).

2.2. *IN VITRO* ASSAY CONDITIONS FOR PHAGOCYTOSIS

For phagocytosis to occur effectively in an *in vitro* system, a number of assay conditions must be met. Major considerations in developing an *in vitro* phagocytic assay include such factors as the chemical composition and buffering capacity of the incubation medium, gas environment, and the presence of opsonic factors. Although many different balanced salt solutions and culture media preparations have been employed in phagocytic studies, a number of criteria are common to all. For example, the maintenance of a physiological pH, the presence of divalent cations and metabolizable substrates for cell utilization are essential ingredients of the incubation medium. Serum is the usual source of macromolecules and, in addition to containing opsonic factors for the promotion of phagocytosis, serum also supplies additional metabolites and adds greater buffering capacity to the medium. Normal serum contains immune (Lennox and Cohn, 1967; Hirsch and Strauss, 1964; Gigli and Nelson, 1968) as well as nonimmune (Saba, 1975) factors that can promote phagocytosis depending on the particle chosen for uptake by phagocytic cells. If serum opsonins are to be avoided, bovine serum albumin can be substituted as a source of protein. It should be noted, however, that proteins are known to have a profound effect on the metabolism of macrophages (Cohn and Benson, 1965b).

Phagocytosis by macrophages has been measured *in vitro* using either cell suspensions or monolayer cultures. The use of cell suspension for quantifying phagocytosis offers several advantages, including that a known number of cells can be added to each test sample and also that cells can be centrifuged following incubation for removal of extracellular particles or for further metabolic analysis of the macrophages or of the incubation medium. One major drawback in using cell suspensions for phagocytic studies is the relatively short incubation interval (usually several hours) before macrophages begin either to lose viability or to adhere to the incubation vessel.

The use of monolayer cultures for phagocytic assays has substantially increased the length of time macrophages can be maintained *in vitro* and has allowed greater flexibility in manipulating culture conditions prior to the measurement of particle uptake. Since PMNL do not normally survive in culture for more than 24 hr, long-term monolayer techniques for "professional" phagocytes are generally limited to mononuclear phagocytes. Macrophages suspended in either balanced salt solution or culture medium, with or without serum, will settle and adhere tenaciously to plastic or glass surfaces at 37°C. Following this adherence step, the medium, along with nonadhering cells, can be removed resulting in monolayers of macrophages. One major advantage of this monolayer technique is that quantitation of phagocytosis can be made with fewer cells per sample than in cell suspension phagocytic assays. Furthermore, removal of extracellular particles is usually rapid and more effective than by centrifugation of cell suspensions. However, problems of particles attaching to cell surfaces or to the culture dish itself do occur. Therefore, appropriate controls and additional procedures are generally needed to ensure that only phagocytic particles are being quantified.

3. TEST PARTICLES USED FOR PHAGOCYTOSIS *IN VITRO*

Although any particle that is visible under light or phase microscopy can be utilized for the measurement of *in vitro* phagocytosis, two major categories of particles are in general use today: artificial and biological particles. Table 1 lists a

TABLE 1. ARTIFICIAL AND BIOLOGICAL PARTICLES COMMONLY USED FOR *IN VITRO* PHAGOCYTOSIS

Type	Particle
Artificial	Polystyrene latex
	Paraffin oil emulsion
	Radiolabeled lipid emulsion
	Carbon
	Asbestos
Biological	Erythrocytes
	Microorganisms
	Immune complexes
	Zymosan
	Denatured serum albumin
	Starch
	Protein-coated zymogens

number of such particles that have been commonly used for measuring phagocytosis by *in vitro* methods. It should be apparent from this representative list that these particles span a range of particle sizes. Although the demarcation between pinocytosis and phagocytosis in terms of particle size is not clearly defined, most investigators regard particles that are visible under light or phase microscopy to be internalized by a phagocytic process. It should also be noted that the type of particle chosen will also govern the type of humoral factors that may accelerate its uptake by phagocytic cells. In some instances, ingestion in the absence of serum opsonins occurs. For example, carbon particles, zymosan, and certain rough strains of bacteria can be phagocytized without serum opsonins (Griffin and Silverstein, 1974). Rather than provide an exhaustive catalog of particles used in phagocytic assays, this section will analyze a select group of artificial and biological particles that have been used extensively to measure macrophage phagocytosis.

3.1. ARTIFICIAL PARTICLES

Polystyrene latex beads are metabolically inert particles frequently employed in *in vitro* phagocytic studies. These particles offer such advantages as ready availability in a number of sizes, long-term stability, and ease in handling. Evaluation of latex ingestion by phagocytic cells has generally been performed either by direct counting of ingested beads by microscopic methods (Korn and Weisman, 1967; Ulrich and Zilversmit, 1970) or by quantifying cell-associative polystyrene (Roberts and Quastel, 1963). Latex beads containing carboxyl groups are now commercially available, thus permitting chemical cross-linking of latex particles with a variety of proteins to their surfaces (Hallgren *et al.*, 1977; Check *et al.*, 1979). Recently, radiolabeled, gelatin-coated latex particles have been utilized to measure the influence of serum and plasma fibronectin on their uptake by macrophage monolayers (Gudewicz *et al.*, 1980; Doran *et al.*, 1980). A major disadvantage in the use of latex beads for phagocytic measurements is their tendency to adhere to either cell or culture dish surfaces. In addition, it is difficult to separate extracellular particles from phagocytic cells either by simply washing monolayers or by centrifugation of cell suspensions (Gardner *et al.*, 1973). Our laboratory has been successful in removing surface-bound gelatin-coated latex beads by treating macrophage monolayers with trypsin after phagocytosis has been completed.

Stossel *et al.* (1971) have used paraffin oil containing oil red O emulsified with serum proteins as a test particle for obtaining kinetic data on the initial rate of ingestion by PMNL and macrophage populations. One advantage of these particles is that they are easily removed from the cell suspension by centrifugation so that a negligible zero-time value can be obtained. These particles offer another advantage in that, unlike latex beads, they do not aggregate or bind nonspecifically to phagocytic cells. While using AM, but not PMNL, the ingestion of albumin-coated paraffin oil particles has been shown to be enhanced in the presence of fresh serum and divalent cations (Mason *et al.*, 1973).

Another lipid emulsion particle has been used extensively *in vivo* and in liver slice preparations to measure phagocytosis by organ-localized macrophages. This particle was originally developed by DiLuzio and Riggi (1964) to measure reticuloendothelial cell function *in vivo*. This lipid emulsion can easily be radiolabeled with ^{131}I- or ^{14}C-labeled fatty acids and has an average particle size between 0.5 and 3.0 μm. Several investigators (Saba *et al.*, 1966; Molnar *et al.*, 1977; Allen *et al.*, 1973) have utilized a gelatinized, lipid emulsion particle to measure uptake by liver slices *in vitro* and have demonstrated the ability of nonimmune opsonic factors to markedly enhance particle uptake. Plasma fibronectin has recently been identified as the nonimmune opsonic factor in serum responsible for the enhanced uptake by macrophages of this colloid (Blumenstock *et al.*, 1978; Saba *et al.*, 1978; Molnar *et al.*, 1979). Although these gelatinized lipid emulsion particles have been used extensively to measure reticuloendothelial cell activity in liver slice preparations, these particles have not been used successfully to quantitate phagocytosis by isolated cell preparations of macrophages.

3.2. BIOLOGICAL PARTICLES

By far the most commonly used particles for the quantitation of phagocytosis have been live and heat-killed microorganisms. In view of the important role assigned to "professional" phagocytes in the defense against infectious agents invading the host, preparations of either bacteria, yeast, or fungi have frequently been employed to measure the uptake and fate of microorganisms within phagocytic cells (Cohn and Morse, 1959; Lehrer and Cline, 1969). Large numbers of bacteria can easily be grown in suitable culture broth, centrifuged, washed in physiological salt solution, and finally resuspended to the desired concentration by turbidimetric techniques (Mackaness, 1960). Viable bacterial counts are also usually determined from known dilutions by pour-plate methods. Viable bacterial suspensions are readily radiolabeled during their log phase of growth by the incorporation of ^{14}C- or ^{3}H-labeled substrates such as amino acids or thymidine (Downey and Diedrick, 1968) or by the incorporation of ^{32}P (Michell *et al.*, 1969). Other methods have also been utilized to tag microorganisms. For example, Viken (1974) has reported a method for the ^{131}I electrolytic labeling of killed *Candida albicans*. A major drawback in the use of microorganisms as test particles for phagocytosis is the difficulty in discerning attached and ingested microorganisms by whatever method is used to quantitate phagocytosis. In addition, the use of killed organisms or radiolabeling introduces the possibility of altered surface antigens. Another disadvantage in using microorganisms in phagocytic assays is the inability of obtaining kinetic data regarding the rate of uptake of the microorganisms due to the large variability in phagocytic uptake from one experiment to another.

Red blood cells (RBC) are another commonly used biological particle in the quantitation of phagocytosis by macrophages. RBC have become a convenient test particle due to their uniform size, ease of radiolabeling, visibility under light or phase microscopy, and defined antigenic composition. Although normal, homologous RBC are not avidly ingested by macrophages, their uptake may be induced by modification of the cell membrane surface. For example, glutaraldehyde-treated RBC are excellent test particles for the discrimination of attached versus ingested RBC by macrophage monolayers (Rabinovitch, 1967). RBC pretreated with antibody and/or complement have been used extensively in the study of particle binding as well as determining the receptors mediating particle recognition (Gigli and Nelson, 1968; Rabinovitch and DeStefano, 1973; Holland *et al.*, 1972; Mantovani *et al.*, 1972). RBC are also easily radiolabeled with ^{51}Cr or ^{59}Fe (Gray and Sterling, 1950) and have been used not only to measure phagocytosis but, more recently, in the determination of antibody-dependent cellular cytotoxicity by mononuclear cells (Poplack *et al.*, 1976; Hersey, 1973). Probably the major methodological advantage in using RBC for measuring phagocytosis is that noningested RBC can be thoroughly removed from the sample by a hypotonic lysis procedure, thereby leaving only those RBC that have been internalized to be counted or measured.

4. METHODS TO QUANTIFY PHAGOCYTOSIS *IN VITRO*

In general, quantitation of phagocytosis *in vitro* is determined by one of two basic approaches, namely, microscopic observation of ingested particles or the uptake process can be measured by chemical or radioactive counting techniques. In either case, it is usually preferable to measure the accumulation of particles within the phagocyte as opposed to the disappearance of particles from the incubation medium. One advantage in measuring intracellular particle uptake is that quantitation of the initial rates of ingestion can be made. Since a major problem in all phagocytic assays is the separation of extracellular from internalized particles, certain criteria must be established for each assay system in order to avoid errors in interpretation of the data. These criteria have been outlined by Stossel (1975) and include such determining factors as (1) the ingestion rate should be proportional to particle concentration at less than saturating doses of particles; (2) the ingestion rate should be independent of particle concentration at high particle-to-cell ratios; and (3) the inhibition of phagocytosis at ice-bath temperatures or in the presence of metabolic inhibitors. Since the variety of methods employed to study phagocytosis in the past century seems almost endless, this section will limit its discussion to only major approaches that have been applied to the measurement of phagocytosis by isolated macrophages.

4.1. MORPHOLOGICAL AND BACTERIAL COUNTING METHODS

Since the initial observations of phagocytosis by isolated phagocytes nearly a century ago, there have been a number of now classic microscopic studies devoted to examining this process in detail (Fenn, 1921; Hanks, 1940; Mudd *et al.*, 1934). The assessment of phagocytosis by light microscopy (e.g., Blake and

Swanson, 1975) and by phase-contrast miscroscopy (Gibbs and Roberts, 1975) have depended upon the laborious counting of phagocytes that have ingested particles and the number of particles per phagocyte to arrive at an index of the phagocytic process for the population examined (North, 1968). Quantitatively, this method has been improved by including not only the number of cells that have ingested particles but also the proportion of cells containing a specified number of particles (Sbarra and Karnovsky, 1959). Several disadvantages of the light microscopic approach to the measurement of phagocytosis are readily apparent. For example, this technique is quite tedious and very time-consuming and cannot be used effectively when large numbers of samples are to be examined in a single experiment. In addition, these methods are relatively insensitive for deriving any kinetic data of phagocytosis (Smith and Wood, 1958). More recently, the electron microscopic approach to the study of phagocytosis has added a new dimension to our understanding of the ultrastructural components of this process and has elegantly demonstrated the fine structure of the cellular elements that interact with the ingested particle. In addition, electron microscopic studies have aided our understanding of how certain facultative parasites are able to survive and multiply within phagocytic cells (Pearson *et al.*, 1963; Leake *et al.*, 1971). With the use of scanning electron microscopic tehniques, macrophage phagocytosis can be observed three-dimensionally, as extensions of macrophage cell membranes reach out to surround the particle to be ingested (Walters *et al.*, 1976; Tizard and Holmes, 1974). A procedure originally described by Rabinovitch (1967) has been used effectively in examining the attachment and ingestion phases of erythrocyte phagocytosis by macrophages in culture. This morphological approach has revealed that ingestion of aldehyde-treated RBC was dependent on temperature, divalent cations, and serum factors while attachment was not dependent on these factors.

Suspensions of bacteria or yeast are frequently used, not only for the microscopic examination of phagocytosis, but also for determining the microbicidal activity of the phagocyte population under study. By using centrifugation techniques to separate the phagocyte and noningested microbes, it is possible to estimate and compare phagocytic rates of ingestion and postphagocytic killing of the microorganism in question (Cohn and Morse, 1959; Hirsch and Strauss, 1964). Several dilutions of both extracellular and cell-associated microorganisms are usually prepared and disrupted from cells (Baughn and Bonventre, 1975; Hoff, 1975) and counted by standard pour-plate techniques to obtain the total number of microbes that have been ingested and the number of viable microbes remaining within the phagocytes. The resulting uptake data are usually plotted semilogarithmically over time of incubation in order to obtain killing activity. When using this methodology, appropriate controls must be conducted to rule out any extracellular microbicidal activity in the incubation medium as well as to be able to distinguish precisely between microorganisms attached to or ingested by the phagocytes. When the test organisms are rapidly growing bacteria, antimicrobicidal agents may be added to suppress the growth of noningested bacteria, on the assumption that these agents do not penetrate phagocytic cells (Tan *et al.*, 1971). One advantage in measuring microbicidal activity of a given phagocyte population is that, in addition to uptake data, another physiological endpoint of the phagocytic process is also being quantified. However, the major limitations include the now familiar problems of separation of ingested microbes from nonspecific adherence and also the clumping of bacteria and yeast when using pour-plate techniques. In general, these methods are usually reliable when the data will yield large differences between control and experimental values.

4.2. RADIOACTIVE AND CHEMICAL METHODS

The use of uniformly radiolabeled particles is a convenient quantitative technique for studying *in vitro* phagocytosis and intracellular particle degradation by phagocytic cells. For example, radiolabeled live and heat-killed bacteria have been widely used to measure endocytosis by phagocyte cell suspensions and monolayer cultures (Michell *et al.*, 1969; Cohn, 1963; Root *et al.*, 1972). When cell suspensions of phagocytes are employed, differential and density gradient centrifugation are commonly used to separate noningested from internalized particles (Solomkin *et al.*, 1978). Once again, appropriate controls must be established in order to verify that the radioactivity associated with the cells is in fact due to phagocytosis. It is usually advisable to use as many criteria as possible to establish the mechanism of particle uptake. In certain procedures, it is possible to minimize the attachment of particles to the surfaces of macrophages as when using radiolabeled RBC as described previously. The problem of surface adherence has been investigated by Hallgren and Stalenheim (1976) with complexes of IgG and protein A from *Staphylococcus aureus*. These aggregates interact with Fc receptors on macrophages and are internalized while the extracellular complexes can be solubilized by using an excess of protein A. Another recent modification developed to minimize nonspecific radioactivity not associated with phagocytosis has been the inhibition of [^{3}H]uridine incorporation into *C. albicans* (Yamamura *et al.*, 1977) and *S. aureus* (Lam and Mathison, 1979) following their phagocytosis by PMNL. This method is reported to be a sensitive index of yeast and bacteria phagocytosis since the rapid incorporation of [^{3}H]uridine into the particles stopped abruptly following their phagocytosis. Since the PMNL suspension did not incorporate [^{3}H]uridine to any significant degree during the short incubation period, the reduction in uridine incorporation in the particle and cell mixture was taken as a quantitative measure of phagocytosis as long as there was a linear correlation between uridine incorporation and yeast number. Forsgren *et al.* (1977) have described a method, which is a modification of the spectrophotometric method described by Stossel *et al.* (1972), that measures phagocytosis of a radiolabeled emulsion of lipopolysaccharide-coated oil particles. The authors describe the advantages of this method as (1) phagocytosis can be assessed with fewer phagocytes than most cell suspension assays and (2) quantitation of opsonic activity can be performed with as little as 10 μl of serum. Jones and Buchanan (1978) recently described a dual-labeling method for quantitatively assessing phagocytosis of [^{3}H]uracil-labeled gonococci by [^{14}C]leucine-labeled PM monolayers. This method offers an advantage not normally present in most *in vitro* phagocytic assays in that not only phagocytosis can be measured but the number of macrophages remaining after the incubation period can also be assessed. Therefore, both macrophage ingestion and attachment functions can be performed simultaneously. Finally, with regard to the problem of surface attachment of radiolabeled particles to macrophage monolayers, our laboratory has observed that treatment of macrophage monolayers with trypsin following the phagocytic assay is a mandatory step in order to eliminate adherent latex particles. Nonspecific binding of radioactive particles to macrophage monolayers and culture dishes can create just as many difficulties in the quantitation of phagocytosis as are thought to exist with the use of cell suspension assays.

Roberts and Quastel (1963) developed an *in vitro* phagocytic assay to measure the uptake of polystyrene latex particles by PMNL by using *p*-dioxane to extract the ingested polystyrene from the phagocytes for spectrophotometric analysis. This technique has been applied to macrophage monolayers (Tsan and Berlin, 1971) and has been used to characterize the kinetics of latex phagocytosis by the protozoan *Acanthamoeba* (Weisman and Korn, 1967). The ingestion of paraffin oil droplets containing oil red O by PMNL and macrophages can also be quantitated spectrophotometrically following extraction with *p*-dioxane (Stossel *et al.*, 1972). Depending on the origin of the phagocytes and the type of phagocyte examined, the ingestion rate of emulsified oil particles will differ. For example, guinea pig PMNL will phagocytize these particles without serum opsonic factors (Stossel *et al.*, 1972) whereas rabbit AM (Stossel *et al.*, 1973) and human PMNL (Stossel *et al.*, 1973) phagocytosis of these particles was enhanced by serum factors.

Phagocytosis is an integrated series of complex events that is accompanied by alterations in metabolism which have been termed *metabolic concomitants of phagocytosis*. Sbarra and Karnovsky (1959) were the first to observe that phagocytosis of a variety of particles markedly stimulated the rate of [1-^{14}C]glucose oxidation in phagocytes. Since PMNL and PM have low rates of Krebs cycle activity and therefore low basal rates of glucose oxidation, the large increments in the initial rate of [1-^{14}C]glucose oxidation are directly proportional to the initial rate of ingestion (Stossel *et al.*, 1972). The oxidation of glucose is more complex in AM (Romeo *et al.*, 1973) and in activated macrophages (Karnovsky and Lazdins, 1978), since these cells have significant glucose oxidation via aerobic pathways. Caution must be used in the interpretation of metabolic data in these macrophage populations as an indicator of phagocytic activity. Furthermore, in studies utilizing glucose oxidation as a measure of phagocytosis, it is important to keep in mind that certain particles, such as endotoxin, can stimulate macrophage glucose metabolism without being phagocytized (Cline *et al.*, 1968).

5. SUMMARY

This chapter has reviewed methodology currently being applied to the quantitation of phagocytosis *in vitro* by macrophages. Methods for obtaining viable macrophage populations for *in vitro* phagocytic studies from tissues containing a significant reticuloendothelial cell component have also been briefly described. In addition, major advantages and disadvantages of various artificial and biological particles employed to measure phagocytosis *in vitro* have been discussed. Also a survey of major *in vitro* assays that have been applied to the *in vitro* measurement of macrophage phagocytosis has been given in order to provide an awareness of the various limitations each assay intrinsically possesses. The continual expansion of our knowledge of phagocytosis and of the mechanisms regulating this fundamental biological process will rest not only on future advances in methodology but ultimately on the fascination and imagination of those who endeavor to unlock its many mysteries.

REFERENCES

Allen, C., Saba, T. M., and Molnar, J., 1973, Isolation, purification and characterization of opsonic protein, *J. Reticuloendothelial Soc.* **13**:410.

Baughn, R. E., and Bonventre, P. F., 1975, Phagocytosis and intracellular killing of *Staphylococcus aureus* by normal mouse peritoneal macrophages, *Infect. Immun.* **12**:346.

Blake, M., and Swanson, J., 1975, Studies on gonococcus infection. IX. *In vitro* decreased association of pilated gonococci with mouse peritoneal macrophages, *Infect. Immun.* **11**:1402.

Blumenstock, F. A., Saba, T. M., Weber, P., and Laffin, R., 1978, Biochemical and immunological characterization of human opsonic α_2SB glycoprotein: Its identity with cold-insoluble globulin, *J. Biol. Chem.* **253**:4287.

Bonventre, P. F., and Oxman, E., 1965, Phagocytosis and intracellular disposition of viable bacteria by the isolated perfused rat liver, *J. Reticuloendothelial Soc.* **2**:313.

Brodersen, M. P., and Burns, C. P., 1973, The separation of human monocytes from blood including biochemical observations, *Proc. Soc. Exp. Biol. Med.* **144**:941.

Check, I. J., Wolfman, H. C., Coley, T. B., and Hunter, R. L., 1979, Agglutination assay for human opsonic factor using gelatin-coated latex particles, *J. Reticuloendothelial Soc.* **25**:351.

Cline, M. J., Melmon, K. L., Davis, W. C., and Williams, H. E., 1968, Mechanism of endotoxin interaction with human leucocytes, *Br. J. Haematol.* **15**:539.

Cohen, A. B., and Cline, M. J., 1971, The human alveolar macrophage: Isolation, cultivation *in vitro*, and studies of morphologic and functional characteristics, *J. Clin. Invest.* **50**:1390.

Cohn, Z., 1963, The fate of bacteria within phagocytic cells. I. The degradation of isotopically labelled bacteria by polymorphonuclear leucocytes and macrophages, *J. Exp. Med.* **117**:27.

Cohn, Z. A., and Benson, B., 1965a, The differentiation of mononuclear phagocytes: Morphology, cytochemistry and biochemistry, *J. Exp. Med.* **121**:153.

Cohn, Z. A., and Benson, B., 1965b, The *in vitro* differentiation of mononuclear phagocytes. II. The influence of serum on granule formation, hydrolase production, and pinocytosis, *J. Exp. Med.* **121**:835.

Cohn, Z. A., and Morse, S. I., 1959, Interactions between rabbit polymorphonuclear leucocytes and staphylococci, *J. Exp. Med.* **110**:419.

DiLuzio, N. R., and Riggi, S. J., 1964, The development of a lipid emulsion for the measurement of reticuloendothelial function, *J. Reticuloendothelial Soc.* **1**:136.

Doran, J. E., Mansberger, A. R., and Reese, A. C., 1980, Cold insoluble globulin-enhanced phagocytosis of gelatinized targets by macrophage monolayers: A model system, *J. Reticuloendothelial Soc.* **27**:471.

Downey, R. J., and Diedrick, B. F., 1968, A new method for assessing particle ingestion by phagocytic cells, *Exp. Cell Res.* **50**:483.

Fenn, W. O., 1921, The phagocytosis of solid particles. I. Quartz, *J. Gen. Physiol.* **3**:439.

Filkins, J. P., and Smith, J. J., 1965, Plasma factor influencing carbon phagocytosis in the isolated perfused rat liver, *Proc. Soc. Exp. Biol. Med.* **119**:1181.

Fogelman, A. M., Seager, J., Hokom, M., and Edwards, P. A., 1979, Separation of and cholesterol synthesis by human lymphocytes and monocytes, *J. Lipid Res.* **20**:379.

Forsgren, A., Schmeling, D., and Zettervall, O., 1977, Quantitative phagocytosis by human polymorphonuclear leucocytes: Use of radiolabelled emulsions to measure the rate of phagocytosis, *Immunology* **32**:491.

Gardner, D. E., Graham, J. A., Miller, F. J., Illing, J. W., and Coffin, D. L., 1973, Technique for differentiating particles that are cell-associated or injested by macrophages, *Appl. Microbiol.* **25**:471.

Gibbs, D. L., and Roberts, R. B., 1975, The interaction *in vitro* between human polymorphonuclear leukocytes and *Neisseria gonorrhoeae* cultivated in the chick embryo, *J. Exp. Med.* **141**:155.

Gigli, I., and Nelson, R. A., Jr., 1968, Complement dependent immune phagocytosis, *Exp. Cell Res.* **51**:45.

Gray, S. J., and Sterling, K., 1950, The tagging of red cells and plasma proteins with radioactive chromium, *J. Clin. Invest.* **29**:1604.

Griffin, F. M., and Silverstein, S. C., 1974, Segmental response of the macrophage plasma membrane to a phagocytic stimulus, *J. Exp. Med.* **139**:323.

Gudewicz, P. W., Molnar, J., Lai, M. Z., Beezhold, D. W., Siefring, G. E., Credo, R. B., and Lorand, L., 1980, Fibronectin-mediated uptake of gelatin-coated latex particles by peritoneal macrophages, *J. Cell Biol.* **87**:427.

Hallgren, R., and Stalenheim, G., 1976, Quantification of phagocytosis by human neutrophils: The use of radiolabelled staphylococcal protein A–IgG omplexes, *Immunology* **30**:755.

Hallgren, R., Jansson, L., and Verge, P., 1977, Kinetic studies of phagocytosis of IgG-coated latex particles with a thrombocyte counter, *J. Lab Clin. Med.* **90**:786.

Hanks, J. H., 1940, Quantitative aspects of phagocytosis as influenced by the number of bacteria and leukocytes, *J. Immunol.* **38**:159.

Hart, I. R., and Fidler, I. J., 1979, The collection, purification and characterization of canine peripheral blood monocytes, *J. Reticuloendothelial Soc.* **26**:121.

Hersey, P., 1973, Macrophage effector function: An *in vitro* system of assessment, *Transplantation* **15**:282.

Hirsch, J. G., and Strauss, B., 1964, Studies on heat-labile opsonin in rabbit serum, *J. Immunol.* **92**:145.

Hoff, R., 1975, Killing *in vitro* of *Trypanosoma cruzi* by macrophages from mice immunized with *T. cruzi* or BCG, and absence of cross-immunity on challenge *in vivo*, *J. Exp. Med.* **142**:299.

Holland, P., Holland, N. H., and Cohn, Z. A., 1972, The selective inhibition of macrophage phagocytic receptors by anti-membrane antibodies, *J. Exp. Med.* **135**:458.

Jones, R. B., and Buchanan, T. M., 1978, Quantitative measurement of phagocytosis of *Neisseria gonorrhoeae* by mouse peritoneal macrophages, *Infect. Immun.* **20**:732.

Karnovsky, M. L., and Lazdins, J. K., 1978, Biochemical criteria for activated macrophages, *J. Immunol.* **121**:809.

Kavet, R. I., and Brain, J. D., 1980, Methods to quantify endocytosis: A review, *J. Reticuloendothelial Soc.* **27**:201.

Koenig, M. G., Heyssel, R. M., Melly, M. A., and Rogers, D. E., 1965, The dynamics of reticuloendothelial blockade, *J. Exp. Med.* **122**:117.

Korn, E. D., and Weisman, R. A., 1967, Phagocytosis of latex beads by *Acanthamoeba*. II. Electron microscopic study of the initial events, *J. Cell Biol.* **34**:219.

Lam, C., and Mathison, G. E., 1979, Phagocytosis measured as inhibition of uridine uptake: A method that distinguishes between surface adherence and ingestion, *J. Med. Microbiol.* **12**:459.

Leake, E. S., Evens, D. G., and Myrvik, Q. N., 1971, Ultrastructural patterns of bacterial breakdown in normal and granulomatous rabbit alveolar macrophages, *J. Reticuloendothelial Soc.* **9**:174.

Lehrer, R. I., and Cline, M. J., 1969, Interaction of *Candida albicans* with human leukocytes and serum, *J. Bacteriol.* **98**:996.

Lennox, E. S., and Cohn, M., 1967, Immunoglobulins, *Annu. Rev. Biochem.* **36**:365.

Mackaness, G. B., 1960, The phagocytosis and inactivation of staphylococci by macrophages of normal rabbits, *J. Exp. Med.* **112**:35.

Mantovani, B., Rabinovitch, M., and Nussenzweig, V., 1972, Phagocytosis of immune complexes by macrophages: Different roles of the macrophage receptor sites for complement (C3) and for immunoglobulin (IgG), *J. Exp. Med.* **135**:780.

Mason, R. J., Stossel, T. P., and Vaughan, M., 1973, Quantitative studies of phagocytosis by alveolar macrophages, *Biochim. Biophys. Acta* **304**:864.

Melly, M. A., Duke, L. J., and Koenig, M. G., 1972, Studies on isolated cultured rabbit Kupffer cells, *J. Reticuloendothelial Soc.* **12**:1.

Michell, R. H., Pancake, S. J., Noseworthy, J., and Karnovsky, M. L., 1969, Measurement of rates of phagocytosis: The use of cellular monolayers, *J. Cell Biol.* **40**:216.

Molnar, J., McLain, S., Allen, C., Laga, H., Gara, A., and Gelder, F., 1977, The role of an α_2-macroglobulin of rat serum in the phagocytosis of colloidal particles, *Biochim. Biophys. Acta* **493**:37.

Molnar, J., Gelder, F. B., Lai, M. Z., Siefring, G. E., Credo, R. B., and Lorand, L., 1979, Purification of opsonically active human and rat cold-insoluble globulin (plasma fibronectin), *Biochemistry* **18**:3909.

Mudd, S., McCutcheon, M., and Lucke, B., 1934, Phagocytosis, *Physiol. Rev.* **14**:210.

Munthe-Kaas, A. C., 1976, Phagocytosis in rat Kupffer cells *in vitro*, *Exp. Cell Res.* **99**:319.

Munthe-Kaas, A. C., Berg, T., Seglen, P. O., and Seljelid, R., 1975, Mass isolation and culture of rat Kupffer cells, *J. Exp. Med.* **141**:1.

Musson, R. A., and Henson, P. M., 1979, Humoral and formed elements of blood modulate the response of peripheral blood monocytes. I. Plasma and serum inhibit and platelets enhance monocyte adherence, *J. Immunol.* **122**:2026.

Myrvik, Q. N., Leake, E. S., and Fariss, B., 1961, Studies on pulmonary alveolar macrophages from the normal rabbit: A technique to procure them in a high state of purity, *J. Immunol.* **86**:128.

Myrvik, Q. N., Leake, E. S., and Oshima, S., 1962, A study of macrophages and epithelioid-like cells from granulomatous (BCG-induced) lungs of rabbits, *J. Immunol.* **89**:745.

North, R. J., 1968, The uptake of particulate antigens, *J. Reticuloendothelial Soc.* **5**:203.

Pearson, G. R., Freeman, B. A., and Hines, W. D., 1963, Thin-section electron micrographs of monocytes infected with *Brucella suis*, *J. Bacteriol.* **86**:1123.

Pisano, J. C., Filkens, J. P., and DiLuzio, N. R., 1968, Phagocytic and metabolic activities of isolated rat Kupffer cells, *Proc. Soc. Exp. Biol. Med.* **128**:917.

Poplack, D. G., Bonnard, G. D., Holiman, B. J., and Blaese, R. M., 1976, Monocyte-mediated antibody-dependent cellular cytotoxicity: A clinical test of monocyte function, *Blood* **48**:809.

Rabinovitch, M., 1967, The dissociation of the attachment and injestion phases of phagocytosis by macrophages, *Exp. Cell Res.* **46**:19.

Rabinovitch, M., and DeStefano, M. J., 1973, Particle recognition by cultivated macrophages, *J. Immunol.* **110**:695.

Reed, P. W., and Tepperman, J., 1969, Phagocytosis-associated metabolism and enzymes in the rat polymorphonuclear leukocyte, *Am. J. Physiol.* **216**:223.

Roberts, J., and Quastel, J. H., 1963, Particle uptake by polymorphonuclear leucocytes and Ehrlich ascites-carcinoma cells, *Biochem. J.* **89**:150.

Romeo, D., Zabucchi, G., Marzi, T., and Rossi, F., 1973, Kinetic and enzymatic features of metabolic stimulation of alveolar and peritoneal macrophages challenged with bacteria, *Exp. Cell Res.* **78**:423.

Root, R. K., Rosenthal, A. S., and Balestra, D. J., 1972, Abnormal bactericidal, metabolic, and lysosomal functions of Chediak–Higashi syndrome leukocytes, *J. Clin. Invest.* **51**:649.

Saba, T. M., 1975, Aspecific opsonins, in: *The Immune System and Infectious Diseases* (E. Nester and F. Milgrom, eds.), pp. 489–504, Karger, Basel.

Saba, T. M., Filkins, J. P., and DiLuzio, N. R., 1966, Properties of the "opsonic system" regulating *in vitro* hepatic phagocytosis, *J. Reticuloendothelial Soc.* **3**:398.

Saba, T. M., Blumenstock, F. A., Weber, P., and Kaplan, J. E., 1978, Physiologic role for cold-insoluble globulin in systemic host defense: Implications of its characterization as the opsonic α_2SB glycoprotein, *Ann. N.Y. Acad. Sci.* **312**:43.

Sbarra, A. J., and Karnovsky, M. L., 1959, The biochemical basis of phagocytosis. I. Metabolic changes during the ingestion of particles by polymorphonuclear leukocytes, *J. Biol. Chem.* **234**:1355.

Silverstein, S. C., Steinman, R. M., and Cohn, Z. A., 1977, Endocytosis, *Annu. Rev. Biochem.* **46**:669.

Silverstein, S. C., Michl, J., and Sung, S.-S.J., 1978, Phagocytosis, in: *Transport of Macromolecules in Cellular Systems* (S. C. Silverman, ed.), pp. 245–264, Dahlem Konferenzen, Berlin.

Smith, M. R., and Wood, W. B., 1958, Surface phagocytosis: Further evidence of its destructive action upon fully encapsulated pneumococci in the absence of type-specific antibody, *J. Exp. Med.* **107**:1.

Solomkin, J. S., Mills, E. L., Giebink, G. S., Nelson, R. D., Simmons, R. L., and Quie, P. G., 1978, Phagocytosis of *Candida albicans* by human leukocytes: Opsonic requirements, *J. Infect. Dis.* **137**:30.

Stossel, T. P., 1974, Phagocytosis, *N. Engl. J. Med.* **290**:833.

Stossel, T. P., 1975, Phagocytosis: Recognition and ingestion, *Semin. Hematol.* **12**:83.

Stossel, T. P., Polland, T. D., Mason, R. J., and Vaughan, M., 1971, Isolation and properties of phagocytic vesicles from polymorphonuclear leukocytes, *J. Clin. Invest.* **50**:1745.

Stossel, T. P., Mason, R. J., Hartwig, J., and Vaughan, M., 1972, Quantitative studies of phagocytosis by polymorphonuclear leukocytes: Use of emulsions to measure the initial rate of phagocytosis, *J. Clin. Invest.* **51**:615.

Stossel, T. P., Alper, C. A., and Rosen, F. S., 1973, Serum-dependent phagocytosis of paraffin oil emulsified with bacterial lipopolysaccharide, *J. Exp. Med.* **137**:690.

Tan, J. S., Watanakunakorn, C., and Phari, J. P., 1971, A modified assay of neutrophil function: Use of lysostaphin to differentiate defective phagocytosis from impaired intracellular killing, *J. Lab. Clin. Med.* **78**:316.

Tizard, I. R., and Holmes, W. L., 1974, Phagocytosis of sheep erythrocytes by macrophages: Some observations under the scanning electron microscope, *J. Reticuloendothelial Soc.* **15**:132.

Tsan, M. F., and Berlin, R. D., 1971, Effect of phagocytosis on membrane transport of nonelectrolytes, *J. Exp. Med.* **134**:1016.

Ulrich, F., and Zilversmit, D. B., 1970, Release from alveolar macrophages of an inhibitor of phagocytosis, *Am. J. Physiol.* **218**:1118.

Viken, K. E., 1974, ^{125}I-labelling of *Candida albicans* by electrolysis, *Acta Pathol. Microbiol. Scand. Sect. B* **82**:219.

Walters, M. N. I., and Papadimitriou, J. M., 1978, Phagocytosis: A review, *CRC Crit. Rev. Toxicol.* **5**:377.

Walters, M. N. I., Papadimitriou, J. M., and Robertson, T. A., 1976, The surface morphology of the phagocytosis of micro-organisms by peritoneal macrophages, *J. Pathol.* **118**:221.

Wardley, R. C., Lawman, M. J., and Hamilton, F., 1980, The establishment of continuous macrophage cell lines from peripheral blood monocytes, *Immunology* **39**:67.

Weisman, R. A., and Korn, E. D., 1967, Phagocytosis of latex beads by *Acanthamoeba*. I. Biochemical properties, *Biochemistry* **6**:485.

Yamamura, M., Boler, J., and Valdimarsson, H., 1977, Phagocytosis measured as inhibition of uridine uptake by *Candida albicans*, *J. Immunol. Methods* **14**:19.

Zuckerman, S. H., and Douglas, S. D., 1979, Dynamics of the macrophage plasma membrane, *Annu. Rev. Microbiol.* **33**:267.

Zuckerman, S. H., Ackerman, S. K., and Douglas, S. D., 1979, Long-term human peripheral blood monocyte cultures: Establishment, metabolism and morphology of primary human monocyte-macrophage cell cultures, *Immunology* **38**:401.

4

Kinetics of Vascular Clearance of Particles by Phagocytes

SIGURD J. NORMANN

1. INTRODUCTION

There are several good reasons for studying the kinetics of vascular clearance of intravenously administered particles. The primary objectives are to identify those parameters which limit the rate of vascular clearance, show how these parameters interact, and determine at which points control of vascular clearance occurs. At present, the kinetics of phagocytosis are best studied *in vivo* and the descriptions of the process have been largely achieved based on *in vivo* measurements. *In vivo,* there is an adequate number of cells to ensure measurable uptake over a sustained period of time, cell-associated particles are readily separable from those in the fluid phase, measurement of removal is readily achieved, and the conditions of flow, particle availability, and media are biologically relevant. The insights gained from the study of vascular clearance with model particles are presumably relevant to the basic issues of how macrophages recognize nonself and of how things get into cells.

The fixed phagocytes of the liver and spleen are the cells primarily responsible for recognizing and removing from the circulation potentially injurious materials (Benacerraf *et al.,* 1957). Among these substances are microorganisms, fibrin aggregates that form normally and in increased amounts during disseminated intravascular coagulation, immune complexes, enzymes, and denatured autologous proteins. The process contributes to normal protein turnover and failure to remove injurious substances seems to be a major factor in the perpetuation of traumatic and hypovolemic shock (Altura and Hershey, 1968).

Vascular clearance involves several sequential steps: possible interaction with serum constituents (opsonins), binding to a phagocyte surface, and endocytosis. Accordingly, control of clearance could exist at several levels including availability of opsonin, concentration of particles, number of membrane recep-

SIGURD J. NORMANN • Department of Pathology, College of Medicine, University of Florida, Gainesville, Florida 32610.

tors, affinity of association between particle and cell membranes, and the velocity of endocytosis. In this review, we will develop the kinetic theory of vascular clearance, show how factors of particle and host origin influence clearance velocity, and conclude with a presentation of how different sets of particles may interact to reduce the clearance velocity of both.

2. KINETICS OF VASCULAR CLEARANCE: A ZERO- OR FIRST-ORDER RATE EQUATION?

Blood clearance of intravenously injected particles can be used to quantitate the phagocytic activity of macrophages lining vascular channels. Biozzi *et al.* (1953) working with carbon particles and aggregated albumin found that the clearance velocity (dC/dt) of a given particle was directly proportional to the particle concentration (C) in the circulation:

$$dC/dt = -kC \tag{1}$$

Many investigators have used this relationship to determine the rate constant (k) of phagocytosis; k has units of minutes^{-1}. This constant is usually determined experimentally by measuring the concentration of particles in the circulation at several different time intervals after injection. As shown in Fig. 1, a straight line

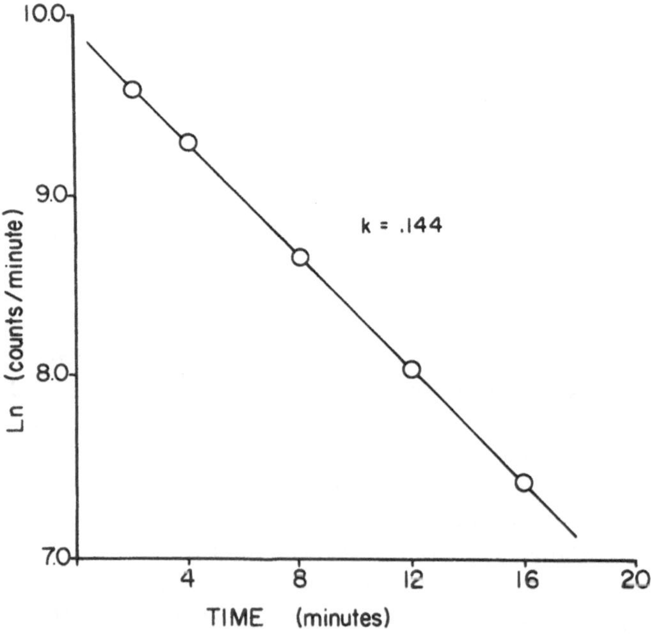

FIGURE 1. Typical example of the vascular clearance of 100 μg/100 g body wt of acetylated C-14 denatured bovine albumin in rats. (Reproduced from data of Normann, 1973b, *J. Reticuloendothelial Soc.* **14**:587, with permission of Academic Press, Inc.)

usually emerges when the ln concentration is plotted against time. The slope of the line is equivalent to k and is determined by linear regression methods. No uniformity exists in whether or not the data are plotted using logarithms to the base 10 or natural logarithms. However, the latter is preferred for kinetic studies and should be adopted in future published work.

It is generally recognized that the dose of particles influences the rate at which they are removed from the circulation. In 1953, Biozzi and co-workers injected increasingly large colloid doses and concluded that the clearance rate was inversely proportional to the concentration of injected particles:

$$k(\text{dose}) = \text{constant} \tag{2}$$

But this relationship should not be true if vascular clearance strictly follows a first-order rate equation as suggested by equation (1). Indeed, this relationship is not true when examined over a range of particle concentrations; in particular, it is not accurate for either small or large particle doses. When large particle doses were examined, the data obtained by Parker and Finney (1960) suggested that the clearance equation might be either zero order (independent of particle concentration) or first order (dependent on particle concentration) at various segments of the clearance curve. This observation was confirmed by Fred and Shore (1967) and Fred *et al.* (1967) leading them to propose that phagocytosis involves a rate-limiting interaction between the particle and the phagocytic cell surface. In their model, the rate-limiting factor for phagocytosis was the number of binding sites for particles on the surface of the phagocyte.

In a further analysis of this problem, Normann (1974) measured the removal from the circulation of a series of particle concentrations. When plotted on a linear scale, a curvature was observed in the plot at low particle concentrations for these particles were being removed in an exponential manner; that is, clearance velocity was proportional to blood concentration and followed a first-order rate equation. As the particle concentration increased, however, the initial clearance data conformed to a straight line on both linear and logarithmic scales indicating that at very high particle concentration zero-order kinetics would be observed.

The above observations suggested certain similarities to enzyme kinetics wherein the order of the reaction is dependent on substrate concentration. The controversy as to whether the clearance equation was zero or first order was resolved by Normann (1973b) when he plotted the initial clearance velocity against the dose of particles injected. Typical results are diagrammed in Fig. 2 and support the following conclusions.

1. As the particle concentration increases, clearance velocity approaches a maximum phagocytic velocity (V_{max}).
2. In analogy to enzyme kinetics, the order of the reaction varies with the particle concentration.
3. A particle–membrane constant (K_p) exists which corresponds to the Michaelis–Menten constant (K_m). This constant is related to the affinity by which the particles bind to the surface of the phagocyte.

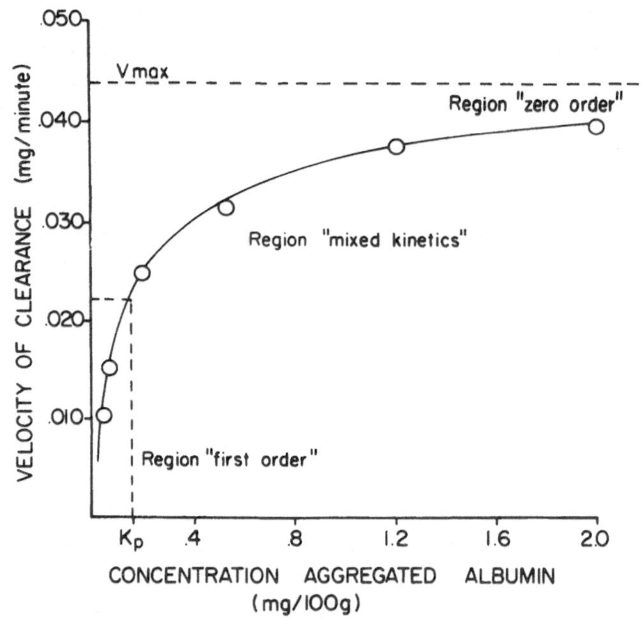

FIGURE 2. Effect of aggregated albumin concentration on the clearance velocity of acetylated C-14 denatured bovine albumin. Clearance velocity increases with increasing albumin concentration toward a maximum rate of clearance. K_p is the particle–membrane constant. Vascular clearance will exhibit zero-order, first-order, or mixed kinetics depending on the concentration of aggregated albumin injected. (Reproduced from data of Normann, 1973b, *J. Reticuloendothelial Soc.* **14**:587, with permission of Academic Press, Inc.)

For practical purposes, zero-order kinetics will be observed when the particle concentration is 100 times greater than K_p and, conversely, first-order kinetics will be observed when the particle concentration is below $0.01K_p$. Between these extremes, mixed zero- and first-order kinetics will be observed. For example, removal of carbon particles in rats will be zero order when concentrations are equal to or greater than 460 mg/100 g body wt and first-order reactions will be observed at or below 0.05 mg carbon/100 g body wt. The larger concentration of particles cannot be given. Indeed, the concentrations of particles generally administered are closer to those concentrations yielding first-order rather than zero-order kinetics. For this reason, an approximation of clearance velocity can be made by assuming first-order kinetics. The initial clearance velocity is then estimated as the product of k and the dose of particle injected:

$$\text{initial clearance velocity} = k(\text{dose}) \tag{3}$$

Initial clearance velocity will vary for different doses of particles. Thus, the product of k(dose) is not constant as assumed by Biozzi and co-workers [equation (2)]. Indeed, their own data are consistent with this conclusion. To avoid confusion and for accurancy, the formulation of Biozzi and co-workers [equation

(2)] should be replaced with the concept of maximum phagocytic velocity or V_{max}.

3. VASCULAR CLEARANCE, SURFACE-SATURATION KINETICS, AND PHAGOCYTOSIS

Meredith (1961) postulated that vascular clearance of intravenously administered colloidal gold was analogous to an enzyme–substrate reaction. Although Meredith pointed out that phagocytosis is a nonenzymatic process, he argued that it should proceed and be limited by the formation of an intermediate particle–membrane complex. This concept was further developed by Iio and Wagner (1963) and applied to the clearance of aggregated albumin in dogs and humans. Subsequently, Caro and Radicella (1970) and Cohen *et al.* (1968) presented a similar analysis for the phagocytic kinetics of radiogold colloid in the rat. Analogy to enzyme kinetics suggests that specific surface receptors for particles may be involved. Such receptors have been demonstrated on monocytes and macrophages for the Fc portion of immunoglobulins and for complement (Lay and Nussenzweig, 1968; Huber and Fudenberg, 1970; Bianco *et al.*, 1975; Buglio *et al.*, 1967).

Alternatively, phagocytosis may involve less-specific surface adsorption followed by engulfment. Gosselin (1956, 1967) observed that the rate of ingestion at any time was proportional to the amount of surface-bound particles and that attachment of gold colloid to rabbit peritoneal macrophages was similar to a Langmuir adsorption isotherm. van Oss and Gillman (1972b) observed that thermodynamic surface energy considerations underlie the phagocytosis of bacteria and that a correlation exists between the hydrophobicity of bacteria or latex and the degree to which they are engulfed. This concept is further developed by van Oss and colleagues in Chapter 1 of this volume.

Regardless of whether attachment of particles to the phagocyte surface involves reaction with specific receptors or chemisorption to less-specific surface loci, the essential feature is the formation of a particle–membrane complex. This conclusion led Normann (1973b) to postulate that phagocytosis is a surface-initiated reaction and that attachment loci on the surface of the cell become progressively saturated with particles as their concentration increases. If so, vascular clearance conforms to surface-saturation kinetics and obeys a rate equation determined by the formation of a particle–membrane intermediate.

The vascular clearance of a variety of particles including carbon, radiogold colloid, latex, aggregated albumin, foreign or formaldehyde-fixed red blood cells, lipid emulsion, and bacteria has been shown to conform to surface-saturation kinetics. Evidence to support this hypothesis has been obtained using a double-reciprocal plot of clearance velocity versus the particle concentration (Lineweaver–Burk plot) (Lineweaver and Burk, 1934). In all instances examined, the reciprocal of the clearance velocity ($1/V$) was directly related to the reciprocal of the particle concentration ($1/C$). A typical experiment is presented in Fig. 3. Such a plot of $1/V$ versus $1/C$ yields an ordinate intercept equivalent to the

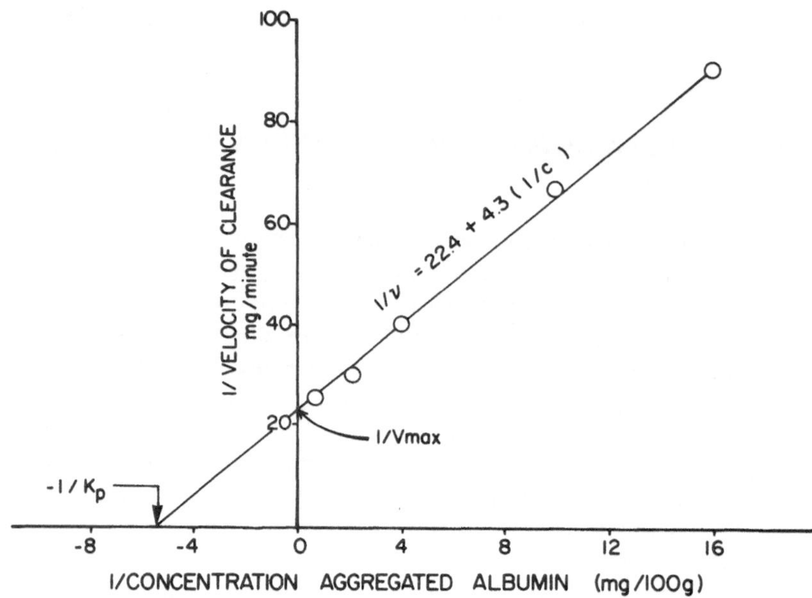

FIGURE 3. Lineweaver–Burk plot of the clearance of acetylated C-14 aggregated bovine albumin from the circulation of rats.

inverse of the maximum phagocytic velocity ($1/V_{max}$) and an abscissa intercept equivalent to the negative reciprocal of the particle–membrane constant K_p. These two parameters are exceedingly important determinants of the functional status of the phagocytes and should be determined in describing fully the clearance of a given particle from the circulation under normal and experimental conditions.

4. DERIVATION OF THE PARTICLE–MEMBRANE CONSTANT (K_p) AND MAXIMUM PHAGOCYTIC VELOCITY (V_{max})

It is generally recognized that phagocytosis consists of a surface attachment phase followed by a phase of engulfment (Rabinovitch, 1967; Jones, 1975):

$$\begin{array}{c} \text{particle} \\ + \\ \text{phagocyte} \end{array} \underset{k_2}{\overset{k_1}{\rightleftharpoons}} \begin{array}{c} \text{particle–} \\ \text{membrane} \\ \text{complex} \end{array} \xrightarrow{k_3} \begin{array}{c} \text{phagosomal} \\ \text{particle} \end{array} \tag{4}$$

attachment phase engulfment phase

Evidence to support this contention has been obtained from experiments using temperature or chemicals to block cellular ingestion while permitting attachment to occur (Rabinovitch, 1967). Ultrastructural studies emphasize adherence to the cell surface as an early event in the clearance of inert particles (carbon, thorium dioxide, iron oxide, and latex) as well as cells (bacteria, altered or damaged mammalian cells) (Horn *et al.*, 1969; Casley-Smith and Reade, 1965; Karrer, 1960). The process is relatively selective because the intensity of binding to peroxidase-positive Kupffer cells is vastly greater than to peroxidase-negative endothelial cells or parenchymal cells (Widman *et al.*, 1973). Particles of different types can be observed attached to the surface membranes of the same phagocyte (Normann *et al.*, 1968). Such surface-bound proteins and particles can be eluted from the membranes *in vitro* either by removing radiolabeled materials from the surface by exchange with nonradiolabeled colloids, by repetitive washings after binding, or by enzymatic treatments to remove surface-held molecules (Gosselin, 1956, 1967; Ulrich and Zilversmit, 1970; Unanue *et al.*, 1969). The amount of substance released varies considerably with the agent used, the temperature of the experiment, and the time period after exposure at which recovery is attempted. For instance, the amount recovered from phagocytes relative to the total amount associated with the cell has been reported to be as high as 25% for radiogold colloid (Gosselin, 1956, 1967), 12–21% for polystyrene (Ulrich and Zilversmit, 1970), and 3–6% for hemocyanin and horseradish peroxidase (Schmidtke and Unanue, 1971). Greater amounts of nonaggregated than aggregated albumin can be recovered indicating that the aggregate binds more strongly to the membrane.

In the development which follows, the approach will be from the point of view of surface phenomena. As observed in equation (4), the reversible nature of the attachment phase is indicated by the individual rate constants of k_1 for attachment and k_2 for detachment. The ratio of k_1/k_2 expresses the affinity of the particle for the cell surface. It seems reasonable to assume that the attachment phase involves multiple receptor loci (specific or nonspecific) on the surface, each of which can be occupied by only one particle. If X represents that fraction of the surface occupied by particles, then the rate of particle addition onto the surface is proportional to that fraction of the surface not occupied by particles $(1 - X)$ and to the concentration of unbound particles (C). The equilibrium expression between the rate of attachment and detachment becomes

$$k_1(1 - X)(C) = k_2X \tag{5}$$

or

$$X = \frac{K_aC}{1 + K_aC} \quad \text{where } K_a = k_1/k_2$$

The above equation is the familiar Langmuir adsorption isotherm originally used by Gosselin (1956) to describe the attachment of radiogold colloid to rabbit peritoneal macrophages. The affinity with which particles bind to the cell is expressed by the association constant K_a. The assumption that the attachment phase of phagocytosis conforms to a Langmuir adsorption isotherm does not take into account either heterogeneity of receptor sites or the influence of a bound particle on an adjacent unbound site.

The second phase of phagocytosis or the phase of engulfment must follow attachment and represent a surface-initiated reaction. This does not imply that attachment alone will initiate a reaction in all instances. It does imply that attachment is a requisite condition for phagocytosis and that the intensity of the engulfment process is proportional to the numbers of particles held on the surface. Different particles have different rates of engulfment independent of their affinity for the cell surface. If E_o represents the total number of surface loci that can be occupied by particles, then E_oX represents the number of particles attached to the surface. If the rate of particle engulfment (phagocytosis) is proportional to the number of surface-held particles, the initial velocity V of phagocytosis becomes

$$V = k_3E_oX \tag{6}$$

If it is assumed that following engulfment there is reappearance of unoccupied sites, then the rate of change of X at equilibrium becomes

$$dX/dt = k_1(1 - X)C - (k_2 + k_3)X = 0 \tag{7}$$

or

$$X = \frac{C}{C + K_p} \quad \text{where } K_p = \frac{k_2 + k_3}{k_1} \tag{8}$$

and substituting equation (8) into equation (6)

$$\frac{1}{V} = \frac{1}{k_3E_o} + \frac{K_p}{k_3E_o}\left(\frac{1}{C}\right) \tag{9}$$

Equation (9) predicts that the reciprocal velocity of phagocytosis should be directly related to the reciprocal of the particle concentration (Lineweaver–Burk plot). Further, the maximum phagocytic velocity as noted in equation (6) should occur when all membrane sites are occupied or

$$V_{max} = k_3E_o \tag{10}$$

As noted in equation (8), the particle–membrane constant is defined as follows:

$$K_p = \frac{k_2 + k_3}{k_1} \tag{11}$$

From a double-reciprocal plot of $1/V$ versus $1/C$, it is apparent from equation (9) that the intercept on the abscissa represents the negative reciprocal of K_p and the ordinate intercept represents the inverse of V_{max}.

5. AFFINITY OF PARTICLES FOR THE PHAGOCYTE SURFACE: THE ASSOCIATION CONSTANT K_a

The interaction between particles and the cell surface is an essential and potentially rate-limiting step in the process of vascular clearance. Methods to measure the affinity of binding of macromolecules to cell surfaces have been developed for cells *in vitro*. While most studies have concerned the association of antigen with immunocytes, a few studies have been reported using phagocytes. Mehl and Lagunoff (1975) examined the association of aggregated albumin complexes with isolated rat peritoneal macrophages while several studies have been performed using monomeric IgG and either rabbit alveolar macrophages (Arend and Mannik, 1973) or mouse peritoneal macrophages (Unkeless and Eisen, 1975).

Estimation of the number of binding sites per cell and the average affinity of association between particles and the cell surface is performed using Scatchard plots (Scatchard, 1949). In order to perform these studies, it is essential to deter-

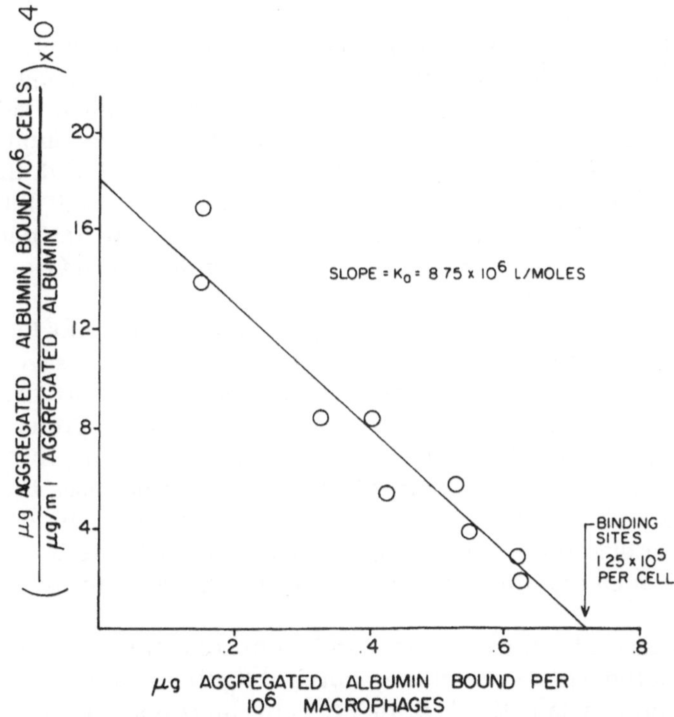

FIGURE 4. Scatchard plot of the binding of aggregated albumin (52 monomer units) to rat peritoneal macrophages. The slope of the line is equivalent to the affinity of binding of the aggregated albumin to the macrophages whereas the intercept on the abscissa is equivalent to the number of binding sites per cell. (Reproduced from data of Mehl and Lagunoff, 1975, *J. Reticuloendothelial Soc.* **18**:125, with permission of Dr. Mehl and the Reticuloendothelial Society.)

mine the number of molecules bound per cell (r) and the molar concentration (c) of molecules available for binding. A series of concentrations are examined from which a plot is made of the ratio of r/c versus r. The slope of the resulting line represents the apparent association constant K_a. Extrapolation to an infinite concentration of molecules—e.g., as r/c approaches zero—yields an absicissa intercept equivalent to the maximum number of binding sites per cell. An example is provided in Fig. 4.

In performing these experiments with phagocytes, two technical problems must be overcome. First, the cell-bound particles must be separated from the nonbound particles. This step is usually achieved by either multiple washings or by centrifugation through a gradient of suitable density. In his study on aggregated albumin, Mehl (1975) preferred gelatin gradients to multiple washings because the separations were cleaner and there was decreased cell manipulation with a concomitant increase in cell recovery and viability. Second, endocytosis must be reduced or blocked so that the cell-associated particles are held on the cell membrane. *In vitro*, endocytosis is blocked either by low temperature or by chemicals (Cohn, 1966; de Terra and Rustad, 1959; Mehl, 1975; Berken and Benacerraf, 1966). The determination of bound particles is usually made by isotopic measurements.

In the study reported by Mehl (1975), bovine serum albumin (BSA) was aggregated using glutaraldehyde cross-linking and the aggregates separated according to size. An aggregate composed of 15 BSA monomers had an apparent association constant of 5.12×10^6 liter/moles and a maximum number of binding sites per cell of 8.0×10^4. As the size of the aggregate increased, the association constant increased but the number of binding sites decreased. Table 1 summarizes the data on the binding of aggregated albumin to rat peritoneal macrophages as well as data on the Fc receptor binding of IgG to alveolar and peritoneal macrophages.

The data tabulated in Table 1 for the binding of immunoglobulins, monomeric and aggregated albumins to macrophages *in vitro* have several interesting implications for the vascular clearance of proteins and particulates *in vivo*. For instance, the data suggest that nondenatured monomeric proteins may be removed slowly from the circulation because of their low affinity for the phagocyte surface. Aggregation and presumably denaturation increases the affinity and may be the dominant factor promoting clearance. However, the effect of increasing aggregate size on affinity is opposed by the decrease in the number of binding sites such that the clearance rate may become relatively independent of particle size.

If dissociation from the cell is marginal in comparison to association, then the inverse of the particle membrane constant K_p should approximate the apparent association constant K_a. A K_a of 1.8×10^6 liter/moles as observed by Mehl (1975) for aggregates of 15 BSA monomers has a reciprocal value $1/K_a$ of 1.18×10^{17} molecules/liter. A rat has a blood volume of about 7.2 ml/100 g so this value of $1/K_a$ is equal to 8.5×10^{14} molecules/100 g. It is interesting that this value lies in the same general range as the K_p for heat-aggregated BSA in rats of 1.8×10^{14} molecules/100 g. This similarity exists despite differences due to *in vivo* and *in*

TABLE 1. SUMMARY OF APPARENT ASSOCIATION CONSTANT AND SITES PER CELL FOR DIFFERENT SUBSTRATES AND CELLS

Cell types	Substrate	Apparent association constant (liter/moles)	Sites/cell	References
Rabbit alveolar macrophage				
Resident	Monomeric IgG	7.6×10^5	1.2×10^6	Arend and Mannik (1973)
Elicited (Freund's adjuvant)	Monomeric IgG	9.0×10^5	2.2×10^6	Arend and Mannik (1973)
Rat peritoneal macrophage				
Resident	Monomeric BSA	2.5×10^5	9.0×10^5	Mehl (1975)
Elicited (oyster glycogen)	Aggregated BSA			
	4 units	1.8×10^6	2.1×10^5	Mehl (1975)
	15 units	5.1×10^6	8.0×10^4	Mehl (1975)
	52 units	2.2×10^7	3.0×10^4	Mehl (1975)
Mouse peritoneal macrophage				
Resident	Monomeric IgG2a	1.4×10^8	1.1×10^5	Unkeless and Eisen (1975)
	Monomeric IgG2b	6.8×10^6	5.0×10^4	Unkeless and Eisen (1975)
Elicited (thioglycollate)	Monomeric Ig2a	1.3×10^8	4.4×10^5	Unkeless and Eisen (1975)
	Monomeric IgG2b	9.0×10^6	2.9×10^5	Unkeless and Eisen (1975)
Lymphocyte from immunized guinea pig	DNP-guinea pig	6.2×10^8	6.8×10^4	Davie and Paul (1972)

vitro measurements and the possibility of significant dissociation from the cell surface. It underscores the value of both *in vivo* and *in vitro* measurements in determining kinetic constants important to the vascular clearance of particles.

Fc Receptors

It is interesting that the affinity of the Fc receptor for rabbit IgG on rabbit alveolar macrophages is similar to that of mouse IgG2b for mouse peritoneal macrophages (Table 1). While the binding of these immunoglobulins is greater than for monomeric BSA, it is considerably less than that of mouse IgG2a for murine macrophages. This observation suggests that a specific high-affinity receptor exists on murine macrophages for IgG2a in contrast to a low-affinity receptor for IgG2b. Recent experiments confirm the fact that two Fc receptors exist on murine macrophages: Fc receptor I is trypsin sensitive and binds IgG2a while Fc receptor II is trypsin resistant and binds IgG1 and IgG2b. Monoclonal antibody prepared against the glycoprotein of Fc receptor II does not block Fc receptor I (Unkeless, 1979).

The magnitude of the association constant of IgG2a for Fc receptor I is approximately equal to that of an antigen combining with its antibody which raises the question as to why such immunoglobulins are not permanently cytophilic. The lower affinity of rabbit IgG and of murine IgG2a may in fact be more relevant to vascular clearance of immune complexes. By cooperative (not necessarily additive) enhancement of K_a by immune aggregate formation, they may promote clearance in a manner analogous to nonimmune aggregation or denaturation of proteins. This notion is consistent with the observation that a single IgG molecule combined with antigen is not sufficient to promote clearance (Mannik *et al.*, 1971; Arend and Mannik, 1971; Normann, 1972a; Rowley and Turner, 1966). On the other hand, Fc receptors may not play a significant role in clearance of immune complexes formed with complement-fixing IgG since such complexes coated with complement do not bind staphylococcal protein A (Scharfstein *et al.*, 1979). Because staphylococcal protein A binds to the Fc portion of IgG (Kronvall and Frommel, 1970), these experiments suggest that complement components mask Fc segments. *In vivo* where complement is abundant, it would be anticipated that the Fc segments of complement-fixing IgG in immune complexes would be unavailable for binding to the Fc receptors on macrophages.

6. PARTICLE SELECTION AND THE CLEARANCE BEHAVIOR OF DIFFERENT PARTICLES

Wilkins and Myers (1964) listed several criteria which particles should possess when used as probes of macrophage function, among which were (1) particle stability, (2) monodispersity, (3) reproducibility, and (4) some feature by which they can be recognized such as isotopic labeling. The more important characteristics of particles that modify clearance are discussed below.

6.1. PARTICLE STABILITY

Two types of particle stability are recognized: (1) intrinsic stability implying that the particles are neither destroyed nor catabolized except after their removal from the circulation and (2) extrinsic stability implying that the particles remain in suspension: that is, they do not promote excessive intravascular coagulation, clump, or embolize to the pulmonary bed.

Gelatin has often been used to confer suspension stability on a variety of inert colloids (Halpern *et al.*, 1953). Of these preparations, colloidal carbon has been the most frequently used. As the amount of gelatin increases, the rate of particle clearance decreases. Dobson (1957) observed that the rate of removal of chromium phosphate was inversely proportional to the amount of gelatin added. Normann and Benditt (1965a) demonstrated a similar effect for colloidal carbon and Murray and Katz (1955) for radiogold colloid. The observed effect of gelatin is due to a changed relationship between the particle and the plasma. One effect of gelatin is to diminish deposition of fibrinogen around the particles making them more stable in blood (Halpern *et al.*, 1953). While reducing the pronounced hydrophilic nature of the colloid and its capacity for coagulation (Lyman *et al.*, 1965), gelatin is nonetheless hydroscopic. Thus, a second effect is to coat the particles with a moderately hydrophilic surface which may directly interfere with phagocytosis (van Oss and Gillman, 1972a). A third effect may involve interaction between gelatin and blood proteins that influence particle clearance (Filkins and DiLuzio, 1966). The nature of this substance may be a natural antibody to gelatin since such has been described in rats and dogs but not rabbits (Maurer, 1954; Murray, 1963a). However, it is more likely that this substance is fibronectin, an α_2-macroglobulin also identified as LETS protein or cold-insoluble globulin (Blumenstock *et al.*, 1977b). Fibronectin is a glycoprotein with a molecular weight of 450,000 whose concentration in plasma is about 300 μg/ml. It binds with high affinity to denatured collagen and moderately to fibrin but not to native collagen, fibrinogen, or most bacteria (Saba *et al.*, 1978; Bevilacqua *et al.*, 1980). Since gelatin is a solubilized form of collagen which binds fibronectin, the effective depletion of fibronectin by binding onto excess nonparticle gelatin could slow the clearance rate of the gelatin-stabilized colloid.

The relationship between particle stability and vascular clearance has been examined also for the removal of immune complexes. While it is generally appreciated that soluble complexes formed in antigen excess circulate for prolonged periods of time, soluble complexes formed in antibody excess also have a slow removal rate (Normann, 1972a). Horse antiserum which fixes complement poorly was used to form soluble complexes in antibody excess (prozone effect). The clearance of complexes formed in antigen excess was slow and independent of antiserum concentration until 80% of the antigen had been bound to antibody. At equivalence, clearance was rapid and cellular localization revealed the majority of complexes within fixed macrophages of liver and spleen. In antibody excess, high-molecular-weight nonprecipitating complexes were formed yet the clearance rate was exceedingly slow. Thus, the rate of vascular clearance *in vivo* precisely reflected the precipitin curve measured *in vitro* (Fig. 5). This experiment was important as it demonstrated that (1) combination of antigen with antibody alone is not sufficient to promote rapid removal, (2) the size of the complex may not always correlate with clearance velocity, and (3) particle stability may be a most critical determinant of removal.

In conclusion, particle compatibility with blood is a major determinant of clearance velocity. This conclusion is supported by data derived from carbon (Halpern *et al.*, 1953; Normann and Benditt, 1965b), colloidal chromium phosphate (Dobson, 1957), colloidal radiogold (Murray and Katz, 1955), immune complexes (Normann, 1972a), and denatured proteins (Benacerraf *et al.*, 1957; Iio and Wagner, 1963; Mehl, 1975; Normann, 1974). If a substance aggregates in blood by immune or nonimmune mechanisms, its removal rate is increased. This statement emphasizes interactions with fibrinogen and other plasma proteins such as fibronectin as well as immunoglobulins. Another critical factor appears to be the degree of surface free energy that determines the particle interaction with water (van Oss and Gillman, 1972). It is well known that bacteria with hydrophilic capsules resist phagocytosis in contrast to those with hydrophobic capsules. In addition, denaturation of proteins involves disruption

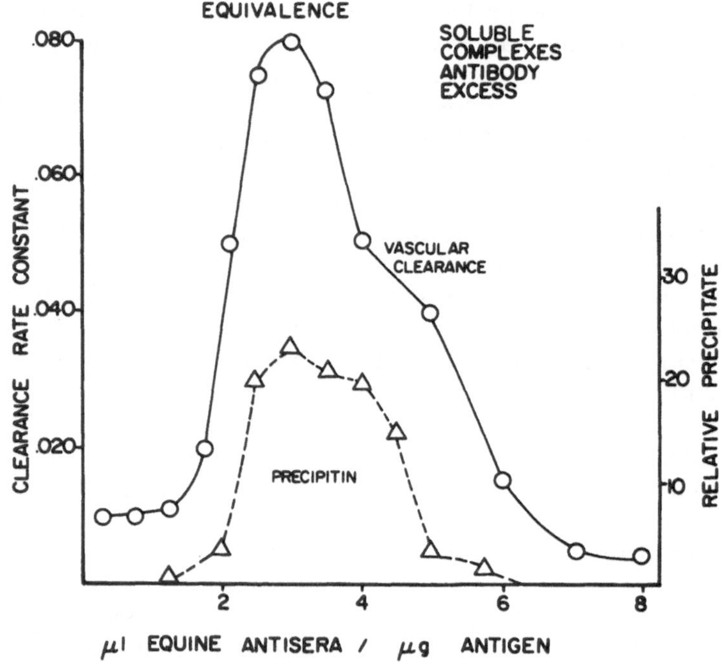

FIGURE 5. Correlation between the clearance rate of antigen from the circulation of rats in the presence of equine antisera and the precipitin curve of the antigen–antibody complexes. Equine antisera form soluble complexes in antibody excess; although these complexes averaged about 1 million molecular weight, they were only slowly removed from the circulation. (Reproduced from data of Normann, 1972a, *Immunology* **108**:521, with permission of Williams and Wilkins.)

of internal bonds resulting in increased hydrophobicity. This may be a critical factor in removal of monomeric denatured proteins as well as those in aggregate form.

6.2. PARTICLE CHARGE

Charge modification of particles can alter their removal rates and their sites of deposition. For instance, Wilkins and Myers (1966) observed that positively charged latex particles were cleared slower than particles of identical size but negative charge. While the former particles were localized mainly in the lung, the latter were directed to the liver. This observation *in vivo* is hard to reconcile with the observation of several investigators that *in vitro* making a particle more positive relative to the negative cell surface enhances attachment of the particle to the cell surface (Weiss *et al.*, 1972; Danon *et al.*, 1972; Gasic *et al.*, 1968). One possible explanation advanced by Wilkins (1971) relates to his observation that electrophoretic mobility measured *in vitro* may have no relationship to *in vivo* conditions since particles with different charge in the absence of serum have similar mobilities in the presence of serum. Wilkins finally concluded that charge alone is not enough to explain recognition, clearance rates, or sites of deposition.

6.3. MONODISPERSITY

Particles of different size may localize in different anatomical sites. For instance, Dobson *et al.* (1949) prepared stable colloids of yttrium, zirconium, columbium, and lanthanum and showed that particle size was critical to their bodily distribution. Gabrieli (1952) and Jallut *et al.*, (1955) studied the cellular distribution of labeled chromium phosphate and concluded that the coarser particles (2–4 μm) tended to localize in the lung whereas the finer particles (about 1 μm) concentrated in the liver. Vascular clearance involving the lung is generally faster than by the liver or spleen but such clearance may be due to embolization rather than phagocytosis. Any general conclusion relating body distribution to particle size is clearly unwarranted. Homogeneous particles with varying sizes such as formalinized red blood cells (7 μm), lipid emulsion (1 μm), and carbon particles (0.08 μm) all localize readily to the fixed phagocytes of the liver and spleen.

Controversy exists as to the effect of particle size on clearance velocity. Investigations using metal colloids generally support the conclusion that large particles are removed faster than small particles (Gofman, 1949; Dobson *et al.*, 1949; Dobson, 1957). However, Schoenberg *et al.* (1961) found no correlation between the size of polystyrene spheres and their rates of clearance. A similar conclusion was reached by Sheppard *et al.*, (1951) working with colloidal gold. It is important to distinguish changes in clearance rate due to size from those which may be attributed to other factors. For instance, aggregation of proteins by immunologic or nonimmunologic means increases clearance velocity but this may be due to factors other than size. This conclusion is suggested by the studies on antigen–antibody complexes presented above (Normann, 1972a). Further protein aggregation by heat and other physical or chemical methods is generally associated with denaturation and an increase in hydrophobicity. Whatever the mechanism, a suspension of polydispersed particles may have a clearance curve composed of more than one component. To avoid such double-exponential clearance curves, it is highly desirable that the particles be monodispersed and of nearly identical surface characteristics.

6.4. SUMMARY

A variety of particles are available with which to measure phagocytosis *in vivo*. A summary of particle size, and the K_p and V_{max} values determined in rats for several different commonly used particles is provided in Table 2. The data for carbon, aggregated albumin, and lipid emulsion are recorded both in milligram units, commonly used experimentally, and in particle units to facilitate comparison with latex particles and formalinized red blood cells. When expressed as cells, particles, or molecules, the data range from 10^8 to 10^{13} particles/100 g rat body wt. The magnitude of this value particularly for V_{max} underscores the tremendous potential for clearing the blood of foreign colloids and particles.

TABLE 2. SUMMARY OF MAXIMUM PHAGOCYTIC VELOCITY (V_{max}) AND THE PARTICLE–MEMBRANE CONSTANT (K_p) FOR THE VASCULAR CLEARANCE OF SEVERAL DIFFERENT PARTICLES IN RATS[a]

Particle	Approximate size (Å)	K_p (U/100 g)	V_{max} (U/min/100 g)	Units
Carbon		4.71	1.22	mg
Monomeric bovine albumin		88.34	ND	mg
Aggregated bovine albumin		0.192	0.044	mg
Lipid emulsion		33.49	18.98	mg
Carbon	800	2.36×10^{13}	6.10×10^{12}	Particles
Monomeric bovine albumin	45	7.71×10^{17}	ND	Molecules
Aggregated bovine albumin	120	1.79×10^{14}	4.13×10^{13}	Molecules
Lipid emulsion	10,000	7.80×10^{9}	4.42×10^{9}	Particles
Formalized rabbit red cells	75,000	1.76×10^{9}	4.03×10^{8}	Cells
Latex spheres	2,300	3.65×10^{8}	1.40×10^{12}	Particles
Gelatin-stabilized radiogold colloid	600	4.97×10^{12}	1.40×10^{12}	Particles

[a]Data from Normann (1974).

7. HOST FACTORS MODULATING VASCULAR CLEARANCE

A variety of host factors influence the clearance rates of intravenously administered particles. The most important ones are discussed below.

7.1. LIVER BLOOD FLOW

Hepatic sinusoidal blood flow is an important determinant of vascular clearance rates. However, the contribution of flow to clearance is dependent on the number of particles injected. It is generally agreed that tracer colloids are removed from the circulation at rates proportional to the rate of hepatic sinusoidal blood flow (Dobson and Jones, 1952; Benacerraf *et al.*, 1955, 1957). On the other hand, large colloid doses which are not removed in a single passage through the liver are assumed to be limited more by the activity of the phagocytic cells than by perfusion rates. Theoretically, the clearance of large particle doses could be influenced by liver blood flow if the delivery rate of the particles relative to the activity of the macrophages was such that individual Kupffer cells engulfed all particles in their immediate vicinity.

The extraction ratio (Asiddao *et al.*, 1964) expresses the difference in particle concentration across the liver and is related to the amount of phagocytosis as shown below:

$$\text{extraction ratio} = \frac{\text{amount of phagocytosis}}{\text{particle concentration (flow)}} \qquad (12)$$

or

$$E = P/CF$$

It is apparent that the extraction ratio (E) at constant flow (F) should vary with the dose and nature of the particles administered. By analyzing data obtained in the perfused liver, Normann (1972b) concluded that the critical extraction ratio for rats was 0.136. Tracer colloids have E approaching 0.80 (Dobson and Jones, 1952) and are essentially removed in a single passage through the liver. For this reason, their clearance velocity is directly related to liver blood flow. As the dose increases, E decreases eventually reaching the critical number (0.136) wherein adequate numbers of particles are being delivered to the cells. If E is reduced further, variations in flow have little effect on phagocytosis.

Application of these principles has been used to measure liver blood flow and to explain certain changes in clearance velocity. In rats, Benacerraf *et al.* (1957) were able to estimate liver blood flow as 1.35 ml/min per g liver by using tracer concentrations of aggregated albumin and assuming an extraction ratio of 0.84. This estimate compares favorably to that obtained by Normann (1972b) of 1.69 ml/min per g liver as measured with an electromagnetic flow meter. Benacerraf *et al.* (1957) noted that clearance rates decrease with age yet portal flow increases both in total amount and in ml/min per g liver (Normann, 1973a). With the concentration of particles used, the extraction ratio was low and an increase in portal flow did not enhance clearance. Since the removal of particles per gram liver is constant with age, the decreased clearance in elderly subjects probably relates to an increase in total vascular volume greater than the increase in liver weight. In hypovolemic shock or in intestinal excision, there is decreased phagocytosis and decreased portal blood flow. In these conditions, the extraction ratio rises and often approaches or even exceeds the critical ratio. These conditions illustrate that decreased clearance for a given particle dose can occur when portal flow is reduced.

7.2. SPECIES

It is often assumed that identical particles injected into different species are cleared at different rates. While the absolute numbers may differ, this conclusion

TABLE 3. COMPARATIVE DATA FOR THE MAXIMUM PHAGOCYTIC VELOCITY (V_{max}) AND THE PARTICLE–MEMBRANE CONSTANT (K_p) FOR THE VASCULAR CLEARANCE OF AGGREGATED ALBUMIN IN DIFFERENT MAMMALIAN ORDERS[a]

Mammalian order	Ratio liver weight to body weight	K_p (molecules/kg)	V_{max} (molecules/kg per min)
Mouse	0.052	3.83×10^{14}	21.39×10^{14}
Rat	0.044	1.79×10^{14}	10.23×10^{14}
Rabbit	0.037	ND	5.81×10^{14}
Guinea pig	0.037	ND	11.35×10^{14}
Dog	0.033	3.14×10^{14}	7.37×10^{14}
Man	0.022	2.33×10^{14}	9.95×10^{14}

[a]Compiled from data of Iio and Wagner (1963) and Normann (1974).

fails to emphasize the extraordinary similarity with which the macrophages of different species are functionally identical in terms of maximum phagocytic velocity and the particle membrane constant K_p. Table 3 compiles data for aggregated albumin injected into six different mammalian orders. It is observed that V_{max} varies over a very narrow range of (5.8 to 21.4) × 10^{14} molecules/kg per min and that for mouse, rat, man, and dog K_p is nearly identical. Similar data have been obtained in animals for carbon particles. Both the magnitude of the number for maximum phagocytic velocity and the very narrow range of values for K_p and V_{max} underscore the remarkable constancy of the phagocytic ability of different mammals relative to their body size.

7.3. KUPFFER CELL NUMBER

The principal organs concerned with phagocytosis of intravenously administered particles are the liver and spleen. Since most colloids are removed principally by the liver (> 80%), we will concentrate attention on the hepatic Kupffer cell.

The number of liver phagocytes can be determined by multiplying the total number of liver cells by the nuclear ratio of phagocytes to nonphagocytes. The latter determination is made by examining histological sections in which appropriate means are taken to delineate the presence of phagocytic particles. Carbon particles are excellent because they can be observed readily in routine histological sections. The number of nuclei associated with particles is counted as well as all other nuclei. These numbers are then corrected for nuclear size and for visibility of nuclear fragments using Floderus formula (Marrable, 1962). For mouse liver, I have estimated the number of phagocytes that ingest carbon particles as 4.6 × 10^7/g liver. This figure is in agreement with that obtained for rat liver using peroxidase staining reactions and electron microscopy (Widman *et al.*, 1973). The number of phagocytes ingesting carbon particles in liver corresponds exactly to the number ingesting fluorescein-labeled heat-aggregated albumin (Normann *et al.*, 1968).

As observed in Table 3, normal liver weight as a function of body weight is remarkably constant for different mammalian orders. Furthermore, the percentage of liver cells which are actively phagocytic is also remarkably constant averaging 15% of all cells. For these reasons, the number of liver phagocytes under normal conditions is proportional to the body weight. Thus, correction of the clearance velocity for body weight and/or the weight of the liver and spleen is neither necessary nor desirable. Yet many authors report a corrected phagocytic index:

$$\frac{\text{corrected phagocytic}}{\text{index}} = \left(\frac{\text{body weight}}{\begin{matrix}\text{liver}\\ \text{weight}\end{matrix} + \begin{matrix}\text{spleen}\\ \text{weight}\end{matrix}} \right) k^{1/3} \tag{13}$$

This formula is misleading since by taking the cube of the clearance velocity the variance of the number of most interest is minimized. The rational and general acceptance of this formula has not been properly justified.

The number of Kupffer cells is an important determinant of the number of available phagocytic sites. This observation leads to three important conclusions. First, if it is assumed that the number of sites per cell for a given particle is constant and whereas the number of Kupffer cells per killigram body weight is constant, then the number of available sites is also constant per killigram body weight for different mammalian orders. Second, if the number of sites is constant and V_{max} is constant, then the endocytic velocity for a given particle is the same for different mammalian orders. Third, experimental or clinical conditions which alter the number of Kupffer cells and hence the number of binding sites may well alter the rates of vascular clearance.

7.4. OPSONINS

An opsonin is any biological substance which interacts with a particle to enhance its rate of phagocytosis. The best known opsonins are antibodies (IgG or IgM) and their function *in vivo* has been readily demonstrated in several different systems. Complement as an opsonin has been demonstrated *in vitro* (Bianco *et al.*, 1975; Griffin *et al.*, 1975; Ehlenberger and Nussenzweig, 1977) but its function *in vivo* is less certain (Mannik *et al.*, 1971). Opsonins unrelated to the immune system also exist. For example, serum fibronectin may be important in clearance of gelatin-coated particles, collagen fragments, and microaggregates of fibrin. Under certain circumstances, even fibrin could be considered an opsonin because its coating around certain particles enhances their clearance (Wilkins, 1971). Obviously, the availability of opsonins may influence vascular clearance rates possibly by promoting aggregation or in some way altering the surface of the particle to increase its affinity for the phagocyte.

8. CLEARANCE INHIBITION BY PARTICLE INJECTION

The clearance rate of an observed particle may decrease following injection of the same or different particle (Biozzi *et al.*, 1957). This phenomenon is interesting because of the possible relationship between phagocyte malfunction and an increased susceptibility to infection and shock.

In early studies a depression in clearance rates due to particle injection was interpreted as "blockade" of the RES, implying a paralysis of function due to a physical stuffing of the cells with particles (Benacerraf *et al.*, 1957; Biozzi *et al.*, 1953). This interpretation is not correct and belies the tremendous capacity of the fixed phagocytes for removal of injected particles. Indeed, repeated injection of either inert or metabolizable particles usually leads to enhanced not inhibited clearance (Dobson *et al.*, 1967; Normann, 1970). These findings are the antithesis of cellular satiation. In 1968, Normann and colleagues measured simultaneously the vascular clearance of two dissimilar particles (carbon and aggregated albumin) from the circulation of rats and observed that the clearance rate of both particles was inhibited. This finding suggested that one phase of inhibition arose from the presence of the particles in the circulation. Meanwhile, Parker and

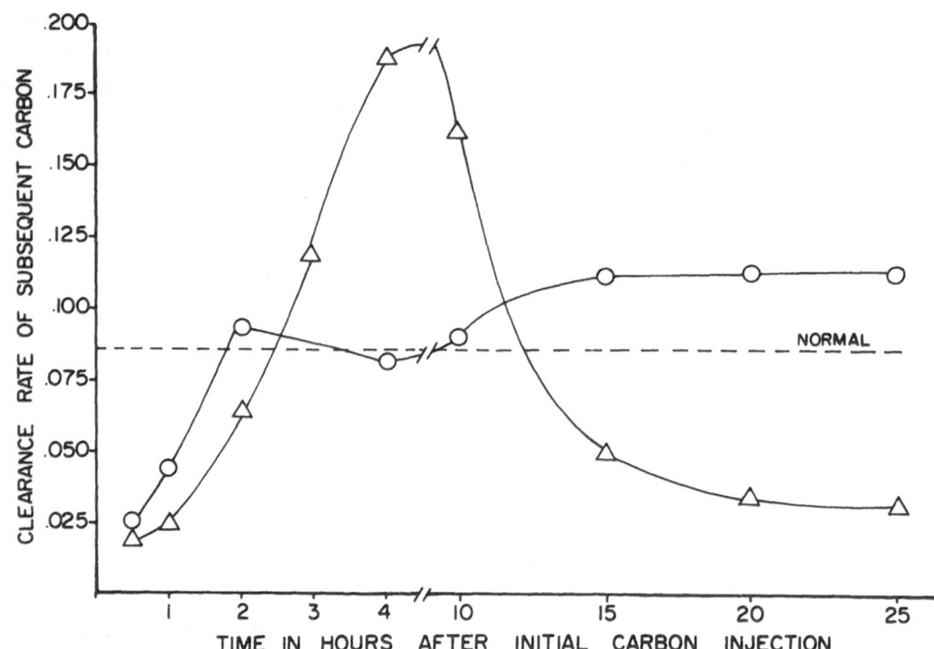

FIGURE 6. The influence of an initial injection [(○-○) 10 mg carbon/100 g; (△-△), 30 mg carbon/100 g] of carbon on the clearance rate of a subsequent carbon injection. There are two phases of inhibition: the first phase begins with the initial carbon injection and lasts only a few hours; the second phase is observed only after a large carbon injection and is separated from the first phase by a period of augmented clearance rates. The second phase begins about 12 hr after the initial carbon injection and lasts several days. (Reproduced from data of Normann, 1970, *Infect. Immun.* **1**:327, with permission of the American Society for Microbiology.)

Finney (1960) observed in mice that a period of normal clearance may follow particle injection and exist prior to the onset of RES paralysis. This observation suggested that a second period of RES depression may follow particle clearance and represent some form of cellular derangement induced by colloid ingestion. From such studies it was concluded that the usual description of RES blockage as a single, sustained period of depression originating with particle ingestion may not be correct. This idea was further developed by Normann (1970) who showed that the phenomenon was composed of two phases which could be dissociated in time and mechanism (Fig. 6).

8.1. IMMEDIATE-ONSET RES PARALYSIS

Under appropriate experimental conditions this type of inhibition can be observed between nearly all types of particles. There does not seem to be a requirement for similarity of surface or for a specific opsonin. Inhibition is observed with metabolizable as well as inert particles. The duration of inhibition is determined by the length of time that an adequate concentration of particles remains in the circulation. Accordingly, the duration is usually short (minute to hours) but may be long if the clearance rate of the particles is exceptionally slow or if an exceedingly large dose is administered. The probable mechanism is a competitive inhibition between the particles probably for receptor domains on the surface of the phagocytes. In some instances wherein particles of similar surface are injected sequentially, inhibition may arise due to competition for or depletion of opsonins required for optimal phagocyte binding and ingestion.

8.2. DELAYED-ONSET RES PARALYSIS

This type of inhibition is observed following the injection of inert particles such as carbon, or thorotrast and toxic particles such as silicon dioxide (Pearsall and Weiser, 1968), endotoxin (Heilman, 1965; Benacerraf and Sebestyen, 1957; Levy and Ruebner, 1968) or certain lipid emulsions (Stuart, 1962; Stuart *et al.*, 1960). It is not observed following injection of aggregated albumins (Normann, 1970). Following onset of inhibition, recovery is slow and may take several days. Since this type of paralysis occurs after the particles have left the circulation, it is most likely due to some form of cellular derangement. The abnormality in the cell might represent a failure to synthesize opsonin but opsonin depletion arising from particle binding and ingestion is unlikely since the paralysis can be preceded by a period of normal or even enhanced particle removal (Parker and Finney, 1960; Normann, 1970). Recovery may involve cellular proliferation of new phagocytes (Benacerraf *et al.*, 1954; Kelly *et al.*, 1960, 1962).

Historically, phagocytosis inhibition produced by colloid injection was believed to represent a single period of RES paralysis due either to depletion of blood opsonin (Jenkins and Rowley, 1961; Jenuet and Good, 1969; Pisano *et al.*, 1968) or to a cellular limitation on the amount of particles ingested (Benacerraf *et al.*, 1954; Biozzi *et al.*, 1953, Biozzi *et al.*, 1963). Because the phenomenon consists of two phases as outlined above, cellular satiation is an unlikely explanation. On the other hand, the requirement for opsonins even to gelatin-stabilized lipid emulsions argues for the hypothesis that opsonin depletion limits particle clearance rates (Pisano *et al.*, 1968).

8.3. OPSONIN DEPLETION

Opsonin depletion that arises from particle injection should produce a sustained depression in clearance rate corrected only as the opsonin level is restored. Specificity in the inhibition should be observed; that is, only particles with identical opsonin requirements should be affected. The responsible opsonin should be present in limited quantity so that it can be depleted by particle injection. Finally, the opsonin should be a rate-limiting requirement for phagocytosis.

That opsonin can be rate limiting has been demonstrated for bacteria by injecting opsonic antisera during clearance and noting an acceleration in the removal rate of the bacteria (Benacerraf and Miescher, 1960). Recently, fibronectin with a binding affinity for denatured collagen, certain bacteria, and fibrin has been demonstrated in human and rat sera (Blumenstock *et al.*, 1977,a,b). However, the significance of this binding protein to particle removal is unclear. *In vitro*, Bevilacqua *et al.* (1980) reported that fibronectin receptors on phagocytes increased attachment but not ingestion of gelatin-coated latex beads and tanned erythrocytes. In contrast, Doran *et al.* (1981) reported that fibronectin promoted both attachment and ingestion. However, they cautioned that such studies must be interpreted with care since fibronectin is divalent for gelatin binding and causes aggregation of gelatin-coated latex beads contributing to bead retention. *In vivo*, injection of serum containing fibronectin during clearance of gelatin-stabilized colloids failed to augment the particle clearance rate (Koenig *et al.*, 1965; Normann and Benditt, 1965b). Moreover, gelatin itself was slowly removed from the circulation (Koenig *et al.*, 1965; Normann, 1970) implying a limited effect of any opsonin in its removal. Persistence of circulating gelatin might complicate interpretation of the clearance rate of a subsequently injected colloid, particularly if the gelatin bound to the surface of the particles. On the other hand, serum levels of fibronectin are decreased by injection of a gelatin-stabilized lipid emulsion (Blumenstock *et al.*, 1977b) and after trauma (Kaplan and Saba, 1977) and septic shock (Saba et al., 1978). In the latter two conditions, fibronectin levels might be reduced by binding to collagen fragments or circulating fibrin aggregates. Under these conditions, removal of a test lipid emulsion was reduced and could be increased by infusion of the purified fibronectin (Blumenstock *et al.*, 1977b; Saba *et al.*, 1978). The role of opsonin in vascular clearance is discussed in greater detail in Part B of this volume.

8.4. COMPETITIVE INHIBITION BETWEEN PARTICLES

In a study of heat-denatured bovine and human albumins, Normann (1973b) showed that the particles were two similar but distinct populations which acted as competitive inhibitors of each other. The double-reciprocal plot of the velocity of clearance versus particle concentration (Lineweaver–Burk plot) showed that the presence of the second particle altered the particle–membrane constant K_p but did not alter the maximum phagocytic velocity. Representative data from this experiment are presented in Fig. 7. Furthermore, the degree of inhibition increased with increasing dose of inhibiting colloid. These two facts are best explained by a fully competitive inhibition between the particles. Subsequently, Normann (1974) showed that a similar competitive inhibition existed between various combinations of rabbit red cells, carbon particles, latex spheres, lipid emulsion, aggregated albumins, and bacteria. The variety of particles that produce competitive inhibition argues that the inhibition is for receptor domains on the cell surface rather than for opsonin.

If a particle acting as an inhibitor is itself a substrate for the cell, then the degree of inhibition in the measured particle will be determined by the concentration of the inhibitor (I) and the particle–membrane constant (K_p^i) of the

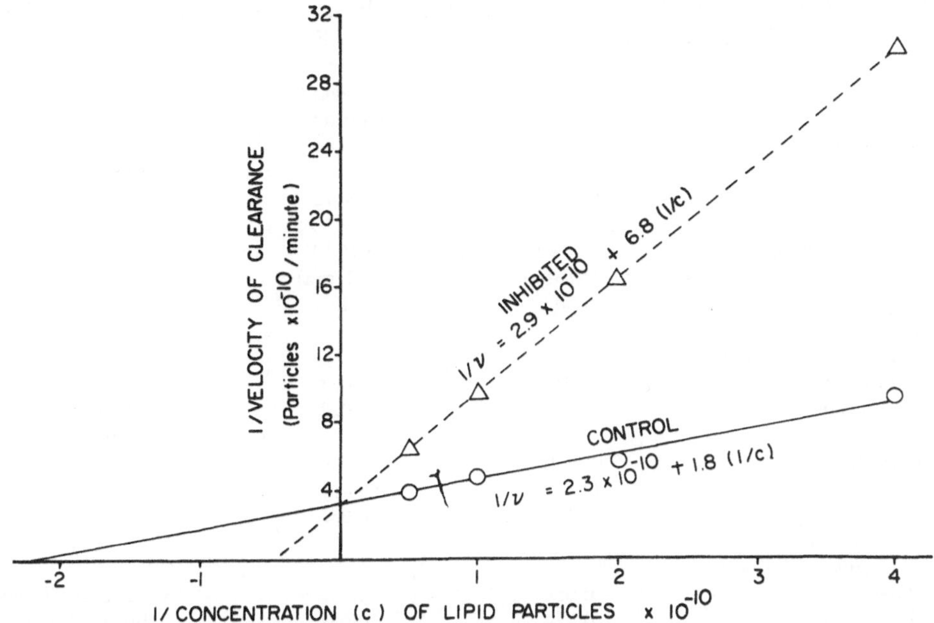

FIGURE 7. Lineweaver–Burk plot of the competitive inhibition of lipid emulsion clearance by latex particles. Control is the clearance of lipid emulsion in the absence of latex whereas inhibited is the clearance in the presence of 2.1 × 10⁹ latex particles. (Reproduced from data of Normann, 1974, *Lab. Invest.* **31**:286, with permission of Williams and Wilkins.)

inhibitor (Normann, 1974). The relationship is as follows:

$$V = \frac{V_{max}}{1 + \frac{K_p}{C}\left(1 + \frac{I}{K_p^i}\right)} \tag{14}$$

Thus, if one knows the V_{max} and K_p of the measured particle and the K_p^i of the inhibitor, it is possible to predict the degree of inhibition given the relative concentrations of the two particles. An example is provided in the data of Table 4. Observable inhibition between different sets of particles requires that certain conditions be fulfilled: specifically, sufficient concentrations and adequate affinities of the particles for the phagocyte. The reason for these conditions is explained by equation (14) and illustrated by the data of Table 4. Equation (14) also provides an explanation for the conclusion of some investigators that a specificity exists in the interaction between particles in the circulation; i.e., only certain particles are inhibitory (Murray, 1963b; Schapiro *et al.*, 1966; Drutz *et al.*, 1967). These studies often employed tracer concentrations of particles and it would be predicted that had adequate particle concentrations been used the apparent specificity would not have been observed.

It is apparent from equation (14) that anything which alters K_p, V_{max}, or K_p^i will alter the degree of inhibition. Opsonins are one class of substances known

TABLE 4. PREDICTION OF DUAL PARTICLE INHIBITION BASED ON CONCENTRATIONS AND THE PARTICLE–MEMBRANE CONSTANTS OF THE PARTICLES INJECTED

Measurement	Carbon particles	Aggregated albumin molecules	Inhibition
V_{max}	6.1×10^{12}	4.1×10^{13}	
K_p	2.4×10^{13}	1.8×10^{14}	
Concentration/100 g	5.0×10^{13} (10 mg)	9.3×10^{13} (0.1 mg)	
Rate constant k	0.083	0.152	
k carbon and aggregated albumin			
Predicted	0.071	0.063	Aggregated albumin and carbon inhibited
Observed	0.068	0.072	
Concentration/100 g	2.0×10^{13} (4 mg)	9.3×10^{14} (1 mg)	
Rate constant k	0.139	0.037	
k carbon and aggregated albumin			
Predicted	0.036	0.032	Carbon inhibited but not aggregated albumin
Observed	0.034	0.034	

to influence these constants. Therefore, it is not surprising that the degree of inhibition between particles may be altered by coating the particles with opsonin prior to injection (Normann and Benditt, 1965b) or by supplying excess opsonin. In some instances, the degree of inhibition may be sufficiently reduced that the effect of a second particle becomes statistically insignificant.

The phenomenon of competitive inhibition between particles was observed by Benacerraf *et al.* (1957) and interpreted by them as a preferential phagocytosis of some particles over others. These workers used this phenomenon to demonstrate the vast variety of particles removed from the circulation by the RES. The subsequent experiments by Normann (1974) confirmed these original findings, provided the mathematical formulation of the nature of the inhibition, and demonstrated that the inhibition was fully competitive in nature.

9. CODA

This review has concerned the kinetics of vascular clearance of particles by phagocytes. The emphasis has been on the nature of the clearance equation and the importance of the particle–membrane intermediate as a rate-limiting step in the phagocytic process. The two key parameters in describing particle clearance from the circulation are the particle–membrane constant K_p and the maximum clearance velocity V_{max}. The particle–membrane constant includes the affinity K_a of the particle for the membrane of the phagocyte. The application of these concepts has demonstrated that different sets of particles in the circulation simultaneously interact in a predictable manner to produce a fully competitive type of inhibition.

REFERENCES

Altura, B. M., and Hershey, S. G., 1968, RES phagocytic function in trauma and adaptation to experimental shock, *Am. J. Physiol.* **215:**1414.

Arend, W. P., and Mannik, M., 1971, Studies on antigen–antibody complexes. II. Quantification of tissue uptake of soluble complexes in normal and complement-depleted rabbits, *J. Immunol.* **107:**63.

Arend, W. P., and Mannik, M., 1973, The macrophage receptor for IgG: Number and affinity of binding sites, *J. Immunol.* **110:**145.

Asiddao, C. B., Filkins, J. P., and Smith, J. J., 1964, Metabolic and surface factors governing phagocytosis in the perfused rat liver, *J. Reticuloendothelial Soc.* **1:**393.

Benacerraf, B., Halpern, B. N., Biozzi, G., and Benos, S. A., 1954, Quantitative study of the granulopectic activity of the reticuloendothelial system. III. The effect of cortisone and nitrogen mustard on the regenerative capacity of the RES after saturation with carbon, *Brit. J. Exp. Path.* **35:**97.

Benacerraf, B., Biozzi, G., Cuendet, A., and Halpern, B. N., 1955, Influence of portal blood flow and of partial hepatectomy on the granulopectic activity of the reticuloendothelial system, *J. Physiol. (London)* **128:**1.

Benacerraf, B., and Sebestyen, M. M., 1957, Effect of bacterial endotoxins on the reticuloendothelial system, *Fed. Proc.* **16:**860.

Benacerraf, B., Biozzi, G., Halpern, B., and Stiffel, C., 1957, Physiology and phagocytosis of particles by the R.E.S., in: *Physiopathology of the Reticuloendothelial System* (B. N. Halpern, ed.), Blackwell, Oxford.

Benacerraf, B., and Miescher, P., 1960, Bacterial phagocytosis by the reticuloendothelial system *in vivo* under different immune conditions, *Ann. N.Y. Acad. Sci.* **88:**184.

Berken, A., and Benacerraf, B., 1966, Properties of antibodies cytophilic for macrophages, *J. Exp. Med.* **123:**119.

Bevilacqua, M. P., Mosesson, M. W., and Bianco, C., 1980, Fibronectin receptors on mononuclear phagocytes, *Clin. Res.* **28:**340.

Bianco, C., Griffin, F. M., and Silverstein, S. C., 1975, Studies of the macrophage complement receptor: Alteration of receptor function upon macrophage activation, *J. Exp. Med.* **141:**1278.

Biozzi, G., Benacerraf, B., and Halpern, B. N., 1953, Quantitative study of the granulopectic activity of the reticuloendothelial system. II. A study of the kinetics of the granulopectic activity of the R.E.S. in relation to the dose of carbon injected: Relationship between the weight of the organs and their activity, *Br. J. Exp. Pathol.* **34:**441.

Biozzi, G., Benacerraf, B., Halpern, B. N., and Stiffel, C., 1957, The competitive effect of certain colloids on the phagocytosis of other colloids and the phenomenon of phagocytic preference, *RES Bull.* **3:**3.

Biozzi, G., Stiffel, C., Halpern, B. N., and Mouton, D., 1963, Lack of action of serum opsonins in phagocytosis of inert particles by cells of reticuloendothelial system, *Proc. Soc. Exp. Biol. Med.* **112:**1017.

Blumenstock, F., Weber, P., and Saba, T., 1977a, Isolation and biochemical characterization of opsonic glycoprotein from rat serum, *J. Biol. Chem.* **252:**7156.

Blumenstock F., Weber, P., Saba, T. M., and Laffin, R., 1977b, Electro-immunoassay of alpha-2-opsonic protein during reticuloendothelial blockade, *Am. J. Physiol.* **232:**R80.

Buglio, A. F., Cotran, R. S., and Jandl, J. H., 1967, Red cells coated with immunoglobulin G: Binding and sphering by mononuclear cells in man, *Science* **158:**1582.

Caro, R., and Radicella, R., 1970, A theoretical analysis of the kinetics of the phagocytosis of radiogold colloids in the rat, *Acta Physiol. Lat. Am.* **20:**45.

Casley-Smith, J. R., and Reade, P. C., 1965, An electron microscopical study of the uptake of foreign particles by the livers of foetal and adult rats, *Br. J. Exp. Pathol.* **46:**473.

Cohen, Y., Ingrand, J., and Caro, R., 1968, Kinetics of the disappearance of gelatin protected radiogold colloids from the blood stream, *Int. J. Appl. Radiat. Isot.* **19:**703.

Cohn, Z. A., 1966, The regulation of pinocytosis in mouse macrophages. I. Metabolic requirements as defined by the use of inhibitors, *J. Exp. Med.* **124:**557.

Danon, D., Goldstein, L., Marikovsky, Y., and Skutelsky, E., 1972, Use of cationized ferritin for a label of negative charges on cell surfaces, *J. Ultrastructure Res.* **38:**500.

Davie, J. M., and Paul, W. E., 1972, Receptors on immunocompetent cells. IV. Direct measurement of avidity of cell receptors and cooperative binding of multivalent ligands, *J. Exp. Med.* **135**:643.

de Terra, N., and Rustad, R. C., 1959, The dependence of pinocytosis on temperature and aerobic respiration, *Exp. Cell Res.* **17**:191.

Dobson, E. L., Gofman, J. W., Jones, H. B., Kelly, L. S., and Walker, L. A., 1949, Studies with colloids containing radioisotopes of yttrium, zirconium, columbium, and lanthanum. II. The controlled selective localization of radioisotopes of yttrium, zirconium, and columbium in the bone marrow, liver, and spleen, *J. Lab. Clin. Med.* **34**:305.

Dobson, E. L., 1957, Factors controlling phagocytosis, in: *Physiopathology of the Reticuloendothelial System* (B. N. Halpern, ed.), p. 80, Blackwell, Oxford.

Dobson, E. L., and Jones, H. B., 1952, The behavior of intravenously injected particulate material, *Acta Med. Scand. (Suppl.)* **273**:1.

Dobson, E. L., Kelly, L. S., and Finney, C. R., 1967, Kinetics of the phagocytosis of repeated injection of colloidal carbon: Blockade, a latent period or stimulation? A question of timing and dose, in: *The Reticuloendothelial System and Atherosclerosis* (N. R. DiLuzio and R. Paoletti, eds.), *Adv. Exp. Med. Biol.* **1**:63.

Doran, J. E., Mansberger, A. R., Edmondson, H. T., and Reese, A. C., 1981, Cold insoluble globulin and heparin interactions in phagocytosis by macrophage monolayers: Mechanism of heparin enhancement, *J. Reticuloendothelial Soc.* **29**:285.

Drutz, D. J., Koenig, M. G., and Rogers, D. E., 1967, Further observations of the mechanism of reticuloendothelial blockade, *J. Exp. Med.* **126**:1087.

Ehlenberger, A. G., and Nussenzweig, V., 1977, The role of membrane receptors for C3b and C3d in phagocytosis, *J. Exp. Med.* **145**:357.

Filkins, J. P., and DiLuzio, N. R., 1966, Mechanism of gelatin inhibition of reticuloendothelial function, *Proc. Soc. Exp. Biol. Med.* **122**:177.

Fred, R. K., Harris, J. G., Parker, H. G., and Shore, M. L., 1967, A mathematical model of RES phagocytic function, *J. Reticuloendothelial Soc.* **4**:524.

Fred, R. K., and Shore, M. L., 1967, Application of a mathematical model to the study of RES phagocytosis in mice, in: *The Reticuloendothelial System and Atherosclerosis,* (N. R. Diluzio and R. Paoletti, eds.), *Adv. Exp. Med. Biol.* **1**:1.

Gabrieli, E. R., 1952, Distribution of radioactive chromium phosphate in tissues of rodents following intravenous injection, *Acta Path. Microbiol. Scand.* **31**:195.

Gabrieli, E. R., 1961, The velocity at which radioactive colloids disappear from the blood: Studies on the function of the reticuloendothelial system, *Acta Physiol. Scand.* **23**:283.

Gasic, G. J., Berwick, L., and Sorrentino, M., 1968, Positive and negative colloidal iron as cell surface electron stains, *Lab. Invest.* **18**:63.

Gofman, J. W., 1949, Studies with colloids containing radioisotopes of yttrium, zirconium, colombium, and lanthanum. I. The chemical principles and methods involved in preparations of colloids of yttrium, zirconium, and lanthanum, *J. Lab. Clin. Med.* **34**:297.

Gosselin, R. E., 1956, The uptake of radiocolloids by macrophages *in vitro:* A kinetic analysis with radioactive colloidal gold, *J. Gen. Physiol.* **39**:625.

Gosselin, R. E., 1967, Kinetics of pinocytosis, *Fed. Proc.* **26**:987.

Griffin, F. M., Bianco, C., and Silverstein, S. C., 1975, Characterization of the macrophage receptor for complement and demonstration of its functional independence from the receptor for the Fc portion of immunoglobulin G, *J. Exp. Med.* **141**:1269.

Halpern, B. N., Benacerraf, B., and Biozzi, G., 1953, Quantitative study of granulopectic activity of the reticuloendothelial system. I. The effect of the ingredients present in India ink and of substances affecting blood clotting *in vivo* on the fate of carbon particles administered intravenously in rats, mice, and rabbits, *Br. J. Exp. Pathol.* **34**:426.

Heilman, D. H., 1965, The selective toxicity of endotoxin for phagocytic cells of the reticuloendothelial system, *Int. Arch. Allergy* **26**:63.

Horn, R. G., Koenig, M. G., Goodman, J. S., and Collins, R. D., 1969, Phagocytosis of *Staphylococcus aureus* by hepatic reticuloendothelial cells: An ultrastructural study, *Lab. Invest.* **21**:406.

Huber, H., and Fudenberg, H. H., 1970, The interaction of monocytes and macrophages with immunoglobulins and complement, *Ser. Haematol.* **3**:160.

Huber, H., Polley, M., Linsocott, W., Fudenberg, H., and Müller-Eberhard, H., 1968, Human monocytes: Distinct receptor site for the third component of complement and for immunoglobulin G, *Science* **162:**1281.

Iio, M., and Wagner, H. N., 1963, Studies on the reticuloendothelial system (RES). I. Measurement of the phagocytic capacity of the RES in man and dog, *J. Clin. Invest.* **42:**417.

Jallut, O., Penguidon, L., Lerch, S., Neukom, S., and Feissly, R., 1955, Speed of disappearance from the blood and distribution in different organs of radio-active colloidal $CrPO_4$ as a function of the injected particle size, *RES Bull.* **1:**70.

Jenkins, C. R., and Rowley, D., 1961, The role of opsonins in the clearance of living and inert particles by cells of the reticuloendothelial system, *J. Exp. Med.* **114:**363.

Jenuet, F. S., and Good, R. A., 1969, Reticuloendothelial function in the isolated perfused liver. II. Phagocytosis of heat aggregated bovine serum albumin. Demonstration of two components in the blockade of the reticuloendothelial system, *J. Reticuloendothelial Soc.* **6:**94.

Jones, T. C., 1975, Attachment and ingestion phases of phagocytosis, in: *Mononuclear Phagocytes in Immunity, Infection, and Pathology* (R. van Furth, ed.), p. 269, Oxford: Blackwell Scientific.

Kaplan, J. E., and Saba, T. M., 1977, α-2-glycoprotein opsonic deficiency after trauma, *Proc. Soc. Exp. Biol. Med.* **156:**14.

Karrer, H. E., 1960, Electron microscopic study of the phagocytosis process in lung, *J. Biophys. Biochem. Cytol.* **7:**357.

Kelly, L. S., Dobson, E. L., Finney, C. R., and Hirsch, J. D., 1960, Proliferation of the reticuloendothelial system in the liver, *Am. J. Physiol.* **198:**1134.

Kelly, L. S., Brown, B. A., and Dobson, E. L., 1962, Cell division and phagocytic activity in liver reticulo-endothelial cells, *Proc. Soc. Exp. Biol. Med.* **110:**555.

Koenig, M. G., Heyssel, R. M., Melly, M. A., and Rogers, D. E., 1965, The dynamics of reticuloendothelial blockade, *J. Exp. Med.* **122:**117.

Kronvall, G., and Frommel, D., 1970, Definition of staphylococcal protein A reactivity for human immunoglobulin G fragments, *Immunochemistry* **7:**124.

Lay, W., and Nussenzweig, V., 1968, Receptors for complement on leukocytes, *J. Exp. Med.* **128:**991.

Levy, E., and Ruebner, B. H., 1968, Hepatic changes induced by a single dose of endotoxin in germfree mice, *Am. J. Path.* **52:**97.

Lineweaver, H., and Burk, D., 1934, The determination of enzyme dissociation constants, *J. Am. Chem. Soc.* **56:**658.

Lyman, D. J., Muir, W. M., and Lee, I. J., 1965, The effect of chemical structure and surface properties of polymers on the coagulation of blood. I. Surface free energy effects, *Trans. Am. Soc. Artif. Intern. Organs.* **11:**301.

Mannik, M., Arend, W. P., Hall, A. P., and Gilliland, B. C., 1971, Studies on antigen–antibody complexes. I. Elimination of soluble complexes from rabbit circulation, *J. Exp. Med.* **133:**713.

Marrable, A. W., 1962, The counting of cells and nuclei in microtome sections, *Q. J. Miscros. Sci.* **103:**331.

Maurer, P. H., 1954, Antigenicity of oxypolygelatin and gelatin in man, *J. Exp. Med.* **100:**497.

Mehl, T. D., 1975, Uptake of aggregated albumin by rat macrophages *in vitro:* Affinities of cells for monomeric and aggregated BSA, Doctoral thesis, University of Washington.

Mehl, T. D., and Lagunoff, D., 1975, Uptake of aggregated albumin by rat macrophages *in vitro:* Affinities of cells for monomeric and aggregated bovine serum albumin, J. Reticuloendothelial Soc. **18:**125.

Meredith, O. M., Jr., 1961, Kinetics of reticuloendothelial phagocytic response to intravenously administered Au colloid gold in rabbits, *Arch. Biochem. Biophys.* **95:**352.

Murray, I. M., 1963a, Clearance rate in relation to agglutinins for gelatin-stabilized colloid in the rat, *Am. J. Physiol.* **204:**655.

Murray, I. M., 1963b, The mechanism of blockade of the reticuloendothelial system, *J. Exp. Med.* **117:**139.

Murray, I. M., and Katz, M., 1955, Factors affecting the rate of removal of gelatin-stabilized radiogold colloid from the blood. I. The retardation of the radiogold disappearance rate by gelatin, *J. Lab. Clin. Med.* **46:**262.

Normann, S. J., 1970, Function of the reticuloendothelial system. IV. Evidence for two types of particle-induced reticuloendothelial paralysis, *Infect. Immun.* **1**:327.

Normann, S. J., 1972a, Clearance of equine antitoxin–toxin complexes by the reticuloendothelial system, *J. Immunol.* **108**:521.

Normann, S. J., 1972b, Function of the reticuloendothelial system. V. Studies on the correlation between phagocytic rate and liver blood flow, *J. Reticuloendothelial Soc.* **12**:473.

Normann, S. J., 1973a, Reticuloendothelial system function. VI. Experimental alterations influencing the correlation between portal blood flow and colloid clearance, *J. Reticuloendothelial Soc.* **13**:47.

Normann, S. J., 1973b, The kinetics of phagocytosis. I. A study of the clearance of denatured bovine albumin and its competitive inhibition by denatured human albumin, *J. Reticuloendothelial Soc.* **14**:587.

Normann, S. J., 1974, Kinetics of phagocytosis. II. Analysis of *in vivo* clearance with demonstration of competitive inhibition between similar and dissimilar foreign particles, *Lab. Invest.* **31**:161.

Normann, S. J., and Benditt, E. P., 1965a, Function of the reticuloendothelial system. I. A study on the phenomenon of carbon clearance inhibition, *J. Exp. Med.* **122**:693.

Normann, S. J., and Benditt, E. P., 1965b, Function of the reticuloendothelial system. II. Participation of a serum factor in carbon clearance, *J. Exp. Med.* **122**:709.

Normann, S. J., Lagunoff, D., and Benditt, E. P., 1968, Function of the reticuloendothelial system. III. Simultaneous measurement of two particles during clearance inhibition, *Lab. Invest.* **19**:353.

Parker, H. G., and Finney, C. R., 1960, Latent period in the induction of reticuloendothelial blockade, *Am. J. Physiol.* **198**:916.

Pearsall, N. N., and Weiser, R. S., 1968, The macrophage in allograft immunity. I. Effects of silica as a specific macrophage toxin, *J. Reticuloendothelial Soc.*, **5**:107.

Pisano, J. C., Patterson, J. T., and DiLuzio, N. R., 1968, Reticuloendothelial blockade: Effect of puromycin on opsonin-dependent recovery, *Science* **162**:565.

Rabinovitch, M., 1967, The dissociation of the attachment and ingestion phases of phagocytosis by macrophages, *Exp. Cell Res.* **46**:19.

Rowley, D., and Turner, K. J., 1966, Number of molecules of antibody required to promote phagocytosis of one bacterum, *Nature (London)* **210**:496.

Saba, T. M., Blumenstock, F. A., Scovill, W. A., and Bernard, H., 1978, Cryoprecipitate reversal of opsonic surface binding glycoprotein deficiency in septic surgical and trauma patients, *Science* **201**:622.

Scatchard, G., 1949, The attraction of proteins for small molecules and ions, *Ann. N.Y. Acad. Sci.* **51**:660.

Schapiro, R. L., MacIntyre, W. J., and Schapiro, D. I., 1966, The effect of homologous and heterologous carrier on the clearance of colloidal material by the reticuloendothelial system, *J. Lab. Clin. Med.* **68**:286.

Scharfstein, J. E., Correa, B., Gallo, G. R., and Nussenzweig, V., 1979, Human C4 binding protein: Association with immune complexes *in vitro* and *in vivo*, *J. Clin. Invest.* **63**:276.

Schmidtke, J. R., and Unanue, E. R., 1971, Macrophage–antigen interaction: Uptake, metabolism and immunogenicity of foreign albumin, *J. Immunol.* **197**:331.

Schoenberg, M. D., Gilman, P., Mumaw, V. R., and Moore, R. D., 1961, The phagocytosis of uniform polystyrene latex particles (PLP) by the reticuloendothelial system (RES) in the rabbit, *Br. J. Exp. Pathol.* **42**:486.

Sheppard, C. W., Jordan, G., and Hahn, P. F., 1951, Disappearance of isotopically labeled gold colloids from the circulation of the dog, *Am. J. Physiol.* **164**:345.

Stuart, A. E., Biozzi, G., Stiffel, C., Halpern, B. N., and Mouton, D., 1960, The stimulation and depression of reticulo–endothelial phagocytic function by simple lipids, *Brit. J. Exp. Path.* **41**:599.

Stuart, A. E., 1962, Effect of cholesterol oleate on the phagocytic function of the reticulo-endothelial system, *Nature (London)* **196**:78.

Ulrich, F., and Zilversmith, D. B., 1970, Release from alveolar macrophages of an inhibitor of phagocytosis, *Am. J. Physiol.* **218**:1118.

Unanue, E. R., Cerottini, J. C., and Beford, M., 1969, Persistence of antigen on the surface of macrophages, *Nature (London)* **222**:1193.

Unkeless, J., 1979, Characterization of monoclonal antibody directed against mouse macrophage and lymphocyte Fc receptors, *J. Exp. Med.* **156**:580.

Unkeless, J., and Eisen, H., 1975, Binding of monomeric immunoglobulins to Fc receptors of mouse macrophages, *J. Exp. Med.* **142**:1520.

van Oss, C. J., and Gillman, C. F., 1972a, Phagocytosis as a surface phenomenon. I. Contact angles and phagocytosis of non-opsonized bacteria, *J. Reticuloendothelial Soc.* **12**:283.

van Oss, C. J., and Gillman, C., 1972b, Phagocytosis as a surface phenomenon. II. Contact angles and phagocytosis of encapsulated bacteria before and after opsonization by specific antiserum and complement, *J. Reticuloendothelial Soc.* **12**:497.

Weiss, L., Ziegel, R., Jung, O. S., and Bross, I. D. J., 1972, Binding of positively charged particles to glutaraldehyde fixed human erythrocytes, *Exp. Cell Res.* **70**:57.

Widman, J. J., Cotran, R. S., and Fahimi, H. D., 1973, Mononuclear phagocytes (Kupffer cells) and endothelial cells: Identification of two functional cell types in rat liver sinusoids by endogenous peroxidase activity, *J. Cell Biol.* **52**:159.

Wilkens, D. J., and Myers, P. A., 1964, Design of colloids for RES testing and therapy, *J. Reticuloendothelial Soc.* **1**:344.

Wilkins, D. J., and Myers, P. A., 1966, Studies on the relationship between the electrophoretic properties of colloids and their blood clearance and organ distribution in the rat, *Brit. J. Exp. Path.* **47**:568.

Wilkins, D. J., 1967, Interaction of charged colloids with the RES, in: *The Reticuloendothelial System and Atherosclerosis* (N. R. DiLuzio and R. Paoletti, eds.), *Adv. Exp. Med. Biol.* **1**:25.

Wilkins, D. J., 1971, A possible relationship between blood clotting and the recognition by the body of foreign particles, in: *The Reticuloendothelial System and Immune Phenomena* (N. R. DiLuzio and K. Flemming, eds.), *Adv. Exp. Med. Biol.* **15**:77.

5

Vascular Clearance of Microorganisms

R. J. MOON

1. INTRODUCTION

Approximately 15 to 20 years ago there was a substantial outpouring of information regarding the nature and mechanisms of vascular clearance of microorganisms and inert particles from the circulation. The single most substantial influence which spurred these investigations seemed to be the studies by Biozzi *et al.* (1955) describing the use of carbon to study the kinetics of the clearance of microorganisms from the blood. There ensued numerous studies aimed at describing the nature of RES clearance by using a variety of microorganisms and inert colloids as tools (Benacerraf and Miescher, 1960; Benacerraf *et al.*, 1957, 1959; Biozzi *et al.*, 1955, 1957, 1960; Jenkin and Rowley, 1961; Rowley, 1962) and review articles are available which describe the early development of this research field with respect to microbial clearance (Rowley, 1962; Merigan, 1974) and inert particle clearance (Saba, 1970). More recently, the appearance of new information relating to the clearance of microorganisms has been rather sporadic. The objective of this chapter is to discuss some of the more recent contributions relating to clearance of a variety of microorganisms *in vivo* as well as in perfused organ models and to suggest some future directions for this research area.

2. *IN VIVO* STUDIES

2.1. VIRUSES

There have been a number of recent studies on the clearance of viruses from the bloodstream of animals. These have been largely confined to the clearance of

R. J. MOON • Division of Basic Sciences, School of Medicine, Mercer University, Macon, Georgia 31207.

togaviruses. A series of publications by Jahrling and co-workers (1973, 1975, 1976, 1977) evaluate vascular clearance rates of virulent and avirulent Venezuelan equine encephalitis (VEE) viruses. A 1975 report describes the selective clearance of an avirulent small plaque clone of VEE by hepatic reticuloendothelial cells. Thirty minutes after intracardiac administration of large (virulent) and small (avirulent) plaque variants, more than 99% of the small plaque organisms were removed from the blood with almost 50% of these concentrated in the liver. By contrast, less than 1% of the large plaque variants were cleared from the blood in the same time period. Since these viruses do not replicate in hepatic sinusoidal cells, the net effect of hepatic trapping was postulated to effectively decrease the infectious dose and hence the virulence of the viruses for the host. Likewise, an attenuated TC-83 Trinidad donkey strain of VEE was rapidly cleared by the RES while the parent strain was cleared much more slowly (Jahrling and Gorelker, 1975). During the course of the disease, virulent strains could exceed 10^6 PFU/ml of blood while the vaccine strain seldom exceeded 10^3 PFU/ml, again correlating RES clearance with virulence. In a later study in guinea pigs, Jahrling reported that three of six benign virus strains of VEE maintained high viremias similar to six virulent VEE strains (Jahrling *et al.*, 1976). Hence, it became clear that an absolute correlation between virulence and clearance does not hold for all organisms. Conceivably some of the various avirulent viruses lack the macromolecules necessary for binding to hepatic receptors (see below).

Examples can also be cited where clearance of virus by the RES leads to subsequent multiplication of virions within RES tissues as an integral part of the pathogenesis of the disease. For example, Wolinsky and Stroop (1978) have shown that following intraperitoneal injection of neuroadapted mumps virus, the viruses are rapidly distributed among numerous host organs including the liver and spleen and multiply there for approximately 7 to 10 days until antibody appears. Once antibody appears, viruses persist only in brain and kidney and are undetectable among other organs. In this instance, RES organs seem to play an integral role in development of viremia apparently necessary for development of the disease. No evidence as to whether multiplication occurred in RES cells per se prior to viremia is presented.

Sigel *et al.* (1968) performed an interesting study on clearance of T_2 bacteriophage by the lemon shark. Following intravenous or intramuscular injection, the viruses were cleared relatively rapidly, showing a 4 log reduction after 2 days and complete clearance by 5 to 7 days. Sharks were free of antibody at the time of administration but developed antibody starting by day 5 after virus exposure.

2.2. BACTERIA

The number of studies on vascular clearance of bacteria from blood has declined dramatically in recent years as more attention is being paid to the interactions of bacteria with surfaces of isolated cells *in vitro*. An area of some research interest of late has been attempts to correlate tissue trophisms of bacteria with diseases such as pyelonephritis and endocarditis. In studies on pyelonephritis, Johnson and Latta (1978) have demonstrated that injection of 10^6 to 10^9 *E. coli* into renal arteries results in sufficient trapping of organisms in the kidney to conduct electron microscopic studies. Upon initial exposure, bacteria adhered to both glomerular and intertubular capillaries. Ruthenium red staining showed a close connection between the largely carbohydrate microcapsule and the sialoglycoprotein endothelial surface coat. By 10 min PMNL and monocytes appeared and phagocytosed the trapped bacteria resulting in acute inflammation. The authors suggest the inflammation and subsequent thrombosis may be mediated by complement activated through the alternate pathway. With respect to development of endocarditis, the mechanisms involved in trapping of bacteria on heart valves have received some attention in recent years. Many details by which bacteria establish themselves on heart valves are still poorly understood. Some time ago Durach and Beeson (1972) demonstrated that while normal animals show rapid clearance of bacteria (*Streptococcus viridans*), a persisting bacteremia ensued in animals with preexisting sterile vegetation caused by insertion of a polyethylene catheter into the right side of the heart. In such animals substantial numbers of injected bacteria adhered to the vegetation and multiplied rapidly. By 2 days the organisms reached stationary phase and were probably the source of the persistent bacteremia. Apparently bacteria can reach the lesions either by direct inoculation from blood (*Proteus, Staphylococcus*) or by adhesion of phagocytes (PMNL) laden with bacteria to the primary vegetation (Durach, 1975). Subsequent bacterial growth and the biological consequences have been described in detail (Durach *et al.*, 1978; Angrist *et al.*, 1967).

In related studies, Hamburger *et al.* (1971) have shown that intravenously injected staphylococci do not localize on heart valves immediately but primarily reside in liver, spleen, etc. While normal dogs tended to dispose completely of the bacteria, dogs with surgically induced aortic insufficiency (AI) had staphylococci colonizing the aortic and mitral valves by 18 hr. While the bacterial numbers progressively decreased with time in normal animals, in dogs with AI they increased in all tissues. AI dogs subsequently developed endocarditis. The American opossum is the only experimental animal that regularly develops bacterial endocarditis (BE) spontaneously. Postulating that rates of bacterial clearance from the bloodstream might be a significant variable in explaining this occurrence, Musher and Richie (1974) studied the clearance rates of *Staphylococcus epidermidis* and *Streptococcus faecalis* during a 30-min period of observation. No differences in early clearance rates were observed while 7 of the 26 animals died of endocarditis within 3 weeks. Six of the seven dead animals had BE at autopsy not caused by the particular challenge variant. The authors conclude that congenital problems or other host factors rather than clearance rates of bacteria are most likely involved in the spontaneous onset of BE.

Among studies performed *in vitro* on the adherence of bacteria to heart valves, Gould *et al.* (1975) reported on the ability of 14 strains of aerobic gram-positive cocci and gram-negative bacilli to adhere to human or canine aortic valve leaflets. Strains of enterococci, viridans streptococci, staphylococci (coag$^+$

ar$^-$), and *Pseudomonas aeruginosa* adhered more readily than did *Escherichia coli* or *Klebsiella pneumoniae*. A positive correlation existed between the ability to adhere to aortic tissue and the likelihood to cause endocarditis. In a more detailed study, Scheld *et al.* (1978) investigated the role of dextran on the surface of streptococci in adherence to the constituents of nonbacterial thrombotic endocarditis (NBTE) (fibrin and platelets) *in vitro* and *in vivo*. These authors found that dextran on the surface of streptococci substantially increased the likelihood of binding to a fibrin–platelet matrix *in vitro*. *In vivo* correlates on the importance of dextran were consistent with the *in vitro* data. The authors conclude that dextran production is important not only in adherence of oral streptococci to teeth but also to the constituents of NBTE and may play a role in the overall pathogenesis of the disease. A concern that circulating antibody to dextran as a result of immunization with streptococcal vaccines for prevention of caries seems unwarranted since Durach *et al.* (1978) have recently shown that the frequency of development of endocarditis in humorally immune rabbits is, in fact, significantly less than in normal rabbits.

2.3. FUNGI

Data concerning the mechanisms by which the RES traps and kills fungi is, at best, sketchy. A number of reports demonstrate that the distribution of *C. albicans* among RES organs follows a fairly typical pattern with the possible exception that the lungs initially remove a larger proportion of the circulating yeast than might be expected (Trnovec *et al.*, 1978; Sawyer *et al.*, 1976; Leunk and Moon, 1979). Trnovec *et al.* (1978) using killed radiolabeled yeast suggest that uptake in the lungs may be transitory while the persistence of label in the lungs is long-lasting. The initial vascular clearance of yeast does not correlate well with the persistence of organisms in those tissues. For example, Leunk and Moon (1979) have shown that while less than 2% of an initial inoculum localizes in the kidneys, it is in these organs that the yeast eventually multiply and are responsible for most of the pathogenesis observed in systemic candidosis. As the infection proceeds, yeast trapped in the liver, lungs, etc. are effectively destroyed by the natural host defense systems.

3. LIVER PERFUSION STUDIES

The liver perfusion model offers a unique opportunity to study the interactions of cellular and humoral factors involved in microbial trapping and killing by the liver. Many years ago Manwaring and Coe (1916) used hepatic perfusions to describe clearance of pneumococci. Encapsulated bacteria suspended in Ringer solution with or without normal sera were not cleared while over 80% of the organisms suspended in 1% immune sera were cleared. The serum component involved was heat stable at 60°C for 30 min and was called *endothelial opsonin*. Manwaring and Fritschen (1923) compared clearance rates of a number of organisms including *Staphylococcus aureus, E. coli,* and *Bacillus anthracis* in various organs and showed high clearance rates in the liver and spleen but substantially lower rates in lungs, intestine, and central nervous system tissue. Little research interest in perfused organs as a method for studying vascular clearance of microorganisms existed until 1958 when Howard and Wardlaw (1958) published studies showing that normal human, rat, and mouse sera all increased clearance of *E. coli* by perfused rat liver. Their studies showed that the opsonic activity of sera was reduced by heating at 56°C for 30 min, absorption with homologous strains of bacteria or with antigen–antibody complexes. From their studies they concluded that serum factors involved in clearance of bacteria by the liver were specific antibody, complement, and probably properdin. In these experiments the actual number of viable bacteria in the perfused livers was not determined. Hence, the effect that these various serum treatments had on an actual hepatic bacterial killing was not determined. Experiments were also done using other strains of gram-negative and gram-positive bacteria (Howard, 1961). The presence of serum enhanced bacterial clearance of all gram-negative bacteria. The gram-positive bacteria were cleared well in the absence of serum and its presence actually reduced clearance.

Bonventre and Oxman (1965) studied the role of humoral and cellular factors on clearance and killing of *Staphylococcus aureus* and *Salmonella enteritidis* by perfused rat livers. Their results indicate that the immunological status of the liver or serum had no effect on clearance or killing of *Staphylococcus aureus* but that immune serum substantially enhanced clearance rates of *Salmonella enteritidis.* We have confirmed this latter observation (Friedman and Moon, 1980). Jeunet *et al.* (1968, 1969) evaluated the ability of continuously perfused rat livers to phagocytose *Salmonella typhosa* and *Brucella melitensis.* These authors showed that both organisms were recognized and phagocytosed well by perfused rat livers in the absence of antibodies or opsonins. When sufficient numbers of bacteria had been exposed to the liver via perfusion, an apparent RES blockade occurred whereby successive exposures were cleared more slowly. The "blockaded" liver could still clear colloidal carbon at normal rates, suggesting some sort of cellular specificity to blockade rather than simply a nonspecific saturation of the phagocytic system.

Recent data from our laboratory are essentially consistent with those described above but extend the observations in a number of critical areas. First, scanning electron microscopy has clearly revealed that endothelial as well as Kupffer cells play a critical role in removal of *Salmonella typhimurium* (Moon *et al.*, 1975; Friedman and Moon, 1977) or *Candida Albicans* (Sawyer *et al.*, 1976) from perfusion media. Second, livers from mice pretreated with DQ12 silica, a specific macrophage toxin which severely depletes hepatic macrophage populations, still retain approximately 50% of their ability to remove bacteria from the perfusion medium (Friedman and Moon, 1977). Cell-associated pili are particularly important in removal of *S. typhimurium* SR-11. Data in Table 1 from a recent study by Mr. Robert Leunk in our laboratory clearly show that 60 to 70% of type 1 piliated organisms are removed from perfusion media on a single pass through mouse liver whereas only 1.2% of a nonpiliated spontaneous mutant of SR-11 is

TABLE 1. PERFUSION OF BROTH AND AGAR-CROWN *S. TYPHIMURIUM* SR-11 AND A NONPILIATED MUTANT STRAIN

	MHD[b] for guinea pig RBC	% Recovery in[a]		
		Liver	Perfusate	Total recovery
Broth-grown *S. typhimurium* (piliated phase)	8.6×10^7	66.7 ± 10.6	25.8 ± 7.7	92.5 ± 8.8
Nonpiliated mutant	Negative[c]	1.2 ± 0.6	98.7 ± 9.2	99.8 ± 9.2

[a]Means ± S.D. of at least 7 experiments.
[b]Minimum hemagglutinating dose (bacteria/ml).
[c]Bacteria did not cause hemagglutination at doses $\leq 2.2 \times 10^{10}$/ml.

cleared. Further, Table 2 shows that mannose and α-methyl-D-mannoside can inhibit trapping while related saccharides cannot. Whether or not such specificity for removal of given organisms can be related to specific saturable receptors on the liver cell surfaces remains to be established but the available data would favor such a hypothesis.

In the absence of fresh plasma, livers can trap *S. typhimurium* very efficiently but killing of the organisms is minimal (Moon *et al.*, 1975). Serum or plasma does not enhance trapping substantially but does stimulate the bactericidal activity of the organ *in situ*. Complement seems to play an important role in this process, since any manipulation which depletes complement activity such as heating plasma to 60°C for 30 min, treatment with EDTA, or specific immune absorption of C3 results in loss of its bactericidal catalyzing properties (Friedman and Moon, 1980). Data have been obtained suggesting that complement activation of the alternate complement pathway by lipopolysaccharide is a control step in this process.

TABLE 2. SACCHARIDE INHIBITION OF *S. TYPHIMURIUM* STRAIN SR-11 BY PERFUSED LIVERS

Saccharide[b]	% Recovery in[a]		
	Liver	Perfusate	Total recovery
None	66.7 ± 10.6	25.8 ± 7.7	92.5 ± 8.8
Mannose	16.2 ± 9.2[c]	82.3 ± 16.6[c]	98.5 ± 9.5
α-Methyl-D-mannoside	23.2 ± 6.1[c]	74.8 ± 10.9[c]	98.0 ± 9.1
Fructose	40.4 ± 15.3[c]	50.1 ± 13.4[c]	90.7 ± 11.6
Galactose	73.4 ± 8.1	25.3 ± 2.3	98.9 ± 7.2
Glucose	69.7 ± 7.7	36.4 ± 6.5	106.6 ± 5.4
Lactose	82.8 ± 2.2	22.6 ± 4.7	105.4 ± 5.9
Mannitol	62.5 ± 6.3	33.8 ± 5.0	96.2 ± 6.5
Ribose	63.1 ± 12.2	27.5 ± 4.3	90.5 ± 14.1
Sucrose	69.9 ± 6.9	25.3 ± 7.9	95.2 ± 8.8

[a]Mean ± S.D. of at least 6 experiments.
[b]All sugars were present at 1% (w/v) in M199 for the entire perfusion.
[c]$p < 0.001$ vs. control by Student's *t* test.

In contrast to the data of Jeunet *et al.* (1969), in our hands immune plasma substantially enhanced removal of *S. typhiumurium* from a single-pass perfusion system. Livers from immune mice were not enhanced in their clearance capabilities in the absence of plasma.

Ruggerio *et al.* (1977) have also used the perfused liver to study *Salmonella* clearance. Using continuous perfusion rather than single pass, they found that about 95% of *S. typhi* could be cleared by 1 hr compared with 65% for *S. paratyphi* B. The presence of immune serum resulted in 1 to 2 times as much killing of *S. paratyphi* B or *S. typhi* than would have occurred with normal serum alone. We have shown that serum devoid of anti-*Salmonella* antibody will not kill *S. typhimurium in vitro* (Moon *et al.*, 1975).

4. FUTURE DIRECTIONS

As a major organ for vascular clearance of microorganisms, the mechanisms involved in attachment of foreign molecules to hepatic cells become critically important in understanding its antimicrobial action. In recent years there has been increasing evidence accumulating which describes specific cellular receptors important in removal of materials from the blood. Steer (1980) has recently reviewed the carbohydrate recognition systems of hepatic sinusoidal cells. Growing in part out of the innovative studies by Ashwell and Morell (1977) on the specific elimination of desialylated glycoproteins by the liver, a specific binding protein for recognition and clearance of galactose-terminated glycoproteins was isolated and characterized from rabbit liver (Hudgin *et al.*, 1974; Kawasaki and Ashwell, 1976; Prior and Ashwell, 1976). Subsequent studies both *in vivo* and *in vitro* have documented not only the existence of galactose receptors on hepatocytes but also receptors on sinusoidal cells which recognize glycoproteins terminating in *N*-acetylglucosamine, mannose (Kawasaki *et al.*, 1978), and fucose (Prieels *et al.*, 1978). Hubbard *et al.* (1979) have confirmed the *in vivo* and *in vitro* studies by transmission electron microscopy. Using ^{125}I-labeled test glycoproteins these authors have definitively demonstrated that galactosyl-terminated macromolecules bind exclusively to hepatocytes while mannose- and *N*-acetylglucosamine-terminated macromolecules bind almost exclusively to sinusoidal cells.

Other types of hepatic cell receptors have also been demonstrated. For example, Munthe-Kass (1977) showed that Kupffer cells have both Fc and complement receptors. The carbohydrate component of immunoglobulins also seems uniquely involved in vascular clearance of immune complexes (Thornburg *et al.*, 1980). Apparently concealed while unreacted with antigen in the circulation, the conformational change induced by antigen coupling with antibody appears to expose carbohydrates in such a way as to enhance removal by the liver. It is interesting to note that the carbohydrate associated with IgM contains mannose-rich oligosaccharides and hence tends to be removed by sinusoidal cells while IgG complexes apparently contain galactose-rich glycoproteins and are primarily taken up by hepatocytes. The precise relationships between Fc, complement, and carbohydrate receptors need further study but it is becoming apparent that a number of different carbohydrate receptors are present on hepatic cells. Some data from our laboratory extend in a preliminary way the thinking on hepatic receptors to trapping *S. typhimurium* by the liver. As noted above, type 1 pili seem crucial to the removal of *Salmonella* from the circulation in nonimmune mice. The bacteria bind to both Kupffer and endothelial cells and trapping can be inhibited by mannose present in the perfusion medium (Tables 1 and 2). We have extended these observations to show that if bacteria are mixed with 1% mannose and then washed to remove unbound mannose, such treatment also results in inhibition of hepatic trapping (data not shown) suggesting that mannose binds to pili and inhibits association with sinusoidal cells. These data not only suggest that some type of receptor for pili exists on sinusoidal cells but also that reaction of pili with mannose changes the conformation of the ligand in such a way that it can no longer bind to the appropriate liver cell receptor(s). Further studies on these complex interactions are under way.

5. CONCLUSIONS

Following the flurry of research activity in the 1950s on the clearance of microorganisms from the blood there has been a relative lull in the literature with regard to substantially new insight into the mechanisms by which reticuloendothelial organs trap microorganisms. While a number of studies have enhanced the breadth of our understanding with respect to selected organisms, fundamental advances in understanding new mechanisms are scant. By contrast, the ways in which macromolecules interact with a variety of reticuloendothelial cells as well as PMNL and red blood cells have substantially increased and it appears now that the time is right to approach a more detailed understanding of the molecular mechanisms by which microbial cells interact with cells lining the vasculature. These studies will be complex because one not only has to identify the important adherence antigens on the microbial surface and the appropriate receptor ligands on the vasculature cells but one also has to keep in mind the interaction with humoral elements including specific antibody which can radically alter the surface which the microbe presents to host tissue. It is also becoming increasingly apparent that cells of the mononuclear phagocyte system are not the only cell type involved in vascular clearance. Hepatic endothelial cells have already been clearly implicated and other cell types will be soon to follow. The exciting advances on the mechanisms of adhesion of microbial cells to mucous membranes of the oral, intestinal, and GI systems have far surpassed our understanding of the mechanisms by which microbes adhere to the systemic vasculature. The success in understanding the adherence mechanisms in enteric disease highlights the value of such approaches being applied to infections of deeper tissues. Studies on the host cell interaction with the vasculature should not only enhance our understanding of host defense against microbial invasion but also give new insight into how a microbe can avoid host defenses and establish systemic disease. This field is ripe for contributing new fundamental insight into the nature of the host–parasite interaction and the future looks bright.

REFERENCES

Angrist, A., Oka, M., and Nakao, K., 1967, Vegetative endocarditis, in: *Pathology Annual* (S. C. Sommers, ed.), pp. 155–180, Appleton–Century–Crofts, New York.

Ashwell, G., and Morell, H. G., 1977, Membrane glycoproteins and recognition phenomena, *Trends Bio Sci.* **2:**76.

Benacerraf, B., and Miescher, L., 1960, Bacterial phagocytosis by the reticuloendothelial system *in vivo* under different immune conditions, *Ann. N.Y. Acad. Sci.* **88:**184.

Benacerraf, B., Biozzi, G., Halpern, B. N., Stiffel, C., and Mouton, D., 1957, Phagocytosis of heat-denatured human serum and albumin labeled with ^{131}I and its use as a means of investigating liver blood flow, *Br. J. Exp. Pathol.* **38:**35.

Benacerraf, B., Sebestyen, M. M., and Schlossman, S., 1959, A quantitative study of the kinetics of blood clearance of P^{32} labeled *Escherichia coli* and staphylococci by the reticuloendothelial system, *J. Exp. Med.* **110:**27.

Biozzi, G., Benacerraf, B., and Halpern, B. N., 1955, The effect of *Salmonella typhi* and its endotoxin on the phagocytic activity of the reticuloendothelial system in mice, *Br. J. Exp. Pathol.* **31:**226.

Biozzi, G., Halpern, B. N., Benacerraf, B., and Stiffel, C., 1957, Phagocytic activity of the reticuloendothelial system in experimental infections, in: *Physiopathology of the Reticuloendothelial System*, Blackwell, Oxford.

Biozzi, G., Benacerraf, B., and Halpern, B. N., 1958, Quantitative study of granulopectic activity of reticuloendothelial system. II. The study of kinetics of granulopectic activity of the reticuloendothelial system in relation to dose of carbon injected: Relationship between weight of organs and their activity, *Br. J. Exp. Pathol.* **34:**441.

Biozzi, G., Howard, J. G., Halpern, B. N., Stiffel, C., and Mouton, D., 1960, The kinetics of blood clearance of isotopically labeled *Salmonella enteritidis* by the reticuloendothelial system in mice, *Immunology* **3:**74.

Bonventre, P. F., and Oxman, E., 1965, Phagocytosis and intracellular decomposition of viable bacteria by the isolated perfused rat liver, *J. Reticuloendothelial Soc.* **2:**313.

Durach, D. T., 1975, Experimental bacterial endocarditis. IV. Structure and evaluation of very early lesions, *J. Pathol.* **115:**81.

Durach, D. T., and Beeson, P. B., 1972, Experimental bacterial endocarditis. I. Colonization of a sterile vegetation, *Br. J. Exp. Pathol.* **53:**44.

Durach, D. T., Gilliland, B. C., and Petersdorf, R. G., 1978, Effect of immunization on susceptibility to experimental *Streptococcus mutans* and *Streptococcus sanguis* endocarditis, *Infect. Immun.* **22:**52.

Friedman, R. L., and Moon, R. J., 1977, Hepatic clearance of *Salmonella typhimurium* in silica-treated mice, *Infect. Immun.* **16:**1005.

Friedman, R. L., and Moon, R. J., 1980, Role of Kupffer cells, complement, and specific antibody in the bactericidal activities of perfused livers, *Infect. Immun.* **29:**152.

Gould, K., Raminez-Ronda, G. H., Holmens, R. K., and Sanford, J. P., 1975, Adherence of bacteria to heart valves *in vitro*, J. Clin. Invest. **56:**1364.

Hamburger, M., Gall, E. A., and Scott, N. C., 1971, Distribution of intravenously inoculated staphylococci in heart valves and viscera, *Arch. Intern. Med.* **127:**496.

Howard, J. G., 1961, The reticuloendothelial system and resistance to bacterial infection, *Scott. Med. J.* **6:**60.

Howard, J. G., and Wardlaw, A. C., 1958, The opsonic effect of normal serum on the uptake of bacteria by the RES: Perfusion studies with isolated rat liver, *Immunology* **1:**338.

Hubbard, A. L., Wilson, G., Ashwell, G., and Stukenbrok, H., 1979, An electron microscope autoradiographic study of the carbohydrate recognition systems in rat liver. I. Distribution of ^{125}I-ligands among the liver cell types, *J. Cell Biol.* **83:**47.

Hudgin, R. L., Pricer, W. E., and Ashwell, G., 1974, The isolation and properties of a rabbit liver binding protein specific for asialoglycoproteins, *J. Biol. Chem.* **249:**5536.

Jahrling, P. B., 1976, Virulence heterogeneity of a predominantly avirulent Western equine encephalitis virus population, *J. Gen. Virol.* **32:**121.

Jahrling, P. B., and Gorelker, L., 1975, Selective clearance of a benign clone of Venezuelan encephalitis virus from hamster plasma by hepatic reticuloendothelial cells, *J. Infect. Dis.* **13:**667.

Jahrling, P. B., and Scherer, W. F., 1973, Growth curves and clearance rates of virulent and benign Venezuelan encephalitis viruses in hamsters, *Infect. Immun.* **8:**456.

Jahrling, P. B., Heisey, G. B., and Hesse, R. A., 1977, Evaluation of vascular clearance as a marker for virulence of alphaviruses: Disassociation of rapid clearance with low virulence of Venezuelan encephalitis virus strains in guinea pigs, *Infect. Immun.* **17:**356.

Jenkin, C. R., and Rowley, D., 1961, The role of opsonins in the clearance of living and inert particles by cells of the reticuloendothelial system, *J. Exp. Med.* **114:**363.

Jeunet, F. S., Cain, W. A., and Good, R. A., 1968, Differential recognition of *Brucella* organisms by Kupffer cells: Studies with isolated, perfused liver, *Proc. Soc. Exp. Biol. Med.* **129:**187.

Jeunet, F. S., Cain, W. A., and Good, R. A., 1969, Reticuloendothelial function in the isolated perfused liver. III. Phagocytosis of *Salmonella typhosa* and *Brucella melitensis* and the blockade of the RES, *J. Reticuloendothelial Soc.* **6:**391.

Johnson, W. H., and Latta, H., 1978, Acute hematogenous pyelonephritis in the rabbit (electron microscopic study of *E. coli* localization and early acute inflammation), *Lab. Invest.* **38:**439.

Kawasaki, T., and Ashwell, G., 1976, Chemical and physical properties of an hepatic membrane protein that specifically binds asialoglycoprotein, *J. Biol. Chem.* **251:**1296.

Kawasaki, T., Etoh, R., and Yamashima, I., 1978, Isolation and characterization of a mannan-binding protein from rabbit liver, *Biochem. Biophys. Res. Commun.* **81:**1018.

Leunk, R. D., and Moon, R. J., 1979, Physiological and metabolic alterations accompanying systemic candidiasis in mice, *Infect. Immun.* **26:**1035.

Manwaring, W. H., and Coe, H. C., 1916, Endothelial opsonins, *J. Immunol.* **1:**401.

Manwaring, W. H., and Fritschen, W., 1923, Study of microbic–tissue affinity by perfusion methods, *J. Immunol.* **8:**83.

Merigan, T. C., 1974, Host defenses against viral disease, *N. Engl. J. Med.* **290:**323.

Moon, R. J., Vrable, R. A., and Broka, J. A., 1975, *In situ* separation of bacterial trapping and killing functions of the perfused liver, *Infect. Immun.* **12:**411.

Munthe-Kass, A. C., 1977, Endocytosis studies on cultured rat Kupffer cells, in: *Kupffer Cells and Other Liver Sinusoidal Cells* (E. Wisse and D. L. Knook, eds.), p. 325, Elsevier/North-Holland, Amsterdam.

Musher, D. M., and Richie, Y., 1974, Bacterial clearance and endocarditis in American opossums, *Infect. Immun.* **9:**1126.

Prior, W. E., and Ashwell, G., 1976, Subcellular distribution of a mammalian hepatic binding protein specific for asialoglycoproteins, *J. Biol. Chem.* **251:**7539.

Prieels, J. P., Pizzo, S. V., Glascow, L. R., Paulson, J. C., and Hill, R. L., 1978, Hepatic receptor that specifically binds oligosaccharides containing fucosyl $\alpha 1 \rightarrow 3$ *N*-acetylglycosamine linkages, *Proc. Natl. Acad. Sci. USA* **75:**2215.

Rowley, P., 1962, Phagocytosis, *Adv. Immunol.* **2:**241.

Ruggerio, G., Utili, R., and Andreana, A., 1977, Clearance of viable *Salmonella* strains by isolated, perfused rat livers: A study of serum and cellular factors involved and of the effect of treatments with carbon tetrachloride or *Salmonella enteritidis* lipopolysaccharide, *J. Reticuloendothelial Soc.* **21:**79.

Saba, T. M., 1970, Physiology and physiopathology of the reticuloendothelial system, *Arch. Intern. Med.* **126:**1031.

Sawyer, R. T., Moon, R. J., and Beneke, E. S., 1976, Hepatic clearance of *Candida albicans* in rats, *Infect. Immun.* **14:**1348.

Scheld, W. M., Valone, J. A., and Sande, M. A., 1978, Bacterial adherence in the pathogenesis of endocarditis, *J. Clin. Invest.* **61:**1394.

Sigel, M. M., Acton, R. T., Evans, E. E., Russell, W. J., Wells, T. G., Painter, B., and Lucas, A. H., 1968, T_2 bacteriophage clearance in the lemon shark, *Proc. Soc. Exp. Biol. Med.* **128:**977.

Steer, C. J., 1980, Carbohydrate recognition systems of sinusoidal liver cells: An update, *Kupffer Cell Bull.* **3**:15.

Thornburg, R. W., Day, J. F., Baynes, J. W., and Thorpe, S. R., 1980, Carbohydrate-mediated clearance of immune complexes from the circulation, *J. Biol. Chem.* **255**:6820.

Trnovec, T., Schel, D., Lemaneh, M., Faberova, V., Bezek, S., Gajdasih, A., and Kopruda, V., 1978, The distribution in mice of intravenously administered labeled *Candida albicans*, *Sabouraudia* **16**:299.

Wolinsky, J. S., and Stroop, W. G., 1978, Virulence and persistence of three prototype strains of mumps viruses in newborn hamsters, *Arch. Virol.* **57**(4):355.

6

RES Control of Endotoxemia

HENRY GANS

1. INTRODUCTION

The classic definition of endotoxin is a material that is freed from the cell wall upon death of gram-negative microorganisms,* in contrast to exotoxins, which are toxins produced and excreted by certain bacteria. Presently, endotoxins, a term used interchangeably with lipopolysaccharides (LPS), are regarded as the cell wall components derived from the gram-negative bacteria, obtained by various extraction procedures. Hence, when working with endotoxins, the manner in which the material is obtained should be specified. Their chemical composition consists mainly of lipoproteins and polysaccharides with minor impurities, predominantly proteins and nucleic acids. The LPS–protein† complexes thus obtained vary considerably in size. This is reflected in differences in molecular weight which can range from one-half to several million daltons (Beer *et al.*, 1966). Relatively heat stable, they are destroyed only upon prolonged heating at temperatures above 100°C. Different LPS preparations vary in toxicity; however, many effects are similar to and to a degree interchangeable and independent of the source of the organism. Thus, certain endotoxin-related phenomena, as for instance the Shwartzman‡ phenomenon which relies upon two temporally

*Gram-negative bacteria from which endotoxin has been isolated include the pathogenic organisms *Salmonella typhi, Shigella dysenteriae, Brucella melitensis, Vibrio cholera, Neisseria gonorrhoeae,* and *N. meningitidis,* and the relatively avirulent species *Escherichia coli.*

†The protein fraction and a phospholipid fraction can be separated from the LPS by phenol extraction.

‡The *local* Shwartzman reaction develops when endotoxins are injected s.c. into rabbits in a sublethal dose resulting in a mild inflammatory skin response. A second dose of the *same* or another endotoxin administered i.v. in the same amount as the first injection but 24 hr later results in hemorrhagic necrosis of the originally injected skin site. Shwartzman, who observed this phenomenon in 1928, noted that the changes were due to leukocyte–platelet thrombi in the venules. The *generalized* reaction, which occurs if both injections are given i.v., 24 hr apart, results in bilateral cortical necrosis in the rabbit due to fibrinoid occlusion of the renal cortical vessels (Good and Thomas, 1953; McKay *et al.*, 1959; Bohle *et al.*, 1959; Horn and Collins, 1968).

HENRY GANS • Surgical, Research, and Pathology Services of the Danville Veterans Administration Medical Center, and University of Illinois School of Basic Medical Science and Clinical Medicine, and Division of Nutritional Sciences, College of Agriculture, University of Illinois at Urbana, Urbana, Illinois 61801.

spaced endotoxin injections, a challenging and a provocative dose, can be elicited with endotoxins derived from different gram-negative bacteria.

To repeat, endotoxin has two basic chemical components, a lipid and a carbohydrate. The lipid portion is the toxic and mitogenic, hence the biologically active, moiety. The polysaccharide fraction, or the O or somatic antigen, constitutes the antigenic component. The latter is resistant to prolonged heating at 100°C, to alcohol, and to dilute acids, features that are applied in its preparation. As an antigen, it agglutinates slowly with anti-O antibodies in finely granular masses. Because the toxic and the antigenic moieties are associated with distinctly different components of endotoxin, antisera prepared against endotoxin usually fail to protect against its effects. The antisera react predominantly with the polysaccharide fraction, normally leaving the toxic lipid fraction unaltered.

Endotoxin is active over a very wide dosage range. Endotoxin, administered *in vivo* in amounts of less than 0.001 μg/kg, elicits a marked pyrogenic response in man and rabbit. The quantities required to cause shock, diarrhea, and numerous other changes are much larger; with these larger doses, marked congestion, hemorrhage, and necrosis, chiefly due to damage to the vascular endothelium, are found in several organs. Different species of animals show great variation in susceptibility to endotoxin—the cat is the most sensitive while the rat is the most resistant animal. Zweifach compiled the following relative LD_{75} for circulatory collapse: cat, 1; rabbit, 2.5; dog, 8; guinea pig, 20; and rat, 300.

Animals subjected to repeated injections of one endotoxin become not only refractory to its pyrogenic and other biological effects but also to the action of other endotoxins, traumatic shock, and radiation-induced injury. This state is designated *tolerance* or *refractoriness*, a condition that subsides a few weeks after its last exposure to endotoxin (Beeson, 1947a,b; Reichard, 1967; Greisman and Woodward, 1970; Reichard, 1972; Moreau and Skarnes, 1973). Tolerance to endotoxin is unrelated to immunity, a condition that persists much longer. It is accompanied by a number of interesting changes. One is the markedly accelerated clearance of endotoxin from blood. Agents which block RES function, such as thorotrast and carbon particles, interfere with blood clearance of endotoxin, thereby eliminating tolerance (Beeson, 1947a,b). This observation implicates the RES as an important factor in control of endotoxemia.

2. DETECTION OF ENDOTOXIN

The progress of endotoxin studies has been seriously hampered by the lack of expedient, reliable techniques for the detection and determination of endotoxin at very low levels. Initially, biological assays such as the chick embryo method, the endotoxin-mediated necrotizing action of intradermally injected epinephrine in rabbits, and the pyrogenic response in the rabbit were widely used. More recently applied biological assays include the mouse lethality test, the tumor-necrotizing effect, the Shwartzman reaction, bone marrow necrosis, the adrenalectomized or actinomycin D-treated mouse, and the lead acetate-sensitized rat. Differences in sensitivity and reproducibility continue to plague most of these tests, despite continuing efforts toward their improvement. The latest developments in this area are sensitive chemical assays that include complement-inactivation, radioimmunoassay for O antigen and the limulus amebocyte lysate gelation assay which demonstrate and quantitate endotoxins reproducibly in minute quantities in biological fluids and tissues.

Toxicity to chick embryos is determined by i.v. injection of small quantities of the test substance dissolved in saline into the chick embryo at 39.5°C on day 10 of incubation (Finkelstein, 1964). Timing is crucial since susceptibility to endotoxin decreases with increasing embryonic maturity. Mortality is recorded after 24 hr. Death is due to an endotoxin-induced marked hypoglycemia, a response much less pronounced and sustained in older embryos (Smith and Thomas, 1956).

The pyrogenic response is determined in New Zealand white rabbits of standard weight (Chapman, 1942). The average normal temperature of each animal is initially established with an electric recording thermometer and only those animals with temperatures ranging between 39 and 39.5°C are selected. The test substance is dissolved in pyrogen-free saline and injected into the marginal ear vein. The temperature is recorded and a temperature curve is prepared. The rise in temperature is compared to reference curves obtained following injection of standard doses of endotoxin.

The tumor-necrotizing effect of endotoxin is applied in the tumor hemorrhagic assay. It is performed on mice of standard weights bearing a standard number of tumor cells (e.g., sarcoma 37 cells) implanted s.c. 7 days prior to the assay. The presence of endotoxin is demonstrated if hemorrhage is observed 24 hr after challenge by i.v. injection of 0.1–0.2 ml of a test substance into the lateral tail vein.

The limulus amebocyte lysate (LAL) test uses amebocytes, the circulating blood cells of the hemolymph of horseshoe crabs (*Limulus polyphemus*) (Levin *et al.*, 1970, 1971). This lysate gels in the presence of minute, nanogram, quantities of endotoxin. A highly sensitive technique when applied for pyrogen detection, it can be tempermental when utilized for the determination of endotoxin presence in biological fluids. Because endotoxins in biological fluids are associated with and reversibly inactivated by certain plasma proteins (Oroszlon *et al.*, 1966), several methods have been developed to free it from this association. Chloroform extraction, dilution with distilled water, and the heating of biological fluid are among the more common procedures applied today. Unfortunately, the inference drawn from a positive LAL test is not always correct because false-positive results have been obtained with it in the presence of thrombin, thromboplastin, and other materials, thus limiting its usefulness (Elin and Wolff, 1973; Stumacher *et al.*, 1973; Suzuki *et al.*, 1977).

The radioimmunoassay is far less sensitive than the LAL test. Also, it is only specific for the particular O antigen used for antibody production. Hence, its usefulness is presently rather limited. Actinomycin D, 12.5–25 μg/C.F.W. mouse (s.c.), enhances endotoxin toxicity (injected i.p.) more than 10,000-fold. It thus serves as an excellent assay to demonstrate endotoxin in this species (Pieroni *et al.*, 1970).

Lead acetate, 3.5–5 mg/100 g body wt, administered into the dorsal vein of the penis, has been shown to sensitize rats to endotoxin more than 100,000-fold by mechanisms that, as yet, have not been fully elucidated (Selye *et al.*, 1966; Filkins, 1970). Using commercial endotoxin preparations, administration of 1 ng endotoxin/100 g body wt in a lead acetate-sensitized rat results in a mortality rate of 20%. The mortality associated with 10 ng is 30%; with 100 ng (0.1 μg), 60%; and after 1 μg, 80% (Gans and Matsumoto, 1974). Because the mortality rates vary little between different laboratories using the same batch of endotoxin, its use allows for a semiquantitative determination of endotoxin levels in biological fluids. Increase in the lead acetate dose or addition of adrenalectomy enhances the test's sensitivity even more. The sensitizing effect of lead acetate for *E. coli* endotoxin is greatest when lead and endotoxin are present stimultaneously. However, considerable sensitization persists when endotoxin is injected from 1 hr before to 7 hr after lead administration (Selye *et al.*, 1966). Besides lead, several other agents can induce a hypersensitivity to endotoxin, e.g., actinomycin D, zymosan, glucan, BCG, CCl_4, and mitomycin C (Seyberth *et al.*, 1972). One common denominator in endotoxin hyperreactivity, once considered solely due to RES depression, was found to be a marked, frequently lethal hypoglycemia, inadequate glyconeogenesis, and depletion of hepatic and other tissue glycogen stores (Filkins, 1973). Also, lead salts were recently found to inhibit plasma enzymes that ordinarily inactivate considerable amounts of endotoxin. Considering the amount of endotoxin inactivated in this manner, this may represent another mode of action of lead-induced sensitization to endotoxin (unpublished observation).

3. SOURCES OF INFECTION

Although the gut constitutes a major reservoir for gram-negative organisms and the oral–fecal route a common pathway of contamination, gram-negative infections, in our moderate climate, are not uncommonly confined to the urinary tract or as a local and systemic infection in the compromised, malnourished, or debilitated host. In contrast, healthy individuals can readily resist the effects of the common gram-negative organisms and their endotoxins, Hence, endotoxin shock develops more commonly in elderly, debilitated, and chronically ill patients. Leukemia, lymphoma, sarcoma, and carcinoma seem to predispose to infections by gram-negative organisms, not unlike malnutrition or diabetes. Immune deficienty disorders, primary as well as those induced by chemotherapy or immunosuppression, also seem to sensitize the patient. In tropical climes, especially those with limited public health facilities, enteric infections by *Shigella*, *Salmonella*, and *Vibrio* organisms remain common health hazards.

Virulence, invasiveness, pathogenicity, and the size of the infecting dose determine the outcome of gram-negative infection. Equally important, however, is the host's response to the invading organism or its endotoxin. The local and systemic manifestations of nonenteric gram-negative infections may derive from the bacterial infection per se, the endotoxemia alone, or a combination of the two. Endotoxins share a number of common biological properties irrespective of the bacterial species from which they derive. These include pyrexia, leukopenia, diarrhea, prostration, and, in the extreme form, shock and death. Depending on which clinical features develop, pathological effects of gram-negative infections are subject to enormous variations.

In any case, the organisms residing in the lower gastrointestinal tract ordinarily cause very little harm to the host. Avirulent strains of *E. coli* and of other gram-negative microorganisms of the normal gut contain a great deal of endotoxin. Yet, this fails to affect the host adversely. Even in the compromised individual, nosocomial infections with gram-negative organisms are usually transmitted diseases rather than the result of a local breakdown of constraining mechanisms as recent epidemic outbreaks of *Serratia marcescens* and other opportunistic organisms would suggest. It should be mentioned here that to implicate gram-negative commensals as a cause for clinical disease requires more than their demonstration in culture, particularly since they are easily mistaken for contaminants. Hence, serotyping has been recently applied to positively identify the organism as the cause of the disease and in establishing a definite diagnosis since a rise in antibody titer can be anticipated to develop during such an infection.

Clinical infection usually develops because the organism can enter the host by surviving and penetrating mucosal surfaces, multiply in the nutritional environment of the host tissues, inhibit or injure host defense mechanisms, and inflict injury to the host. *Endotoxemia* may occur *early* during a gram-negative infection. Following effective mobilization of the host defenses, leading to control of infection and killing of the bacteria with or without the aid of antibiotics, a *late* endotoxin release (as in a Herxheimer type of reaction) can also occur which may be extremely detrimental, if not lethal.

4. EFFECT OF ENDOTOXIN AT THE CELLULAR LEVEL

LPS, after its extraction from the outer membrane of gram-negative bacteria and its partial purification, resembles membrane fragments which in the electron micrograph appear as discs, ribbons, vesicles, and lamellae (Beer *et al.*, 1966). These consist of bilayer structures, oriented with fatty acids (lipid A) on the interior and hydrophilic polysaccharides on the outer surface. Hence, the latter are exposed to the surrounding water and thus participate in the suspension of endotoxin in aqueous media.

LPS show great affinity for biological membranes, especially for those to which they are initially exposed, such as endothelial cells, blood cells (Brunning *et al.*, 1964), etc. Shands proposed the concept of "edge attachment" between LPS and mammalian membranes. The lipid moiety, which is buried deep within the LPS particle, appears to be essential for this coupling. He proposed that "the particle must attach by their edges, whereby some fatty acids may be solubilized into the membrane" (Shands, 1973). This is thought to be an important characteristic significant enough to account for most of the actions of LPS since endotoxin without lipid A exerts few, if any, noxious effects. Alkali treatment of the preparation enhances this particular effect, e.g., such endotoxin lyses erythrocytes *in vitro* presumably as a result of conformational changes induced in the red cell membrane.

Endotoxin has a number of stimulating effects on immune responses, particularly macrophage- and lymphocyte-mediated responses. These effects, thought to be membrane transmitted, include a rise in intracellular levels of cAMP (Gimpel *et al.*, 1974; Braun, 1975). Presumably the underlying mechanism is the activation of the membrane-associated enzyme, adenylate cyclase, which converts intracellular ATP into cAMP.* cAMP, in combination with an intracellular binding protein, is associated with activation of certain kinases (e.g., phosphorylating enzymes) that affect the biosynthesis of macromolecules, e.g., the synthesis of nucleic acid, and thus cell proliferation. Also, it activates preexisting enzymes. However, once formed, cAMP is short-lived; it is rapidly converted by phosphodiesterase into inactive AMP, a reaction that is inhibited by theophylline, caffeine, and papaverine. Presently, formation of cAMP is believed to also play an important role in endotoxin induced macrophage and B-cell activation.

Interestingly, several adverse effects of endotoxin are thought to occur in an identical manner. Thus, injury to endothelial cells (Gaynor *et al.*, 1970), granulocytes (Horn and Collins, 1968; Lerner *et al.*, 1968, 1971), erythrocytes (Shands, 1973), and platelets (Des Prez *et al.*, 1961; Des Prez, 1967; Horowitz *et al.*, 1962; Das *et al.*, 1973; Harviger *et al.*, 1975) results presumably from the direct effect of endotoxin on the cell membrane. In addition, immunocytotoxic effects are believed to play a role in endotoxin-induced thrombocytopenia and granulocytopenia, changes that are absent in complement-deficient animals (Fong and Good, 1971). Since endotoxin activates complement (Gewurz *et al.*, 1968; May *et al.*, 1972), its adverse effect on certain blood and blood vessel lining cells is presently thought to be due, at least in part, to complement activation.

5. ENDOTOXIN TRANSFER ACROSS THE NORMAL GUT WALL

In certain animal species, the neonate's gut wall is fairly permeable to macromolecules; in others, however, it is far less so. Colostral immunoglobulins, for instance, readily traverse the intestinal wall of the calf (Krahlenbuhl and Campiche, 1969). They enter the columnar epithelial cells that cover the villi of the small intestine, especially those cells of the distal ileum by endocytosis. Escaping the action of lysosomal enzymes, they traverse the gut wall intact to enter the ileal venules and lymphatics (Halliday, 1955). These immunoglobulins play a significant role in preventing neonatal infections, e.g., coli-sepsis in calves. Whereas immunoglobulins can traverse the gut wall intact in certain animals, this is not the case in others, notably in man (Brambell, 1970). Ruminants lose this property shortly after birth, while in rodents it persists for 2 to 3 weeks. In

*Stimulators of adenylatecyclase action are the catecholamines (epinephrine, norepinephrine, and isoproterenol).

the rabbit, on the other hand, the major share of immunoglobulins reach the neonate transplacentally. In these animals, neither ferritin, horseradish peroxidase, nor hemocyanin escape intact across the gut wall to any significant degree (Krahlenbuhl and Campiche, 1969).

The transfer of intact macromolecules (Walker *et al.*, 1972, 1973; Wands *et al.*, 1976; Volkheimer *et al.*, 1968; Warshaw *et al.*, 1971, 1974) and of gut antigens (Bernstein and Orary, 1968; Kagnoff, 1978; Mattingly and Wachsman, 1978) has been demonstrated conclusively. The escape of endotoxins across the gut wall of mammals during the postneonatal period, on the other hand, has met with considerable controversy. This represents an important issue, especially in shock, as first proposed by Ravin *et al.* (1960), and in chronic liver disease (Gans and Matsumoto, 1974; Liehr *et al.*, 1975) where endotoxins are thought to play a significant role in shaping the clinical manifestations of the disease. Unfortunately, this issue has never been resolved to everyone's satisfaction. Thus, it remains to be established whether endotoxin is absorbed intact across the gut wall and if so the circumstances and the manner in which this takes place, the amounts involved, the conditions that promote it, and the defenses mobilized against it. This is interesting since up until recently the gut was generally considered to be impervious to macromolecules because of its protective epithelial covering, its mucous secretion, the intestinal motility that removes noxious agents rapidly, distally, and because of several local defense mechanisms present in the gut such as gastric acid, intestinal enzymes, coproantibodies (secretory IgA Swarbrick *et al.*, 1979), and the handling of antigens by macrophages.

This much is clear: normal intestinal epithelium forms an extensive cover that extends from the crypts to the tufts of villi. Before weaning, however, the vacuolated columnar epithelial cells referred to above can readily be seen throughout the small intestine, particularly in the distal ileum, an area that is now regarded to be especially associated with the transfer of immunoglobulins from colostrum and milk (Moon *et al.*, 1973). In the suckling mammal these cells disappear after weaning. In rats and mice this occurs between 15 and 20 days after birth, in the pig 20–22 days postnatally, and in ferrets somewhat later, at approximately 35 days (Clarke and Hardy, 1970, 1971a,b). In germfree animals these cells are seen to persist much longer and to show much slower rates of migration from the crypt outward, suggesting that changes in the gut flora associated with weaning of the animal are somehow accompanied by a much faster replacement of the gut epithelium (Clarke and Hardy, 1971a,b).

In adult animals small food particles and carbon have been seen to pass across the gut wall (Joel *et al.*, 1978). They are retained by macrophages in the gut wall, especially at sites of desquamation. This process of mechanical transfer of particulate matter across the intestinal wall appears to be magnified by certain drugs that enhance intestinal motility and is observed predominantly at the tips of the intestinal villi (Volkheimer *et al.*, 1968b; Volkheimer *et al.*, 1968a,b).

In contrast to the normal gut epithelium, the mucosa overlying Peyer's patches has an interesting and different appearance. It contains specialized cells, designated by Owen and Jones (1974) and by Owen (1977) as the membranous or "M" cell, a very thin epithelial cell with only rudimentary villi. Rather than villi, these cells have micropits and microfolds. Also, they lack the thick glycocalyx layer normally adherent to the cell surface facing the intestinal lumen. At the level of the "M" cell, gut-derived antigens, escaping enzymatic or other forms of degradation, are believed to gain access to the lymphoid tissue of the small bowel wall. The lymphoid cells are present in the interstices; they approach the gut lumen in this area to within 0.3 μm (Owen and Jones, 1974). Antigens, such as horseradish peroxidase (Cornell *et al.*, 1971), are seen in the micropits of "M" cells shortly after their introduction into the gut lumen. Somehow, later they occur in the vesicles inside these cells (but not in the adjacent columnar epithelium) and subsequently in the intercellular space located between "M" cell and lymphocyte from where they gain access to Peyer's patches (Owen, 1977). Hence, antigen can interact with T and B lymphocytes in this area to initiate an immune response. This response can be summarized as follows: precursor cells sensitized by antigens migrate to the mesenteric lymph nodes where they mature. Here they enter the thoracic duct and reach the systemic circulation. As mature plasma cells they selectively home in on the lamina propria and into other secretory tissues as IgA-secreting B cells.

Presently, the transfer of intact gut-derived antigens across the gut wall has been firmly established; however, considerable controversy remains concerning the molecular size "cut-off," and especially concerning the amount that can escape. Most investigators believe only relatively small molecules can escape as such; they regard the quantities to be minute—nanograms per milliliter (Brambell, 1970; Gans and Matsumoto, 1974). Others, however, have found that much larger molecules escape in far greater quantities and, for some food antigens, arrive at amounts ranging from a few percent to as much as 20% of the ingested material to be absorbed undigested (Hemmings and Williams, 1978). If larger quantities of antigen were to escape as such, some would enter into the lymphatic as well as into the venous circulation. Such material could possibly exert profound adverse effects.

It remains intriguing that macromolecules can traverse the gut wall unaltered considering the number of protective mechanisms present within the lumen of the intestine. For instance, in addition to the barrier function of the endothelium, secretory IgA is produced and secreted into the gut lumen in response to oral immunization by gut-associated antigens both food-derived and bacterial in origin. Produced by plasma cells in the lamina propria of the bowel wall, this material provides antibacterial and antigenic protection by several mechanisms: first, its action on microorganisms causes their immobilization, agglutination, and/or prevention of adhesion to the mucous membrane; second, its association with bacterial products, e.g., endotoxins and other gut-derived antigens, causes their neutralization. Secretory IgA, in contrast to some of the other immunoglobulins, is quite resistant to the local activity of proteases presumably because of its association with the secretory piece, a glycoprotein molecule with a molecular weight of 70,000. Secretory IgA prevents the absorption of considerable amounts of gut- and food-derived antigens because these antigens are readily absorbed, unaltered, into the circulation of individuals with selective IgA deficiency where they are demonstrated either as such or as circulating immune complexes (Cunningham-Rundles *et al.*, 1978; Kaufman and Hobbs, 1970).

Bile and digestive enzymes also play a major protective role against the absorption of intact macromolecules, including endotoxins. Bile diversion in the rat has been shown to be associated with the transfer of active endotoxin from the gut lumen into the systemic circulation (Koczar *et al.*, 1969). Presently, we do not know whether this is due to lack of bile salts, to absence of biliary IgA, or to a diminished emulsifying action induced by bile. Particularly interesting is the amount of endotoxin absorbed from the intact gut under normal conditions and during disease states (especially liver disease), a feature of which we know very little indeed.

6. ENDOTOXIN ABSORPTION IN DISEASE

Portal vein occlusion causes increased endotoxin release from the gut of monkeys (Olcay *et al.*, 1974), possibly also of the dog (Gans *et al.*, 1971b), but not in rats (Gans, 1974). Intestinal permeability for macromolecules to which the normal bowel wall is impervious is presumably also enhanced in inflammatory bowel disease (Colin *et al.*, 1979; Tabagehai *et al.*, 1977), after irradiation (Smith *et al.*, 1963; Wilson *et al.*, 1970), and following alcohol intake (Worthington *et al.*, 1978). Chronic alcoholism, and the pancreatic insufficiency it engenders, also interferes with IgA-mediated protection against the absorption or transfer of macromolecular antigens across the gut wall. In addition, alcohol depresses phagocytosis (Blalock, 1918; Nolan and Ali, 1973; Liu, 1975). Hence, it is expected to interfere with the elimination of those agents that escape into the portal vein and with their inactivation by the liver. Although alcoholic liver disease has been ascribed to the direct, adverse effect of alcohol and its derivatives on the liver, it is generally conceded that other mechanisms probably also play a role. Finding elevated antibody titers against gut-associated organisms such as *E. coli* and bacteroides in patients with alcoholic liver disease suggested the possibility that gut-derived organisms and their endotoxins also participate in alcohol-mediated hepatotoxicity. Interestingly, antigen–antibody complexes of low titers against common food antigens and enteric endotoxins in patients with alcoholic liver cirrhosis have been described (Triger *et al.*, 1972; Fink and Schultze, 1982). However, the actual role of endotoxin in the pathogenesis of alcoholic liver disease, despite numerous claims, remains to be determined, and if substantiated, its mechanism of action remains to be established (see below).

We became involved in this issue when we postulated in 1972 the release of enteric endotoxin into the systemic circulation of patients with liver cirrhosis to explain the many, diverse and apparently unrelated clinical manifestations and laboratory findings (Gans *et al.*, 1972). Previously we had found the release of gram-negative organisms from the gut in dogs during acute hepatic failure and their spillover into the systemic circulation (Gans *et al.*, 1971a), observations validating those made by Hume *et al.* (1971) and subsequently also in humans in acute fulminating hepatic failure (Wilkinson *et al.*, 1974). Since the normal liver can readily eliminate endotoxin it remained to be demonstrated that under normal circumstances endotoxin is able to escape from the gut into the portal circuit. This was established first *in vivo* by Gans and Matsumoto using bile-free Thiry Vella fistulae in normal and portacaval-shunted rats (1974), and subsequently *in vitro* by Nolan and co-workers with the use of the everted isolated gut sac (1977). However, neither condition is physiologic and the actual spillover of endotoxin into portal vein blood and thoracic duct lymph in the *intact* animal remains to be established. In liver cirrhosis, Prytz *et al.* (1976) and subsequently many others (Jacob *et al.*, 1977; Clemente *et al.*, 1977; Triger *et al.*, 1978; etc.) demonstrated endotoxins in the systemic circulation of patients with portacaval shunts.* The amount of endotoxin that can traverse the gut wall into the portal vein and/or thoracic duct was also assessed in lead-sensitized portacaval-shunted rats. In these animals approximately 5 ng endotoxin/ml blood was demonstrated (Gans, 1978). This is a minute amount. However, it appears adequate to cause macrophage activation (D. Morrison, personal communication, 1981). If Kupffer cell activation occurs due to absorbed endotoxin, its implications certainly require further evaluation.

Subsequently, the demonstration of endotoxin in the ascites of patients with liver cirrhosis indicated that endotoxin also crossed the gut wall (Tarao *et al.*, 1979). The frequency of this occurrence remains to be established; however, in those instances where it was present, a positive correlation could be established between its level in the ascitic fluid and the albumin concentration of the ascitic fluid, a rise attributed to an enhanced vascular permeability induced by the endotoxin. The presence of endotoxin in ascitic fluid can cause serious problems in patients treated by peritoneo-venous shunt (LeVeen shunt) (Moeschl *et al.*, 1978).

The long-standing controversy concerning the escape of enteric endotoxin during shock (Greene *et al.*, 1961; Ravin *et al.*, 1960) and following mesenteric artery occlusion (Cuevas and Fine, 1971) has resolved itself with the aid of the limulus lysate assay. Presently, it would appear that endotoxins can escape under these circumstances in increased amounts. Whether other toxins are also released remains to be established. Another condition associated with the escape of endotoxins across the intestinal wall is intestinal (jejuno-ileal) bypass surgery, a form of treatment applied for morbid obesity. Blind intestinal loops are created during this procedure. Occasionally these show an overgrowth of gram-negative bacteria. This is thought to account for profound, at times lethal, changes in the liver and for the development of arthritis, intravascular hemolysis, thrombocytopenia, and neutropenia (see below). The latter symptoms appear to correlate closely to the presence of immune complexes that contain antigens of *E. coli* and *B. fragilis* (Shagrin *et al.*, 1971; Peters *et al.*, 1975; Moake *et al.*, 1977; Rose *et al.*, 1977). These symptoms have subsided following intestinal reanastomosis or treatment with broad-spectrum antibiotics suggesting that the antigens responsible for the immune complex disease derive from intestinal

*It should be mentioned here that Fulenwider *et al.* (1980) were unable to substantiate the finding of endotoxemia in hepatic cirrhosis.

microorganisms which are presumably absorbed from the surgicially created blind intestinal loop (Hollenbeck *et al.*, 1975; Simmonds *et al.*, 1975).

7. ENTERIC ENDOTOXIN'S POTENTIATION OF VARIOUS HEPATOTOXINS

Long before the issue of endotoxin escape from the intestine became a point of controversy, a number of studies appeared presenting evidence for its participation in the hepatotoxic effects of certain toxins (carbon tetrachloride and galactosamine) and of choline deficiency. The first evidence suggesting that carbon tetrachloride acts indirectly on the liver was provided by Leach and Forbes who in 1941 showed that sulfonamides protect against its hepatotoxic effect.

Subsequently György demonstrated in 1954 that aureomycin and streptomycin effectively prevent the hepatic necrosis and cirrhosis associated with choline deficiency, a finding confirmed by Rutenburg *et al.* (1957). These investigators showed that nonabsorbable antibiotics are superior in this respect, thus implicating the gut flora in the observed changes. Subsequently, Broitman *et al.* (1964) demonstrated that neomycin's protection against the choline deficiency-induced hepatic defects could be abrogated by adding purified endotoxin to the animal's drinking water. This implied that endotoxin contributes to these changes.

More recently, liver necrosis was observed in patients undergoing intestinal bypass to treat morbid obesity. This was initially thought to be due to weight loss from chronic diarrhea and to malabsorption; later, however, it could also be attributed to products derived from gut microorganisms. The bacterial overgrowth observed in the blind gut loop syndrome was found to be indistinguishable from those observed in alcoholic liver disease (Peters *et al.*, 1975). Hence, it is assumed that the hepatic changes observed in these patients are due to the effect of increased absorption of enteric endotoxins (Hollenbeck *et al.*, 1975). Unfortunately, direct evidence to buttress this point is not available.

The amounts of endotoxin that presumably escape from the normal gut of the rat are minute indeed (Gans and Matsumoto, 1974; Gans, 1978). Were we to assume that no marked species differences exist in this regard, and that under normal circumstances intestinal absorption of endotoxin remains constant, the evidence presented would suggest that the action of very small quantities of endotoxin, acutely or chronically, can significantly amplify liver injury from different causes. If this point can be confirmed, the issue of what the circumstances are and the frequency with which this mechanism plays a role remains to be established. Claims that this is the modus operandi of galactosamine hepatitis (Gruen *et al.*, 1977), Reye's syndrome (Cooperstock *et al.*, 1975), and sudden neonatal deaths (DiLuzio and Friedmann, 1973) have also been made.

Enteric endotoxins have also been implicated as a possible mechanism of alcohol-induced liver disease. Alcohol, because of its ability to enhance intestinal permeability for macromolecules (Worthington *et al.*, 1978) and because of its depressant effect on the RES (Ali and Nolan, 1967; Liu, 1975), is thought to facilitate the escape of endotoxin across the gut wall. Decreased clearance is believed to prolong its presence in the circulation. Whether endotoxin indeed plays such a role (Peters *et al.*, 1975) has to the best of our knowledge never been established. For instance, there is a clear-cut need for studies of the effect of alcohol intake in germfree animals. Also, the endotoxin-amplified effect of hepatotoxins, alcohol, choline deficiency, etc. needs further confirmation. The overwhelming evidence obtained so far challenges the blithe notion that the gut flora is an inert entity.

Finally, the mechanism of endotoxin-induced hepatotoxicity remains to be elucidated. Endotoxin may exert a direct effect, a potentiating effect, or an indirect effect. Presently it is not known how much endotoxin is required to exert a direct effect; most, if not all, studies with endotoxin use doses far larger than those normally found in the portal circulation. Nolan and Ali have dealt with the endotoxin potentiating effects. Presently, we do not know whether it exists as such or merely reflects an indirect effect.

The last effect, namely its indirect effect, may well turn out to be the most significant one. Endotoxin is readily cleared from blood by macrophages, predominantly Kupffer cells (see below). This process is presumably associated with macrophage activation. Macrophage activation is accompanied by release of numerous substances including proteolytic enzymes which Adams and his associates have shown to exert profound cytotoxicity. Interesting in this regard is Ferluga and Allison's (1978) study demonstrating that endotoxin-induced macrophage activation results in a highly lethal form of hepatitis in mice; this condition is associated with high serum transaminase levels, glycogen depletion, and hypoglycemia. Hepatocellular damage by proteolytic enzymes released by activated macrophages (e.g., Kupffer cells) is not a new concept. It was previously postulated as a possible mechanism of liver disease to explain the development of a rapidly progressing form of liver cirrhosis observed in children with homozygous α_1-antitrypsin deficiency (Gans *et al.*, 1969b; Sharp *et al.*, 1969). This involves the absence of a low-molecular-weight serum α_1-globulin that inhibits the different serine proteases (Gans, 1972). Unfortunately, not enough is known presently about macrophage activation, especially about the conditions that elicit protease release, the types and amounts of enzymes released and their local effects, to make any rational judgement as to its true significance. But it would appear from the evidence available that endotoxin-induced macrophage activation could well result in serious liver damage. In fact, results of Bradfield and Souhami's studies (1980) suggest that endotoxin may not even be required. Phagocytosis by Kupffer cells in general is sufficient in this regard since they found it to be associated with enhanced hepatocyte turnover rates.

8. THE ROLE OF THE KUPFFER CELL IN PROCESSING ENTERIC ENDOTOXINS

Endotoxins or other gut-derived antigens that spill over into the portal vein are normally sequestered and degraded by Kupffer cells. This process interferes with their possible escape into the systemic circulation, thus preventing their reaching antibody-producing sites. Chase (1946) demonstrated that oral ingestion of a hapten which is absorbed unaltered will induce immunological unresponsiveness or tolerance; little or no immunologic response occurs to its subsequent parenteral administration. Neither specific antibodies are formed nor can a delayed-type hypersensitivity be elicited against it. Cantor and Dumont (1967) confirmed this in the dog by demonstrating that following oral feeding of dinitrochlorobenzene, circulating antibodies to it are absent. Subsequent s.c. administration also elicits no effect. The tolerance is abolished in the portacaval-shunted animal. This suggests that clearance of antigen by the Kupffer cell from the portal vein blood renders it nonimmunogenic. In contrast, Kupffer cell bypass, e.g., following portacaval shunting, allows the antigen to reach antibody-producing sites with development of antibodies (Thomas *et al.*, 1973; Thomas and Parrott, 1974; Keraan *et al.*, 1974; Simjee *et al.*, 1975). The mechanism of orally induced tolerance, applied for centuries in folk medicine, is unknown, although several possible explanations for it have been advanced (Kagnoff, 1978; Mattingly and Wachsman, 1978). Evidence that macrophages can degrade endotoxin has been clearly established from *in vitro* and *in vivo* studies. Filkins (1971) in *in vitro* experiments with macrophages, Mori *et al.* (1973) in our laboratory in *in vivo* experiments, and more recently Yamaguchi *et al.* (1982) demonstrated the ability of rat Kupffer cells to sequester and detoxify endotoxin, and in the latter study to determine the amount of endotoxin inactivated during a single passage through the liver.

Hence, liver disease, particularly the advanced liver cirrhosis that is associated with portal hypertension and development of porta-systemic shunts, can be expected to be associated with circulating antibodies against gut-derived antigens. Interestingly, hypergammaglobulinemia is frequently observed in these patients and in experimental animals with liver disease or porta caval shunt. Bjoerneboe *et al.* (1972) and Triger *et al.* (1972) showed that immunoglobulin response is related to gut-derived antigens, including *E. coli* (Keraan *et al.*, 1974, Thomas *et al.*, 1974; Simjee *et al.*, 1975) and certain food antigens (Triger *et al.*, 1972). The latter process is significant in that the extent of hypergammaglobulinemia which accompanies chemically induced liver injury in conventionally reared rats is not different from that observed in germfree rats.

The assumption that the Kupffer cell plays a central role in the development of immunological tolerance for absorbed enteric antigens is undoubtedly too simplistic, particularly if one considers the fact that besides entry into the portal vein, certain antigens tend to escape predominantly into the mesenteric lymphatics which bypass Kupffer cell function altogether. The mechanisms implicated in the development of immunological unresponsiveness, that is, a reduced systemic antibody response against parenterally administered antigens, are presently thought to include the appearance of T suppressor cells in Peyer's patches (Asherson *et al.*, 1977), inhibition from those antigens combined with IgA (André *et al.*, 1975), and a low zone tolerance due to the adsorption of very tiny amounts of antigens (Mitchison, 1964).

9. ENDOTOXIN-INDUCED HOST RESPONSES

The biological effects of endotoxin are many. The most obvious one is fever and leukocytosis. The most severe change is progressive hypotension leading to shock and death. The responses are determined by the quantity of endotoxin present within the circulation. Much, if not all, of our information has been obtained with the use of very large amounts of endotoxin. Unfortunately, very little is known about the effect of intermittent or continuous administration of small quantities of endotoxin. Also, some of the responses, their nature, mechanism of action, and actual effect remain to be more clearly defined.

Hypotension has been ascribed, in part, to the release of kinins, a group of polypeptides with profound vasodilatory properties, which are present in high concentration in man and primate during the early phase of septic shock (Fox *et al.*, 1961). They are thought to account for the warm skin, bounding pulse, and decreased peripheral resistance so commonly observed under those circumstances. And, indeed, endotoxin infusion is associated with abnormally high bradykinin levels in man and monkey. It is doubtful, however, that this is the only fact responsible for these changes, since cats and rabbits have normal kinin systems that fail to respond to endotoxin. Yet these animals are subject to the same cardiovascular manifestations following endotoxin administration as other species. Hence, agents besides bradykinin play a role in inducing these changes as well. Other humoral factors such as histamine, serotonin, catecholamines, and angiotensin are also implicated; however, their roles in this condition remain to be determined.

The release of catecholamines, evidenced by its rise in blood levels, is pronounced following exposure to large doses of endotoxin. They derive from granulated vesicles of preterminal sympathetic nerves (Wolfe *et al.*, 1962). Normally discharged by stress, these vesicles are also depleted by amphetamines, reserpine, and cocaine, drugs that tend to potentiate the development of shock (Potter and Axelrod, 1962). This release is a significant response enabling the animal to preserve blood flow to the most vital areas, the heart and brain. It causes inotropic and chronotropic effects on the heart, vasoconstriction of skin, renal, and visceral arterioles, vasodilation of arteries of the striated muscles and myocardium, and constriction of most of the veins (Kaiser *et al.*, 1964). Prolonged and extensive vasoconstriction of the renal and visceral arteriolar vessels results in a severe hypoxia, anerobic glycolysis, and ultimately damage to the tissue, not uncommon findings of endotoxin shock. How this response relates to those observed following smaller amounts of endotoxin remains to be established.

Histamine protects against lethal doses of endotoxin, anaphylaxis, and scalding (Fox and Lasker, 1961). Whether it is released by the mast cell, as some contend, or is formed locally through the action of histidine decarboxylase, as has been implied by others, it is involved in the vasodilation and increased capillary permeability that is associated with a postcapillary venular spasm (Shayer, 1960). However, a great deal remains to be learned about the relationship of histamine, especially concerning the amounts released during endotoxemia.

Angiotensin, a vasoactive octapeptide produced by renin, is a response to renal ischemia. It raises the blood pressure by inducing severe arterial vasoconstriction. Serotonin, released especially through the action of endotoxin on platelets, acts as a vasoconstrictor on the pulmonary vasculature (Kobold *et al.*, 1964).

Endotoxin also induces *disseminated intravascular coagulation* (a consumption coagulopathy) that under certain well-defined circumstances can lead to deposition of fibrin which occludes the microvascular circulatory beds. This enhances tissue ischemia. The ensuing hypoxia results in profound metabolic changes and ultimately serious tissue injury. Because fibrin formation is obvious, the Shwartzman reaction has served as a model in experiments designed to unravel the mechanisms involved in the development of endotoxin-induced intravascular coagulation (Cluff and Berthrong, 1953; Good and Thomas, 1953; Stetson and Good, 1957; Kliman and McKay, 1958; Shapiro and McKay, 1958; Bohle *et al.*, 1959; Gans and Kirivit, 1960, 1961; Spaet *et al.*, 1961; Horowitz *et al.*, 1962; Lee, 1962; McKay, 1965; Rappaport and Hjort, 1967; Horn and Collins, 1968; Phillips *et al.*, 1968; Filkins and DiLuzio, 1968; Lerner *et al.*, 1969; McKay *et al.*, 1969; Gaynor *et al.*, 1970; Fong and Good, 1971; Mueller-Berghaus and Schneberger, 1971; Bergstein and Michael, 1973; Gans, 1975; Alving *et al.*, 1979).

Several mechanisms have at one time or another been implicated for the initiation of the intravascular clotting process; yet, what at one time seemed to be the cause frequently turned out to be one of the effects. As mentioned above, endotoxin in sufficient concentrations ultimately affects the endothelial cell, the platelet, and the granulocyte adversely. Within a few minutes after administration of certain endotoxin doses (the initial dose remains to be determined) all these cells exhibit profound changes. For instance, damaged endothelial cells are released in great numbers, so that they appear as free circulating cells in the peripheral blood (Gaynor *et al.*, 1970). This process is a direct one on the cell, it is *not* prevented by anticoagulants, hence the injury is not clotting related, but it is able to induce profound changes in the hemostatic mechanism. The denuded areas and the exposed basement membrane they leave behind in the vascular wall form a nidus for platelet adhesion and aggregation. Also, the vascular injury is associated with a Factor XII activation (McKay *et al.*, 1969). Yet, those anticoagulants that interfere with Hageman factor activation and the use of lysozyme, which also inhibits activation of Factor XII, fail to prevent the development of intravascular clotting (Mueller-Berghaus and Schneberger, 1971). Hence, Factor XII consumption associated with endotoxemia does not appear to be the cause but rather the result of the intravascular clotting. Whether release of tissue thromboplastin, under these circumstances, accounts for activation of the extrinsic clotting process remains to be determined.

Endotoxin, as noted above, binds to a lipid fraction on the platelet membrane. Those platelets whose cell membranes are altered activate Factors XII and XI, thereby sustaining the intrinsic clotting system. The binding of endotoxin to the platelet membrane is also associated with the "release reaction" (release of platelet factor 3, serotonin, and ADP) and with "platelet aggregation" (Des Prez *et al.*, 1961; Des Prez, 1967; Horowitz *et al.*, 1962). Hence, shortly after endotoxin administration, masses of platelets can be seen filling branches of the pulmonary artery, a feature that is accompanied by a pronounced thrombocytopenia. In fact, these platelet aggregates tend to block the pulmonary vessels. This abnormality and the serotonin release that causes vasoconstriction raises pulmonary artery pressure and vascular resistance, thus reducing the return flow of blood to the heart. This, in turn, causes a drop in the systemic blood pressure. Initially this was regarded as a direct effect of endotoxin on the platelet; however, it is absent in complement-deficient or in cobra venom-treated, decomplemented animals. This suggests that besides its direct effect on the platelet membrane, endotoxin can induce thrombocytopenia as a result of complement activation (Fong and Good, 1971).

The release reaction not only occurs *in vivo;* the incubation of platelet-rich plasma with endotoxin also results in platelet aggregation and the release of serotonin and platelet factor 3 into the plasma (Des Prez *et al.*, 1961; Horowitz *et al.*, 1962; Des Prez, 1967). Whereas every cell possesses latent phospholipid procoagulants, e.g., the phosphatidylethanolamine and phosphatidylserine localized inside the cell membrane, platelets, upon stimulation by endotoxin, or after their exposure to thrombin, serotonin, basic polymers, epinephrine, ADP, collagen, and antigen–antibody complexes, uniquely extrude these membrane components as platelet factor 3 activity. The fact that substitution of phosphatidyl lipids or platelet factor 3 for platelets prior to the administration of a second dose of endotoxin fails to induce a Shwartzman reaction (Mueller-Berghaus *et al.*, 1967) or the administration of purified platelet factor 3 activity fails to initiate intravascular clotting in rabbits suggests, however, that disseminated intravascular coagulation (DIC) is not directly induced by platelets. Yet, platelets are required for the development of an endotoxin-induced formation of microclots as the absence of thrombi in thrombocytopenic animals would indicate (McKay, 1965).

Besides platelets, other blood cells and endothelial cells play a role in endotoxin-related coagulation abnormalities. Thomas and Good (1952) observed that the Shwartzman phenomenon is absent in the granulocytopenic rabbit, suggesting the significant role that granulocytes play in the development of this phenomenon (Horn and Collins, 1968). Interestingly, the sensitivity of this cell for endotoxin exceeds that of the platelet manyfold; leukopenia in rabbits occurs with endotoxin doses of 0.1 μg/kg. This dose fails to elicit a drop in the number of circulating platelets (Alving *et al.*, 1979). Rabbit leukocytes are especially sensitive to minute quantities of endotoxin; hence, it has been suggested that endotoxin-induced leukopenia may even be a more sensitive test for detecting the presence of minute amounts of endotoxin than pyrogenicity (see p. 117).

The *in vitro* addition of endotoxin to whole blood shortens the clotting time. Under those circumstances leukocytes release a procoagulant similar in activity to tissue thromboplastin (Lerner *et al.*, 1968, 1971). Leukocytes, activated by endotoxin after their injection into rabbits, induce clot formation in the lungs. Rabbit leukocytes also activate plasminogen, a feature absent in human and dog leukocytes (Gans and Hanson, 1964). The presence of fibrin as a response to stimuli implies an increase in fibrin formation relative to its breakdown (fibrinolysis). Hence, deficiencies in fibrinolytic activity are expected to promote the persistence of fibrin deposits. In addition to the noted release of coagulant activity, an enhanced activation of plasminogen was observed during the interaction of endotoxin on granulocytes. Clearly, endotoxin favors the release of clotting as well as of lytic activities from the granulocyte. Both activities are confined to the granules (Horn and Collins, 1968) which can replace the whole cell in this regard. Presently, the active clotting and lytic principles of the neutorphil granule fraction have not been isolated. It is assumed that they represent neutral proteases.

Endotoxin reduces not only the blood level of various coagulation factors, but also of antithrombin III activity (Merskey *et al.*, 1964, 1966). If sufficiently profound this can lead to development of a hemorrhagic diathesis (consumption coagulopathy). Similar changes also occur in the fibrinolytic enzyme system with decreased levels of plasminogen, of plasminogen activator activity both systemically and locally from the vascular endothelium, and of antiplasmin. The final result is generation of soluble fibrin (fibrin monomer complexes), of fibrinopeptides, and of fibrin(ogen) split products, which can be demonstrated in the circulating blood during endotoxemia. Fibrin-rich microthrombi, however, are rarely observed (Bohle *et al.*, 1959; McKay, 1965). This is because soluble fibrin, and fibrin monomer, are usually eliminated from the circulation before they have had a chance to polymerize. The precipitation of fibrin and the formation of microthrombi only occur if critical blood concentrations are exceeded and since fibrin is eliminated via the RES if the cells of the RES are saturated or blocked. Whereas fibrin monomer production is inhibited by heparin, the endotoxin-induced polymerization of preformed fibrin monomer is heparin resistant.

To illustrate when fibrin formation occurs, slow infusion of thrombin in quantities sufficiently large to partially defibrinate a dog rarely causes the overt formation of fibrin or microthrombi. However, in animals pretreated immediately prior to thrombin infusion with endotoxin, with inhibitors of fibrinolysis, or with RES-blocking agents, microthrombi are frequently observed (Gans, 1966a,b). Hence, the presence or absence of fibrin clots sufficiently large to occlude small vessels seems to depend upon the concomitant action of several mechanisms. Fortunately, these rarely occur simultaneously.

The *pyrogenic* response to endotoxin is due to release of endogenous pyrogens from both granulocytes (Murphy, 1967) and macrophages (Page and Good, 1957). The latter are much more important than the former in eliciting this reaction, for febrile responses to endotoxin are not reduced in granulocytopenic patients. In addition, refractoriness or tolerance to endotoxin, defined as the absence of a febrile response after repeated exposures to endotoxin, is observed in the complete absence of neutrophils, again indicating that the macrophage constitutes the major source of pyrogen (Page and Good, 1957).

Considering the significant role that Kupffer cells play in sequestering endotoxin (see above), these cells are expected to contribute significantly to endotoxin's febrile response. This fact was confirmed by Dinarello *et al.* (1968) using isolated Kupffer cells and by Greisman and Woodward (1970) using *in vivo* hepatic perfusion studies in humans. Conversely, "tolerance" to endotoxin is associated with a greatly enhanced clearance of endotoxin from the blood by the Kupffer cell and with diminished release of pyrogens, causing a diminished or absent febrile response (Greisman and Woodward, 1970).

10. HOST IMMUNE RESPONSES TO ENDOTOXIN

The outcome of an infection with a gram-negative organism or the endotoxin derived from it ultimately depends to a significant extent upon the responses the host can muster toward its defense. The mechanisms of host defense by which endotoxin is rendered innocuous have been the subject of extensive investigations. These include antibody production, complement activation, opsonization, the action of natural inhibitors, and phagocytosis.

Endotoxin exerts a marked *adjuvant* effect in that it elicits an enhanced antibody response to antigens. Allison *et al.* (1973) demonstrated that this endotoxin-induced B-cell-stimulating effect is macrophage-mediated. Since endotoxin in appropriate quantities activates the macrophage, it was initially thought that the adjuvant effect of endotoxin was in some way related to the manner of antigen processing by the macrophage prior to antigen transfer to antibody-producing cells. However, this adjuvant effect is absent in thymectomized or nude mice. When endotoxin-treated activated macrophages were found to produce a soluble (secretory) factor that stimulates T cells, this suggested that the endotoxin-induced adjuvant effect is due to an interleukin which stimulates T helper cells. Interestingly, macrophage suppressor activity can also be demonstrated. Recently, evidence has been presented to indicate that the type of response elicited (helper versus suppressor response) depends upon the degree of macrophage stimulation (Babb *et al.*, 1983; Pennline and Herscowitz, 1981). This observation reiterates the significance of trying to define the factors that elicit macrophage activation and their effects, especially those related to endotoxin. Are these also dose dependent, and if so, what effects are elicited by the different amounts of endotoxin?

Endotoxin is also a *mitogen* for B cells, but not for T cells. B cells, particularly those of mice, show increased DNA synthesis, mitosis, division, and differentiation into antibody-producing cells that secrete increased amounts of antibodies, predominantly of the IgM class.

As *immunogens*, both the LPS fraction and lipid A elicit antibody responses. As a T-independent antigen, LPS elicits an IgM response. Its extent depends upon the manner in which endotoxin is prepared. There are several different ways of preparing endotoxin and it is essential that the preparation method be indicated! The lipid A fraction constitutes the core of the endotoxin preparation, hence its hidden location renders it usually less reactive; however, when suitably exposed, or by using preparations derived from polysaccharide-deficient rough strains of gram-negative organisms, antibody responses to lipid A can be elicited. Antibodies against lipid A of one strain commonly cross-react with those of other gram-negative organisms.

Endotoxin activates *complement* via the classical as well as the alternate pathways; hence, C_9 generation is expected to effect cytolysis, a process involving the enzymatic degradation of cell membrane components. This same process incidentally also plays a role in the *in vivo* disintegration of endotoxin itself (Gewurz *et al.*, 1968; May *et al.*, 1972).

The host, through its exposure to the indigenous gram-negative gut flora, acquires an altered susceptibility to gram-negative organisms. Dubos and Schaedler (1960) and Schaedler and Dubos (1962) demonstrated that germfree mice exhibit better growth and *marked resistance to the lethal effects of LPS;* however, their susceptibility to infection is also greatly increased. After exposure to the normal indigenous gut flora, immunocompetent germfree animals, when compared to conventionally reared animals, exhibit a markedly enhanced susceptibility to the lethal effects of endotoxin, suggesting that endotoxin lethality has an immunological basis* (Glode and Rosenstreich, 1976; Rosenstreich *et al.*, 1977). Several mechanisms are involved. First, conventionally reared mice possess suppressor T lymphocytes that reduce B-cell responsiveness to LPS (McGee *et al.*, 1980). Second, as was pointed out above, certain macrophage secretory products elicited by LPS also contribute to endotoxin toxicity.

From certain brief preceding remarks it is quite clear that the RES is heavily involved in the defense against gram-negative sepsis and endotoxemia. Antibody to LPS as such, through its binding to the Fc receptor or as a result of its combination with complement through its binding to the C3 receptor, enhances the uptake of gram-negative organisms by the phagocyte, undoubtedly a significant feature in the defense against gram-negative sepsis. Also, endotoxins bind to cell membranes, including the macrophage membrane. Subsequently these cells undergo specific changes: initially, bacterial endotoxins suppress RES granulopectic activity, but reduced clearance seems to occur only temporarily (Benacerraf and Sebestyen, 1959). Subsequently a stimulation of the macrophage and its activity is observed. The stimulated cells increase in size, show increased ruffled membrane activity, enhanced adhesion to glass, a rise in glucose utiliza-

*The endotoxin resistance in the C3H/HeJ strain of mice, which is genetically unresponsive to LPS, was recently found to involve reactions of endotoxin and phagocytic cells besides a failure of LPS to properly interact with B lymphocytes. Several mediators produced by the macrophage, notably the colony-stimulating activity (CSA), pyrogen, acute-phase serum proteins, and leukocyte endogenous mediator (LEM), appear in the C3H/FeJ mouse, an LPS-responsive closely related strain, but fail to do so in the resistant strain. Subsequently LEM's effect on body temperature, CSA, acute-phase proteins, and fibrinogen present in the endotoxin-sensitive strain appears to be absent in the resistant one (Kampschmidt *et al.*, 1980), suggesting that possibly other manifestations of toxicity may also be mediated by macrophage secretory products or mediators. At the time of writing this chapter, little work had been done with the LPS-resistant mouse strain. Results of these few studies indicate, however, that the use of this strain will eventually provide considerable help to dissect out the different endotoxin effects. Hence, the results of further studies in this area are eagerly awaited.

tion, increased membrane enzyme adenylate cyclase activity, production and release of collagenase, elastase, plasminogen activator, and presently ill-defined cytolytic enzymes, as well as prostaglandins, and an increased number of cytoplasmic granules. These are typical changes associated with increased phagocytosis and pinocytosis, and cause enhanced bacterial and tumor cell killing. In addition, macrophages release a number of mediators (see footnote, p. 133) in response to exposure to endotoxin, notably pyrogen, colony-stimulating activity, luekocyte endogenous mediator, and an insulin-like substance (Filkins, 1980).

The "activated" macrophage is eminently suited for the task of eliminating inordinate amounts of detritis generated *in vivo* by endotoxin, including damaged endothelial cells, platelet aggregates, fibrin monomer, and fibrin(ogen) degradation products. The role of fibronectin (Saba, 1980) in the nonimmunological clearance of fibrin (Lee, 1962; Gans and Lowman, 1967), cell remnants, and other debris (Gans *et al.*, 1966) including thromboplastin (Spaet *et al.*, 1961), thrombin (Gans *et al.*, 1969a; Goodnough and Saito, 1982), etc., by the RES will be discussed in Chapter 12.

Braude *et al.* (1955), using ^{51}Cr-labeled endotoxin, found it to gradually disappear from the blood following its injection. Beeson (1947a,b) was the first to clearly demonstrate that this process occurred by *phagocytosis*. The disappearance rate is quite rapid if small sublethal doses of endotoxin are given. Following administration of larger quantities this process takes place at a much slower pace (Wiznitzer *et al.*, 1959, 1960). A recent quantitative assessment performed in our laboratory demonstrated that the Kupffer cells can inactivate 1.5 μg endotoxin/g liver per hour during a single passage through the rat liver, an amount far in excess of that normally expected to occur during gram-negative sepsis (Yamaguchi *et al.*, 1982).

Details concerning the process of blood clearance are provided by electron microscopic studies that reveal that the macrophage takes up labeled endotoxin by pinocytosis. During the first 15 min, endotoxin is avidly absorbed upon the macrophage membrane; then small membrane invaginations are formed that contain endotoxin. After 45 min, endotoxin is found on the macrophage membrane and in the heterophagosomes; however, at 24 hr, none remains on the membrane (Bona, 1973). The factors that affect the uptake of endotoxin by macrophages are presently unknown; however, we found in heparinized, decomplemented rats, in which we assume that fibrin formation and complement activation did not occur, that endotoxin clearance was essentially the same as that observed in normal rats (Gans and Wendel, 1976).

The critical endotoxin dose, the absolute amount cleared from blood if all phagocytes come into play, is unknown. From previous observations we assume that it is relatively small. After an initial rapid clearance phase, lasting approximately 5–6 min, endotoxin is eliminated slowly from the circulation. This suggests that the actual preempting of these attachment sites on the phagocyte, its subsequent incorporation by the cell, and its intracellular inactivation take place at relatively slow rates. Hence, it has become evident that the rate of endotoxin clearance depends greatly upon the dose of endotoxin present, the status of the RES, the state of tolerance of the individual or animal, in addition to the number of circulating platelets. Thrombocytopenic animals, it should be noted, show a distinct delayed endotoxin clearance for the circulation (Das *et al.*, 1973). Tolerance is characterized by very rapid blood clearance of endotoxin, a process that occurs independent of complement or other serum factors (Gans and Wendel, 1976). Presently it is assumed that it is due to an increased number of receptor sites on macrophages, on a larger macrophage population, or both.

Local proliferation of Kupffer cells has been observed following stimulation by *Corynebacterium parvum* or zymosan. Since phagocytes derive from monocytes (hence, they are bone marrow derived!) (Kelley and Dobson, 1971; Warr and Sljivic, 1974), the question presents itself whether endotoxin, in addition to recruiting granulocytes from bone marrow, also promotes the release of monocytes.

Finally, we present a brief note on the inactivation of endotoxin by serum, a process that can be both physiologic and pharmacologic in origin. Oroszlon *et al.* (1966) demonstrated that the prolonged presence of endotoxin in blood is further enhanced through the endotoxin's association with, and reversible inactivation by certain plasma proteins. Thus, under normal conditions, endotoxin in this form is hard to detect in plasma or serum. Special techniques for its dissociation are required to release the LPS before it can react in certain identification reactions, e.g., the limulus lysate test.

Besides binding to plasma proteins, endotoxins are inactivated by certain plasma/serum enzymes, a process that is greatly enhanced in tolerant rats (Moreau and Skarnes, 1973). These LPS serum inhibitors inactivate more than 1.5 μg *E. coli*/ml serum per hour *in vitro* as determined by the LAL technique (unpublished observation). If this mechanism operates *in vivo* to the same degree as *in vitro*, it must be assumed that strict rules in blood collecting and serum preparation have to be followed to prevent loss of endotoxin activity in blood.

Besides these physiological inhibitors, inhibition of circulating endotoxin also occurs as a result of the action of certain pharmacological agents. Polymyxin B, for instance, a cationic cyclic decapeptide containing both lipophilic and lipophobic groups, provides protection against gram-negative organisms and the endotoxins derived from them. *In vivo* protection by polymyxin B has been demonstrated against endotoxins derived from *E. coli*, *Aerobacter aerogenes*, and *Serratia marcescens*. Interestingly, Cooperstock, who noted the great sensitivity of various endotoxins to as little as 1 μg polymyxin B/ml *in vitro*, also demonstrated that those endotoxins most susceptible were derived from organisms resistant to polymyxin. This suggests that the polymyxin B bactericidal properties and its endotoxin-detoxifying properties are not necessarily directly related.

ACKNOWLEDGMENTS. The author gratefully acknowledges the help received from M. Schillaci, R. Lawlyes, and the members of the Word Processing Center, University of Illinois School of Medicine, Urbana–Champaign.

REFERENCES

Ali, M. V., and Nolan, J. P., 1967, Alcohol induced depression of reticuloendothelial function in the rat, *J. Lab. Clin. Med.* **70:**295.

Allison, A. C., Davies, P. C., and Page, R. C., 1973, Effect of endotoxin on macrophages and other lymphoreticular cells, *J. Infect. Dis.* **128:**S204.

Alving, B. M., Evatt, B. J., Levin, J., Bell, W. R., Ramsey, R. B., and Levin, F. C., 1979, Platelet and fibrinogen production: Relative sensitivities to endotoxin, *J. Lab. Clin. Med.* **93:**437.

André, C., Heremans, J. F., Vaerman, J. P., and Cambiaso, C. L., 1975, A mechanism for the induction of immunological tolerance by antigen feeding: Antigen–antibody complexes, *J. Exp. Med.* **142:**1509.

Asherson, G. L., Zembala, M., Perera, M. A., Mayhew, C. C., and Thomas, W. R., 1977, Production of immunity and unresponsiveness in the mouse by feeding contact sensitizing agent and the role of suppressor cells in the Peyer's patches, mesenteric lymph nodes, and other lymphoid tissues, *Cell. Immunol.* **33:**145.

Babb, J. L., Billing, P. A., Gans, H., and Yamaguchi, Y., 1983, Liver and the immune system: Cellular basis and mechanism of adjuvant and suppressive effects of acute hepatoxin in exposure in inbred mice, *Proc. Soc. Exp. Biol. Med.*

Beer, H., Braude, A. I., and Brinton, C. C., Jr., 1966, A study of particle size, shapes, and toxicities present in Boivin type endotoxin preparation, *Ann. N.Y. Acad. Sci.* **133:**450.

Beeson, P. B., 1974a, Tolerance to bacterial pyrogens. I. Factors influencing its development, *J. Exp. Med.* **86:**29.

Beeson, P. B., 1947b, Tolerance to bacterial pyrogens. II. Role of the reticuloendothelial system, *J. Exp. Med.* **86:**39.

Benacerraf, B., and Sebestyen, M. M., 1959, Effect of bacterial endotoxins on the reticuloendothelial system, *Fed. Proc.* **16:**860.

Bergstein, J. M., and Michael, A. F., Jr., 1973, Renal cortical fibrinolysis activity in the rabbit following one or two doses of endotoxins, *Thromb. Diath. Haemorrh.* **29:**27.

Bernstein, J. D., and Ovary, L., 1968, Absorption of antigens from the gastrointestinal tract, *Int. Arch. Allergy Appl. Immunol.* **33:**521.

Bjoerneboe, M., Prytz, H., and Orskov, F., 1972, Antibodies to intestinal microbes in serum of patients with cirrhosis of the liver, *Lancet* **1:**58.

Bohle, A., Krecke, H. J., Miller, F., and Sitte, J., 1959, Über die Natur des sogenannten Fibrinoids bei der generalisierten Schwartzmanschen Reaktion: Elektronmikroskopische Untersuchungen an Kaninchennieren, in: *First International Symposium on Immunopathology*, p. 339, Schwabe, Basel.

Bona, C. A., 1973, Fate of endotoxin in macrophages, in *Bacterial Lipopolysaccharides* (E. H. Kass and S. M. Wolff, eds.), pp. 63–73, University of Chicago Press, Chicago.

Bradfield, J. W. R., and Souhami, R. L., 1980, Hepatocyte damage secondary to Kupffer cell phagocytosis, in: *The Reticuloendothelial System and Pathogenesis of Liver Disease* (H. Liehr and M. Grun, eds.), p. 165, Elsevier/North-Holland, Amsterdam.

Brambell, F. W. R., 1970, in: *The Transmission of Passive Immunity from Mother to Young*, Elsevier, Amsterdam.

Braude, A. I., Carey, F. J., Sutherland, D., and Zalesky, M., 1955, Studies with radioactive endotoxin. I. The use of Cr^{51} to label endotoxin of *Escherichia coli*, *J. Clin. Invest.* **34:**850.

Braun, W., 1975, Immunologic and antineoplastic effects of endotoxin: Role of membranes and mediation of cyclic adenoisine 3′,5′-monophosphate, *J. Infect. Dis.* **125:**S180.

Broitman, S. A., Gottleib, L. S., and Zamcheck, N., 1964, Influence of neomycin and ingested endotoxin in the pathogenesis of choline deficiency cirrhosis in the adult rat, *J. Exp. Med.* **119:**633.

Brunning, R. D., Woolfrey, B. F., and Schrader, W. H., 1964, Studies with endotoxin. II. Endotoxin localization in the formed elements of the blood, *Am. J. Pathol.* **44:**401.

Cantor, H. M., and Dumont, A. E., 1967, Hepatic suppression of sensitization to antigen absorbed into the portal system, *Nature (London)* **215:**744.

Chapman, C. J., 1942, The use of rabbits for the detection of pyrogenic substances in the solutions for intravenous administration, *Q. J. Pharm. Pharmacol.* **15:**361.

Chase, M. W., 1946, Inhibition of experimental drug allergy by prior feeding of the sensitizing agent, *Proc. Soc. Exp. Biol. Med.* **61**:257.

Clarke, R. M., and Hardy, R. N., 1970, Structural changes in the small intestine associated with the uptake of PVP by the young ferret, rabbit, guinea pig, cat, and chicken, *J. Physiol. London* **209**:669.

Clarke, R. M., and Hardy, R. N., 1971a, Histological changes in the small intestine of the young pig and their relation to macromolecular uptake, *J. Anat.* **108**:63.

Clarke, R. M., and Hardy, R. N., 1971b, Uptake of PVP in the young goat, *J. Anat.* **108**:79.

Clemente, C., Bosch, J., Rodes, J., Arroyo, V., Mas, A., and Maragall, S., 1977, Functional renal failure and haemorrhagic gastritis associated with endotoxemia in cirrhosis, *Gut* **18**:556.

Cluff, L. E., and Berthrong, M., 1953, The inhibition of the local Shwartzman reaction by heparin, *Bull. Johns Hopkins Hosp.* **92**:353.

Colin, R., Grancher, T., and Lemeland, J. F., 1979, Endotoxemia in patients with inflammatory enterocolitis, *Gastroenterol. Clin. Biol.* **3**:15.

Cooperstock, M. S., Tucker, R. P., and Baublis, J. V., 1975, Possible pathogenic role of endotoxin in Reye's syndrome, *Lancet* **1**:1272.

Cornell, R., Walker, W. A., and Isselbacher, K. J., 1971, Small intestinal absorption of horseradish peroxidase: A cytochemical study, *Lab. Invest.* **25**:42.

Cuevas, P., and Fine, J., 1971, Demonstration of the lethal endotoxemia in experimental occlusion of the superior mesenteric artery, *Surg. Gynecol. Obstet.* **133**:81.

Cunningham-Rundles, C., Brandeis, W. E., Good, R. A., and Day, N. K., 1978, Milk precipitins, circulating immune complexes, and IgA deficiency, in: *Secretory Immunity and Infection* (J. R. McGee, J. Mastecky, and J. L. Babb, eds.), p. 523, Plenum Press, New York.

Das, J., Schwartz, A. A., and Folkman, J., 1973, Clearance of endotoxin by platelets: Role in increasing the accuracy of the limulus gelatin test and in combating experimental endotoxemia, *Surgery* **74**:235.

Des Prez, R. M., 1967, The effect of bacterial endotoxin on rabbit platelets. V. Heat labile plasma factor, *J. Immunol.* **99**:966.

Des Prez, R. M., Horowitz, H. I., and Hook, E. W., 1961, Effect of bacterial endotoxin on rabbit platelets, *J. Exp. Med.* **114**:857.

DiLuzio, N. R., and Friedmann, T. J., 1973, Bacterial endotoxins in the environment, *Nature (London)* **244**:49.

Dinarello, C. A., Bodel, P. T., and Atkins, E., 1968, The role of the liver in the production of fever and in pyrogenic tolerance, *Trans. Assoc. Am. Physicians* **81**:334.

Dubos, R. J., and Schaedler, R. W., 1960, The effect of the intestinal flora on the growth rate of mice and on their susceptibility to experimental infection, *J. Exp. Med.* **111**:407.

Elin, R. J., and Wolff, S. M., 1973, Non-specificity of the limulus amebocyte lysate test, *J. Infect. Dis.* **128**:349.

Ferluga, J., and Allison, A. C., 1978, Role of mononuclear infiltrating cells in pathogenesis of hepatitis, *Lancet* **2**:610.

Filkins, J. P., 1970, Bioassay of endotoxin inactivation in the lead sensitized rat, *Proc. Soc. Exp. Biol. Med.* **134**:610.

Filkins, J. P., 1971, Comparison of endotoxin detoxification by leukocytes and macrophages, *Proc. Soc. Exp. Biol. Med.* **137**:1396.

Filkins, J. P., 1973, Hypoglycemia and depressed hepatic gluconeogenesis during endotoxicosis in lead sensitized rats, *Proc. Soc. Exp. Biol. Med.* **142**:915.

Filkins, J. P., 1980, Endotoxin: Enhanced secretion of macrophage insulin-like activity, *J. Reticuloendothelial Soc.* **27**:507.

Filkins, J. P., and DiLuzio, N. R., 1968, Heparin protection in endotoxin shock, *Am. J. Physiol.* **214**:1074.

Fink, P., and Schultze, K. D., 1982, The polyethylene glycol precipitation technique and the particle-counting immunoassay for detection of circulating immune complex-like material in liver cirrhosis and septicemia, *J. Lab. Clin. Med.* **99**:852.

Finkelstein, R. A., 1964, Observations on mode of action of endotoxin on chick embryos, *Proc. Soc. Exp. Biol. Med.* **115**:702.

Fong, J. S., and Good, R. A., 1971, Prevention of the localized and generalized Shwartzman reaction by anticomplementary agent, cobra venom factor, *J. Exp. Med.* **134:**642.

Fox, C. L., Jr., and Lasker, S. E., 1961, Protection by histamine and metabolite in anaphylaxis, scalds, and endotoxin shock, *Am. J. Physiol.* **157:**589.

Fox, R. H., Goldsmith, R., Kidd, D. J., and Lewis, G. P., 1961, Bradykinin as vasodilator in man. *J. Physiol. (London)* **157:**589.

Fulenwider, J. T., Sibley, C., Stein, S. P., Nordlinger, B. M., and Ivey, G. L., 1980, Endotoxemia of cirrhosis: An observation not substantiated, *Gastroenterology* **78:**1001.

Gans, H., 1966a, The thrombogenic properties of EACA, *Am. Surg.* **163:**175.

Gans, H., 1966b, Study of mechanisms that preserve blood fluidity and the effect of their inhibition on thrombogenesis, *J. Surg. Res.* **6:**87.

Gans, H., 1972, Considerations on the etiology of the hepatic injury observed in patients with homozygous alpha I-antitrypsin deficiency, in: *Pulmonary Emphysema and Proteolysis* (C. Mittman, ed.), p. 115, Academic Press, New York.

Gans, H., 1974, Effect of ligation of hepatic artery and occlusion of the portal vein on the development of endotoxemia, *Surg. Gynecol. Obstet.* **139:**689.

Gans, H., 1975, Mechanism of heparin protection in endotoxin shock, *Surgery* **77:**602.

Gans, H., 1978, Effect of Kupffer cell deficiency in the rat: The development of systemic endotoxemia and its effect on the liver, *J. Reticuloendothelial Soc.* **24**(Suppl.):25A.

Gans, H., and Hanson, M., 1964, Fibrinolytic properties of proteases derived from human, dog, and rabbit leukocytes, *Thromb. Diath. Haemorrh.* **10:**379.

Gans, H., and Krivit, W., 1960, Effect of endotoxin shock on the clotting mechanism of dogs, *Am. Surg.* **152:**69.

Gans, H., and Krivit, W., 1961, Effect of endotoxin shock on the variation in the response in different species of animals, *Am. Surg.* **153:**453.

Gans, H., and Lowman, J., 1967, The uptake of fibrin and fibrin degradation products by the isolated perfused rat liver, *Blood* **29:**526.

Gans, H., and Matsumoto, K., 1974, Are enteric endotoxins able to escape from the intestine?, *Proc. Soc. Exp. Biol. Med.* **147:**736.

Gans, H., and Wendel, G., 1976, Evaluation of the possible role of serum factors in the clearance of endotoxins from blood, *J. Surg. Res.* **21:**415.

Gans, H., Lowman, J., and Fahr, G., 1966, Uptake of intact and disintegrated platelets by the isolated perfused rat liver, *Fed. Proc.* **25:**2900.

Gans, H., Subramanian, V., and Tan, B. H., 1968, Selective phagocytosis: A new concept in protein catabolism, *Science* **159:**107.

Gans, H., Stern, R., and Tan, B. H., 1969a, On the *in vivo* clearance of thrombin, *Thromb. Diath. Haemorrh.* **22:**1.

Gans, H., Sharp, H., and Tan, B. H., 1969b, Antiprotease deficiency and familial infantile liver cirrhosis, *Surg. Gynecol. Obstet.* **129:**289.

Gans, H., Mori, K., Lindsey, E., Kaster, B., Richter, D., Quinlan, R., Dineen, P., and Tan, B. H., 1971a, Septicemia as a manifestation of acute hepatic failure, *Surg. Gynecol. Obstet.* **132:**783.

Gans, H., Mori, K., Quinlan, R., Richter, D., and Tan, B. H., 1971b, The intestine as a source of plasminogen activator activity, *Proc. Soc. Exp. Biol. Med.* **136:**627.

Gans, H., Matsumoto, K., and Mori, K., 1972, Antibodies and intravascular clotting in liver cirrhosis, *Lancet* **1:**1181.

Gaynor, E., Bouvier, C., and Spaet, T. H., 1970, Vascular lesions: Possible pathogenetic basis of the generalized Shwartzman reaction, *Science* **170:**986.

Gewurz, H., Mergenhagen, S. E., Nowotny, A., and Phillips, J. R., 1968, Interactions of the complement system with native and chemically modified endotoxins, *J. Bacteriol.* **95:**397.

Gimpel, L., Hodgins, D. S., and Jacobson, E. D., 1974, Effect of endotoxin on hepatic adenylate cyclase activity, *Circ. Shock* **1:**31.

Glode, L. M., and Rosenstreich, D. L., 1976, Genetic control of B cell activation by bacterial lipopolysaccharide is mediated by multiple distinct genes or alleles, *J. Immunol.* **117:**2061.

Good, R. A., and Thomas, L., 1953, Studies on the generalized Shwartzman reaction. IV. Prevention of the local and generalized Shwartzman reaction, *J. Exp. Med.* **97:**871.

Goodnough, L. T., and Saito, H., 1982, Specific binding of activated clotting factors from the circulation, *J. Lab. Clin. Med.* **9**:873.

Greene, R., Wiznitzer, T., Rutenburg, S., Frank, E., and Fine, J., 1961, Hepatic clearance of endotoxin absorbed from the intestine, *Proc. Soc. Exp. Biol. Med.* **108**:261.

Greisman, S. E., and Woodward, C. L., 1970, Mechanism of endotoxin tolerance. VII. The role of the liver, *J. Immunol.* **105**:1468.

Gruen, M., Liehr, H., and Rasenack, U., 1977, Significance of endotoxemia in experimental galactosamine hepatitis in rats, *Acta Hepato-Gastroenterol.* **24**:64.

György, P., 1954, Antibiotics and liver injury, *Ann. N.Y. Acad. Sci.* **57**:925.

Halliday, R., 1955, Absorption of antibodies from immune sera by gut of young rat, *Proc. R. Soc. London Ser. B* **143**:408.

Harviger, J., Harviger, A., and Timmons, S., 1975, Endotoxin sensitive membrane component of human platelets, *Nature (London)* **256**:925.

Hemmings, W. A., and Williams, E. W., 1978, Transport of large breakdown products of dietary proteins through the gut wall, *Gut* **19**:715.

Hollenbeck, J. T., O'Leary, J. P., Maher, J. W., and Woodward, F. R., 1975, An etiologic basis for fatty liver after jejunoileal bypass, *J. Surg. Res.* **18**:83.

Horn, R. G., and Collins, R. D., 1968, Studies on the pathogenesis of the generalized Shwartzman reaction: The role of granulocytes, *Lab. Invest.* **18**:101.

Horowitz, H. I., Des Prez, R. M., and Hook, E. W., 1962, Activation of platelet factor 3 by bacterial endotoxin in rabbits, *Fed. Proc.* **21**:66.

Hume, D., Mendez, G., Gayle, W. E., Smith, D. H., Abouna, G. M., and Lee, H. M., 1971, Current methods of support of patients in hepatic failure, *Transplant. Proc.* **3**:1525.

Jacob, H. I., Goldberg, P. K., Bloom, N., Degensheim, G. A., and Kozinn, P. J., 1977, Endotoxin and bacteria in portal blood, *Gastroenterology* **72**:1268.

Joel, D. D., Larssue, J. A., and LeFabre, M. E., 1978, Distribution and rate of ingested carbon particles in mice, *J. Reticuloendothelial Soc.* **24**:477.

Kagnoff, M., 1978, Effects of antigen feeding on intestinal and systemic immune responses, after antigen feeding, *Cell. Immunol.* **40**:186.

Kaiser, G. A., Ross, J., Jr., and Braunwald, E., 1964, Alpha and beta adrenergic receptor mechanisms in systemic venous bed, *J. Pharacol. Exp. Ther.* **144**:156.

Kampschmidt, R. F., Pulliam, L. A., and Upchurch, H. F., 1980, The activity of partially purified leukocytic endogenous mediator in endotoxin-resistant C3H/HeJ mice, *J. Lab. Clin. Med.* **95**:616.

Kaufman, H. S., and Hobbs, J. R., 1970, Immunoglobulin deficiencies in an atopic population, *Lancet* **2**:1061.

Kelley, L. S., and Dobson, E. L., 1971, Evidence concerning the origin of liver macrophages, *Br. J. Exp. Pathol.* **52**:88.

Keraan, M. M., Meyers, O. L., Engelbrecht, G. H., Hickman, R., Saunders, S. J., and Terblanche, J., 1974, Increased serum immunoglobulin levels following portacaval shunt in the normal rat, *Gut.* **15**:468.

Kliman, A., and McKay, D. G., 1958, The prevention of the generalized Shwartzman reaction by fibrinolytic activity, *Arch. Pathol.* **66**:719.

Kobold, E. E., Lucas, R., and Thal, H. P., 1964, Chemical mediators released by endotoxin, *Surg. Gynecol. Obstet.* **118**:807.

Koczar, L. T., Bertok, L., and Varteresz, V., 1969, Effect of bile acids on the intestinal absorption of endotoxin in rats, *J. Bacteriol.* **100**:220.

Krahlenbuhl, J. P., and Campiche, M. A., 1969, Early stages of intestinal absorption of specific antibodies in the newborn: An ultrastructural, cytochemical, and immunological study in the pig, rat, and rabbit, *Cell Biol.* **42**:345.

Leach, B. E., and Forbes, J. C., 1941, Sulfonamide drugs as protective agents against carbon tetrachloride poisoning, *Proc. Soc. Exp. Biol. Med.* **48**:361.

Lee, L., 1962, Reticuloendothelial clearance of circulating fibrin in the pathogenesis of the generalized Shwartzman reaction, *J. Exp. Med.* **115**:1065.

Lerner, R. G., Rappaport, S. I., and Spitzer, J. M., 1968, Endotoxin induced intravascular clotting: The need for granulocytes, *Thromb. Diath. Haemorrh.* **20**:430.

Lerner, R. G., Rappaport, S. I., Siemens, J. K., and Spitzer, J. M., 1969, Disappearance of fibrinagen I^{131} after endotoxin: Effect of a first and second injection, *Am. J. Physiol.* **214**:532.

Lerner, R. G., Goldstein, R., and Cummings, G., 1971, Stimulation of human leukocyte thromboplastic activity by endotoxin, *Proc. Soc. Exp. Biol. Med.* **138**:145.

Levin, J., Poore, T. E., Lander, N. P., and Oser, R. S., 1970, Detection of endotoxin in the blood of patients with sepsis due to gram negative bacteria, *N. Engl. J. Med.* **283**:1313.

Levin, J., Tomasulo, P. A., and Oser, R. S., 1971, Detection of endotoxin in human blood and the demonstration of an inhibitor, *J. Lab. Clin. Med.* **75**:903.

Liehr, H., Grun, M., Brunswig, D., and Sautter, T. H., 1975, Endotoxemia in liver disease: Treatment with polymyxin B, *Lancet* **1**:810.

Liu, Y. K., 1975, Phagocytic capacity of reticuloendothelial system in alcoholics, *J. Reticuloendothelial Soc.* **25**:605.

McGee, J. R., Kiyono, H., Michalek, S. M., Babb, J. L., Rosenstreich, D. L., and Mergenhagen, S. E., 1980, Lipopolysaccharide regulation of the immune response: T lymphocytes from normal mice suppress mitogenic and immunogenic responses to L.P.S., *J. Immunol.* **124**:1603.

McKay, D. G., 1965, Disseminated intravascular coagulation: An intermediary mechanism of disease, Harper & Row, New York.

McKay, D. G., and Shapiro, S. S., 1958, Alterations in the blood coagulation system induced by bacterial endotoxin I *in vivo* (generalized Shwartzman reaction), *J. Exp. Med.* **107**:353.

McKay, D. G., Gitlin, D., and Craig, J. M., 1959, Immunochemical demonstration of fibrin in the generalized Shwartzman reaction, *Arch. Pathol.* **67**:270.

McKay, D. G., Mueller-Berghaus, G., and Cruse, V., 1969, Activation of Hageman factor by ellagic acid in the generalized Shwartzman reaction, *Am. J. Pathol.* **54**:393.

Mattingly, J. A., and Wachsman, B. H., 1978, Immunologic suppression after oral administration of antigen I specific suppressor cells formed in rat Peyer's patches, after oral administration, after sheep erythrocytes, and their systemic migration, *J. Immunol.* **121**:1878.

May, J. E., Kane, M. A., and Frank, M. M., 1972, Host defense against bacterial endotoxemia: Contribution of the early and late components of complement to detoxification, *J. Int. Dis.* **128**:(1):168.

Merskey, C., Johnson, A. J., Pert, J. H., and Wohl, H., 1964, Pathogenesis of fibrinolysis in defibrination syndrome: Effect of heparin administration, *Blood* **24**:701.

Merskey, C., Kleiner, G. J., and Johnson, A. J., 1966, Quantitative estimation of split products of fibrinogen in human serum: relation to diagnosis and treatment, *Blood* **28**:1.

Mitchison, N. A., 1964, Induction of immunological paralysis in two zones of dosages, *Proc. R. Soc. London Ser. B* **61**:275.

Moake, J. L., Kageler, W. V., and Cimo, P. L., 1977, Intravascular hemolysis, thrombocytopenia, leukopenia, and circulating immune complexes after jejunalileal bypass surgery, *Ann. Intern. Med.* **86**:576.

Moeschl, P., Lubee, G., Keiler, A., and Kreuzer, W., 1978, Is endotoxin containing ascites a contradiction for peritoneo-venous shunt?, *International Research Communic. Service* **6**:454.

Moon, H. W., Kohler, E. M., and Whipp, S. C., 1973, Vacuolation: A function of cell age in porcine ileal absorptive cells, *Lab. Invest.* **28**:23.

Moreau, S. C., and Skarnes, R. C., 1973, Host resistance to bacterial endotoxemia: Mechanisms in endotoxin-tolerant animals, *J. Infect. Dis.* **128**:114.

Mori, K., Matsumo, K., and Gans, H., 1973, On the *in vivo* clearance and detoxification of endotoxin by lung and liver, *Ann. Surg.* **177**:159.

Mueller-Berghaus, G., and Schneberger, R., 1971, Hageman factor activation in the generalized Shwartzman reaction induced by endotoxin, *Br. J. Haematol.* **21**:513.

Mueller-Berghaus, G., Goldfinger, D., Margaretten, W., and McKay, D. G., 1967, Platelet factor 3 and the generalized Shwartzman reaction, *Thromb. Diath. Haemorrh.* **18**:726.

Murphy, P. A., 1967, Quantitative aspects of the release of leukocyte pyrogen from rabbit blood incubated with endotoxin, *J. Exp. Med.* **126**:763.

Nolan, J. P., and Ali, M. V., 1973, Endotoxin and the liver. II. Effect of tolerance on carbon tetrachloride-induced injury, *J. Med.* **4**:28.

Nolan, J. P., Hare, D. K., McDevitt, J. J., and Ali, M. V., 1977, *In vitro* studies of intestinal endotoxin absorption. I. Kinetics of absorption in the isolated gut sac, *Gastroenterology* **72:**434.

Olcay, I., Kitahama, A., Miller, R. M., Drupanas, T., Trejor, A., and DiLuzio, N. R., Reticuloendothelial dysfunction and endotoxemia following portal vein occlusion, *Surgery* **75:**64.

Oroszlon, S., McFarland, V. W., Mora, P. T., and Shear, M. J., 1966, Reversible inactivation of an endotoxin by plasma proteins, *Ann. N.Y. Acad. Sci.* **133:**622.

Owen, R. L., 1977, Sequential uptake of horseradish peroxidase by lymphoid follicle epithelium of Peyer's patches in the normal unobstructed mouse intestine: An ultrastructural study, *Gastroenterology* **72:**440.

Owen, R. L., and Jones, A. L., 1974, Epithelial cell specialization within human Peyer's patches: An ultrastructural study of intestinal lymphoid follicles, *Gastroenterology* **66:**89.

Page, A. R., and Good, R. A., 1957, Studies on cyclic neutropenia, *Am. J. Dis. Child.* **94:**623.

Pennline, K. J., and Herscowitz, H. B., 1981, Dual role of alveolar macrophages in humoral and cell mediated immune responses: Evidence for suppressor and enhancing functions, *J. Reticuloendothelial Soc.* **30:**205.

Peters, R. L., Gay, T., and Reynolds, T. B., 1975, Post-jejunoileal-bypass hepatic disease: Its similarity to alcoholic hepatic disease, *Am. J. Clin. Pathol.* **63:**318.

Phillips, L. L., Margaretten, W., and McKay, D. G., 1968, Changes in the fibrinolytic enzyme system following intravascular coagulation induced by thrombin and endotoxin, *Am. J. Obstet. Gynecol.* **100:**319.

Pieroni, R. E., Broderick, E. J., Bundeally, A., and Levine, L., 1970, A simple method for the quantitation of submicrogram amounts of bacterial endotoxin, *Proc. Soc. Exp. Biol. Med.* **133:** 790.

Potter, L. T., and Axelrod, J., 1962, Properties of norepinephrine storage particles of rats, *Pharmacol. Exp. Ther.* **142:**299.

Prytz, H., Holst-Christensen, J., Korner, B., and Liehr, H., 1976, Portal venous and systemic endotoxemia in patients without liver disease and systemic endotoxemia in patients without liver disease and systemic endotoxemia in patients with cirrhosis, *Scand. J. Gastroenterol.* **11:**857.

Rappaport, S. I., and Hjort, P. F., 1967, The blood clotting properties of rabbit peritoneal leukocytes *in vitro, Thromb. Diath. Haemorrh.* **17:**222.

Ravin, H. A., Rowley, D., Jenkins, C., and Fine, J., 1960, On the absorption of bacterial endotoxin from the gastro-intestinal tract of the normal and shocked animal, *J. Exp. Med.* **112:**783.

Reichard, M. S., 1967, Factors that diminish radiation lethality, *Radiology* **89:**501.

Reichard, M. S., 1972, R.E.S. stimulation and transfer of protection against shock, *J. Reticuloendothelial Soc.* **12:**604.

Rose, E., Espinoza, L. R., and Osterland, C. K., 1977, Intestinal bypass arthritis: Association with circulating immune complexes and HLAB27, *J. Rheumatol.* **4:**129.

Rosenstreich, D. L., Glode, L. M., and Mergenhagen, S. E., 1977, Action of endotoxin on lymphoid cells, *J. Infect. Dis.* **136:**S239.

Rutenburg, A. M., Sonninblick, E., Koven, I., Aphrahamiam, H. A., Reiner, L., and Fine, J., 1957, The role of intestinal bacteria in the development of dietary cirrhosis in rats, *J. Exp. Med.* **106:**1.

Saba, T. M., 1980, Plasma fibronectin (opsonic glycoprotein): Its synthesis by vascular endothelial cells and role in cardiopulmonary integrity after trauma as related to reticuloendothelial function, *Am. J. Med.* **68:**577.

Schaedler, R. W., and Dubos, R. J., 1962, The fecal flora of various strains of mice, its bearing on their susceptibility to endotoxin, *J. Exp. Med.* **115:**1149.

Selye, H., Tuchweber, B., and Bertok, L., 1966, Effect of lead acetate on the susceptibility of rats to bacterial endotoxins, *J. Bacteriol.* **91:**884.

Seyberth, H. W., Schmidt-Gayk, H., and Hackethal, E., 1972, Toxicity, clearance, and distribution of endotoxin in mice influenced by actinomycin D, cycloheximide, amantin, and lead acetate, *Toxicon* **10:**491.

Shagrin, J. W., Frame, B., and Duncan, H., 1971, Polyarthritis in obese patients with intestinal bypass, *Ann. Intern. Med.* **75:**377.

Shands, J., 1973, Affinity of endotoxin for membrane, *J. Infect. Dis.* **128:**S189.

Shapiro, S. S., and McKay, D. G., 1958, The prevention of generalized Shwartzman reaction with sodium warfarin, *J. Exp. Med.* **107**:377.

Sharp, H. L., Bridges, R. A., Krivit, W., and Freier, E. F., 1969, Cirrhosis associated with alpha 1-antitrypsin deficiency: A previously unrecognized inherited disorder, *J. Lab. Clin. Med.* **73**:934.

Shayer, R. W., 1960, Relationship of induced histidine decarboxylase activity and histamine synthesis to shock from stress or from endotoxin, *Am. J. Physiol.* **198**:1187.

Shwartzman, J., 1928, Studies on B typhosus toxic substances. 1. Phenomenon of local skin reactivity to B typhosus culture filtrate, *J. Exp. Med.* **48**:247.

Simjee, A. E., Hamilton-Miller, J. M., Thomas, H. C., and Brumfitt, W., 1975, Antibodies to *E. coli* in chronic liver diseases, *Gut* **16**:871.

Simmonds, D. J., Hyland, G., Lesker, P. A., Cohen, M., Stein, T., and Wise, L., 1975, The effect of small bowel resection or bypass on the rat skeleton, *Surgery* **78**:460.

Smith, R. T., and Thomas, L., 1956, The lethal effect of endotoxin in chick embryos, *J. Exp. Med.* **104**:217.

Smith, W. W., Alderman, I. M., Schneider, C., and Cornfield, J., 1963, Sensitivity of irradiated mice to bacterial endotoxin, *Proc. Soc. Exp. Biol. Med.* **113**:778.

Spaet, T. H., Horowitz, H. J., Zucker-Franklin, D., Citron, J., and Bierenski, J. J., 1961, Reticuloendothelial clearance of blood thromboplastin by rats, *Blood* **17**:196.

Stetson, C. A., and Good, R. A., 1957, Studies on the mechanisms of the Shwartzman phenomenon: Evidence for the participation of polymorphonuclear leukocytes in the phenomenon, *J. Exp. Med.* **93**:49.

Stumacher, R. J., Kovnat, M. J., and McCabe, W. R., 1973, Limitations of the usefulness of the limulus assay for endotoxin, *N. Engl. J. Med.* **288**:1201.

Suzuki, M., Mikani, T., and Suzuki, S., 1977, Gelation of limulus lysate by synthetic dextran derivatives, *Microbiol. Immunol.* **21**:419.

Swarbrick, E. T., Stokes, C. R., and Soothill, J. F., 1979, Absorption of antigens after oral immunization and the simultaneous induction of specific systemic tolerance, *Gut* **20**:121.

Tabagehai, S., O'Donoghue, D. P., and Bettelheim, K., 1977, *Escherichia coli* antibodies in patients with inflammatory bowel disease, *Gut* **19**:108.

Tarao, K., Moror, T., Nagakura, Y., Ikevchi, T., Suyama, T., Endo, O., and Fukushima, K., 1979, Relationship between endotoxemia and protein concentration of ascites in cirrhotic patients, *Gut* **20**:205.

Thomas, H. C., and Parrott, D. M. V., 1974, The induction of tolerance to a soluble protein antigen by oral administration, *Immunology* **27**:631.

Thomas, H. C., MacSween, R. N. M., and White, R. G., 1973, Role of the liver in controlling the immunogenicity of commensal bacteria in the gut, *Lancet***1**:1288.

Thomas, L., and Good, R. A., 1952, Studies on generalized Shwartzman reaction: General observations concerning phenomenon, *J. Exp. Med.* **96**:605.

Triger, D. R., Alp, M. H., and Wright, R., 1972, Bacterial and dietary antibodies in liver disease, *Lancet* **1**:60.

Triger, D. R., Boyer, T. D., and Levin, J., 1978, Portal and systemic bacteraemia and endotoxemia in liver disease, *Gut* **19**:935.

Volkheimer, G., Schultz, F. H., and Hoffman, I., 1968a, The effects of drugs on the rate of persorption, *Pharmacol.* **1**:8.

Volkheimer, G., Wendlandt, H., and Wagemann, W., 1968b, Beobachtungen zum Persorption-Mechanismus, *Pathol. Microbiol.* **31**:51.

Volkheimer, G., Schultz, F. H., Aurich, I., Strauch, S., Beuthin, K., and Wendlandt, 1968c, Persorption of particles, *Digestion* **1**:78.

Walker, W. A., and Isselbacher, K. J., 1974, Uptake and transport of macromolecules by the intestine: Possible role in clinical disorders, *Gastroenterology* **67**:531.

Walker, W. A., Cornell, R., Davenport, L. M., and Isselbacher, K. J., 1972, Macromolecular absorption: Mechanism of horseradish peroxidase uptake and transport in adult and neonatal rat intestine, *J. Cell Biol.* **54**:195.

Wands, J. R., LaMont, J. T., Mann, E., and Isselbacher, K. J., 1976, Arthritis associated with intestinal bypass procedure for morbid obesity: Complement activation and characterization of circulating cryoproteins, *N. Engl. J. Med.* **294**:121.

Warr, G. W., and Sljivic, V. S., 1974, Origin and division of liver macrophages during stimulation of mononuclear phagocyte system, *Cell Tissue Kinet.* **7**:559.

Warshaw, A. L., Walker, W. A., Cornell, R., and Isselbacher, K. J., 1971, Small intestinal permeability to macromolecules: Transmission of horseradish peroxidase into mesenteric lymph and portal blood, *Lab. Invest.* **25**:675.

Warshaw, A. L., Walker, W. A., and Isselbacher, K. J., 1974, Protein uptake by the intestine: Evidence for absorption of intact macromolecules, *Gastroenterology* **66**:987.

Wilkinson, S. O., Arroyo, V., Gazzard, B. G., Moodie, H., and Williams, R., 1974, Relationship of renal impairment and hemorrhagic diathesis to endotoxemia in fulminant hepatic failure, *Lancet* **1**:521.

Wilson, R., Barry, T. A., and Bealmear, P. M., 1970, Evidence for a toxic substance of bacterial origin in the blood of irradiated mice, *Radiat. Res.* **41**:89.

Wiznitzer, T., Schweinfurt, F. B., Atkins, N., and Fine, J., 1959, On the relation of the size of intraintestinal pool of endotoxin to the development of irreversibility in hemorrhagic shock, *Ann. N.Y. Acad. Sci.* **78**:315.

Wiznitzer, T., Better, N., and Rachlin, W., 1960, *In vivo* detoxification of endotoxin by the reticuloendothelial system, *J. Exp. Med.* **1112**:1157.

Wolfe, D. E., Potter, L. T., Richardson, K. C., and Axelrod, J., 1962, Localizing tritiated norepinephrine in sympathetic axons by electron microscopic autoradiography, *Science* **138**:440.

Worthington, B. S., Meserole, L., and Shyrotuck, J. A., 1978, Effect of daily ethanol ingestion on intestinal permeability to macromolecules, *Dig. Dis.* **23**:23.

Yamaguchi, Y., Yamaguchi, K., Babb, J., and Gans, H., 1982, *In vivo* quantitation of the rat liver's ability to eliminate endotoxin from portal vein blood, *J. Reticuloendothelial Soc.* **32**:409.

II

Regulatory Interactions with the Blood Elements

7

The Reticuloendothelial System and Erythropoiesis

BRIAN A. NAUGHTON and ALBERT S. GORDON

1. INTRODUCTION: GENERAL ROLE OF THE RES IN ERYTHROPOIESIS

A wide diversity of functions has been ascribed to the cells of the reticuloendothelial system (RES) since that term was first coined by Aschoff and Kiyono in 1913. These cells were identified by their ability to phagocytize foreign particles and stain with vital dyes. Although these attributes were recognized by Metchnikoff as early as 1905, it was the association of these phagocytic cells with reticulum (argentophilic) fibers which first led to their definition as a reticuloendothelial *system*. The terms *mononuclear phagocytic system* and *reticulohistiocytic system* are synonymous with *RES*.

The RES forms a significant percentage of the cell populations of all of the hematopoietic organs and has been linked functionally to the destruction of damaged red cells. Most of the erythrophagocytosis in normal animals occurs in RES cells which line the tortuous sinusoids of the spleen (Rous, 1923; Wagner *et al.*, 1962; Rifkind, 1966; Pictet *et al.*, 1969). The relatively slow blood flow through this organ allows the sinusoidal phagocytes to "recognize" and destroy physically damaged erythrocytes (Wagner *et al.*, 1962), erythrocytes sensitized with metal cations (Jandl and Simmons, 1957), antibody-damaged red cells (Cutbush and Mollison, 1958), and erythrocytes containing inclusions such as Heinz bodies (Slater *et al.*, 1968) or Howell–Jolly bodies (Chen-Li and Weiss, 1973). The functional and morphological condition of the erythrocyte is the determining factor in erythrophagocytosis (Pictet *et al.*, 1969). The splenic macrophages possess the metabolic machinery to lyse the old or damaged red cells so that their iron can be recycled and used for the synthesis of new hemoglobin (Crosby, 1959; Finch *et al.*, 1970). Storage and release of iron by these and other RE cells

BRIAN A. NAUGHTON and ALBERT S. GORDON • Department of Biology, New York University, New York, New York 10003.

modulating hemoglobin synthesis is recognized as an important function of this system (Crosby, 1959; Noyes *et al.*, 1960; Rifkind, 1965; Finch *et al.*, 1970).

In addition to the spleen, the bone marrow and liver are also capable of erythrophagocytosis (Muir and Niven, 1935; von Ehrenstein and Lockner, 1958; Hughes-Jones, 1961). In the bone marrow, this function is mainly confined to reticular cells at the center of erythroid islets (Bessis and Breton-Gorius, 1959, 1962). These central cells were believed to transfer products from the digestion of phagocytized, damaged, or defective red blood cells to the developing erythroblasts which surround it by a process termed *rhopheocytosis* (Policard and Bessis, 1958) although this view has not as yet reached universal acceptance. In contrast to the spleen which can phagocytize erythrocytes displaying only slight defects, hepatic macrophages (Kupffer cells) ingest only severely damaged red cells under normal conditions (Rifkind, 1966), although erythrophagocytosis by Kupffer cells may occur after Friend virus infection (McGarry and Mirand, 1973; Orlic and Mirand, 1977), following splenectomy (Skamene *et al.*, 1978), and after pheylhydrazine treatment (Hertzberg and Orlic, 1980).

In addition to the hematologic function of the RES in iron metabolism, these cells have been implicated in the production of humoral substances which affect the development of blood cells. Blood monocytes and hepatic Kupffer cells produce colony-stimulating factor (CSF) which promotes the growth of granulocytes *in vitro* (Joyce and Chervenick, 1975; Naughton *et al.*, 1982c). Alveolar macrophages have been shown to produce interferon (Kono and Ho, 1965; Acton and Myrvik, 1966) and both Kupffer cells and blood monocytes can produce prostaglandins (Chervenick, and LoBuglio, 1972; Kurland *et al.*, 1979). Kupffer cells are also believed to be the cellular site of origin of hepatic erythropoietin (Ep), the hormone which regulates red cell development (Peschle *et al.*, 1976; Naughton *et al.*, 1977a). Several previous studies have supported this contention (Gruber *et al.*, 1977; Naughton *et al.*, 1978; Rich *et al.*, 1980) and recent experiments have provided direct evidence that the Kupffer cell is the cellular site of origin of Ep in the liver (G. Naughton *et al.*, 1984).

A survey of the current literature concerning extramedullary erythropoiesis and extrarenal Ep production and their relation to the RES will be discussed in this chapter. Details of recent experiments performed in our laboratory dealing with the RES and erythropoiesis will also be included.

2. THE ROLE OF THE MACROPHAGE IN ERYTHROPOIESIS

2.1. RETICULAR CELLS OF THE BONE MARROW: MEDULLARY ERYTHROPOIESIS

Although it may be argued that these cells are not RE cells in the classical definition of the term (Aschoff and Kiyono, 1913; Wisse, 1970), reticular cells are fixed members of the hematopoietic organs. With their associated fibers they form hematopoietic organ stroma, an essential requirement for the hematopoietic inductive microenvironment (Trentin, 1970). In the bone marrow, groups of erythroblasts and other developing blood cells surround a centrally located reticular cell (Bessis, 1956; Bessis and Breton-Gorius, 1959; Berman, 1967; Ben-Ishay, 1977). The reticular or nurse cell has been envisioned to support erythropoiesis by transferring iron to the surrounding erythroblasts (Bessis and Breton-Gorius, 1959), and by phagocytizing defective erythroblasts (Bessis and Breton-Gorius, 1959; Ben-Ishay, 1975) and pyknotic nuclei extruded from mature erythroid elements (Cottier *et al.*, 1963; Tavassoli, 1974). Although it was suggested that cells of the erythroid islets may act as stem cells for further hematopoiesis (Zamboni, 1965), evidence has been provided to the contrary (Rifkind *et al.*, 1969). The hypothesis has been suggested that the central reticular cell produces Ep (Naughton *et al.*, 1979a–c; Unanue, 1976) which causes the differentiation of hematopoietic precursor cells into the erythroid line (Stohlman *et al.*, 1968), although more direct evidence for this is required. According to some investigators, erythrophagocytosis by reticular cells is indicative of ineffective erythropoiesis (Sjögren and Brandt, 1974; Cavallin-Stahl *et al.*, 1974) rather than a "nurse cell" function. Intramedullary destruction of developing red blood cells (Goodman *et al.*, 1968), particularly polychromatophilic erythroblasts (Cavallin-Stahl *et al.*, 1974), is an important aspect of megaloblastic anemia, where basophilic erythroblasts are the predominant cell of the erythroid islet. The number of medullary erythroid islets increases as does the phagocytic activity of the bone marrow RE cells after the administration of erythropoietic stimulatory agents such as phenylhydrazine (Ploemacher and van Soest, 1977; Hertzberg and Orlic, 1980) and antineoplastic drugs such as mechlorethamine or vincristine sulfate (Henry *et al.*, 1974). Bone marrow RE activity increased after acute hemorrhage (Okuyama *et al.*, 1975), a primary erythropoietic stimulus. Although no augmentation of the labeling indices of medullary reticular cells was noted after hypoxia (Yoffey and Yaffe, 1980), the erythrophagocytic activity of these cells increased as did the cellular area, possibly due to elevated ferritin accumulation (Ben-Ishay and Yoffey, 1971a,b). In contrast, depression of erythropoiesis by hypertransfusion did little to alter the appearance or activity of bone marrow RE cells (Nelp *et al.*, 1970).

2.2. RETICULAR CELLS OF THE LIVER AND SPLEEN: EXTRAMEDULLARY ERYTHROPOIESIS

Although the bone marrow is the most important site of erythropoiesis in the adult mammal, red blood cell production occurs in the yolk sac (Sorenson, 1961), spleen (Orlic *et al.*, 1965), and liver (Rifkind *et al.*, 1969) of the fetal mammal. As the bone marrow assumes an erythropoietic function during the neonatal–adolescent stage of maturation, extramedullary erythropoiesis gradually disappears (Bessis, 1956; Metcalf and Moore, 1972), although some hematopoietic foci can be found in the adult spleen (Glew *et al.*, 1973). The reinitiation of hepatic erythropoiesis has been demonstrated during various pathological and experimental conditions such as methylcellulose-induced splenomegaly (Joyce *et al.*, 1976), ionizing radiation treatment (Testa and Hendry, 1977), hypoxia (Bozzini *et al.*, 1970), during Friend virus disease in splenectomized (Metcalf *et al.*, 1959) and intact mice (Orlic and Mirand, 1977), hemangioendothelioma in humans (Alpert and Benisch, 1970), and in liver regenerating after partial hepatectomy in the adult rat (Naughton *et al.*, 1978, 1979a). Developing erythroblasts were associated with macrophages or reticular cells in all of these instances.

In our experiments, adult rats were perfused with a 2.5% Millonig-buffered gluteraldehyde solution from 24 to 96 hr after 80–90% subtotal hepatectomy (hepx) (Naughton *et al.*, 1977a). After mincing of the liver tissue, fixation was concluded in a 1% Millonig-buffered OsO_4 solution prior to sectioning, staining in aqueous 0.4% uranyl acetate, and poststaining with citrated lead hydroxide. By 24 hr after hepx, numerous collagen and reticular fibers were noted (Fig. 1A), apparently a repair phenomenon. Parenchymal cell necrosis occurred in the area near the ligature and numerous Kupffer cells were seen to infiltrate this region (Bucher, 1963; Naughton *et al.*, 1978). The parenchymal cells in the perinecrotic area exhibited some loss of cell architecture, especially in the rough endoplasmic reticulum (Fig. 1B). Smooth endoplasmic reticulum, however, was abundant in these regenerating cells (Fig. 1C). Glycogen, which is abundant in the hepatic parenchyma (Stenger and Confer, 1965), was also drastically reduced in these cells. The number of nucleoli per nucleus and nuclei per cell also increased in the perinecrotic parenchymal cells. In contrast, the phagocytic Kupffer cells lining the hepatic sinusoids (Benacerraf, 1964) showed no evidence of degeneration or loss of cellular architecture at the ultrastructural level (Naughton *et al.*, 1978). The endoplasmic reticulum (predominantly rough) of the Kupffer cells is augmented following hepx (Fig. 1D) and secretion droplets which are located in the cytoplasm of perisinusoidal parenchymal cells appear in close apposition to Kupffer cells (Fig. 1E). Under normal conditions, the Kupffer cell population

FIGURE 1. (A) Electron micrograph of two parenchymal cells in the perinecrotic region at 48 hr after hepx. Collagenous fibers (CF) are noted at the junction of these cells in longitudinal as well as cross-section. Numerous ribosomes (R) are present, some of these associated with a disaggregated rough endoplasmic reticulum. Two large ovoid mitochondria (M) are also identified. 47,000 ×. (B) Light photograph of section of tissue in the portal zone of a liver regenerating 48 hr post-hepx. The loss of cytoplasmic basophilia (after H & E staining) in the parenchymal cells (p) to the right has been attributed to loss of endoplasmic reticulum and cytoplasmic RNA (Bucher, 1963). The sinusoids (S) appear dilated and numerous Kupffer cells (k) are found in the necrotic (N) areas. Some fat (f) deposition is also observed in this region. 440 ×. (C) Electron micrograph of the junction of three parenchymal cells in the centrolobular area of liver regenerating 48 hr after hepx. Smooth endoplasmic reticulum (SER) is abundant and numerous mitochondria (M) with well-defined cristae are noted. Cell membrane (CM) repair is completed in these cells and many desmosomes (d) are observed. 30,000 × (D) Electron micrograph denoting the relationship between a hepatic parenchymal cell (P), a Kupffer cell (k), and the space of Disse (D). The portion of the Kupffer cell in contact with the sinusoid (S) contains numerous mitochondria (M) and rough endoplasmic reticulum (RER). Ribosomes (R) occur both singly and in clusters in the cytoplasm of this cell. A mature erythrocyte (E) is found in the sinusoid. 26,500 ×. (E) Electron micrograph of the junction between a Kupffer cell (k) and a hepatic parenchymal cell at 72 hr following hepx. Collagen fibers (CF) are seen in cross-section in the intercellular space. Numerous polyribosomes (R) are observed in the Kupffer cell cytoplasm and associated with a dilated rough endoplasmic reticulum (RER). A secretion droplet (SD) is seen in the parenchymal cell (P) cytoplasm in close apposition to the Kupffer cell. 28,000 ×.

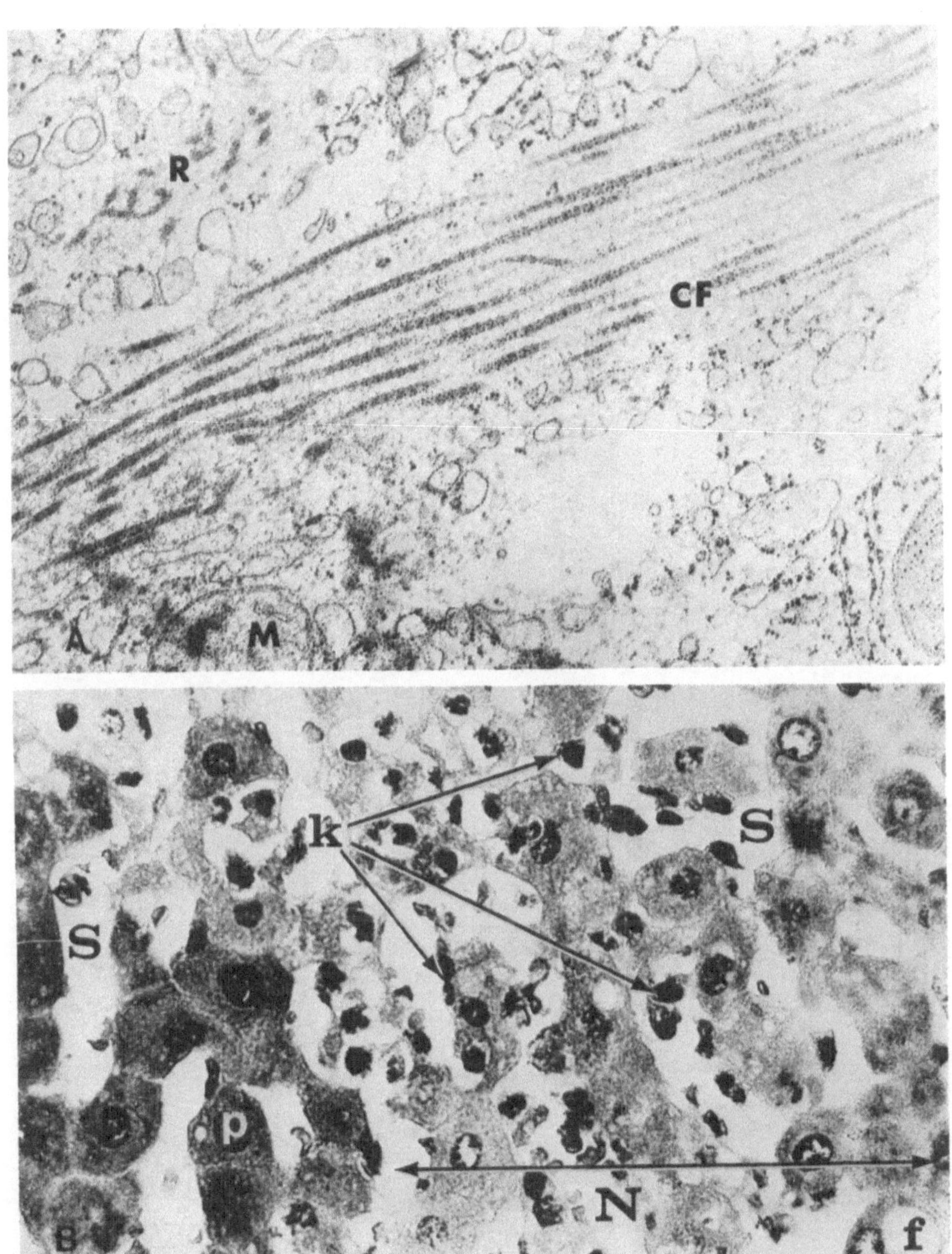
R
CF
A
M
k
S
S
p
N
f

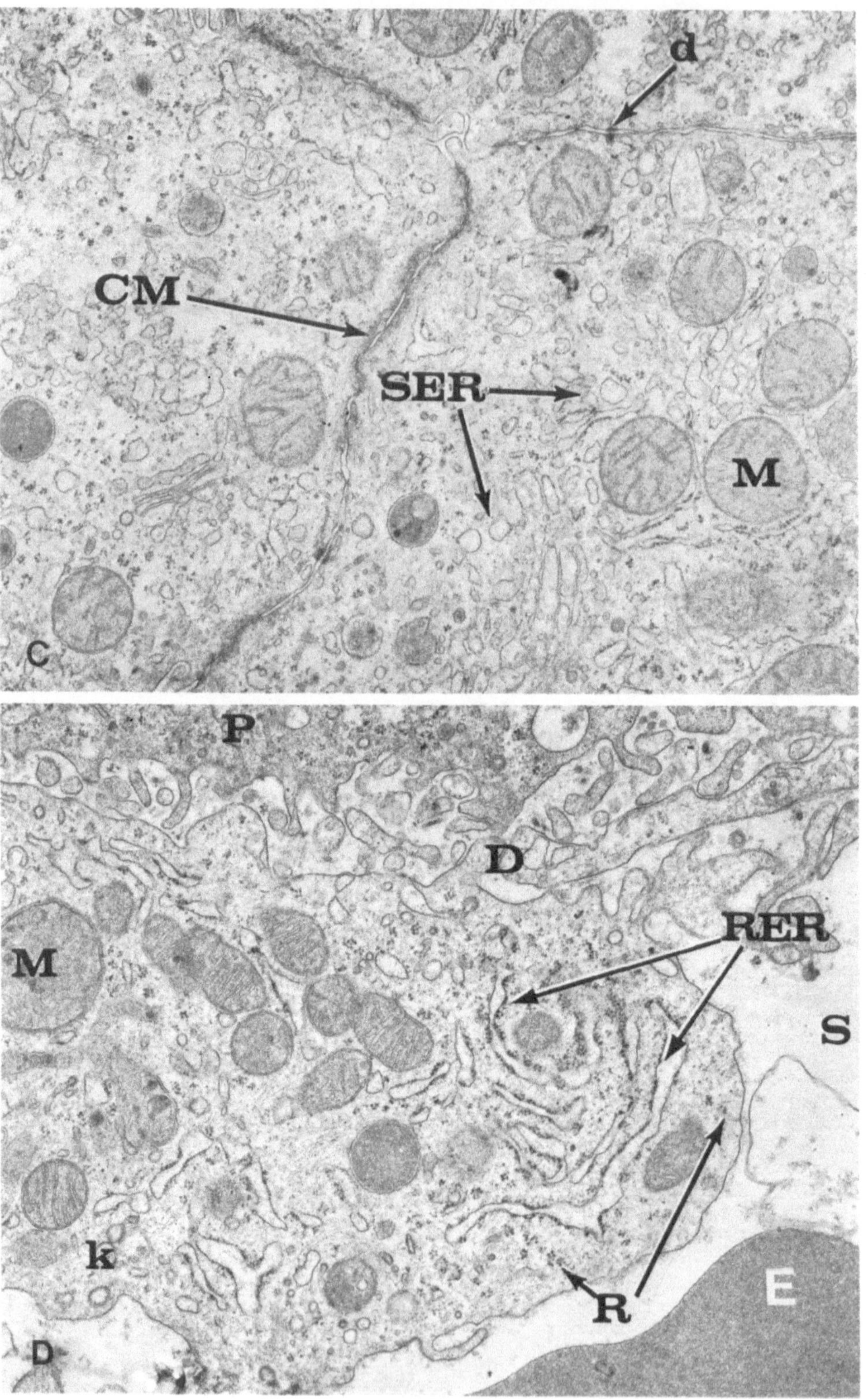

FIGURE 1. (*Continued*.)

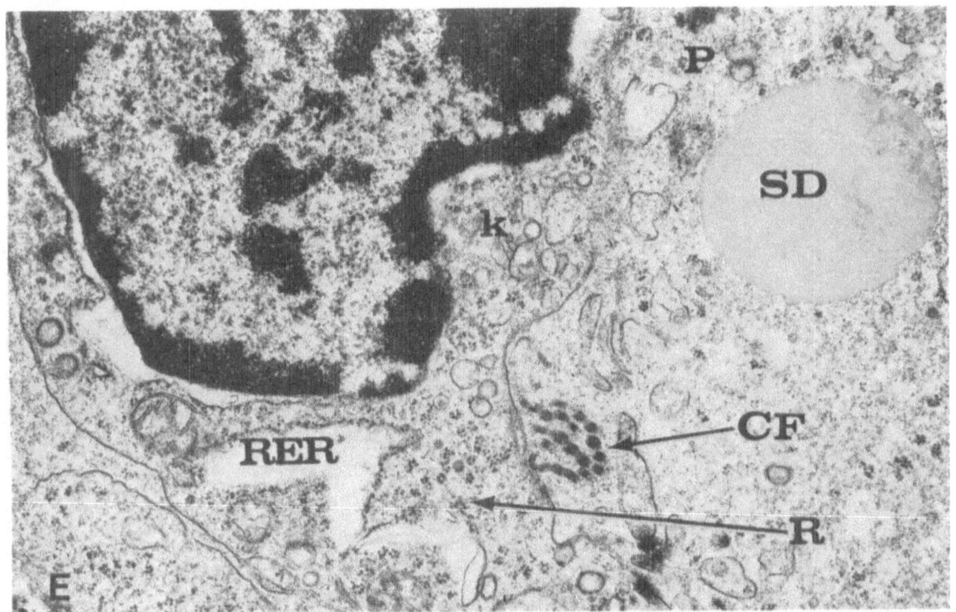

FIGURE 1. (*Continued.*)

only comprises about 15–30% of the total liver cells (Gates *et al.*, 1961) but the ratio of the number of Kupffer to hepatic parenchymal cells increases considerably after hepx (Leong *et al.*, 1959; Bucher, 1963; Naughton *et al.*, 1978, 1979b). The ability of Kupffer cells to phagocytize radioactive chromium-phosphate colloid (Leong *et al.*, 1959) or ^{99m}Tc sulfur colloid (Naughton *et al.*, 1978, 1979b) increases following hepx and remains elevated until hepatic regeneration is near completion. Since hepatic RES clearing capacity is proportional to blood flow at a subblockade dose (Benacerraf *et al.*, 1957), the elevated clearance of radiopharmaceutical colloids may be attributed to the increased blood flow rate per gram of tissue which was noted after hepx (Saba, 1970). However, the rate of flow through the splanchnic circulation diminishes somewhat to compensate for the loss of up to 70% of the functional liver mass (MacKenzie *et al.*, 1975). It seems unlikely that altered blood flow accounts totally for the increased activity of the Kupffer cells after hepx since hypoxia had little effect on ^{99m}Tc sulfur colloid uptake in these studies (Naughton *et al.*, 1978, 1979b,c). Splenectomy, which results in elevated hepatic blood flow, increased Kupffer cell phagocytic activity (Crosby, 1959; Petrelli and Marsila, 1975; Skamene *et al.*, 1978), but this has also been reported in hepatitis (Hymer *et al.*, 1976) and infection (Biozzi *et al.*, 1963; Stiffel *et al.*, 1971) which do not significantly affect hepatic hemodynamics.

In addition to these findings, we discovered hepatic erythropoiesis in the region of the perinecrotic parenchymal cells (Naughton *et al.*, 1979a). This was, in general, extravascular (Fig. 2A) as has been reported in fetal liver (Rifkind *et al.*, 1969) although some instances of sinusoidal erythropoiesis were noted (Fig. 2B). Erythroid islets were prominent in these erythropoietic foci (Fig. 2C) with clusters of erythroblasts surrounding a central reticular cell (Bessis and Breton-

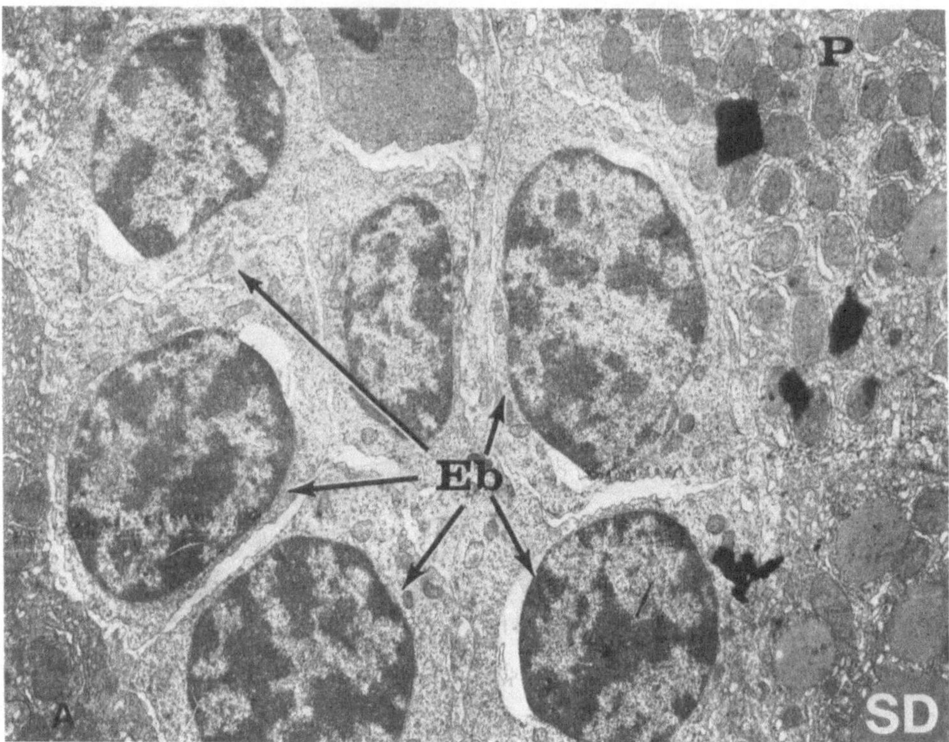

FIGURE 2. (A) Electron micrograph denoting clusters of erythroblasts (Eb) in the extravascular region of liver regenerating 24 hr after hepx. Numerous mitochondria and some secretion droplets (SD) are located in the parenchymal cell (P) cytoplasm. 2300 ×. (B) Electron micrograph of a portion of an erythroid islet located in a sinusoid (S) of liver at 72 hr post-hepx. Cytoplasmic projections of the reticular cell (RC) partially surround an erythroblast (Eb). These cells are associated with a fixed sinusoidal macrophage (Mp), probably a Kupffer cell, 16,000 ×. (C) Low-power electron micrograph of erythropoiesis in liver regenerating at 36 hr after hepx. A reticular cell (RC) with numerous cytoplasmic projections is surrounded by groups of erythroblasts (Eb) at various stages of maturation. Some granulocytic elements (Gr) are noted, an indication that regenerating liver can support granulopoiesis as well. 1000 ×, reproduced at 82½%. (D) Electron micrograph depicting extravascular erythropoiesis in rat liver 48 hr post-hepx. A portion of an erythroid islet is shown with an erythroblast (Eb) and an associated macrophage (Mp), probably a reticular cell. A secretion droplet (SD) and some mitochondria (M) are observed in the parenchymal cell (PC) cytoplasm. 18,500 ×.

Gorius, 1959). In the regenerating liver, satellite-like erythroblasts were found surrounding reticular cells and near the hepatic sinusoids in close proximity to cells resembling mature Kupffer cells. In some cases, the association of erythroblasts with Kupffer cells resembled the extravascular erythroblast– reticular cell conformation (Fig. 2D). Reticular cells of the rat regenerating liver showed positive staining with ferritin-labeled anti-Ep (Fig. 3A). The highly purified antibody used in this study was produced in guinea pigs after immunization with a purified preparation of rat Ep (G. Naughton *et al.*, 1984). Prior to conjugation to ferritin, the guinea pig anti-rat Ep was purified and tested for efficacy by standard immunological techniques including neutralization, immunoelectrophore-

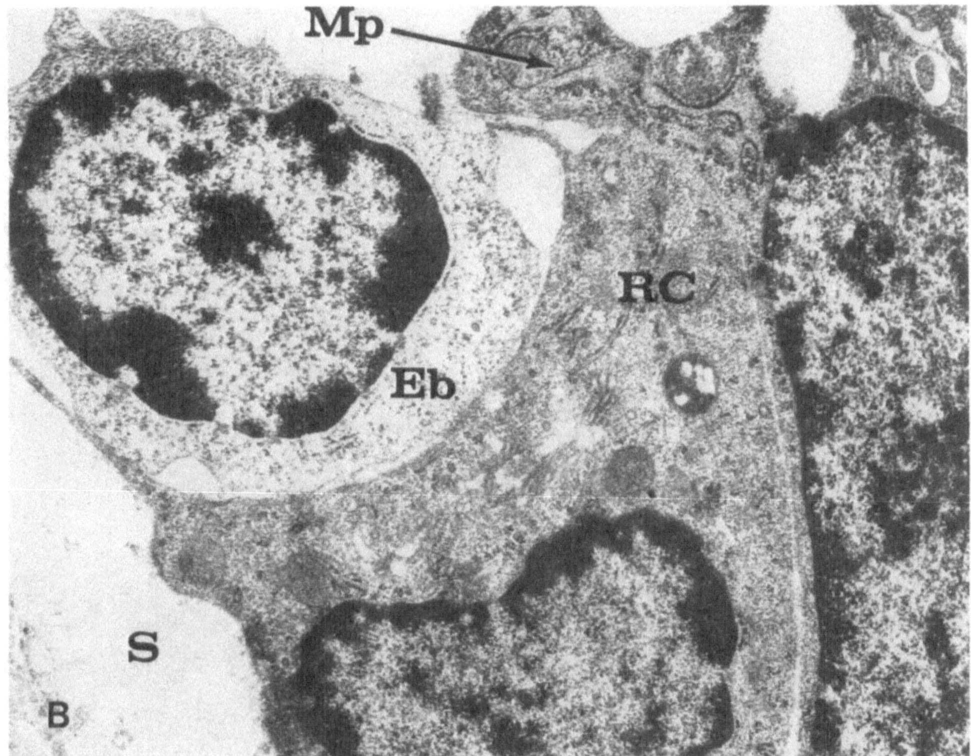

FIGURE 2. (*Continued.*)

sis, and agar gel diffusion. The positive staining reaction indicates that the reticular cell synthesizes and/or stores Ep. An alternate, but unlikely, explanation would be that these cells have phagocytized portions of defective erythroblasts containing bound Ep, e.g., membranes. An intermediary erythroblast with ferritin-labeled anti-Ep deposition on the membrane is seen in Fig. 3B. A more mature erythroblast, located further from the central reticular cell in the "nest," does not appear to have Ep bound to its membrane, since no ferritin deposition is noted (Fig. 3C) (G. Naughton, 1982).

Parenchymal cell secretion droplets were noted in close apposition to the macrophages at the center of erythroid islets (Fig. 2D), similar to that observed for perisinusoidal parenchymal cells and Kupffer cells in the hepx animal (Fig. 1E). Although Kupffer cells can become erythrophagocytic in conditions where spleen function is altered (Petrelli and Marsila, 1975; Stang and Boggs, 1977; Hertzberg and Orlic, 1980), the erythroblast-associated Kupffer cells in this study did not contain any evidence of red blood cell destruction, such as ferritin accumulation or extruded nuclei. The regenerating liver was found to be a site of granulopoiesis (Naughton *et al.*, 1980), also associated with macrophages, although these foci were situated in nonsinusoidal regions. This is in agreement with the observations of Weiss and Chen (1975) and Mohandas and Prenant (1978) for bone marrow hematopoiesis.

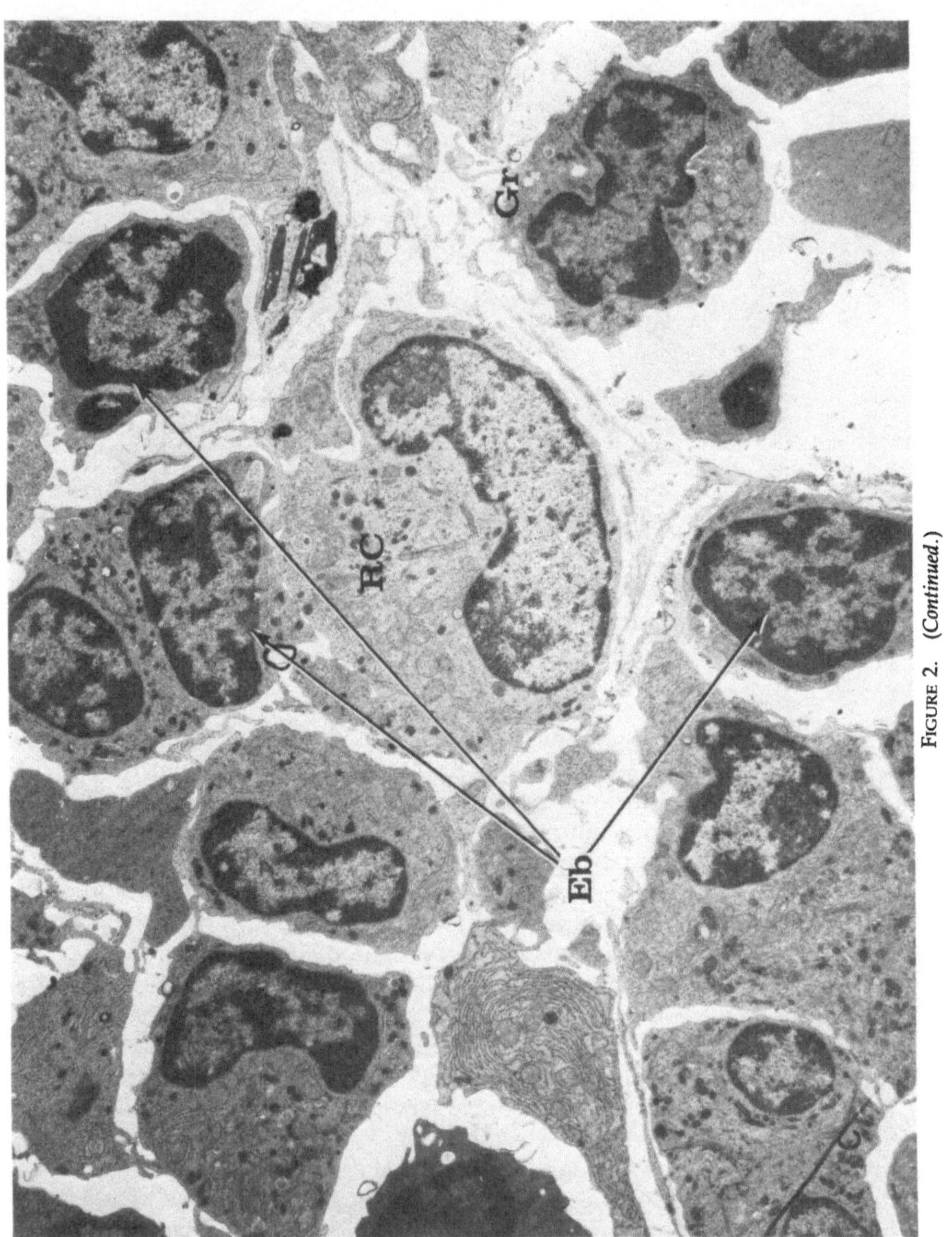

FIGURE 2. (*Continued.*)

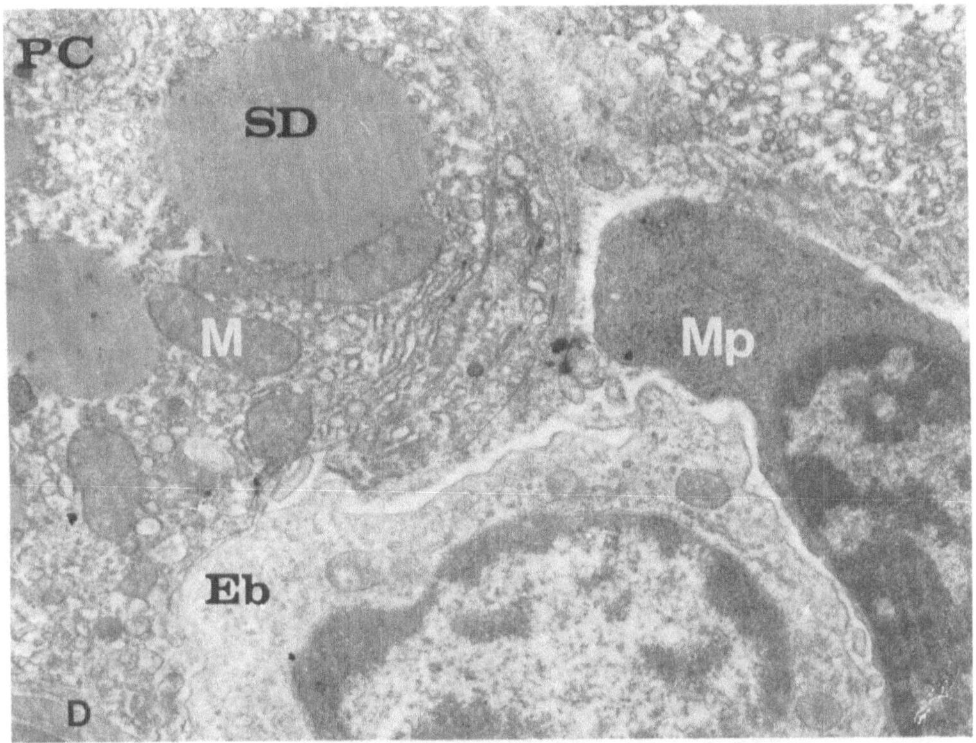

FIGURE 2. (*Continued.*)

Conditions which activate the RE cells appear to favor the reestablishment of extramedullary erythropoiesis in the adult animal. In this regard, the administration of glucan, which activates the RES (Riggi and DiLuzio, 1961), results in the appearance of erythropoietic foci in the adult rat liver (Deimann and Fahimi, 1980) and stimulates *in vitro* hematopoietic cell growth (Niskanen *et al.*, 1978). Activation of the RES with colloidal carbon (Mori and Ito, 1977) or latex particles (Ploemacher and van Soest, 1979) enhances the postirradiation recovery of hematologic cells in spleen as well as bone marrow. Colloidal carbon has also been shown to elevate hepatic Ep production (Peschle *et al.*, 1976) *in vivo* as has silica in cultures of isolated Kupffer cells (Rich *et al.*, 1980). Phenylhydrazine treatment caused an elevation in hepatic and splenic RES activity accompanied by increased erythropoiesis (Hertzberg and Orlic, 1980). Subtotal hepx also activates the RES (Leong *et al.*, 1959; Naughton *et al.*, 1978, 1979b) although the precise reasons for this are as yet unknown. The increased rate of blood flow through the liver of the hepx animal (Saba, 1970) accounts for some of the augmented RES activity. The appearance of degradation products of autophagocytizing hepatic parenchymal cells (Stenger and Confer, 1965) may also

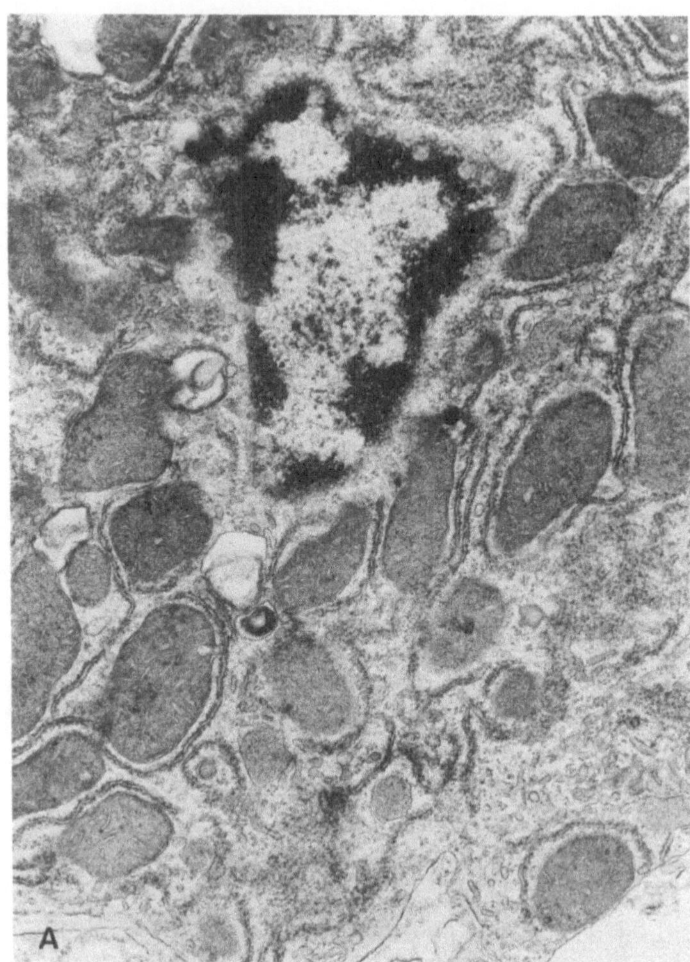

FIGURE 3. (A) Electron micrograph of the central cell of an erythroid islet in the liver of an adult rat at 48 hr after hepx following labeling with guinea pig anti-rat Ep-ferritin. Intracytoplasmic deposition of ferritin is associated with hyperplastic rough endoplasmic reticulum. 20,000 ×, reproduced at 75%. (B) Electron micrograph of the membrane of an early erythroblast, after labeling with ferritin-conjugated guinea pig anti-rat Ep, in the liver of a rat at 48 hr post-hepx. Abundant ferritin deposition on the external part of the cell membrane is noted. 31,920 ×, reproduced at 75%. (C) Electron micrograph of a more mature erythroblast, after labeling with ferritin-conjugated guinea pig anti-rat Ep, in the liver of a rat at 48 hr after hepx. No ferritin deposition is seen on the membrane of this cell which was located at the periphery of an erythroid islet. 12,540 ×, reproduced at 80%.

result in altered RES function. Increased perfusion rate has been associated with elevated erythropoiesis in the spleen (McCuskey *et al.*, 1972) and bone marrow (Trentin, 1970). This phenomenon may be partially responsible for the hepatic erythropoiesis which we noted in adult rats with regenerating livers. The hepatic erythropoiesis in these animals was effective; the absolute reticulocyte count in the peripheral blood increased from 72 to 144 hr following hepx (Naughton *et al.*, 1979a). This is in contrast to the ineffective extramedullary erythropoiesis of

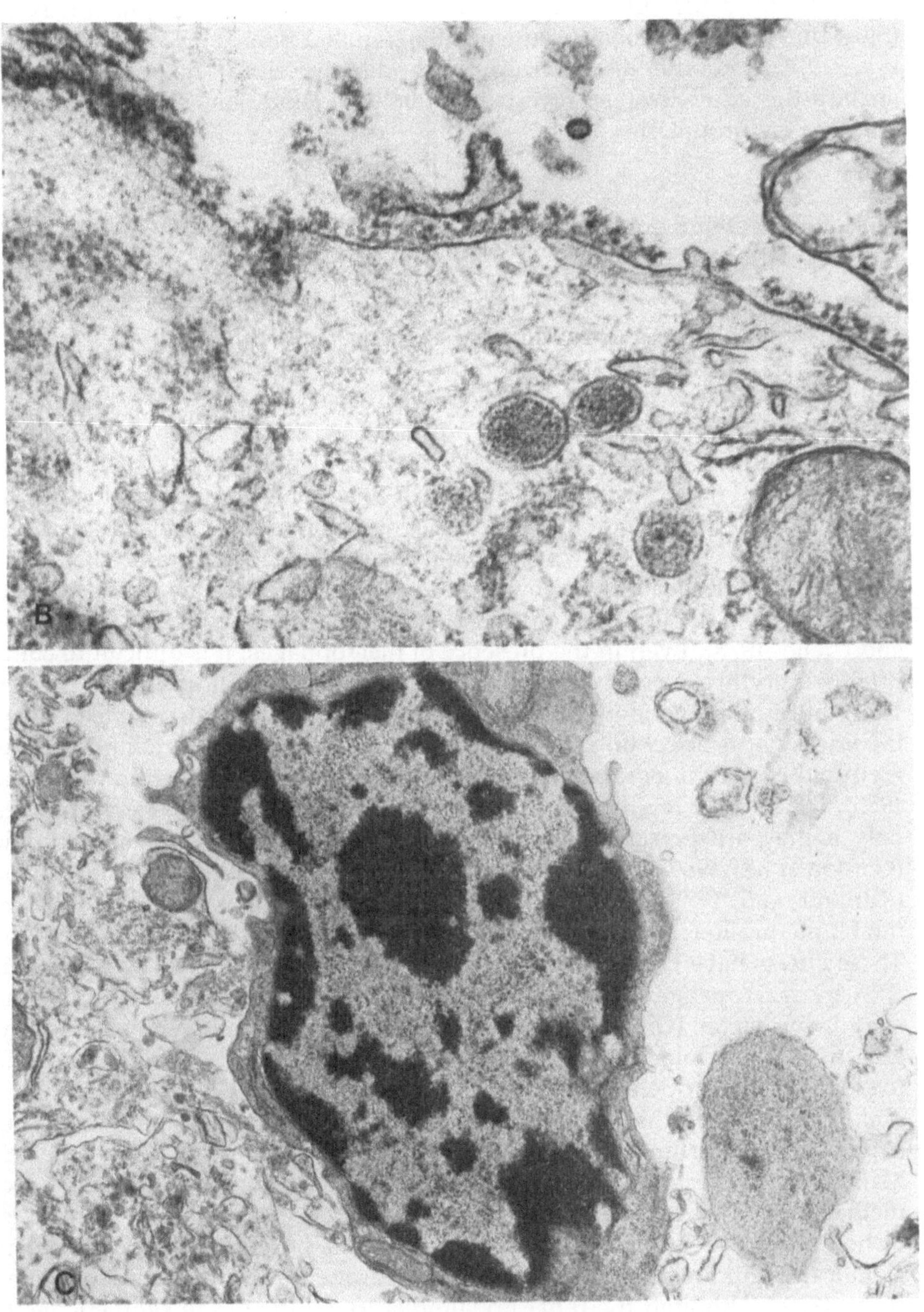

FIGURE 3. (*Continued.*)

pathological conditions where a diminution in bone marrow function is observed (Lord, 1965; Orlic and Mirand, 1977) such as chronic myeloid leukemia (Sjögren, 1976), megaloblastic anemia (Sjögren and Brandt, 1974; Cavallin-Stahl *et al.*, 1974), and other disorders (Metcalf and Moore, 1972). It appears that hepx enables the adult liver to revert to a fetal-like condition and again support effective erythropoiesis.

2.3. HORMONES AND CHEMICAL AGENTS: RELATION OF THE RES TO HEMATOPOIESIS

2.3.1. Macrophages and Erythrocyte Growth *in Vitro*

It is generally acknowledged that erythroblast growth *in vitro* requires the presence of macrophages or factors produced by macrophages (Stephenson *et al.*, 1971; Dunn, 1971; Horland *et al.*, 1977; Murphy and Urabe, 1978). Erythroid colony growth in soft agar inoculated with bone marrow of polycythemia vera patients was dependent on the presence of one or more macrophages per colony (Horland *et al.*, 1977); closely resembling the erythroid islets found *in vivo*. The addition of peritoneal macrophages to cultures of nonadherent bone marrow cells resulted in an elevation of erythroid burst-forming units (BFU-e) (Murphy and Urabe, 1978). Even in the presence of macrophages, it is difficult to culture primitive erythroid cells without adding Ep (Stephenson *et al.*, 1971; Dunn, 1971), but Ep-independent erythroid colony growth has been described following Friend virus infection in mice (Liao and Axelrad, 1975). Although optimum erythroid growth *in vitro* requires macrophages as well as some Ep (Ben-Ishay, 1975), it has been reported that macrophages stimulate *in vitro* erythropoiesis only if they are present in less than 10% of the total cellular concentration (Gordon *et al.*, 1980). In contrast, macrophages inhibit myeloid differentiation (Kurland *et al.*, 1978a), depress lymphocyte transformation (Nelson, 1973), and inhibit the proliferation of neoplastic hematopoietic cells (Kurland *et al.*, 1978a). These effects have been attributed to the production of E series prostaglandins (PG) by macrophages which diminish the response of the CFU-c to colony-stimulating factor (CSF) (Kurland *et al.*, 1978b), the promoter of granulocyte growth *in vitro* (Golde *et al.*, 1972; Joyce and Chervenick, 1975). PG were found to stimulate erythropoiesis (Dukes *et al.*, 1973), especially PGE_2 and PGA_2 (Gross *et al.*, 1976). The erythropoietic effect of PG is apparently mediated through Ep, since anti-Ep abolished their stimulatory action in plethoric mice (Schooley and Mahlmann, 1971). It has been reported that inhibition of PG synthesis by indomethacin resulted in a decreased renal Ep response to hypoxia (Mujovic and Fisher, 1975). To the contrary, inhibition of PG synthesis by aspirin did not significantly diminish the Ep response of hydronephrotic rats to hypoxia (Susic *et al.*, 1979). Although the renal Ep stimulatory effects of the synthetic methylated PG and the naturally occurring E and A series PG have been reported (Lee,

TABLE 1. MEAN EP LEVELS IN THE ANEPHRIC HYPOXIC RAT 24 HR AFTER AN INTRAPERITONEAL INJECTION OF PGA_1, A_2, E_1, E_2, (15s)-15 METHYL E_2, 16-16 DIMETHYL E_2, $F_{2\alpha}$, OR VEHICLE[a]

	Anephric, hypoxic Ep	% Difference from control	Hepx,[b] anephric, hypoxic Ep	% Difference from control
A_1[c]	0.12 ± 0.03[d]	NS	0.18 ± 0.03[d]	NS
A_2	0.28 ± 0.04	+300	0.25 ± 0.04	NS
E_1	0.21 ± 0.03	+200	0.23 ± 0.02	NS
E_2	0.39 ± 0.06	+360	0.28 ± 0.03	+70
(15s)-15 methyl E_2	0.63 ± 0.07	+800	0.33 ± 0.03	+100
16-16 dimethyl E_2	0.48 ± 0.07	+580	0.27 ± 0.02	+70
$F_{2\alpha}$[c]	0.06 ± 0.02	NS	0.09 ± 0.03	NS
Vehicle[e]	0.07 ± 0.03	—	0.16 ± 0.03	—

[a]PG were provided courtesy of Dr. John E. Pike, Upjohn Company, Kalamazoo, Michigan.
[b]Hepx (80–90%) was performed prior to PG injection.
[c]PG were injected in dosages exceeding the minimum effective dose reported by Dukes *et al.* (1975) by 20%. All PG except A_1 and $F_{2\alpha}$ were injected in 125 μg/100 g body wt doses. PGA_1 and $F_{2\alpha}$ were administered in 375 μg/100 g body wt aliquots.
[d]U/ml ± 1 S.E.M.
[e]Olive oil was the vehicle employed (0.5 ml/100 g body wt).
[f]NS = Not significant.

1970; Higgins and Braunwald, 1972; Fisher, 1980), recent evidence from our laboratory indicates that these compounds also stimulate extrarenal Ep production (Table 1), although not to the extent observed in the renal-intact animal (Naughton *et al.*, 1982b). The extrarenal Ep response to PG appears to be mainly hepatic in origin; when 80–90% hepx was performed just prior to PG administration, the extrarenal Ep response was significantly lower than that observed in anephric, hypoxic, PG-injected rats with intact livers (Table 1). Although the physiological mode of action of Ep is affected by PG, it seems unlikely that the production of this principle is mediated solely by these compounds.

Lithium depression of cAMP-mediated PG synthesis (Keighley and Cohen, 1978) blocked the erythropoietic stimulatory action of PGE, and at the same time elevated granulocyte colony formation *in vitro* (Chan *et al.*, 1980). The proposed modulatory effects of PG on hematopoiesis may be due to a feedback action of CSF which has been shown to originate from macrophages (Kurland *et al.*, 1979). In addition, CSF and PGE, which exert opposite effects on erythroid and granuloid hematopoietic stem cells, have been reported to originate from two separate macrophage subpopulations (Kurland *et al.*, 1979). PG may also modulate the cell cycle of cultured macrophages. Treatment of bone marrow cell cultures with indomethacin, a PG synthetase inhibitor, resulted in an elevation in the number of macrophages, although they exhibited a diminished phagocytic capacity (Razin *et al.*, 1980). Several macrophage subpopulations may exist and their role in hematopoiesis may be, at least in part, through their ability to produce various PG which potentiate hematopoietic stem cells to respond to Ep or CSF or stimulate other macrophage subpopulations to produce Ep or CSF.

2.3.2. The RES and Ep

The RES has been implicated as the cellular site of origin of Ep, particularly the Kupffer cells of the liver (Naughton *et al.*, 1976, 1977a, 1979b,c; Peschle *et al.*, 1976; Gruber *et al.*, 1977; Rich *et al.*, 1980). Ep is a glycoprotein hormone which is responsible for the regulation of erythrocyte synthesis (Gordon, 1973). Ep causes the differentiation of hematopoietic precursor cells into the erythroid line (Stohlman *et al.*, 1968) as well as the maturation of early erythroid elements (Wood, 1974) and exerts a dose-dependent effect on erythroid colony growth in bone marrow cultures (Krantz *et al.*, 1963). Radioiron uptake into heme is accelerated by the action of this hormone (Gallien-Lartigue and Goldwasser, 1965) as is the synthesis of some erythrocyte stromal components (Dukes *et al.*, 1964). Ep also increases the growth rate of erythroblasts, as evidenced by elevated tritiated thymidine uptake into DNA of these cells as well as increased mitotic indices after treatment with this principle (Matoth and Kaufmann, 1962; Powsner and Berman, 1967). Early reticulocyte release from the bone marrow is also attributed to the action of Ep (Stohlman, 1961; Gordon *et al.*, 1962). The synthesis of this factor is triggered by a diminished oxygen availability, or hypoxia (Gordon, 1973). The kidney is the major site of synthesis and/or storage of Ep in the adult mammal (Jacobson *et al.*, 1957; Mirand *et al.*, 1959) although extrarenal sources of this principle have been demonstrated (Schooley and Mahlmann, 1972a). The most important site of extrarenal Ep elaboration is the liver (Fried, 1972) which also contains a large population of RE elements, i.e., the Kupffer cells (Gates *et al.*, 1961). Although extrarenal Ep has been reported to account for 10–15% of the total Ep response to hypoxia (Fried, 1972; Schooley and Mahlmann, 1972b), the accuracy of this figure cannot be attested to since the kidneys must be removed prior to this measurement. Renal and extrarenal Ep appear to be immunologically indistinct since treatment of either factor with anti-Ep abolishes their stimulatory effect (Schooley and Mahlmann, 1974).

Although Ep is primarily of renal origin in the adult mammal (Gordon, 1973), the kidneys do not appear to be concerned with fetal (Gordon *et al.*, 1972) or neonatal erythropoiesis (Carmena *et al.*, 1968; Wang and Fried, 1972; Schooley and Mahlmann, 1974). Bilateral nephrectomy (nephrx) does not diminish the Ep response of fetal goats to hypoxia produced by bleeding (Gordon *et al.*, 1972) although a combined hepx and nephrx completely eliminates this Ep appearance (Zanjani *et al.*, 1977). Fetal liver apparently mediates erythropoiesis through Ep production since anti-Ep treatment of final trimester fetuses significantly diminished their red blood cell radioiron incorporation (Zanjani *et al.*, 1974). The liver is not only the primary site of Ep elaboration in the fetus, neonate, and weanling animal but it displays the ability to support erythropoiesis (Rifkind *et al.*, 1969) as well. As the bone marrow assumes an erythropoietic function during the neonatal–adolescent stage of development, extramedullary (hepatic) erythropoiesis disappears (Bessis, 1956) and Ep production becomes progressively renal (Lucarelli *et al.*, 1968). Stimulation of extrarenal Ep sources becomes important in cases of chronic renal insufficiency in man where renal production of this principle has abated (Gordon, 1973; Erslev, 1972) resulting in severe anemia. Detectable levels of Ep have been reported in several anephric human subjects (Naets and Wittek, 1968; Ortega *et al.*, 1977) as well as in anephric animals (Schooley and Mahlmann, 1972a; Gordon, 1973; Naughton *et al.*, 1977a). The action of this Ep, however, may be blocked by an inhibitory factor(s) found in the serum of these patients and animals (Erslev *et al.*, 1971; Ortega *et al.*, 1977). This factor is found not only in renoprival mammals (Neal *et al.*, 1979; Lindemann, 1971; Moriyama *et al.*, 1970), but in rats after hypoxic stimulus (Guduin *et al.*, 1978), following hypertransfusion (Whitcomb and Moore, 1965), and in patients with paroxysmal nocturnal hemoglobinuria (Lewis *et al.*, 1973). An erythrocytic chalone, which specifically inhibits cellular proliferation in the erythron, has been reported as well (Kivilaakso and Rytomaa, 1971; Rytomaa, 1978). Some aspects of the chemical nature of an erythropoiesis inhibitory factor have been reported (Lewis *et al.*, 1973), but it appears that much of the inhibitory effect attributed to the serum of patients with chronic renal disease is due to an effect on erythroid stem cells (CFU-e) by the polyamine, spermine (Radtke *et al.*, 1980). The inhibition due to spermine was dose dependent but did not affect Ep directly. It apparently limits the differentiation of the CFU-e into a more mature stem cell compartment.

Stimulation of extrarenal Ep production has been accomplished mainly by treatment of animals with a variety of agents (Peschle *et al.*, 1976; Ploemacher and van Soest, 1977; Mori and Ito, 1977; Ploemacher and van Soest, 1979; Fried *et al.*, 1979; Deimann and Fahimi, 1980; Liu *et al.*, 1980a) or procedures (Skamene *et al.*, 1978; Naughton *et al.*, 1976, 1977a,b, 1978, 1979a,b; Anagnostou *et al.*, 1977; Flemming, 1963; Liu *et al.*, 1980b; Joyce *et al.*, 1976) which stimulate the RES. Elevated levels of Ep and erythropoiesis were noted after Kupffer cell overload with colloidal carbon (Peschle *et al.*, 1976; Roy *et al.*, 1979), after glucan administration (Deimann and Fahimi, 1980), and after phenylhydrazine treatment (Ploemacher and van Soest, 1977). Phenylhydrazine administration also alters the hepatic contribution to the total Ep response to hypoxia (Dornfest *et al.*, 1983). In this regard, a compensated hemolytic anemia was induced in rats by chronic injection of phenylhydrazine over a 6-week period. Verification of the onset and severity of this anemia was accomplished using standard hematological parameters. A tandem *in situ* perfusion of the livers and kidneys of the experimental rats was performed at different stages of the anemia and the perfusate assayed for Ep. Although the kidney was responsible for most of the total Ep response in the early stages of the anemia, the liver became increasingly more important as the anemia became progressively compensated (after 4–6 weeks of phenylhydrazine treatment) (Fig. 4). In rats administered phenylhydrazine, the Kupffer cells undergo hypertrophy and hyperplasia and erythrophagocytosis is evident (Kent *et al.*, 1969). It seems likely that the increased contribution of the liver to the Ep response is due largely to a stimulatory effect of phenylhydrazine on Kupffer cells. RES activators such as olive oil and triglycerides (Flemming, 1963) caused significant elevation in the hepatic Ep response to hypoxia (Liu *et al.*, 1980a). Alteration of hepatic hemodynamics by splenectomy (Naughton *et al.*, 1977a; Skamene *et al.*, 1978) or splenomegaly induced by methylcelluose administration (Joyce *et al.*, 1976) elevated erythro-

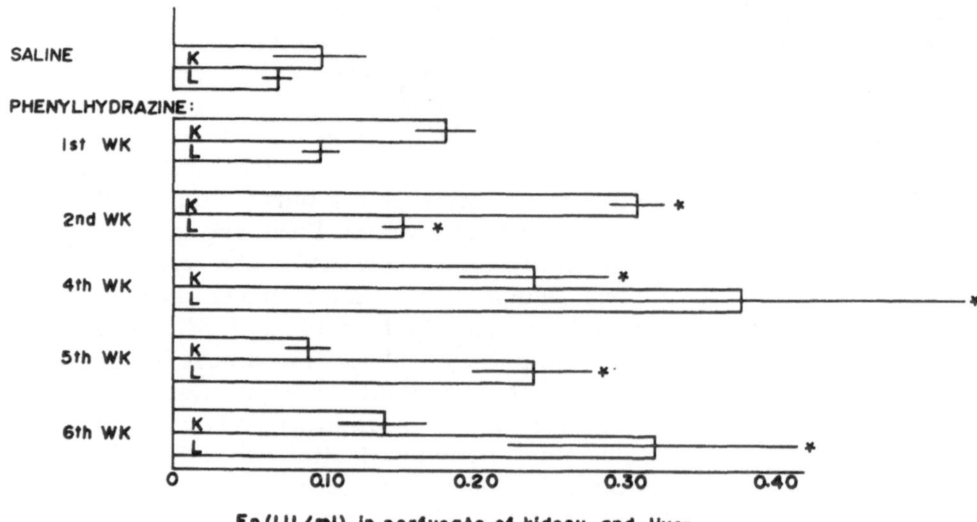

FIGURE 4. Ep levels in the perfusate collected from kidneys (K) and livers (L) of phenylhydrazine-injected rats. Perfusion was performed *in situ* after each organ was excluded from the general circulation (Dornfest *et al.*, 1983). The results were obtained on the third day (72 hr) after the initial phenylhydrazine injection during the first week of treatment and on the first day (24 hr) after each successive injection given during the second, fourth, fifth, and sixth week of treatment. Bars indicate mean values; lines drawn through the top of the bars indicate ± 1 S.E.M. A single asterisk denotes values which are statistically different from comparable control values ($p < 0.05$).

poiesis to some extent. Methylcellulose accumulates in splenic (Glomski, 1969) as well as hepatic (Zuckerman *et al.*, 1984) macrophages, which undergo both hypertrophy and hyperplasia. In the methylcellulose-treated animal, both the liver and the spleen contribute to the extrarenal Ep response to hypoxia (Table 2).

Extrarenal Ep levels were augmented in anephric hypoxic rats after single hepx (Naughton *et al.*, 1976, 1977a; Anagnostou *et al.*, 1977) as well as in rats subjected to repeated partial hepx (Liu *et al.*, 1980b). Hypertrophy and hyperplasia of the Kupffer cells were noted in the livers of hepx rats when compared to sham-operated controls (Leong *et al.*, 1959; Naughton *et al.*, 1977a, 1978, 1979b,c). The cellular and cytoplasmic relative areas of the Kupffer cells increased in direct proportion to the concentration of Ep in the serum of these animals after hepx as measured from histological photographs by planimetry (Eränko, 1955) (Fig. 5; Naughton *et al.*, 1978, 1979b). Kupffer cell areas in the hepx animals decreased significantly after hypoxia, the primary stimulus for Ep production and/or release, was imposed (Fig. 5). This is consistent with the view that hypoxia causes the release of certain substance(s) from these cells; our laboratory has hypothesized that this substance is Ep (Naughton *et al.*, 1978, 1979b,c). After hepx, the increase in Kupffer cell number and size is greatest in cells at the perinecrotic zone nearest the original ligature (Naughton *et al.*, 1979b). The area was also observed to support erythropoiesis (Naughton *et al.*,

TABLE 2. SERUM EP LEVELS IN METHYLCELLULOSE- OR SALINE-TREATED ANIMALS[a]

Group	Ep (IU/ml)
Methylcellulose, nephrx, hypoxia[b]	0.77 ± 0.03[c]
Splenectomy, methylcellulose, nephrx, hypoxia[b]	0.54 ± 0.04
Methylcellulose, simultaneous nephrx and splenectomy, hypoxia[b]	0.53 ± 0.06
Methylcellulose, hypoxia[b]	1.14 ± 0.14
Saline, nephrx, hypoxia[b]	0.08 ± 0.01
Splenectomy, saline, nephrx, hypoxia[b]	0.06 ± 0.01
Saline, nephrx and splenectomy, hypoxia[b]	0.07 ± 0.02
Saline, hypoxia[b]	1.08 ± 0.10

[a]Methylcellulose (20 mg/100 g body wt) was administered chronically (5 injections/week) over a 4-week period.
[b]Hypoxia was applied in a hypobaric chamber at 0.4 atm for 6 hr after which Ep levels were determined using the exhypoxic polycythemic mouse assay.
[c]Mean ± 1 S.E.M.

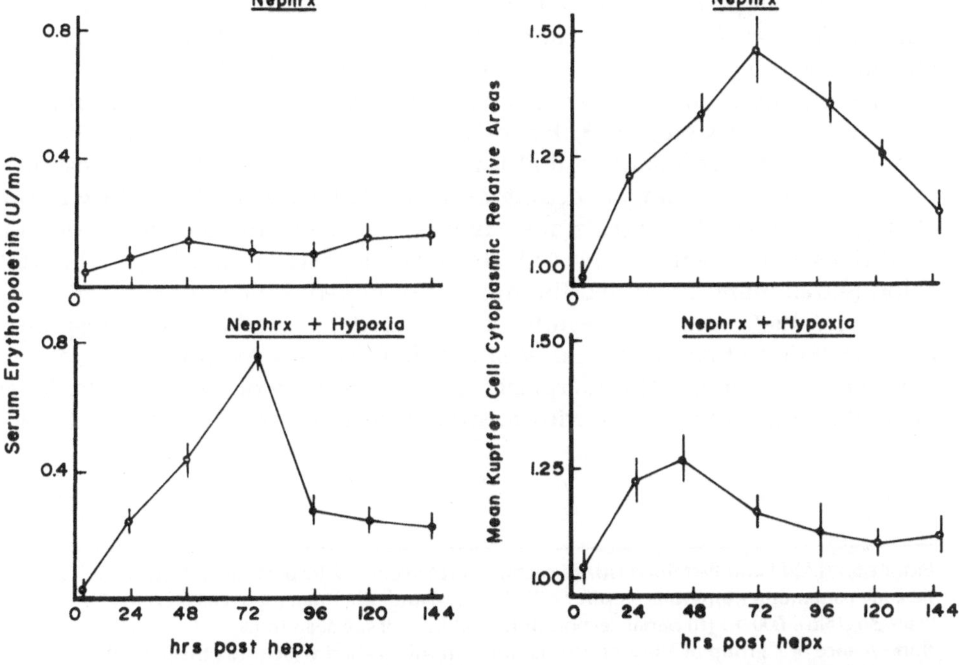

FIGURE 5. Serum Ep and Kupffer cell responses to hepx and nephrx or hepx, nephrx, and hypoxia. Ep levels were determined using the exhypoxic polycythemic mouse assay (Gordon, 1973). Area measurements were performed with a Gelman planimeter. Areas were calculated for equal numbers of cells from experimental and normal animals. The ratio of these measurements was designated as relative area, where relative areas significantly greater than 1 are indicative of hypertrophy. Cytoplasmic relative areas were calculated as the difference between cellular and nuclear relative areas so the data represent only cytoplasmic alterations after the imposed experimental conditions. 1000 cells were counted per point with 5 points per group. Means were calculated from data for 3–5 groups.

1979a) and some granulopoiesis (Naughton *et al.*, 1980). The elevation in Kupffer cell number after hepx (Naughton *et al.*, 1978) cannot be accounted for by mitoses of endogenous Kupffer cells alone since the number of these cells per square millimeter of hepatic tissue is elevated by as early as 18 hr following hepx (Leong *et al.*, 1959; Naughton *et al.*, 1978) which is well below the generation time which has been reported for these cells (Saba, 1970). Although the migration of Kupffer cells within the liver lobule during localized hepatic necrosis has been reported (Parry, 1978) as has Kupffer cell self-proliferation after estrogen stimulation (Kelly and Dobson, 1971), the most likely explanation for this finding is the seeding of the liver with exogenous macrophages or stem cells, presumably from the bone marrow (Mayankii *et al.*, 1978; Gale *et al.*, 1978). Kupffer cells may also be derived from blood monocytes (Kinsky *et al.*, 1969; van Furth, 1980) although this view has not reached unanimity (Wisse, 1970). Monocytes have also been implicated as stem cells for the generation of alveolar macrophages and other fixed tissue cells (Bowden and Adamson, 1980). The hepatic sinusoids apparently offer an ideal microenvironment for macrophages since radiogold-labeled exogenous Kupffer cells (Roser, 1968) as well as peritoneal macrophages (Roser, 1965) seed primarily in the liver of the host. Bone marrow-derived hematopoietic stem cells have been noted in the peripheral blood of humans (Barr *et al.*, 1975; Niskanen *et al.*, 1979) and animals (Lewis *et al.*, 1968). Hematopoietic stem cells have also been seen in fetal liver (Silini *et al.*, 1967) and in regenerating adult rat liver although these are not present in normal adult liver (Varon and Cole, 1966; Hayes *et al.*, 1975). Hepx apparently alters the hepatic microenvironment so that it is favorable to the seeding of these hematopoietic stem cells. Similar seeding of the adult liver with stem cells was noted after ionizing radiation (Testa and Hendry, 1977) and the previously described conditions which activate the RES. Hepatic necrosis, induced by carbon tetrachloride treatment, increased the number of hepatic hematopoietic cells (Varon and Cole, 1966) and also resulted in elevated Ep production in response to hypoxia (Fried *et al.*, 1979; Liu *et al.*, 1980a). These experiments correlate the following conditions: (1) hyperplasia and hypertrophy of Kupffer cells, (2) augmented seeding of the liver with medullary macrophages, (3) elevated seeding

FIGURE 6. (A) H and E-stained frozen section of rat liver at 72 hr post-hepx. Pictured is a perinecrotic zone containing numerous Kupffer cells (arrows) and other macrophages (verified by nonspecific esterase stain). 600 ×. (B) Serial section of the rat liver tissue seen in (A) (72 hr post-hepx). Positive fluorescence of a group of Kupffer cells is noted in this section which was treated with FITC guinea pig anti-rat Ep. This reaction is not positive for all macrophages in this field. 480 ×. (C) H and E-stained frozen section of liver from a rat which was hepx and then nephrx 48 hr later. Pyknosis of some parenchymal cells is evident as is some fatty vacuolization of these cells. Kupffer cells (arrows) are numerous and line the sinusoids. 430 ×. The box outlines the area seen in (D). (D) Guinea pig anti-rat FITC-labeled frozen section of rat liver at 48 hr after hepx and immediately following nephrx. The brightly glowing sinusoidal cells (Kupffer) in this picture are identified by arrows in (C). The large fluorescing area to the right indicates a cluster of Kupffer cells which is also seen in (C). 890 ×.

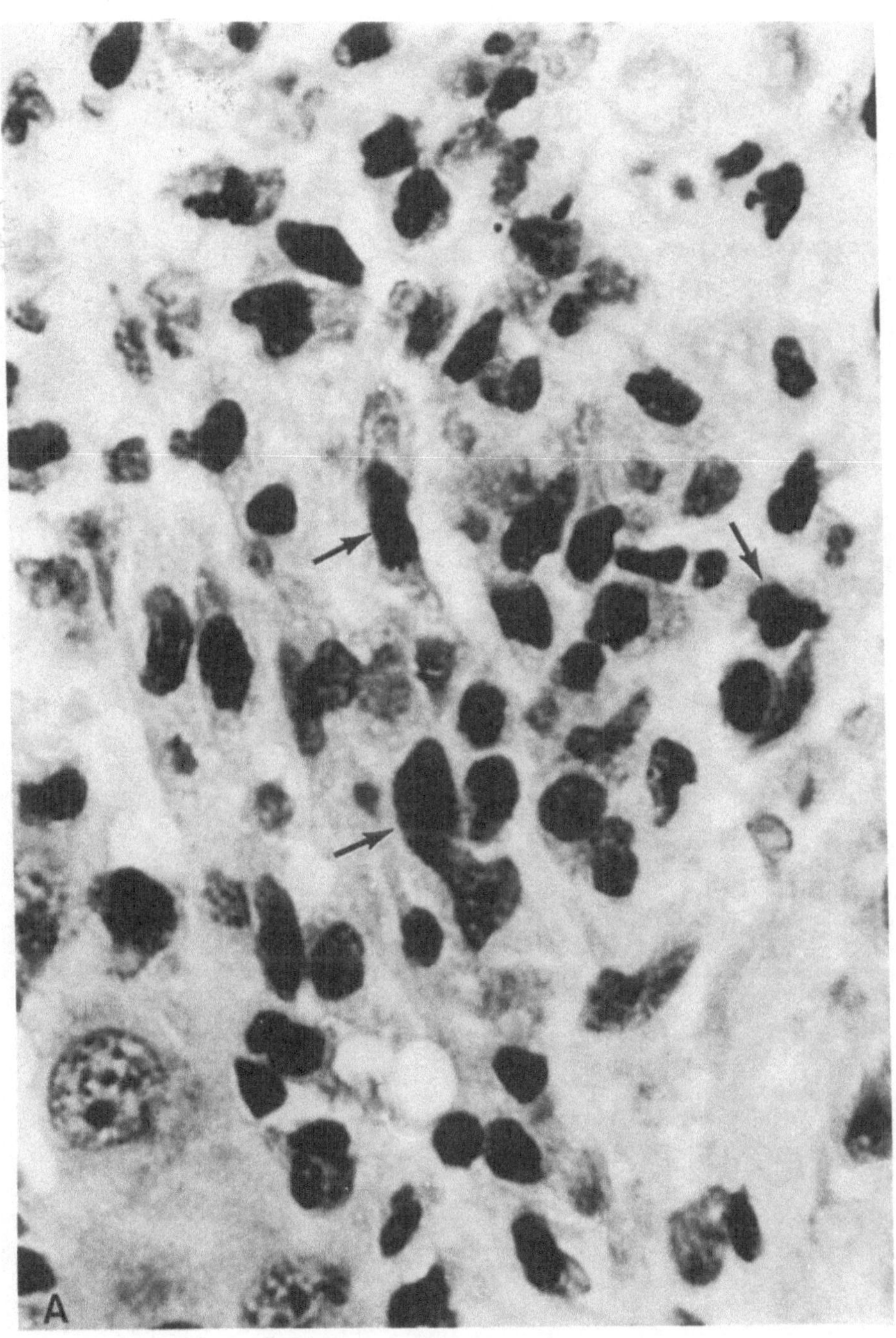
A

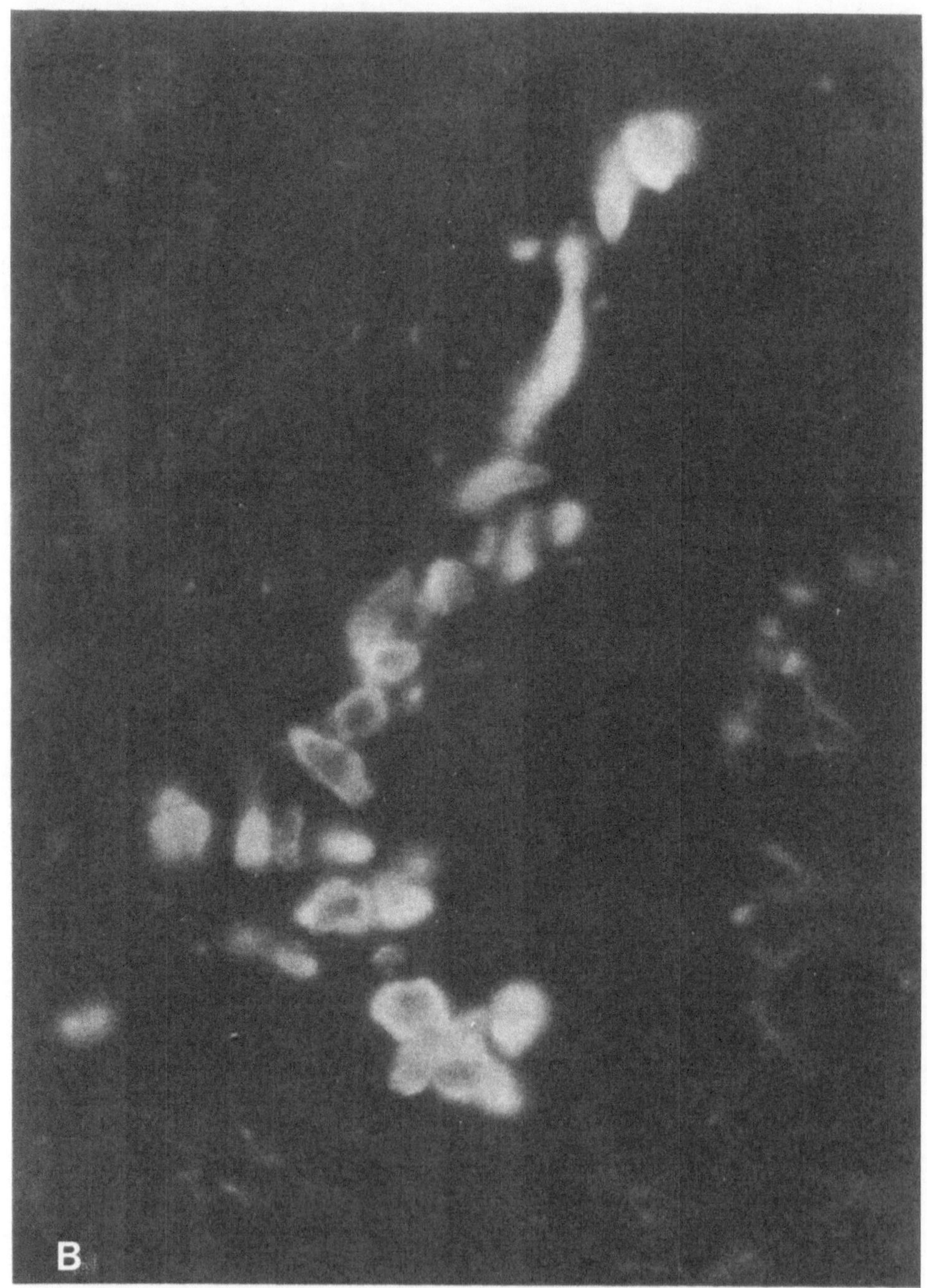

FIGURE 6. (*Continued.*)

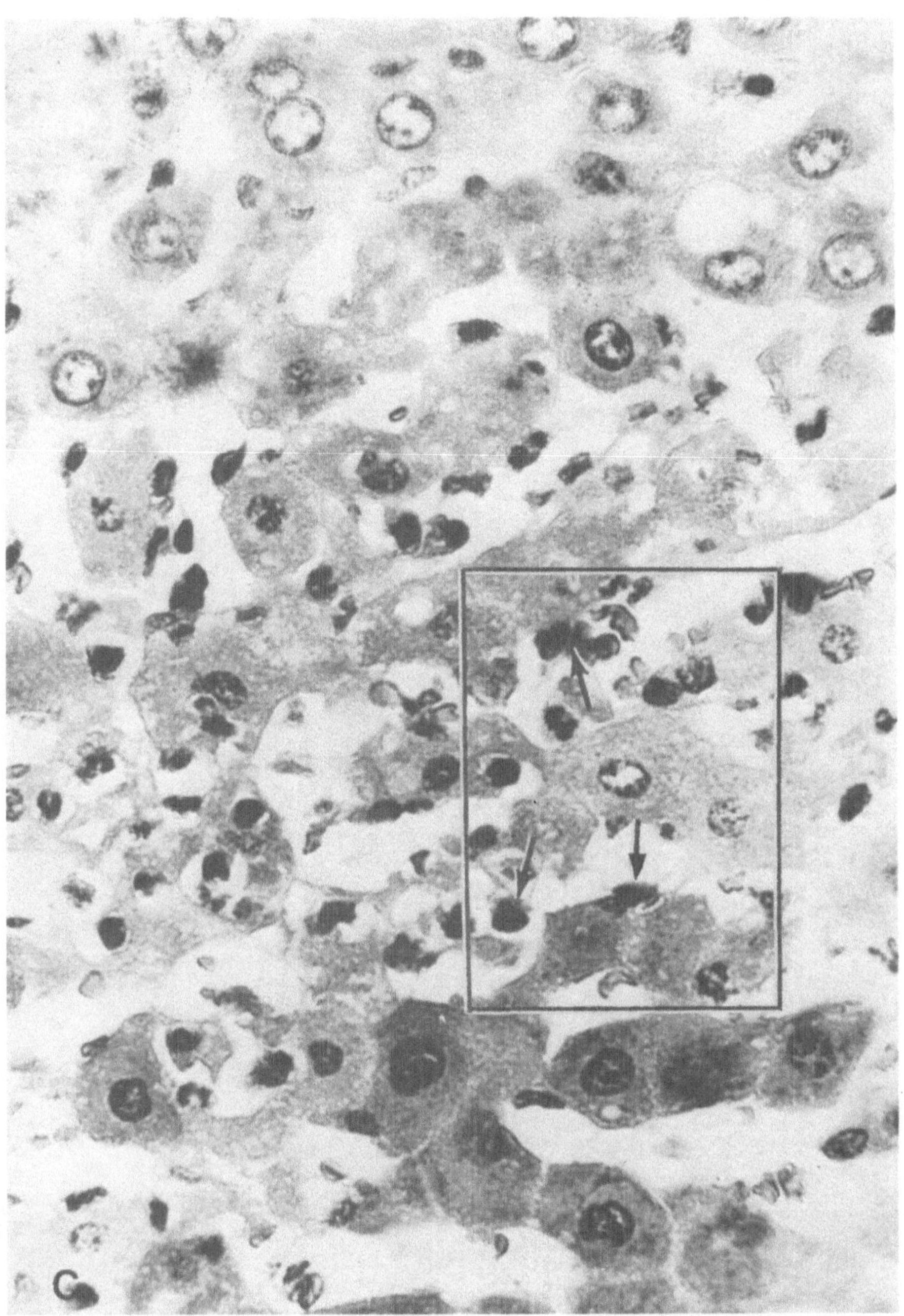

FIGURE 6. (*Continued.*)

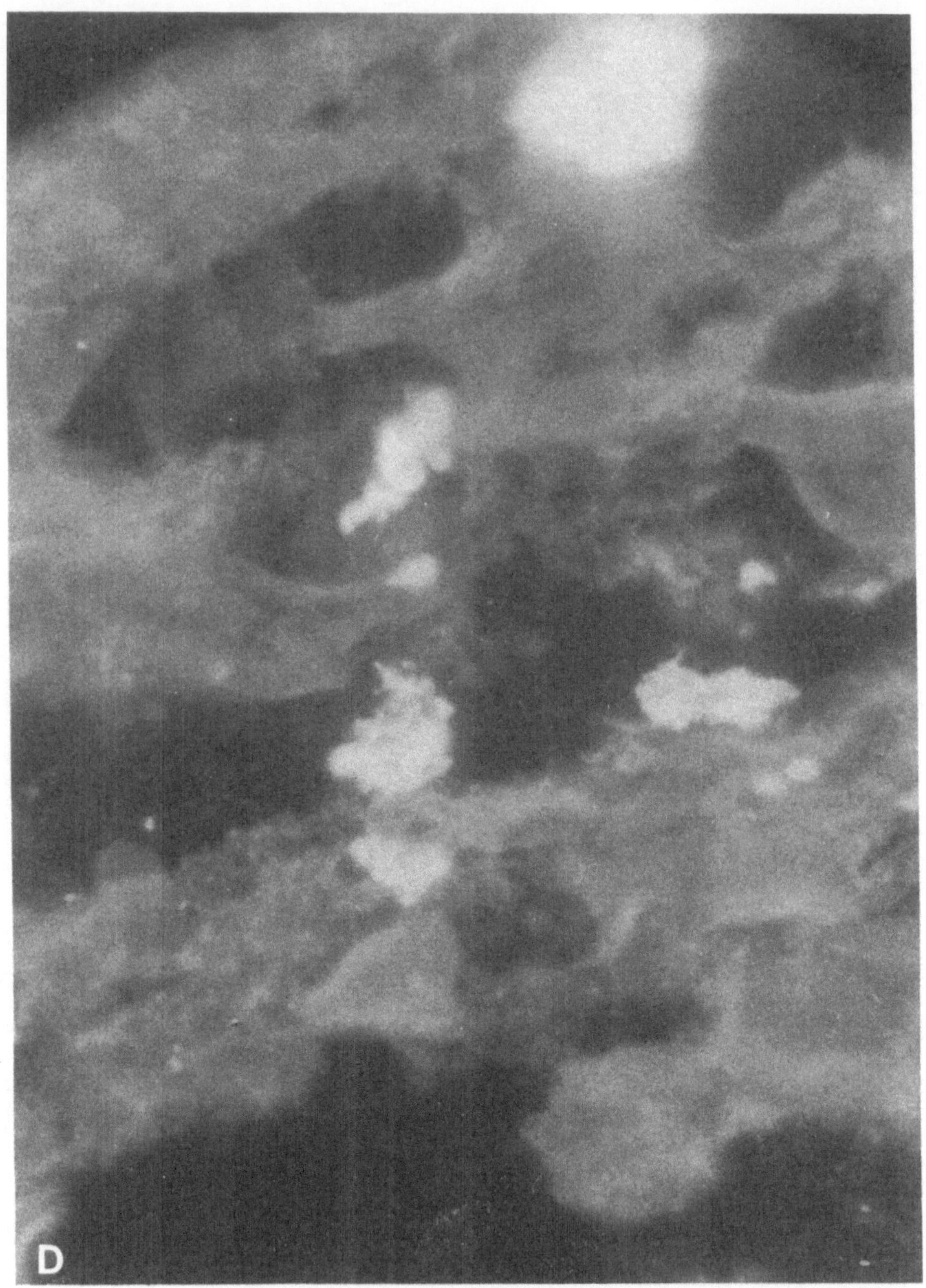

FIGURE 6. (*Continued.*)

of the liver with circulating and bone marrow-derived hematopoietic stem cells, (4) increased production of Ep and CSF, (5) hepatic parenchymal cell necrosis, and (6) a renewed ability of the adult liver to support erythropoiesis and granulopoiesis.

The hypothesis that the Kupffer cell is the site of origin of hepatic Ep (Peschle *et al.*, 1976; Naughton *et al.*, 1976) has been supported by several experiments. Cultured fetal liver Kupffer cells demonstrate positive immunofluorescent staining for Ep (Gruber *et al.*, 1977) using a relatively unpurified anti-Ep. Because of the crude nature of this antibody, further work was necessary to positively identify the cellular site of Ep origin. Similar studies were recently performed in our laboratory. Since rat liver tissue was employed for the localization studies, rat Ep was isolated and purified prior to injection into guinea pigs. This animal is particularly sensitive to rat antigen (Necas *et al.*, 1976). The antibody was then isolated and subjected to purification procedures prior to conjugation with a fluorescent label (FITC). The tissue was serially cryosectioned: sequential sections were labeled with FITC-conjugated guinea pig anti-rat Ep, or treated with hematoxylin and eosin or nonspecific esterase stain (to positively differentiate Kupffer cells from vascular endothelial cells). When liver sections of anephric hypoxic animals with livers regenerating for either 48 or 72 hr post-hepx were treated with the FITC-labeled guinea pig anti-rat Ep, fluorescence of a number of sinusoidal lining cells was seen (Fig. 6B,D). These were identified as Kupffer cells using the nonspecific esterase stain (Koski *et al.*, 1976) (photographs of this reaction are not included here since they do not reproduce well with noncolor plates). However, when compared to the hematoxylin and eosin-stained serial sections (Fig. 6A,C) and the nonspecific esterase-stained sections, not all Kupffer cells in the field exhibited positive fluorescent staining for Ep (Fig. 6A vs. B) (G. Naughton, 1981; G. Naughton *et al.*, 1984). Apparently, not all hepatic macrophages are Ep-producing and/or storing cells. Kupffer cells may be divided into several subpopulations, producing Ep, CSF, and several PG as well. The relative degree of fluorescent staining with the FITC anti-Ep is depicted in Table 3.

TABLE 3. RELATIVE LABELING OF HEPATIC CELLS UPON APPLICATION OF FITC-LABELED GUINEA PIG ANTI-RAT EP[a]

Tissue source	Degree of fluorescence		
	Sinusoidal cells	Parenchymal	Other
Neonate, untreated	0	0	+2 Erythroblasts
Normal rat, untreated	0	0	0
Hepx $\xrightarrow{48\ hr}$ nephrx	+5	0	0
Hepx $\xrightarrow{72\ hr}$ nephrx; 6 hr hypoxia	+4	0	0
Hepx $\xrightarrow{48\ hr}$ nephrx; 6 hr hypoxia	+3	0	0
Hepx $\xrightarrow{72\ hr}$ nephrx; 6 hr hypoxia	+4	0	0
Normal rat; 6 hr hypoxia	+1	0	0
Nephrx; 6 hr hypoxia	+1.5	0	0

[a]0 = negative fluorescence staining reaction; 5 = heavy fluorescence staining reaction.

Fetal mouse liver was found to be capable of producing Ep and CSF although thrombopoietin activity was undetectable (Zucali and Mirand, 1975). Mouse fetal liver macrophages have recently been shown to produce an erythropoietic factor, probably Ep, after stimulation with silica (Rich *et al.*, 1980). Further evidence from our laboratory indicates that Kupffer cells were capable of producing Ep when grown in primary monolayer cultures (Paul *et al.*, 1981) although similar activity was not observed in parenchymal cell cultures. In these experiments, Kupffer cells were characterized by phagocytosis of latex beads, a trypsin-resistant adherence to plastic, and the presence of nonspecific esterase. In contrast, one group has reported that Ep can be produced by the fetal liver parenchymal cell in culture (Congote and Solomon, 1974). Although this view has been stated by others (Erslev *et al.*, 1978), little definitive evidence has been offered in this regard. The reported failure of splenic RE hyperplasia induced by phenylhydrazine to elevate extrarenal Ep levels may not be directly pertinent since the adult spleen produces negligible amounts of Ep (Fruhman, 1970). Indeed, splenic macrophages may not produce significant amounts of Ep under any conditions, either physiological or experimental. Removal of the spleen during hepatic Kupffer cell hypertrophy and hyperplasia induced by hepx had little effect on the Ep response to hypoxia of these animals (Naughton *et al.*, 1977a) although Ep levels were considerably higher than sham-operated controls. Heterogeneous cell populations occur in hepatic as well as other parenchyma (Novikoff, 1959) and have been reported in the littoral cells as well (McPhie, 1979). A similar situation probably exists for RE cells, which synthesize many diverse substances under various physiological conditions and locations. Conditions which result in hyperplasia and/or hypertrophy of hepatic RE cells promote extramedullary erythropoiesis and Ep elaboration whereas activation of splenic RE cells does not appear to alter these parameters. The macrophages in the liver and spleen are morphologically distinct (Rifkind *et al.*, 1969; Wisse, 1970) as are Kupffer cells and peritoneal macrophages (Wisse, 1978). Functional differences have been reported in the erythrophagocytic activities of Kupffer cells and splenic macrophages (Rifkind, 1966; Pictet *et al.*, 1969) as well. The synthesis and/or storage of Ep may be another manifestation of the functional diversity of the hepatic RE versus splenic RE cells.

2.3.3. Hepatic Regeneration: Relation of the RES to Ep Production

Hepatic regeneration has been well documented and the subject of several reviews (Harkness, 1957; Bollman, 1961; Bucher, 1963; Morley, 1974; La Brecque, 1979). It is generally agreed that this process is under humoral control (Moolten and Bucher, 1968; Fisher *et al.*, 1971; Morley, 1974) although some dissenting views have been reported (Islami *et al.*, 1959; Rogers *et al.*, 1961). When normal rats were joined parabiotically to hepx partners, the incidence of liver cell mitoses in the normal parabiont was much higher than those normal animals which were joined to sham-hepx partners (Bucher *et al.*, 1951; Moolten and Bucher, 1968). In addition, serum from hepx animals (Hayes *et al.*, 1969) or humans (Demetriou *et al.*, 1974) elevated the growth of hepatic cells *in vitro*. This serum can stimulate livers which are already regenerating (Adibi *et al.*, 1959), fetal liver DNA synthesis (Paul *et al.*, 1972), and isolated livers when used in the perfusion system (Levi and Zeppa, 1971). Biochemical analysis of this serum has revealed hepatotrophic factor(s) (Morley and Kingdon, 1973; La Brecque, 1979; Goldberg *et al.*, 1980) including α-fetoprotein (Matray *et al.*, 1972). The splanchnic organs which feed the hepatic portal vein have been cited as the origin of the hepatotrophic substances which regulate liver regeneration (Mann *et al.*, 1931; Lee *et al.*, 1968; Starzyl *et al.*, 1973, 1979; Bucher and Swaffield, 1973, 1975). This view, however, is a matter of some dispute (Weinbren, 1978). Liver regeneration in fowl is not dependent on portal blood flow (Thomson and Clarke, 1965) and congenital absence of the portal vein in a human did not diminish the ability of the liver to regenerate (Marois *et al.*, 1979). It appears that these portal vein substances, notably insulin and glucagon, are important metabolic mediators in the process of liver regeneration since in the absence of these substances and/or the organs which produce them, hepatic regeneration is merely slowed, not abolished (Price, 1976; Bucher and Weir, 1976). Experiments performed in our laboratory confirmed these observations. Although insulin treatment resulted in some augmentation of the normal hepatic Ep response to hypoxia, it did not affect Ep elaboration by the regenerating liver (Naughton *et al.*, 1982a). The absence of an action of insulin on the hepx rat was to be expected. Serum insulin titers (Strecker *et al.*, 1980) and the number of hepatic insulin receptors (Bachmann *et al.*, 1979) decrease markedly after hepx. In addition, the rapidly regenerating hepatocytes utilize much of the available blood glucose (Bucher and Malt, 1971) which is the primary stimulus for insulin release (Lacey, 1977). In contrast, glucagon increased the hepatic Ep response by the hepx animal, while reducing Ep synthesis by the normal liver during hypoxia (Naughton *et al.*, 1982a). This may be due, in part, to the reported inhibitory action of glucagon on hepatocellular respiration (Fredlund *et al.*, 1972) which promotes hepatic hypoxia even though it augments blood flow through the hepatic portal vein (Proctor *et al.*, 1980), thereby maximizing oxygenation of other tissues. Recently, the PG system has been hypothesized as a possible stimulator/regulator of liver regeneration (Kanzaki *et al.*, 1979). Although the trophic effect of these compounds on fibroblast synthesis has been reported (Jimnez de Asua *et al.*, 1975; Murota *et al.*, 1977), further substantiation of the influence of PG on hepatic regeneration is required.

Control of liver growth may be due not only to stimulator substance(s) but an inhibitor as well. Inhibition of hepatic regeneration occurs in normal serum- or plasma-treated hepx animals (Smythe and Moore, 1958; Leong *et al.*, 1964), and hepx rats when parabiotically joined to normal partners exhibited lower hepatic cell mitotic indices when compared to nonparabiosed hepx rats (Pechet *et al.*, 1963). The regulation of hepatocellular proliferation by chalones has been hypothesized (Bullough, 1964; Scaife, 1970) and a low-molecular-weight entity capable of inhibiting hepatic cell DNA synthesis has been partially characterized (Sekas *et al.*, 1979). A stimulator–inhibitor system may also act in the induction of renal parenchymal cell hypertrophy following unilateral nephrx (Preuss *et al.*, 1970; Dicker *et al.*, 1977; Radosevic-Stasic *et al.*, 1979). Recent experiments performed in our laboratory have indicated that the production of hepatic Ep following hepx and hypoxia is regulated in a similar way. These are described in the following paragraph.

An elevation of the hepatic Ep response to hypoxia was noted in hepx rats (Naughton *et al.*, 1976, 1977a; Anagnostou *et al.*, 1977). This phenomenon was found to be age-related; the younger animal displayed a higher hepatocellular proliferative response to hepx (Bucher, 1963) than its older counterpart and produced more Ep in response to hypoxia as well (Naughton *et al.*, 1977b). Hepatotrophic substances in the serum of hepx animals were found to stimulate hepatocellular proliferation in normal mammals (Bucher, 1963; Morley, 1974). The possibility that this serum contained factor(s) which could also augment the hepatic Ep response to hypoxia was tested in the following experiments. Rats were exsanguinated from either the hepatic vein or the abdominal aorta at various intervals after hepx or sham hepx. Serum from these animals was administered to normal rats. After a delay of 18–24 hr, these recipient rats were nephrectomized (a necessary procedure for the quantitation of extrarenal Ep) and exposed to hypoxia. In general, serum from rats collected 48–72 hr after hepx was capable of stimulating hepatic Ep in normal recipients when compared to those animals receiving serum from sham hepx donors (Naughton, 1978; Naughton *et al.*, 1978, 1980). In addition, serum from the hepatic vein of hepx animals displayed a greater ability to evoke this response than that derived from the systemic arterial circulation. Furthermore, it was found that renal ablation of the hepx donor rat completely abolished this hepatic venous/systemic arterial variation in the hepatic Ep response to hypoxia in the normal recipients. From these experiments it was concluded that the serum of hepx rats contains a factor(s) which can evoke hepatic Ep elaboration after administration to normal animals. The relationship of this hepatic erythropoietic factor (HEF) to other hepatotrophic factors in this serum is as yet unknown, although preliminary characterization indicates that this substance has a molecular weight of less than 20,000 and is heat stable (Naughton, unpublished observation, 1982). A hepatotrophic substance of similar molecular weight has recently been reported in hepx rats (Goldberg *et al.*, 1980). The HEF is a liver-derived moiety; it has been isolated from the adult regenerating liver via an *in situ* perfusion technique (Dornfest *et al.*, 1981).

The inhibition of the effects of HEF by the kidney could be ascribed to either (1) the elimination of HEF via normal renal excretory function or (2) a specific inhibitor. The first possibility was discounted since bilateral ureteral ligation of the hepx donor did not alter the hepatic venous/systemic arterial variation in the hepatic Ep response of the recipient rat to hypoxia (Naughton, 1978; Naughton *et al.*, 1978, 1980). The buildup of metabolic wastes in either the nephrx or bilaterally ureteral ligated hepx donor animals had little influence on the Ep elaboration of the recipients since serum from normal or sham-operated anephric or bilaterally ureteral ligated controls was incapable of augmenting the hepatic Ep levels of the normal recipients. In further experiments, we found that the renal venous serum of hepx donor animals displayed the ability to diminish the hepatic Ep response to hypoxia when administered to hepx anephric recipients (Naughton *et al.*, 1981). Partial characterization of this serum revealed that the

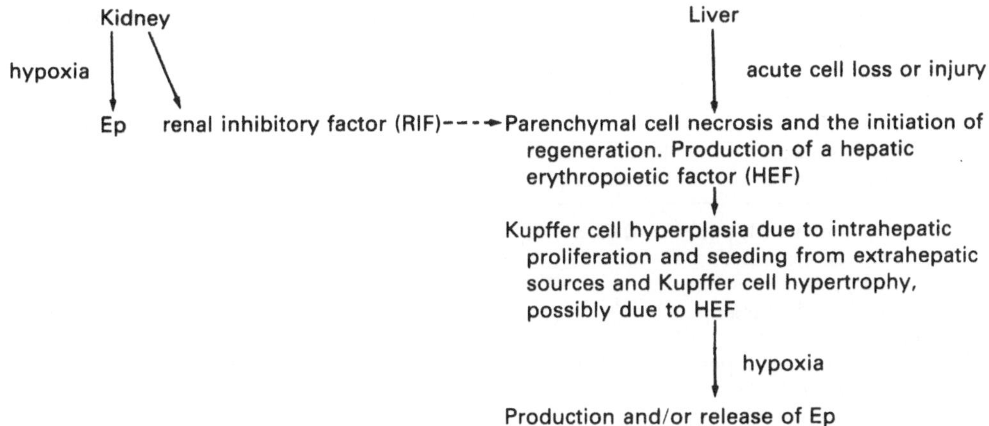

FIGURE 7. Proposed scheme for the renal–hepatic antagonism controlling extrarenal Ep production.

inhibitory component (renal inhibitory factor, RIF) is a high-molecular-weight entity (>100,000). This is similar to the lipid inhibitor postulated by Erslev *et al.* (1971) and in contrast to the low-molecular-weight chalone-type hepato-proliferation inhibitor proposed by Sekas *et al.* (1979). The postulated interaction of these stimulatory and inhibitory factors is presented in Fig. 7.

2.3.4. Significance of HEF and RIF to Problems of Renal Insufficiency in Man

To study the applicability of this renal–hepatic antagonism in hepatic Ep production to problems of the anemia of renal insufficiency in man, serum was collected from several human subjects with hepatic and/or renal disease. This serum was administered to normal rats which were nephrectomized 18–24 hr later and rendered hypoxic. The hepatic Ep levels in normal recipient rats were highest when donor serum was derived from patients which were anephric and suffered from a liver disease characterized by localized hepatic necroses and regenerative foci (Table 4) (Naughton, 1978). Ep has previously been demonstrated in the plasma of cirrhotic liver patients (Mirand and Murphy, 1971) and erythrocytosis has been associated with several hepatic diseases including biliary cirrhosis, viral hepatitis, and hepatocellular carcinoma (McFadzean *et al.*, 1967; Mirand and Murphy, 1971; Sherlock, 1978; Mehrotra *et al.*, 1980). Serum from patients with hepatopathies which were unaccompanied by renal ablation or kidney disease did not evoke recipient rat hepatic Ep levels which were significantly higher than those treated with normal rat or normal human serum. When compared to normal rat or normal human serum-injected animals, rats given serum from an anephric human with no indication of liver disease showed slightly diminished hepatic Ep levels (Table 4). This may be attributed to the presence of an inhibitor to erythropoiesis, possibly spermine (Radtke *et al.*, 1980). Renal dysfunction does not always result in anemia. Several kidney diseases have

TABLE 4. MEAN EP LEVELS IN THE SERUM OF RATS AFTER THE INJECTION OF SERUM OR PLASMA FROM HUMAN SUBJECTS AND FOLLOWED BY NEPHRX AND HYPOXIA (6 HR/0.4 ATM)[a]

Donor serum or plasma source	Ep levels in donor serum or plasma	Hematocrit of human subject prior to dialysis	Ep (U/ml ± S.E.M.) in recipients after nephrx and hypoxia
Anephric human with portal cirrhosis of the liver (Laennec) induced by excessive alcohol consumption	0.08 ± 0.02	26%	0.19 ± 0.03
Anephric human with portal cirrhosis of the liver (Laennec) induced by excessive alcohol consumption	0.07 ± 0.03	29%	0.15 ± 0.03
Anephric human with focal nodular hyperplasia and fibrotic liver associated with heroin addiction	0.07 ± 0.02	28%	0.15 ± 0.04
Anephric human with cavernous hemangioma with accompanying calcification	0.06 ± 0.03	30%	0.22 ± 0.04
Human nephrx due to renal polycystic disease with associated hepatic polycystic disease	0.05 ± 0.02	23%	0.15 ± 0.03
Anephric human with an essentially normal liver	—[b]	12%	—[b]
Human with portal cirrhosis of the liver (Laennec) due to alcoholism No renal involvement	—[b]	42%	0.07 ± 0.02
Normal human	0.05 ± 0.02	45%	0.06 ± 0.02

[a]At 18 hr after injection of serum from human subjects, recipient rats were nephrx and subjected to hypoxia at 0.4 atm for 6 hr. These animals were then exsanguinated and serum prepared. Radioiron uptake into the red blood cells of exhypoxic polycythemic assay mice was determined after the administration of serum from the recipient rats. This datum was compared to values obtained with saline. 0.05 IU Ep, 0.20 IU Ep, and 0.8 IU Ep standards derived from the international reference preparation for human urinary Ep. This was procured by the Department of Physiology, University of Northeast Corrientes, Argentina, and was processed by The Hematology Research Laboratories, Children's Hospital of Los Angeles, under Grant HE-10880 from the National Heart and Lung Institutes of the National Institutes of Health.
[b]Undetectable in the exhypoxic polycythemic mouse assay.

been associated with erythrocytosis and/or elevated Ep elaboration: nephrosclerosis (Sonnenborn *et al.*, 1977), hypernephroma (Waldmann *et al.*, 1968), and renal carcinoma (Jasmin and Riopelle, 1976. A stimulatory response occurs in all cases where hepatic necrosis and repair are indicated and the kidneys are either absent or nonfunctional. No such stimulation is noted in similar cases of hepatic disease where functional kidneys are present. A possible explanation of these findings is as follows: regenerating hepatic cells and possibly a small proportion of normal liver cells have the ability to produce an erythropoietic factor. This factor is regulated by an antagonistic principle, RIF, produced by the normal kidney. In hepatic diseases where the mass of active or regenerating cells is much greater than in normal liver, the HEF elevation is compensated by an augmentation of the activity and/or synthesis of RIF. In such cases where the kidney is absent, the increase in HEF occurs without a concomitant change in RIF levels and serum from these patients becomes capable of stimulating the hepatic Ep response to hypoxia in our normal animal model. The role of the diseased kidney in the onset of liver disease has been discussed in a review (Papper and Vaamonde, 1971). Also, renal failure has reportedly been induced by cirrhosis (Schroeder *et al.*, 1967) and viral hepatitis (Conrad *et al.*, 1964), conditions which promote hepatic parenchymal cell destruction and initiate hepatic regeneration. Although many of the details of the proposed system remain to be elucidated, it appears that it is operative in man as well as in rats (Naughton *et al.*, 1983).

2.3.5. Hormonal Effects on the RE and Erythropoietic System

A summary of the effects of some hormones and other compounds on erythropoiesis and RE activity is given in Table 5. In general, agents which promote either hypertrophy or hyperplasia of the RES also are stimulatory for Ep elaboration and/or erythropoiesis. Indomethacin, which blocks PG synthesis, was also found to diminish the renal Ep response to hypoxia (Mujovic and Fisher, 1975) as well as inhibiting RE activity (Razin and Globerson, 1979). Anabolic hormones such as growth hormone (GH) and adrenocorticotrophic hormone (ACTH) both stimulate the Ep response to hypoxia (Fruhman *et al.*, 1954; Peschle *et al.*, 1971). Likewise, both GH (Rawls *et al.*, 1954) and ACTH (Umehara and Iseki, 1953) have some stimulatory action on the RES. Testosterone, a potent stimulant of erythropoiesis (Gordon, 1973), does not appear to influence the RES (Nicol *et al.*, 1967). Estrogenic compounds, however, are

TABLE 5. THE EFFECTS OF VARIOUS AGENTS ON RES ACTIVITY AND ERYTHROPOIESIS

Substance	Erythropoiesis and/or Ep production	RES activity: hypertrophy and/or hyperplasia
Glucan	↑ (Deimann and Fahimi, 1980)	↑ (Morrow and DiLuzio, 1965)
Testosterone	↑ (Gordon *et al.*, 1970)	← (Nicol *et al.*, 1967)
Estrogen	↓ (Mirand and Gordon, 1966)	↑ (Nicol *et al.*, 1967)
ACTH	↑ (Peschle *et al.*, 1971)	↑ (Umehara and Iseki, 1953)
Colloidal carbon (overload)	↑ (Peschle *et al.*, 1976)	↑ (Benaceraff *et al.*, 1957)
Methylcellulose	↑ (Brabec and Neuwirt, 1975)	↑ (Stang and Boggs, 1977)
Olive oil (and triglycerides)	↑ (Liu *et al.*, 1980a)	↑ (Flemming, 1963)
Indomethacin	↓ (Mujovic and Fisher, 1975)	↓ (Razin and Globerson, 1979)
GH	↑ (Fruhman *et al.*, 1954)	↑ (Rawls *et al.*, 1954)
Phenylhydrazine	↑ (Hertzberg and Orlic, 1980)	↑ (Ploemacher and van Soest, 1977)

potent stimulators of the RES (Nicol *et al.*, 1967; Kinsky *et al.*, 1969) whereas they are inhibitory to both Ep production and erythropoiesis (Mirand and Gordon, 1966). A sexual variation was also noted in the production of HEF by hepx rats (Naughton *et al.*, 1980). Production of this factor was significantly lower in female hepx rats than in their male counterparts. Although the diminished Ep response to hypoxia in female animals has been ascribed to an estrogen suppression of medullary erythropoiesis mediated through direct damage to the CFU-s (Fried *et al.*, 1974), no mechanism has been hypothesized for the low HEF levels in females. Although the effects of the substances in Table 5 on the RES and erythropoiesis may be coincidental rather than causative in some cases, it appears that hypertrophy and/or hyperplasia of the RES, particularly in the liver, favors extramedullary erythropoiesis and is inductive to Ep elaboration.

ACKNOWLEDGMENTS. These studies were supported by Research Grant 2 ROIHLBO3357-21A1 from the National Heart, Lung, and Blood Institute of the National Institutes of Health, and by Grants HRC-877 and 11-105 from the New York State Health Research Council.

REFERENCES

Acton, J. D., and Myrvik, Q. N., 1966, Production of interferon by alveolar macrophages, *J. Bacteriol.* **91:**2300.

Adibi, S., Paschkis, K. E., and Cantarow, A., 1959, Stimulation of liver mitoses by blood serum from hepatectomized rats, *Exp. Cell Res.* **18:**396.

Alpert, L. I., and Benisch, B., 1970, Hemangioendothelioma of the liver associated with microangiopathic hemolytic anemia, *Am. J. Med.* **48:**624.

Anagnostou, A., Schade, S., Barone, J., and Fried, W., 1977, Effects of partial hepatectomy on extrarenal erythropoietin production in rats, *Blood* **50:**457.

Aschoff, L., and Kiyono, T., 1913, Zur Frage der grossed mononuklearen, *Folia Haematologica* **15:**385.

Bachmann, W., Bottger, I., and Haslbeck, M., 1979, The mechanism of insulin resistance in liver disease: Studies in D-galactosamine hepatitis and in partial hepatectomy in rats, *Excerpta Medica* **481:**14.

Barr, R. D., Whang-Peng, J., and Perry, S., 1975, Hemopoietic stem cells in human peripheral blood, *Science* **190:**284.

Benacerraf, B., 1964, Functions of the Kupffer cells, in: *The Liver: Morphology, Biochemistry* (C. L. Rouiller, ed.), Academic Press, New York.

Benacerraf, B., Bilbey, D., Biozzi, G., Halpern, B. N., and Stiffel, C., 1957, The measurement of liver blood flow in partially hepatectomized rats, *J. Physiol. (London)* **136:**287.

Ben-Ishay, Z., 1975, Ultrastructural analysis of erythroid colonies in diffusion chambers in irradiated rats: Effect of cytosine arabinoside administration on donor rats, *Scand. J. Haematol.* **14:**369.

Ben-Ishay, Z., 1977, The erythroid island: Structure and function, in: *Pathobiology Annual* (H. L. Ioachim, ed.), pp. 63–80, Appleton–Century–Crofts, New York.

Ben-Ishay, Z., and Yoffey, J. M., 1971a, Reticular cells of rat bone marrow in different functional states. I. Changes in primary hypoxia, *Isr. J. Med. Sci.* **7:**948.

Ben-Ishay, Z., and Yoffey, J. M., 1971b, Reticular cells of rat bone marrow in hypoxia and rebound, *J. Reticuloendothelial Soc.* **10:**482.

Berman, J., 1967, The ultrastructure of erythroblastic islands and reticular cells in mouse bone marrow, *J. Ultrastruct. Res.* **17:**291.

Bessis, M., 1956, *Cytology of the Blood and Blood Forming Organs*, Grune & Stratton, New York.

Bessis, M., and Breton-Gorius, J., 1959, Nouvelles observations sur l'ilot erythroblastique et la rhophéocytose de la ferritin, *Rev. Hematol.* **14:**165.

Bessis, M., and Breton-Gorius, J., 1962, Iron metabolism in the bone marrow as seen by electron-microscopy: A critical review, *Blood* **19:**635.

Biozzi, G., Stiffel, C., and Mouton, D., 1963, Stimulation et dépression de la fonction phagocytaire du système réticuloendothélial par des émulsions de lipides, *Rev. Eur. Etud. Clin. Biol.* **8:**342.

Bollman, L., 1961, Liver, *Annu. Rev. Physiol.* **23:**183.

Bowden, D. H., and Adamson, I. Y., 1980, Role of monocytes and interstitial cells in the generation of alveolar macrophages. I. Kinetic studies of normal mice, *Lab. Invest.* **42:**511.

Bozzini, C. E., Barrio-Rendo, M. E., Devoto, F. C., and Epper, C. E., 1970, Studies on medullary and extramedullary erythropoiesis in the adult mouse, *Am. J. Physiol.* **219:**724.

Brabec, V., and Neuwirt, J., 1975, Erythropoietin formation in rats with experimental hypersplenism, *Scand. J. Haematol.* **14:**86.

Bucher, N. R., 1963, Regeneration of mammalian liver, *Int. Rev. Cytol.* **15:**245.

Bucher, N. R., Scott, J. E., and Aub, J. C., 1951, Regeneration of the liver in parabiotic rats, *Cancer Res.* **11:**457.

Bucher, N. L. R., and Malt, R. A., 1971, Regeneration of liver and kidney, in: N. Engl. J. Med. Med. Prog. Ser. p. 17.

Bucher, N. L. R., and Swaffield, M. N., 1973, Regeneration of liver in rats in the absence of portal splanchnic organs and a portal blood supply, *Cancer Res.* **33:**3189.

Bucher, N. L. R., and Swaffield, M. N., 1975, Regulation of hepatic regeneration by synergistic action of insulin and glucagon, *Proc. Natl. Acad. Sci. USA* **72:**1157.

Bucher, N. L. R., and Weir, G. C., 1976, Insulin, glucagon, liver regeneration, and DNA synthesis, *Metabolismo* **25:**1423.

Bullough, W. S., 1964, Mitotic and functional homeostasis: A speculative review, *Cancer Res.* **25:**1683.

Carmena, A. D., Howard, D., and Stohlman, F., Jr., 1968, Regulation of erythropoietin. XXII. Production in the newborn animal, *Blood* **32:**376.

Cavallin-Stahl, E., Berg, B., and Brandt, L., 1974, Reticulum cells and erythroblasts in the bone marrow of megaloblastic anemia patients, *Acta Med. Scand.* **195:**185.

Chan, H. S., Sanders, E. F., and Freedman, M. H., 1980, Modulation of human hematopoiesis by prostaglandins and lithium, *J. Lab. Clin. Med.* **95:**125.

Chen-Li, T., and Weiss, L., 1973, The role of the sinus wall in the passage of erythrocytes through the spleen, *Blood* **41:**529.

Chervenick, P. A., and LoBuglio, A. F., 1972, Human blood monocytes: Stimulators of granulocyte mononuclear colony formation *in vitro, Science* **178:**164.

Congote, L. F., and Solomon, S., 1974, On the site of origin of erythropoietic factors in human fetal tissues, *Endocrinol. Res. Commun.* **1:**495.

Conrad, M. E., Schwartz, F. D., and Young, A. A., 1964, Infectious hepatitis: A generalized disease. A study of renal, gastro-intestinal, and hematologic abnormalities, *Am. J. Med.* **37:**769.

Cottier, H., Odartchenko, N., Feinendegen, L. E., Keiser, G., and Bond, V. P., 1963, Autoradiographische untersuchungen uber die entkernung der erythroblasten nach in-vivo-markierung mit thymidine-^{3}H, *Schwiez. Med. Wochenschr.* **93:**1061.

Crosby, W. H., 1959, Normal function of the spleen relative to red blood cells: A review, *Blood* **14:**399.

Cutbush, M., and Mollison, P. L., 1958, Relation between characteristics of blood group antibodies *in vitro* and associated patterns of red cell destruction *in vivo, Br. J. Haematol.* **4:**115.

Deimann, W., and Fahimi, H. D., 1980, Induction of focal hemopoiesis in adult rat liver by glucan, a macrophage activator, *Lab. Invest.* **42:**217.

Demetriou, A., Seifter, E., and Levenson, S., 1974, Effect of sera obtained from normal and partially hepatectomized rats and patients on the growth of cells in tissue culture, *Surgery* **76:**779.

Dicker, S. E., Morris, C. A., and Shipolini, R., 1977, Regulation of compensatory kidney hypertrophy by its own products, *J. Physiol. (London)* **269:**687.

Dornfest, B. S., Naughton, B. A., Kolks, G. A., Liu, P., Piliero, S. J., and Gordon, A. S., 1981, Recovery of an erythropoietin inducing factor from the regenerating liver, *Ann. Clin. Lab. Sci.* **11:**37.

Dornfest, B. S., Naughton, B. A., Johnson, R., and Gordon, A. S., 1983, Hepatic production of erythropoietin in a phenylhydrazine-induced compensated hemolytic state in the rat, *J. Lab. Clin. Med.* **102:**274.

Dukes, P., Takaku, F., and Goldwasser, E., 1964, *In vitro* studies of erythropoietin on ^{14}C-glucosamine incorporation into rat bone marrow cells, *Endocrinology* **74:**960.

Dukes, P. P., Shore, N. A. Hammond, G. D., Ortega, J. A., and Datta, M. C., 1973, Enhancement of erythropoiesis by prostaglandins, *J. Lab. Clin. Med.* **82:**704.

Dukes, P. P., Shore, N. A., Hammond, G. D., and Ortega, J. A., 1975, Prostaglandins and erythropoietin action, in: *Erythropoiesis* (K. Nakao, J. W. Fisher, and F. Takaku, eds.), pp. 3–14, University Park Press, Baltimore.

Dunn, C. D., 1971, The differentiation of haemopoietic stem cells, *Ser. Haematol.* **4:**1.

Eränko, O., 1955, Relative volume measurements, in: *Quantitative Methods in Histology and Microscopic Histochemistry*, (J. P. Gould, ed.), pp. 57–73, Little, Brown, Boston.

Erslev, A. J., 1972, Anemia of chronic disorders, in: *Hematology* (J. W. Williams, E. Beutler, A. J. Erslev, and W. Rundles, eds.), pp. 237–255, McGraw-Hill, New York.

Erslev, A. J., Kazal, L. A., and Miller, O. P., 1971, The action and neutralization of a renal inhibitor of erythropoietin, *Proc. Soc. Exp. Biol. Med.* **138:**1025.

Erslev, A. J., Caro, J., and Kansu, E., 1978, Renal and extrarenal erythropoietin production in anemia, in: *In Vitro Aspects of Erythropoiesis* (M. J. Murphy, ed.), pp. 225–226, Springer-Verlag, Berlin.

Finch, C. A., Deubelbeiss, K., Cook, J. D., Eschbach, J. W., Harker, L. A., Funk, D. D., Marsaglia, G., Hillman, R. S., Slichter, S., Adamson, J. W., Gansomi, A., and Giblett, E. R., 1970, Ferrokinetics in man, *Medicine* **49:**17.

Fisher, B., Szuch, P., and Fisher, E. R., 1971, Evaluation of humoral factors in liver regeneration utilizing liver transplants, *Cancer Res.* **31:**322.

Fisher, J. W., 1980, Prostaglandins and kidney erythropoietin production, *Nephron* **25:**53.

Flemming, K., 1963, Radiation protective effect and pharmacologically changed activity of the reticulo-endothelial system, *Nature (London)* **200:**1117.

Fredlund, P. E., Callum, B., and Tibblin, S., 1972, Influence of glucagon on the ischemic liver in the pig, *Arch. Surg.* **105:**615.

Fried, W., 1972, The liver as a source of extrarenal erythropoietin production, *Blood* **40:**671.

Fried, W., Tichler, T., Dennenberg, J., Barone, J., and Wang, F., 1974, Effects of estrogen on hematopoietic stem cells and on hematopoiesis of mice, *J. Lab. Clin. Med.* **83:**807.

Fried, W., Barone, J., Schade, S., and Anagnostou, A., 1979, Effect of carbon tetrachloride on extrarenal erythropoietin production in rats, *J. Lab. Clin. Med.* **93:**700.

Fruhman, G. J., 1970, Splenic erythropoiesis, in: *Regulation of Hematopoiesis* Vol. 1 (A. S. Gordon, ed.), pp. 339–368, Appleton–Century–Crofts, New York.

Fruhman, G. J., Gerstner, R., and Gordon, A. S., 1954, Effects of growth hormone upon erythropoiesis in the hypophysectomized rat, *Proc. Soc. Exp. Biol. Med.* **85:**93.

Gale, R. P., Sparkes, R. S., and Golde, D. W., 1978, Bone marrow origin of hepatic macrophages (Kupffer cells) in humans, *Science* **201:**937.

Gallien-Lartigue, O., and Goldwasser, E., 1965, On the mechanism of erythropoietin induced differentiation: Effect on RNA synthesis, *Biochim. Biophys. Acta* **103:**325.

Gates, G. A., Henley, K. S., Pollard, H. M., Schmidt, E., and Schmidt, F. W., 1961, The cell population of the human liver, *J. Lab. Clin. Med.* **57:**182.

Glew, R. H., Haese, W. H., and McIntyre, P. A., 1973, Myeloid metaplasia with myelofibrosis: The clinical spectrum of extramedullary hematopoiesis and tumor formation, *Johns Hopkins Med. J.* **132:**253.

Glomski, C. A., 1969, Experimental hypersplenism and tissue iron distribution, *Brit. J. Exptl. Pathol.* **50:**331.

Goldberg, M., Strecker, W., Feeny, D., and Ruhenstroth-Bauer, G., 1980, Evidence for and characterization of a liver cell proliferation factor from blood plasma of partially hepatectomized rats, *Horm. Metab. Res.* **12:**94.

Golde, D. W., Finley, T. N., and Cline, M. J., 1972, Production of colony stimulating factor by human macrophages, *Lancet* **2:**1397.

Goodman, J. R., Wallerstein, R. O., and Hall, S. G., 1968, The ultrastructure of bone marrow histiocytes in megaloblastic anemia and the anemia of infection, *Br. J. Haematol.* **14:**471.

Gordon, A. S., 1973, Erythropoietin, *Vitam. Horm. (N.Y.)* **31:**105.

Gordon, A. S., LoBue, J., Dornfest, B., and Cooper, G., 1962, Reticulocyte and leukocyte release from isolated rat legs and femurs, in: *Erythropoiesis* (L. Jacobson and M. Doyle, eds.), pp. 321–327, Grune & Stratton, New York.

Gordon, A. S., Zanjani, E. D., Levere, R. D., and Kappas, A., 1970, Stimulation of mammalian erythropoiesis by 5B-H steroid metabolites, *Proc. Natl. Acad. Sci. USA* **65**:919.

Gordon, A. S., Zanjani, E. D., Peterson, E. N., Gidari, A. S., LoBue, J., and Camiscoli, J. F., 1972, Studies on fetal erythropoietin, in: *Regulation of Erythropoiesis* (A. S. Gordon, M. Condorelli, and C. Peschle, eds.), pp. 188–201, Il Ponte, Milan.

Gordon, L. I., Miller, W. J., Branda, R. F., Zanjani, E. D., and Jacob, H. S., 1980, Regulation of erythroid colony formation by bone marrow macrophages, *Blood* **55**:1047.

Gross, D. M., Brookins, J., Fink, G. D., and Fisher, J. W., 1976, Effects of prostaglandins A_2, E_2, and $F_{2\alpha}$ on erythropoietin production, *J. Pharmacol. Exp. Ther.* **198**:489.

Gruber, D. F., Zucali, J. R., Wlekinski, J., La Russa, V., and Mirand, E. A., 1977, Temporal transition in the site of rat erythropoietin production, *Exp. Hematol.* **5**:399.

Guduin, V., Michell, A., Scigula, P., Ivanova, V. A., and Gross, J., 1978, Erythropoietin and erythropoiesis inhibitor in normal and hypoxic neonates, *Bull. Exp. Biol. Med.* **86**:1007.

Harkness, R. D., 1957, Regeneration of liver, *Br. Med. Bull.* **13**:87.

Hayes, D. M., Tedo, I., and Matsushima, Y., 1969, Stimulation of *in vitro* growth of rat liver cells with calf serum drawn following partial hepatectomy, *J. Surg. Res.* **9**:133.

Hayes, D. M., Firkin, F. C., Koga, Y., and Hays, E. F., 1975, Formation of colonies in soft agar medium by regenerating liver cells, *Proc. Soc. Exp. Biol. Med.* **148**:596.

Henry, R. E., Warnecke, M. A., and Donati, R. M., 1974, Effect of mechlorethamine or vincristine sulfate on marrow reticuloendothelial and erythropoietic function in rats, *J. Reticuloendothelial Soc.* **15**:406.

Hertzberg, C., and Orlic, D., 1980, An electron microscopic study of erythrophagocytosis in the fetal and neonatal rabbit. *J. Reticuloendothelial Soc.* **28**:15.

Higgins, C. B., and Braunwald, E., 1972, The prostaglandins: Biochemical, physiologic, and clinical considerations, *Am. J. Med.* **53**:92.

Horland, A. A., Walwan, S. R., Murphy, M. J., Jr., and Moore, M. A. S., 1977, Proliferation of erythroid colonies in semi-solid agar, *Br. J. Haematol.* **36**:495.

Hughes-Jones, N. C., 1961, The use of ^{51}Cr and ^{59}Fe as red cell labels to determine the fate of normal erythrocytes in the rabbit, *Clin. Sci.* **20**:315.

Hymer, B., Hobik, H. P., Schaefer, H., Bueltmann, B., Spanel, R., and Haferkamp, O., 1976, Animal experimental studies on chronic granulomatous inflammation and T-lymphocyte-system, *Beitr. Pathol.* **156**:128.

Islami, A. H., Pack, G. T., and Hubbard, J. C., 1959, The humoral factor in regeneration of the liver in parabiotic rats, *Surg. Gynecol. Obstet.* **108**:549.

Jacobson, L. O., Goldwasser, L., and Fried, W., 1957, Role of the kidney in erythropoiesis, *Nature (London)* **179**:633.

Jandl, J. H., and Simmons, R. L., 1957, The agglutination and sensitization of red cells by metallic cations, *Br. J. Haematol.* **3**:19.

Jasmin, G., and Riopelle, J. L., 1976, Renal carcinoma and erythrocytosis in rats following intrarenal injection of nickel subsulfide, *Lab. Invest.* **35**:71.

Jimnez de Asua, L., Clingan, D., and Rudland, P. S., 1975, Initiation of cell proliferation in cultured mouse fibroblasts by prostaglandin $F_{2\alpha}$, *Proc. Natl. Acad. Sci. USA* **72**:2724.

Joyce, R. A., and Chervenick, P. A., 1975, Stimulation of granulopoiesis by liver macrophages, *J. Lab. Clin. Med.* **56**:112.

Joyce, R. A., Pfrimmer, W. J., Turner, A. R., and Boggs, D. R., 1976, Modulation of hepatic hematopoiesis in the adult mouse, *Blood* **164**:129.

Kanzaki, Y., Mahmud, I., Asanagi, M., Fukui, N., and Miura, Y., 1979, Thromboxane as possible trigger of liver regeneration, *Cell. Mol. Biol.* **25**:147.

Keighley, G., and Cohen, N. S., 1978, Stimulation of erythropoiesis in mice by adenosine 3′,5′ monophosphate and prostaglandin E, *J. Med.* **9**:129.

Kelly, L. S., and Dobson, E. L., 1971, Evidence concerning the origin of liver macrophages, *Br. J. Exp. Pathol.* **52**:88.

Kent, G., Minick, O. T., Orfei, E., and Volini, E., 1969, Fine structural changes in the liver in experimental hemolytic anemia, *Am. J. Pathol.* **57**:649.

Kinsky, R. G., Christie, G. H., Elson, J., and Howard, J. G., 1969, Extra-hepatic derivation of Kupffer cells during oestragenic stimulation of parabiosed mice, *Br. J. Exp. Pathol.* **50**:438.

Kivilaakso, E., and Rytomaa, T., 1971, Erythrocytic chalone, a tissue specific inhibitor of cell proliferation in the erythron, *Cell Tissue Kinet.* **4**:1.

Kono, Y., and Ho, M., 1965, The role of the reticuloendothelial system in interferon formation in the rabbit, *Virology* **25**:162.

Koski, I. R., Poplack, D. G., and Blaese, R. M., 1976, A nonspecific esterase stain for the identification of monocytes and macrophages, in: *In Vitro Methods in Cell Mediated and Tumor Immunity*, (B. R. Bloom and J. R. David, eds.), pp. 359–362. Academic Press, New York.

Krantz, S. B., Gallien-Lartigue, O., and Goldwasser, E., 1963, The effect of erythropoietin upon heme synthesis *in vitro*, *J. Biol. Chem.* **238**:4085.

Kurland, J. I., Traganos, F., Darzynkiewicz, A., and Moore, M. A. S., 1978a, Macrophage mediated cytostasis of neoplastic hemopoietic cells: Cytofluorometric analysis of the reversible cell cycle block, *Cell. Immunol.* **36**:318.

Kurland, J. I., Bockman, R. S., Broxmeyer, H. E., and Moore, M. A. S., 1978b, Limitation of excessive myelopoiesis by the intrinsic modulation of macrophage derived prostaglandin E, *Science* **199**:552.

Kurland, J. I., Pelus, L. M., Ralph, P., Bockman, R. S., and Moore, M. A. S., 1979, Induction of prostaglandin E synthesis in normal and neoplastic macrophages: Role for colony stimulating factor(s) distinct from effects on myeloid progenitor cell proliferation, *Proc. Natl. Acad. Sci. USA* **76**:2326.

La Brecque, D. R., 1979, The role of hepatotrophic factors in liver regeneration: A brief review including a preliminary report of the *in vitro* effects of hepatic regenerative stimulator substance (SS), *Yale J. Biol. Med.* **52**:49.

Lacey, P. E., 1977, The physiology of insulin release, in: *The Diabetic Pancreas* (B. W. Volk and K. F. Wellman, eds.), pp. 211–230, Plenum Press, New York.

Lee, J. B., 1970, Prostaglandins, *Physiologist* **13**:379.

Lee, S., Edgington, T. S., and Orloff, M. J., 1968, The role of afferent blood supply in regeneration of liver isografts in rats, *Surg. Forum* **19**:360.

Leong, G. F., Pessotti, R. L., and Brauer, R. W., 1959, Liver function in regenerating rat liver: $CrPO_4$ colloid uptake and bile flow, *Am. J. Physiol.* **197**:880.

Leong, G. F., Grisham, J. W., and Hole, B., 1964, Effect of rapid total exchange transfusion on hepatic DNA synthesis in partially hepatectomized rats, *Cancer Res.* **24**:1496.

Levi, J., and Zeppa, R., 1971, Source of the humoral factor that initiates hepatic regeneration, *Ann. Surg.* **174**:364.

Lewis, J. P., Passovoy, M., Freeman, M., and Trobaugh, F. E., Jr., 1968, The repopulating potential and differentiation capacity of hematopoietic stem cells from the blood and bone marrow of normal mice, *J. Cell. Physiol.* **71**:121.

Lewis, J. P., Neal, W. A., Welch, E. T., Lewis, W. G., DuBose, C. M., Wright, E. T., and Smith, L. L., 1973, The isolation of erythropoiesis regulatory factors by an electro-fractionation technique combined with selective membrane permeability, *Proc. Soc. Exp. Biol. Med.* **142**:845.

Liao, S., and Axelrad, A. A., 1975, Erythropoietin independent erythroid colony formation *in vitro* by hemopoietic cells of mice infected with Friend virus, *Int. J. Cancer* **15**:467.

Lindemann, R., 1971, Erythropoiesis inhibitory factor (EIF). I. Fractionation and demonstration of urinary EIF, *Br. J. Haematol.* **21**:623.

Liu, P., Naughton, B. A., Kolks, G. A., and Gordon, A. S., 1980a, The effect of hepatotoxic agents on hepatic erythropoietin (Ep) production, *Anat. Rec.* **196**:224A.

Liu, P., Naughton, B. A., Kolks, G. A., Kruger, R. E., Piliero, S. J., and Gordon, A. S., 1980b, Hepatic erythropoietin production following double partial hepatectomy in the rat, *J. Surg. Oncol.* **15**:121.

Lord, B. I., 1965, Haematopoietic changes in the rat during growth and during continuous gamma irradiation of the adult animal, *Br. J. Haematol.* **11**:525.

Lucarelli, G., Porcellini, A., Carnevali, C., Carmena, A., and Stohlman, J. F., Jr., 1968, Fetal and neonatal erythropoiesis, *Ann. N.Y. Acad. Sci.* **149**:544.

McCuskey, R. S., Meineke, H. A., and Townsend, S. F., 1972, Studies of the haemopoietic microenvironment. I. Changes in the microvascular system and stroma during erythropoietic regeneration and suppression in the spleens of CF-1 mice, *Blood* **39**:697.

McFadzean, A. J., Todd, D., and Tso, S. C., 1967, Erythrocytosis associated with hepatocellular carcinoma, *Blood* **29**:808.

McGarry, M., and Mirand, E. A., 1973, Incidence of Friend virus induced polycythemia in splenectomized mice, *Proc. Soc. Exp. Biol. Med.* **142**:538.

MacKenzie, R. J., Furnival, C. M., O'Keane, M. A., and Blumgart, L. H., 1975, The effect of hepatic ischaemia on liver function and the restoration of liver mass after 70% partial hepatectomy in the dog, *Br. J. Surg.* **62**:431.

McPhie, J. L., 1979, Peroxidase activity in non-parenchymal cells isolated from rat liver: A cytochemical study, *Acta Hepato-Gastroenterol.* **26**:442.

Mann, F. C., Fishback, F. C., and Gray, J. G., 1931, Experimental pathology of diverting portal blood on the restoration of the liver after partial removal, *Arch. Pathol.* **12**:787.

Marois, D., van Heerden, J. A., Carpenter, H. A., and Sheedy, P. F., II, 1979, Congenital absence of the portal vein, *Mayo Clin. Proc.* **54**:55.

Matoth, G., and Kaufmann, L., 1962, Mitotic activity of erythroblasts pre-exposed to erythropoietin, *Blood* **20**:165.

Matray, F., Sauger, F., Borde, J., Mitrofanoff, P., Grosley, M., Bourg, M., Laumonier, R., and Henet, J., 1972, Présence d'alpha foeto-protéin au cours d'une régeneration hépatique après hépatectomie gauche pour hépatome chez un enfant, *Pathol. Biol.* **20**:353.

Mayankii, D. N., Shcherbakov, V. I., and Mirokhanov, Y. M., 1978, Origin of the Kupffer macrophages in the regenerating liver, *Byull. Eksp. Biol. Med.* **85**:598.

Mehrotra, A. N., Srivastava, A. N., and Mehrotra, R. M. L., 1980, A clinical pathologic study of chronic hepatitis, *Indian J. Med. Res.* **69**:781.

Metcalf, D., and Moore, M. A., 1972, *Hematopoietic Cells,* North-Holland, Amsterdam.

Metcalf, D., Furth, J., and Buffett, R. F., 1959, Pathogenesis of mouse leukemia caused by Friend virus, *Cancer Res.* **19**:52.

Metchnikoff, E., 1905, *Immunity in Infective Disease,* Cambridge University Press, London.

Mirand, E. A., and Gordon, A. S., 1966, Mechanism of estrogen action on erythropoiesis, *Endocrinology* **78**:325.

Mirand, E. A., and Murphy, G. P., 1971, Erythropoietin alterations in human liver diseases, *N.Y. State J. Med.* 860.

Mirand, E. A., Prentice, T. C., and Slaunwhite, W. R., 1959, Current studies on the role of erythropoietin on erythropoiesis, *Ann. N.Y. Acad. Sci.* **77**:677.

Mohandas, N., and Prenant, M., 1978, Three dimensional model of bone marrow, *Blood* **51**:633.

Moolten, F. L., and Bucher, N. R., 1968, Regeneration of rat liver: Transfer of humoral agents by cross circulation, *Science* **158**:272.

Mori, K. J., and Ito, Y., 1977, Recovery of irradiated bone marrow cells in carbon-treated mice, *Radiat. Res.* **69**:134.

Moriyama, Y., Saito, H., and Kinoshita, Y., 1970, Erythropoietin in plasma from patients with chronic renal failure, *Hematologia* **4**:15.

Morley, C. G., 1974, Humoral regulation of liver regeneration and tissue growth, *Perspect. Biol. Med.* **17**:411.

Morley, C. G., and Kingdon, H. S., 1973, The regulation of cell growth. I. Identification and partial characterization of a DNA synthesis stimulating factor from the serum of partially hepatectomized rats, *Biochim. Biophys. Acta* **308**:260.

Morrow, S. H., and DiLuzio, N. R., 1965, The fate of foreign red cells in mice with altered reticuloendothelial function, *Proc. Soc. Exp. Biol. Med.* **119**:647.

Muir, R., and Niven, J. S., 1935, The local formation of blood pigments, *J. Pathol. Bacteriol.* **41**:183.

Mujovic, V. M., and Fisher, J. W., 1975, The role of prostaglandins in the production of erythropoietin (ESF) by the kidney. II. Effects of indomethacin on erythropoietin production following hypoxia in dogs, *Life Sci.* **16**:463.

Murota, S., Morita, I., and Abe, M., 1977, The effects of thromboxane B_2 and 6-ketoprostaglandin $F_{1\alpha}$ on cultured fibroblasts, *Biochim. Biophys. Acta* **479**:122.

Murphy, M. J., and Urabe, A., 1978, Modulatory effect of macrophages on erythropoiesis, in: *In Vitro Aspects of Erythropoiesis* (M. J. Murphy, ed.), pp. 189–191, Springer-Verlag, Berlin.

Naets, J. P., and Wittek, M., 1968, The presence of erythropoietin in the plasma of one anephric patient, *Blood* **31:**249.

Naughton, B. A., 1978, Extrarenal erythropoietin and factors influencing its production, Doctoral dissertation, New York University.

Naughton, B. A., Kaplan, S. M., Roy, M., Piliero, S. J., and Gordon, A. S., 1976, Hepatic regeneration and erythropoietin production in the rat, *Blood* **48:**145A.

Naughton, B. A., Kaplan, S. M., Roy, M., Piliero, S. J., and Gordon, A. S., 1977a, Hepatic regeneration and erythropoietin production in the rat, *Science* **196:**301.

Naughton, B. A., Piliero, J. A., Kruger, R. E., Birnbach, D. J., Roy, M., Piliero, S. J., and Gordon, A. S., 1977b, Age-related variations in hepatic regeneration and erythropoietin production in the rat, *Am. J. Anat.* **149:**431.

Naughton, B. A., Gordon, A. S., Piliero, S. J., and Liu, P., 1978, Extrarenal erythropoietin, in: *In Vitro Aspects of Erythropoiesis* (M. J. Murphy, ed.), pp. 194–217, Springer-Verlag, Berlin.

Naughton, B. A., Kolks, G. A., Arce, J. M., Liu, P., Piliero, S. J., and Gordon, A. S., 1979a, The regenerating liver: A site of erythropoiesis in the adult rat, *Am. J. Anat.* **156:**159.

Naughton, B. A., Birnbach, D. J., Liu, P., Kolks, G. A., Tung, M., Piliero, J. A., Piliero, S. J., and Gordon, A. S., 1979b, Reticuloendothelial system (RES) hyperfunction and erythropoietin (Ep) production in the regenerating liver, *J. Surg. Oncol.* **12:**227.

Naughton, B. A., Birnbach, D. J., Liu, P., Kolks, G. A., Tung, M., Piliero, S. J., and Gordon, A. S., 1979c, Erythropoietin (Ep) production and Kupffer cell alterations following nephrectomy, hypoxia, or combined nephrectomy and hypoxia, *Proc. Soc. Exp. Biol. Med.* **160:**170.

Naughton, B. A., Liu, P., Kolks, G. A., Arce, J. M., Piliero, S. J., and Gordon, A. S., 1980, Evidence for a sexual variation in the production of a hepatic erythropoietic factor by hepatectomized rats, *Am. J. Physiol.* **238:**E245.

Naughton, B. A., Naughton, G. K., Liu, P., DePaola, L. A., Ryan, J. M., LoBue, J., Piliero, S. J., and Gordon, A. S., 1981, A renal inhibitor to hepatic erythropoietin production, *J. Med.* **12:**159.

Naughton, B. A., Liu, P., Naughton, G. K., Zuckerman, G. B., and Gordon, A. S., 1982a, The influence of pancreatic hormones and diabetogenic procedures on erythropoiesis, *J. Surg. Oncol.* **21:**97.

Naughton, B. A., Naughton, G. K., Liu, P., Arce, J. M., Piliero, S. J., and Gordon, A. S., 1982b, The effects of prostaglandins on extrarenal erythropoietin production, *Proc. Soc. Exp. Biol. Med.* **170:**231.

Naughton, B. A., Gamba-Vitalo, C., Naughton, G. K., Liu, P., and Gordon, A. S., 1982c, Granulopoiesis and colony stimulating factor production in the regenerating liver, *Exp. Hematol.* **10:**449.

Naughton, B. A., Liu, P., Naughton, G. K., and Gordon, A. S., 1983, Evidence for an erythropoietin stimulating factor in patients with renal and hepatic disease, *Acta Haematol.* **37:**105.

Naughton, G. K., 1982, Ultrastructural identification of the site of erythropoietin production in the rat liver, Doctoral dissertation, New York University.

Naughton, G. K., Naughton, B. A., Liu, P., Piliero, S. J., and Gordon, A. S., 1981, Localization of erythropoietin in rat liver, *Physiologist* **24:**87A.

Naughton, G. K., Naughton, B. A., Liu, P., Piliero, S. J., and Gordon, A. S., 1984, Localization of the site of erythropoietin production in the rat liver, *Am. J. Anat.*, in press.

Neal, W. A., Lewis, J. P., Welch, E. T., Lutcher, C. L., Moores, R. R., and Wright, C. S., 1979, Inhibition of erythropoiesis by plasma components from sheep, goats, and rabbits, *Am. J. Vet. Res.* **40:**493.

Necas, E., Neuwirt, J., and Gross, J., 1976, Species specificity of guinea pig erythropoietin, *Endocrinol. Exp.* **10:**303.

Nelp, W. B., Gohil, M. N., Larson, S. M., and Bower, R. E., 1970, Long term effects of local irradiation of the marrow on erythron and red cell function, *Blood* **36:**617.

Nelson, D. S., 1973, Production by stimulated macrophages of factors depressing lymphocyte transformation, *Nature (London)* **246:**306.

Nicol, T., Quantock, D. C., and Vernon-Roberts, B., 1967, The effects of steroid hormones on local and general reticuloendothelial activity: Relation of steroid structure to function, in: *The Re-*

ticuloendothelial System and Atherosclerosis (N. R. DiLuzio and R. Paoletti, eds.), pp. 221–242, Plenum Press, New York.

Niskanen, E. O., Burgaleta, C., Cline, M. J., and Golde, D. W., 1978, Effect of glucan, a macrophage activator, on murine hemopoietic cell proliferation in diffusion chambers in mice, *Cancer Res.* **38:**1406.

Niskanen, E., Olofsson, T., and Cline, M. J., 1979, Hemopoietic precursor cells in human peripheral blood, *Am. J. Hematol.* **7:**201.

Novikoff, A. N., 1959, Heterogeneity within the hepatic lobule of the rat, *J. Histol. Cytol.* **7:**240.

Noyes, W. D., Bothwell, T. H., and Finch, C. A., 1960, The role of the reticuloendothelial cell in iron metabolism, *Br. J. Haematol.* **6:**43.

Okuyama, S., Sato, T., Takahashi, K., Ito, Y., and Matsuzawa, T., 1975, Isolated marrow reticuloendothelial activation and erythropoietic response following an acute blood loss, *J. Reticuloendothelial Soc.* **17:**353.

Orlic, D., and Mirand, E. A., 1977, An electron microscopic study of hepatic erythropoiesis in adult mice with Friend virus disease, *Lab. Invest.* **37:**579.

Orlic, D., Gordon, A. S., and Rhodin, J. A., 1965, An ultrastructural study of erythropoietin-induced red cell formation in mouse spleen, *J. Ultrastruct. Res.* **13:**516.

Ortega, J. A., Malekzadeh, M. H., Dukes, P. P., Ma, A., Pennisi, A. V., Fine, R. N., and Shore, N. A., 1977, Exceptionally high serum erythropoietin activity in an anephric patient with severe anemia, *Am. J. Hematol.* **2:**299.

Papper, S., and Vaamonde, C. A., 1971, The kidney and liver disease, in: *Diseases of the Kidney*, Vol. 2 (M. B. Strauss and L. G. Welt, eds.), pp. 1139–1154, Little, Brown, Boston.

Parry, E. W., 1978, Studies on mobilization of Kupffer cells in mice. I. The effect of CCl_4-induced liver necrosis, *J. Comp. Pathol.* **88:**481.

Paul, D., Leffert, H., and Sato, G., 1972, Stimulation of DNA and protein synthesis in fetal rat liver cells by serum from partially hepatectomized rats, *Proc. Natl. Acad. Sci. USA* **69:**373.

Paul, P., Rothmann, S. A., Naughton, B. A., and Gordon, A. S., 1981, Presence of erythropoietin in Kupffer cell conditioned media, *Exp. Hematol.* **9**(Suppl. 9):57A.

Pechet, G. S., Rogers, A. E., and MacDonald, R. A., 1963, Inhibitory humoral factors and liver regeneration, *Fed. Proc.* **22:**192.

Peschle, C., Sasso, G. F., Mastroberardino, G., and Condorelli, M., 1971, The mechanism of endocrine influences on erythropoiesis, *J. Lab. Clin. Med.* **78:**20.

Peschle, C., Marone, G., Genovese, A., Magli, C., and Condorelli, M., 1976, Hepatic erythropoietin: Enhanced production in anephric rats with hyperplasia of Kupffer cells, *Br. J. Haematol.* **32:**105.

Petrelli, F., and Marsila, G., 1975, Consequence of splenectomy of the reticuloendothelial system and on phagocytic activity in rat liver, *Boll. Soc. Ital. Biol. Sper.* **49:**1120.

Pictet, C., Orci, L., Forssmann, W. G., and Girardier, L., 1969, An electron microscopic study of the perfusion fixed spleen. II. Nurse cells and erythrophagocytosis, *Z. Zellforsch. Mikrosk. Anat.* **96:**400.

Ploemacher, R. E., and van Soest, P. L., 1977, Morphological investigation on phenylhydrazine induced erythropoiesis in adult mouse liver, *Cell Tissue Res.* **178:**435.

Ploemacher, R. E., and van Soest, P. L., 1979, Haemopoietic stroma. II. Enhancement of post irradiation recovery of bone marrow stroma by latex particles, *IRCS Med. Sci.* **7:**549.

Policard, A., and Bessis, M., 1958, Sur une mode d'incorporation des macromolécules par la celle, visible au microscope électronique; la rhophéocytose, *C.R. Acad. Sci.* **246:**3194.

Powsner, E., and Berman, L., 1967, Effect of erythropoietin on DNA synthesis by erythroblasts, *Blood* **30:**189.

Preuss, H. G., Terryi, E. F., and Keller, A. I., 1970, Renotropic factor(s) in plasma from uninephrectomized rats, *Nephron* **7:**459.

Price, J. B., Jr., 1976, Insulin and glucagon as modifiers of DNA synthesis in the regenerating rat liver, *Metabolismo* **25:**1427.

Proctor, H. J., Wood, J. J., and Palladino, W. G., 1980, The effect of glucagon on hepatic cellular energetics during a low flow state, *Surgery* **87:**369.

Radosevic-Stasic, B., Cuk, M., and Rukavina, D., 1979, Cellular and humoral mediators in the compensatory renal growth, *Perspect. Biol.* **81:**149.

Radtke, H., Bartos, D., Bartos, F., Rege, A. B., Campbell, R., and Fisher, J. W., 1980, Characterization of an inhibitor of erythropoiesis in chronic renal failure: *In vitro* studies in mouse fetal liver cell cultures, *Exp. Hematol.* **8**(Suppl. 8):298.

Rawls, W. B., Goldzicher, J., Tichner, J. B., and Baker, E., 1954, The effect of (1) hypophysectomy alone and (2) hypophysectomy plus various combinations of steroids and ACTH upon phagocytosis by the reticuloendothelial system of intravenously injected India ink particles, *J. Lab. Clin. Med.* **44**:512.

Razin, E., and Globerson, A., 1979, The effect of various prostaglandins on plasma membrane receptors and function of mouse macrophages, in: *Function and Structure of the Immune System* (W. Muller-Ruchholtz and H. K. Muller-Hermelink, eds.), pp. 415–438, Plenum Press, New York.

Razin, E., Razin, M., and Lohmann-Matthes, M. L., 1980, The role of prostaglandins in the development of macrophages from bone marrow cells, *J. Reticuloendothelial Soc.* **27**:377.

Rich, I. N., Heit, W., and Kubanek, B., 1980, An erythropoietic stimulating factor similar to erythropoietin released by macrophages after treatment with silica, *Blut* **40**:297.

Rifkind, R. A., 1965, Heinz body anemia: An ultrastructural study. 2. Red cell sequestration and destruction, *Blood* **26**:433.

Rifkind, R. A., 1966, Destruction of injured red cells *in vivo, Am. J. Med.* **41**:711.

Rifkind, R. A., Chui, D., and Epler, J., 1969, An ultrastructural study of early morphogenetic events during the establishment of fetal hepatic erythropoiesis, *J. Cell Biol.* **40**:343.

Riggi, S. J., and DiLuzio, N. R., 1961, Identification of a reticuloendothelial stimulating agent in zymosan, *Am. J. Physiol.* **200**:297.

Rogers, A. E., Shaka, J. A., Pechet, G., and MacDonald, R. A., 1961, Absence of a blood borne factor affecting liver regeneration in paired and triplet parabiotic rats, *Fed. Proc.* **20**:287A.

Roser, B., 1965, The distribution of intravenously injected peritoneal macrophages in the mouse, *Aust. J. Exp. Biol. Med. Sci.* **43**:553.

Roser, B., 1968, The distribution of intravenously injected Kupffer cells in the mouse, *J. Reticuloendothelial Soc.* **5**:455.

Rous, P., 1923, Destruction of the red corpuscle in health and disease, *Physiol. Rev.* **3**:75.

Roy, M., Carson, E. J., Burdowski, A., Naughton, B. A., Grubman, S. A., Piliero, S. J., and Gordon, A. S., 1979, The effect of reticuloendothelial overload on extrarenal erythrogenin and erythropoietin production, *J. Reticuloendothelial Soc.* **25**:151.

Rytomaa, T., 1978, The erythrocytic chalone, in: *The Year in Hematology* (R. Silber, J. LoBue, and A. S. Gordon, eds.), pp. 321–339, Plenum Press, New York.

Saba, T. M., 1970, Liver blood flow and intravascular colloid clearance alterations following partial hepatectomy, *J. Reticuloendothelial Soc.* **7**:406.

Scaife, J. F., 1970, Liver homeostasis: An *in vitro* evaluation of a possible specific chalone, *Experientia* **26**:1071.

Schooley, J. C., and Mahlmann, L. J., 1971, Stimulation of erythropoiesis in plethoric mice by prostaglandins and its inhibition by anti-erythropoietin, *Proc. Soc. Exp. Biol. Med.* **138**:523.

Schooley, J. C., and Mahlmann, L. J., 1972a, Erythropoietin production in the anephric rat. I. Relationship between nephrectomy, time of hypoxic exposure, and erythropoietin production, *Blood* **39**:31.

Schooley, J. C., and Mahlmann, L. J., 1972b, Evidence for the *de novo* synthesis of erythropoietin in hypoxic rats, *Blood* **40**:662.

Schooley, J. C., and Mahlmann, L. J., 1974, Extrarenal erythropoietin production by the liver in the weanling rat, *Proc. Soc. Exp. Biol. Med.* **145**:1081.

Schroeder, E. T., Shear, L., Sancetta, S. M., and Gabuzda, G. J., 1967, Renal failure in patients with cirrhosis of the liver, *Am. J. Med.* **43**:887.

Sekas, G., Owen, W. G., and Cook, R. T., 1979, Fractionation and preliminary characterization of a low molecular weight bovine hepatic inhibitor of DNA synthesis in regenerating rat liver, *Exp. Cell Res.* **122**:47.

Sherlock, S., 1978, Primary biliary cirrhosis, *Am. J. Med.* **65**:217.

Silini, G., Pozzi, L. V., and Pons, S., 1967, Studies on the hematopoietic stem cells of mouse fetal liver, *J. Embryol. Exp. Morphol.* **17**:303.

Sjögren, U., 1976, Erythroblastic islands and extramedullary erythropoiesis in chronic myeloid leukemia, *Acta Haematol.* **55**:272.

Sjögren, U., and Brandt, L., 1974, Erythroblastic islands and ineffective erythropoiesis in vitamin B_{12} deficiency, *Acta Med. Scand.* **196**:369.

Skamene, E., Chayasirisobhon, W., and Konshaun, P. A., 1978, Increased phagocytic activity of splenectomized mice challenged with *Listeria monocytogenes, Immunology* **34**:901.

Slater, L. M., Muir, W. A., and Weed, R. I., 1968, Influence of splenectomy on insoluble hemoglobin inclusion bodies in B-thalassemic erythrocytes, *Blood* **31**:766.

Smythe, R. L., and Moore, R. O., 1958, A study of possible humoral factors in liver regeneration in the rat, *Surgery* **44**:561.

Sonnenborn, R., Perez, G. O., Epstein, M., Martello, O., and Pardo, D., 1977, Erythrocytosis associated with the nephrotic syndrome, *Arch. Intern. Med.* **137**:1068.

Sorenson, G. D., 1961, An electron microscopic study of hematopoiesis in the yolk sac, *Lab. Invest.* **10**:178.

Stang, H. D., and Boggs, D. R., 1977, Effect of methylcellulose injection on murine hematopoiesis, *Am. J. Physiol.* **233**:H234.

Starzyl, T. E., Francavilla, A., Halgrimson, C. G., Francavilla, F. R., Brown, T. H., and Putnam, C. W., 1973, The original hormone nature and action of hepatotrophic substances in portal venous blood, *Surg. Gynecol. Obstet.* **137**:179.

Starzyl, T. E., Tereblanche, J., Porter, K. A., Jones, A. F., Usui, S., and Mazzoni, G., 1979, A growth stimulating factor in regenerating canine liver, *Lancet* **2**:127.

Stenger, R. J., and Confer, D. B., 1965, Hepatocellular ultrastructure during liver regeneration after subtotal hepatectomy, *Exp. Mol. Pathol.* **5**:455.

Stephenson, J. R., Axelrad, A. A., McLeod, D. L., and Shreeve, M. M., 1971, Induction of colonies of hemoglobin synthesizing cells by erythropoietin *in vitro, Proc. Natl. Acad. Sci. USA* **68**:1942.

Stiffel, C., Mouton, D., and Biozzi, G., 1971, Role of RES in the defense against invasion by neoplastic, bacterial, and immunocompetent cells, in: *The Reticuloendothelial System and Immune Phenomena* (N. R. DiLuzio and K. Flemming, eds.), pp. 305–314, Plenum Press, New York.

Stohlman, F., Jr., 1961, Shortened survival of erythrocytes produced by erythropoietin or severe anemia, *Proc. Soc. Exp. Biol. Med.* **107**:884.

Stohlman, F., Jr., Ebbe, S., Morse, B., Howard, D., and Donovan, J., 1968, Regulation of erythropoiesis. XX. Kinetics of red cell production, *Ann. N.Y. Acad. Sci.* **149**:156.

Strecker, W., Silz, S., Ruhenstroth-Bauer, G., and Boettger, I., 1980, Insulin and glukagon im teilhepatektomierter ratten, *Z. Naturforsch.* **35**:65.

Susic, D., Milenkovic, P., and Pavlovic-Kentera, V., 1979, The effect of aspirin on erythropoietin formation in the rat, *Proc. Soc. Exp. Biol. Med.* **161**:476.

Tavassoli, M., 1974, Bone marrow erythroclasia: The function of perisinal macrophages relative to the uptake of red cells, *J. Reticuloendothelial Soc.* **15**:163.

Testa, N. G., and Hendry, J. H., 1977, Radiation induced hemopoiesis in adult mouse liver, *Exp. Hematol.* **5**:136.

Thomson, R. Y., and Clarke, A. M., 1965, Role of portal blood supply in liver regeneration, *Nature (London)* **208**:393.

Trentin, J. J., 1970, Influence of hematopoietic organ stroma (hematopoietic inductive microenvironments) on stem cell differentiation, in: *Regulation of Hematopoiesis* Vol. I (A. S. Gordon, ed.), pp. 161–186, Appleton–Century–Crofts, New York.

Umehara, S., and Iseki, T., 1953, Effects of humoral factors on the function of the reticuloendothelial system, *Med. Biol.* **26**:111.

Unanue, E. R., 1976, Secretory function of mononuclear phagocytes: A review, *Am. J. Pathol.* **83**:396.

van Furth, R., 1980, Monocyte origin of Kupffer cells, *Blood Cells* **6**:87.

Varon, M. L., and Cole, L. J., 1966, Hemopoietic colony-forming units in regenerating mouse liver: Suppression by anticoagulants, *Science* **153**:643.

von Ehrenstein, G., and Lockner, D., 1958, Sites of the physiological breakdown of the red blood corpuscles, *Nature (London)* **181**:911.

Wagner, H. N., Razzak, M. A., Gaurtner, R. A., Caine, W. P., and Feagin, O. T., 1962, Removal of erythrocytes from the circulation, *Arch. Intern. Med.* **110**:90.

Waldmann, I. A., Rosse, W. F., and Swarm, R. L., 1968, The erythropoiesis stimulatory factors produced by tumors, *Ann. N.Y. Acad. Sci.* **149**:509.

Wang, F., and Fried, W., 1972, Renal and extrarenal erythropoietin production in male and female rats of various ages, *J. Lab. Clin. Med.* **79**:181.

Weinbren, K., 1978, Other views about the hepatotrophic concept, in: *Hepatotrophic Factors* (R. Porter and J. Whelan, eds.), pp. 139–151, North-Holland, Amsterdam.

Weiss, L., and Chen, L. T., 1975, The organization of hematopoietic cords and vascular sinuses of rat bone marrow, *Blood Cells* **1**:617.

Whitcomb, W. H., and Moore, M. Z., 1965, The inhibitory effect of plasma from hypertransfused animals on erythrocyte iron incorporation in mice, *J. Lab. Clin. Med.* **66**:641.

Wisse, L., 1970, An electron microscopic study of the fenestrated endothelial lining of rat liver sinusoids, *J. Ultrastruct. Res.* **31**:125.

Wisse, E., 1978, Kupffer cells and peritoneal macrophages are different types of cells: A commentary, *Blood Cells* **4**:319.

Wood, W. G., 1974, Erythroid cell proliferation in human bone marrow suspension cultures, *Br. J. Haematol.* **26**:441.

Yoffey, J. M., and Yaffe, P., 1980, The phagocytic central reticular cell of the erythroblastic island in rat bone marrow: Changes in hypoxia and rebound, *J. Reticuloendothelial Soc.* **28**:37.

Zamboni, L., 1965, Electron microscopic studies of blood embryogenesis in humans. I. The ultrastructure of fetal liver, *J. Ultrastruct. Res.* **12**:509.

Zanjani, E. D., Peterson, E. N., Gordon, A. S., and Wasserman, L. R., 1974, Erythropoietin production in the fetus: Role of the kidney and maternal anemia, *J. Lab. Clin. Med.* **83**:281.

Zanjani, E. D., Poster, J., Burlington, H., Mann, L. I., and Wasserman, L. R., 1977, Liver as the primary site of erythropoietin formation in the fetus, *J. Lab. Clin. Med.* **89**:640.

Zucali, J. R., and Mirand, E. A., 1975, Biosynthesis of erythropoietin by mouse fetal liver, *Blood* **46**:85.

Zuckerman, G. B., Naughton, B. A., Gaito, A., and Gordon, A. S., 1984, The effect of methylcellulose on extrarenal erythropoietin production, *Proc. Soc. Exp. Biol. Med.*, in press.

8

Erythroclasia and Bilirubin Metabolism

JOHN C. NELSON, NADER G. IBRAHAM, and RICHARD D. LEVERE

1. ERYTHROCYTE DESTRUCTION

The various components of the RES comprise the principal if not the only graveyard for senescent normal erythrocytes, and a number of its structural features are especially suited for this function. Nevertheless, the principal determinants of the duration of the erythrocyte life span are the intrinsic biochemical and biophysical properties of the erythrocytes. Thus, there is an interplay between erythrocytes which have undergone subtle changes in their properties during aging and a sensitive and highly selective system of erythroclasia; this interaction determines the site, timing, and mechanisms of erythrocyte destruction.

1.1. THE LIFE SPAN OF THE ERYTHROCYTE

1.1.1. Methods

A variety of techniques have been used to measure the life span of circulating erythrocytes and have been important in developing our knowledge of the biochemical and biophysical basis of various hemolytic diseases (Berlin and Berk, 1975; Bentley, 1977). These techniques may be classified as (1) cohort labeling, in which a labeled tracer such as [^{15}N]- or [^{14}C]glycine is physiologically incorporated into developing cells over a limited period; (2) random labeling techniques where circulating erythrocytes are labeled independent of cell age, the most widely used labels being ^{51}Cr and ^{32}P-labeled DFP; and (3)

JOHN C. NELSON, NADER G. IBRAHAM, and RICHARD D. LEVERE • Department of Medicine, New York Medical College, Valhalla, New York 10595.

indirect methods in which measurements such as bilirubin or carbon monoxide production are used to calculate red cell life span. Although much discussion has focused on problems in using these techniques, and it is clear that no perfectly satisfactory means are available, important concepts of transfusion therapy and of the pathogenesis of various hemolytic anemias have emerged.

1.1.2. Finite Life Span

In man the data available from these studies are fairly uniform and indicate the normal potential life span is 115–120 days (Berlin and Berk, 1975). It is clear that in man, during their 1.6×10^5 recirculations (Allison, 1960), normal erythrocytes undergo senescence, and are destroyed almost exclusively in a nonrandom manner after a finite life span. A very low level of random destruction may take place (Eadie and Brown, 1953). The apparent lack of significant random destruction has led investigators to focus on those features of aging erythrocytes which could be important in mediating their destruction, in particular the presence of specific metabolic lesions.

1.2. CHANGES IN THE AGING ERYTHROCYTE

1.2.1. Enzymes and Metabolic Pathways

Reticulocytes have a comparatively high rate of metabolism, consuming some 7.5 times as much glucose as mature erythrocytes, the presence of mitochondria enabling them to utilize the Krebs cycle as well as the Embden–Meyerhof (EM) pathway and hexose monophosphate shunt (HMS). The mature erythrocyte, on the other hand, utilizes 90% of its glucose anaerobically through the EM pathway and about 10% aerobically through the HMS. Anaerobic glycolysis results in the net maximum production of 2 moles of ATP and the reduction of 2 moles of NAD to NADH per mole of glucose consumed with the production of lactate as the end product. The maximum yield of ATP via this pathway is altered by the proportion of 1,3-diphosphoglycerate diverted to the production of 2,3-diphosphoglycerate via a third pathway of glucose metabolism, the Rapoport–Leubering shunt. Glucose metabolized through this pathway is metabolized without a net gain of high-energy phosphate bonds as ATP. This pH-sensitive "energy clutch" pathway (Keitt and Bennett, 1966) serves to regulate both the rate of ADP phosphorylation to ATP and the concentration of 2,3-diphosphoglycerate which reversibly binds to deoxyhemoglobin reducing its oxygen affinity. Although only a small proportion of glucose is metabolized aerobically via the HMS, this pathway is of critical importance in protecting erythrocyte enzymes, membranes, and a small proportion of hemoglobin from oxidative damage through the maintenance of adequate levels of reduced glutathione.

It is clear that significant quantitative or qualitative alterations of key enzymes in each of the major metabolic pathways may result in premature red cell destruction manifested clinically as hemolysis detectable by a shortened red cell survival. Failure of the EM pathway occurs in a variety of congenital enzyme deficiencies including those of hexokinase, glucose phosphate isomerase, phosphofructokinase, aldolase, triose phosphate isomerase, and pyruvate kinase. These disorders are characterized by impaired ATP production and cycling of NAD^+ and NADH with a consequent impaired ability to maintain vital erythrocyte functions. There may be impaired maintenance of cation gradients with calcium gain, potassium and water loss, and increased erythrocyte rigidity reflecting the loss of normal erythrocyte osmotic relationships (Leblond *et al.*, 1978a,b).

The congenital enzymatic disorders of the HMS associated with premature red cell destruction include deficiencies of glucose-6-phosphate dehydrogenase, glutathione peroxidase, glutathione synthetase, and glutathione reductase. These disorders have in common enhanced susceptibility to the denaturation of hemoglobin by noxious environmental oxidative stresses leading to the formation of intracellular precipitates of hemoglobin as Heinz bodies, and possible enhanced susceptibility to other forms of oxidant damage as well.

The activities of many of these same enzymes decrease as the red cell ages. The enzymatic and some other physiochemical changes which accompany erythrocyte aging have been catalogued (Bunn, 1972; Wintrobe *et al.*, 1974). In addition to an overall decrease in the rate of glycolysis (Bernstein, 1959; Chapman and Schaumburg, 1967) and a decrease in red cell ATP (Brok *et al.*, 1966), the activity of specific enzymes of the EM pathway, hexokinase, glucose phosphate isomerase, phosphofructokinase, aldolase, glyceraldehyde-3-phosphate dehydrogenase, and pyruvate kinase, are all decreased. Among the HMS enzymes, the activities of glucose-6-phosphate dehydrogenase and 6-phosphogluconate dehydrogenase both decrease with red cell aging.

While it has been tempting to speculate that the diminished activities of multiple enzymes in aging erythrocytes constitute the biochemical lesion which determines erythrocyte survival, it must be emphasized that most of our understanding of the consequences of these deficits is derived from patients with isolated deficiencies of single enzymes. Even among patients with the same enzyme defect, the clinical severity of the disorder in terms of the apparent degree of hemolysis may be heterogeneous. In pyruvate kinase deficiency, for example, there may be significant disparities between measured enzyme activities and shortened erythrocyte life spans (Zuelzer *et al.*, 1968). Thus, precise characterization of the threshold biochemical lesion in energy metabolism which leads to physiological destruction has remained elusive.

Senescent erythrocytes, similar to other cells, show decreases in other enzymes not directly involved in energy metabolism. The specific activity of superoxide dismutase decreases with cell age suggesting that free radicals could also be important means to other metabolic or structural lesions (Bartosz *et al.*, 1981). The denaturation does not seem to involve H_2O_2, gamma irradiation, or interaction with glucose.

1.2.2. Membranes and Other Constituents

Total erythrocyte lipid, membrane lipid, and other cellular constituents not directly involved in energy metabolism also change during red cell aging. It is generally accepted that total lipid and cholesterol decrease during aging (Westerman *et al.*, 1963; Van Gastel *et al.*, 1965; Winterbourn and Batt, 1970). Significant organelle and membrane lipid is lost from reticulocytes during their maturation (Shattil and Cooper, 1972) and during their early circulation as mature erythrocytes, but little if any during the last 50 to 80% of their life span. While older studies suggested that there is also a decrease in sialic acid content which would account for the decreased negative surface charge number found in older cells (Walter and Selby, 1966) and possibly correlate with red cell survival (Durocher *et al.*, 1975), these findings are doubtful (Bocci, 1981).

During their life span, erythrocytes undergo volume reduction with a loss of hemoglobin and water as well as lipid (Van Gastel *et al.*, 1965; Piomelli *et al.*, 1967). There is initially a progressive increase in the hemoglobin concentration of young erythrocytes probably the result of water and volume loss during the first 25 days of the cell's circulation. Only later is there a progressive decrease in hemoglobin content contributing to the continued decrease in cell volume. This observation is consistent with the concept that there is loss of some hemoglobin along with a portion of cell membrane perhaps indicating progressive fragmentation during aging (Ganzoni *et al.*, 1971).

There is also a change in cation content during cell aging presumably as a consequence of lower cellular ATP. Erythrocyte sodium increases and potassium decreases (Bernstein, 1959). Both calcium and ATP have critical roles in the maintenance of the membrane skeleton through their interaction with spectrin and actin (Lux, 1979) and calcium a major potential role in regulating erythrocyte transglutaminase (Lorand *et al.*, 1979). At this time it is unclear what role these systems play in the aging of normal erythrocytes and in determining normal erythrocyte life span. The combined effects of ATP depletion, a decreased lipid protein ratio, increased oxidant damage, intracellular calcium accumulation and loss of portions of cell membrane as fragments or small vesicles, likely result in modifications of cytoskeletal proteins and possibly their redistribution in the membrane (Sheetz and Singer, 1977; Shotton *et al.*, 1978). Perhaps as a result of these or other processes, neoantigenic determinants may appear on the senescent erythrocyte surface. Kay (1975, 1978) has described a 62,000-dalton cryptic antigen which appears on aged erythrocytes and is capable of binding an IgG autoantibody. The binding of such antibodies, perhaps natural antisugar antibodies, can provide a recognition signal to the RES followed by phagocytic clearance, and a pathway to erythrocyte destruction.

1.2.3. Physical Properties

Whatever the critical biochemical events which lead to normal erythrocyte destruction are, major attention has focused on various physical properties of erythrocytes as a final common pathway of injury. In order for erythrocytes to carry out their expected function of tissue oxygen delivery, they must possess a remarkable degree of deformability. While their diameter averages 8 nm, the capillaries through which they must pass are but 2–3 nm wide. Erythrocytes with reduced deformability are selectively removed by the RES and this has led to the concept that reduced erythrocyte deformability and the RES clearance mechanism is the final pathway which determines erythrocyte survival.

Deformability of red cells has been noted since the work of Van Leeuwenhoek in 1675 and subsequently documented with more modern microscopic and microcinematographic techniques (Krögh, 1930; Wayland, 1973; Branemark and Bagge, 1977). The general factors which regulate erythrocyte deformability have been reviewed and include (1) the viscoelastic properties of the cell membrane, (2) the cell geometry including the surface area to volume ratio, and (3) the viscosity of the intracellular hemoglobin milieu, i.e., the internal viscosity (Mohandas *et al.*, 1979).

In response to various mechanical forces, the erythrocyte behaves as a solid with viscoelastic properties and, until its yield point is reached, can undergo reversible elastic deformations. However, beyond the yield point the membrane behaves as a viscoplastic material and deformations may be irreversible. Numerical values expressing these properties of normal cell membranes have been obtained for the shear modulus of elasticity (deformability), viscosity in the elastic domain, the yield shear, and the plastic viscosity (Hochmuth *et al.*, 1973; Evans and Hochmuth, 1976a,b; Evans *et al.*, 1976).

Reduced erythrocyte deformability accompanying ATP depletion was reported by Weed *et al.* (1969) and is a concomitant of normal erythrocyte aging (La Celle and Arkin, 1970) as well as being characteristic of several types of hemolytic anemia. A close correlation of reduced erythrocyte survival and reduced deformability was found in patients with various erythrocyte disorders (Teitel, 1977). Other studies have been aimed at correlating deformability changes with biochemical studies. Reduced erythrocyte deformability results from heat treatment, sulfhydryl group reacting agents, and by ionophore-mediated calcium entry into erythrocytes which changes the elastic shear modulus and increases plastic deformation.

Erythrocytes, by nature of their biconcave disk shape, can undergo marked deformation and readily traverse passages which are smaller than their own diameter without the loss of surface area. With a signficant loss of cell membrane or an increase in cell water, spherocytes are formed with a resultant minimal surface area per volume and reduced deformability. For such cells, the modulus for area dilation is typically 4 orders of magnitude higher than the elastic shear modulus and substantially greater forces are needed to deform these cells (Evans *et al.*, 1976).

Within normal erythrocytes, hemoglobin in solution at normal concentrations behaves as a Newtonian fluid and, with a viscosity of only 4–6 cP, provides a relatively insignificant contribution to overall erythrocyte rigidity. However, with marked cellular dehydration when the mean cell hemoglobin concentration increases to greater than 38 g/dl, or with various abnormal hemoglobins or with precipitated hemoglobin as Heinz bodies, the internal viscosity may increase to more than 25 cP making this contribution significant (Cokelet and Meiselman, 1968).

A variety of techniques have been used in an attempt to measure erythrocyte deformability and determine the relative contribution of each of the three principal factors. These techniques include viscometry, micropipet aspiration, filtration, shear deformation of attached cells, resistive pulse spectroscopy, and ektacytometry (Chien, 1977; Mohandas *et al.*, 1979).

1.3. SITES OF ERYTHROCYTE DESTRUCTION

In the foregoing sections we have discussed a variety of alterations undergone by all normal erythrocytes during aging which, taken together, constitute a critical prehemolytic lesion. It has been accepted since early in this century that the RES and its mononuclear macrophages constitute the principal graveyard for senescent erythrocytes (Rous and Robertson, 1917; Rous, 1923). Rifkind (1966) has proposed that RES-mediated erythrocyte destruction involves the sequential operation of at least three component process: (1) injury—the development of the threshold prehemolytic lesion; (2) sequestration—resulting in selective filtration of injured cells from the general circulation and their concentration within the vascular spaces of the RES; and (3) hemolysis—constituting the enzymatic degradation of erythrocyte components within the cytoplasm of RE macrophages. In the sections which follow, specific components of the RES are discussed.

1.3.1. Spleen

It has long been proposed that the spleen has a major role in erythrocyte destruction. While this is not disputed in many types of hemolytic anemia where the spleen is indeed the major site of erythrocyte destruction, the evidence that the spleen has a major role in destruction of normally aged cells is surprisingly scanty. Erythrocyte fragments have been described in the spleen, and iron from senescent erythrocytes accumulates in the spleen suggesting that it is indeed such a graveyard (Finch *et al.*, 1950). It is an extraordinarily sensitive filter with unique anatomic features making it highly suitable for this function. Williams *et al.* (1974) attempted to quantitate the relative contribution of the spleen to the destruction of erythrocytes with normal survival. This was noteworthy because of the lack of previous quantitative data. In patients with normal red cell survival, the values for the fraction of destruction taking place in the spleen were typically less than 20%. Pettit (1977) concluded that it is unlikely that the spleen accounts for more than 20% of the total normal erythrocyte destruction. Above all, splenectomy is without significant effect on the normal erythrocyte life span (Tizianello *et al.*, 1961; Ultmann and Gordon, 1965), although it can be argued that the expected effects of splenectomy are offset by the effect of the accumulation of intracellular inclusions in erythrocytes (Bunn, 1972).

1.3.2. Bone Marrow and Liver

Although most attention has been directed toward the spleen as the organ responsible for sequestration and destruction of senescent erythrocytes, recent work has emphasized the potential importance of the bone marrow in this role. In the past there has been reasonable uncertainty in extrapolating from animal studies (where direct sampling is more feasible) because of well-known species differences. The rabbit, for example, has a notably erythroclastic marrow, with macrophages ingesting normal cells (Tavassoli, 1977). The marrow can function as the major site of removal of damaged erythrocytes (Mieschner, 1956; Ehrenstein and Lockner, 1958). Yet in mice and rats the bone marrow is a relatively inefficient trapping system and by some has been considered of major significance only when the hepatic and splenic RE systems are blockaded (Ultmann and Gordon, 1965; Keene and Jandl, 1965). Nevertheless, it has been stated that with the exception of the bone marrow, erythrophagocytic activity has never been documented for any other organ in normal man (Marton, 1975).

A role for the liver in normal erythroclasia as a major RE organ has been accepted even though direct evidence for such a role has been lacking. In fact, a specific subpopulation of sinusoidal cells adapted for this purpose has been proposed (Bissell *et al.*, 1972a,b).

1.4. MECHANISMS OF ERYTHROCYTE DESTRUCTION

1.4.1. Anatomic Correlations of the Spleen

Much interest has been directed to various anatomic features of the spleen which make it an extraordinarily sensitive filtering system. Blood flow through the spleen has both fast and slow components. Rapid transit of erythrocytes takes place through bypassing the cords and entering the sinuses directly, the slow-transit fraction being delayed through percolation through the cords.

Erythrocytes which enter the cords from the marginal zone must pass through the fenestrated wall which separates the cords from the sinuses before they can reach the venous drainage system. Erythrocytes must negotiate this passage by squeezing through fenestrations in the membrane and by squeezing between endothelial cells of the sinuses (Chen and Weiss, 1973; Weiss, 1962, 1974; Weiss and Tavassoli, 1970). The cordal side of this wall includes a network of reticular cells and macrophages which surrounds the cords and separates the sinuses. In passing through the cord walls, erythrocytes come into intimate contact with these cells.

Weiss and Tavassoli (1970) have discussed a variety of stresses besetting erythrocytes during this journey, including mechanical, rheological, and metabolic hazards and the physiologic consequences of stasis. Mechanical hazards include the arterial lumen which may contract, tortuous irregular and narrow passages of the marginal zone and cords, and slow flow through the reticular meshwork of the cords with intimate contact with the macrophages in this network. The rheologic problems include the marked elevation in erythrocyte concentration that results from plasma skimming in the white pulp. The resultant increased viscosity impairs flow leading to impaired oxygen-carrying capacity and a relative anoxic state, with a falling glucose and pH, and elevated lactic acid in the vicinity.

Remarkably, normal erythrocytes negotiate this passage, but reductions in cell deformability such as those which may result from various membrane changes, changes in the cytoplasmic state or from reduction in the surface area to volume ratio, further impede this passage resulting in sequestration and erythrocyte destruction. Based on these relationships, the erythrocyte deformability could be the major determinant of red cell life span in a variety of disorders (Weed, 1970; La Celle and Arkin, 1970; Jacob, 1974).

1.4.2. Anatomic Correlations of the Bone Marrow

The circulatory system through the bone marrow, in contrast to that of the spleen, is best described as a closed system. Circulating blood in the marrow sinusoids is separated by endothelium from the extravascular marrow parenchyma. The marrow macrophages are thus localized to an extravascular compartment separated by endothelium from the circulating blood. Mammalian erythropoiesis takes place extravascularly and cells which are destined for the peripheral circulation via the marrow sinuses must traverse the marrow blood barrier.

Much of our current information regarding this process concerns the means whereby nucleated erythrocytes lose their nuclei to become anucleate reticulocytes and traverse this barrier in the delivery process. The barrier consists of the marrow sinus covered by a thin layer of endothelium whose margins overlap considerably and which lack tight junctions and a discontinuous adventitial layer. The endothelium apparently provides the principal structural basis for the selectivity of the barrier (Tavassoli, 1978). The loss of the erythrocyte nucleus may occur through an active extrusion process coincident with various morphologic and metabolic changes within the cell (Bessis and Bricka, 1952; Simpson and Kling, 1967; Skutelsky and Danon, 1970). In passing through the marrow blood barrier to the marrow sinuses, red cells pass transendothelially via pores rather than through the interendothelial junctions. Macrophages can phagocytize damaged developing erythrocytes resulting in the process of ineffective erythropoiesis which is felt to be quantitatively unimportant under normal circumstances. In fact, in rabbits the perisinal macrophages have been proposed as providing a mechanism to adjust the rate of reticulocyte delivery to the peripheral circulation (Tavassoli, 1977).

It has been observed in rabbits that the bone marrow can function as a major site of removal of damaged erythrocytes. Indeed, as a filter for heat-damaged erythrocytes it can be the equivalant of the spleen, although the latter is more effective in removing Heinz bodies (Ehrenstein and Lockner, 1958; Hughes-Jones, 1961). In man, under various pathological conditions, erythrophagocytosis has been observed with conventional techniques. Such conditions include the aplastic crisis of hereditary spherocytosis (Sansone, 1955), autoimmune hemolytic anemias (Sansone, 1962; Bessis, 1972), and the histiocytosis associated with military tuberculosis (Chandra *et al.*, 1975).

Only recently have direct observations supported the concept that erythrophagocytosis in the bone marrow is a physiological process in man. Although not seen in the usual marrow aspirates from normal human subjects, this process may be demonstrated by special staining techniques and by electron microscopy (Marton, 1970, 1975).

Examination of biopsy specimens which have been embedded in methacrylate or epoxy resin has shown that erythrophagocytosis is seen in patients without bone marrow disease in both intrasinuoidal (vascular) and intraparenchymal (extravascular) compartments. In rabbits macrophages may penetrate the sinus walls to phagocytize erythrocytes within the sinus lumen, this penetration seemingly resulting from the perforation of a single cell rather than passing through the junction between endothelial cells. Such projections are cytoplasmic and continuity of the endothelium is seen in adjacent areas (DeBruyn *et al.*, 1971; Cambell, 1972; Tavassoli and Crosby, 1973). Marton (1975) has derived calculations which indicate that there is sufficient bone marrow erythrophagocytosis to account for a major portion of the destruction of senescent erythrocytes. It must be emphasized, however, that in man there has not yet been a rigorous demonstration that the bone marrow provides the major physiological mechanism for senescent erythrocyte destruction. These observations could represent phagocytosis of newly formed defective cells or a control system for reticulocyte release.

In summary, the bone marrow does not have the specialized cord structure of the spleen. The marrow sinusoids provide a wide vascular bed but with slow erythrocyte passage. The thin sinosoid wall allows passage of macrophages from the marrow to the sinusoid to engulf passing erythrocytes.

1.4.3. Anatomic Correlations of the Liver

The specialized anatomy of the liver provides some similarity to that of the spleen in that there is direct contact between circulating cells and the cordal macrophages. In the liver, which is principally a sinusoidal organ, there is an incomplete layer of sinus endothelial cells and the phagocytic Kupffer cells are in contact with the circulating erythrocytes (Scott and Deane, 1966). It has also been proposed that there is a specific population of sinusoidal cells involved in erythrophagocytosis (Bissell *et al.*, 1972a,b). There is, however, no delayed mixing component. In general, it has been held that the liver is effective in clearing more severely damaged erythrocytes, and special requirements must be met in order for it to have this function (Rifkind, 1966).

It may be possible to infer the importance of various sites in erythrocyte destruction through study of the enzymes involved in hemoglobin catabolism, as detailed in Section 2.2.1. Although the major enzyme system involved in heme catabolism, heme oxygenase, has widespread tissue distribution, its major activities are in the spleen, bone marrow, and liver (Tenhunen *et al.*, 1970; Maines and Kappas, 1974; Ibraham and Levere, 1980).

1.4.4. Erythrocyte Destruction

Despite these various structural features of the spleen, bone marrow, and liver, the exact means by which intact normal aged erythrocytes are destroyed have never been clarified. Several major mechanisms have been proposed, but each is based largely on observations from patients with various sorts of hemolytic anemia. The major mechanisms which may be important are (1) progressive fragmentation, (2) osmotic cell lysis, and (3) erythrophagocytosis.

Based on their failure to directly observe widespread erythrophagocytosis, Rous and Robertson (1917) proposed that erythrocytes undergo progressive fragmentation. In this process, portions of the cell membrane with or without cytoplasmic components, including hemoglobin, are lost. Presumably, progressive microinjury occurs each time followed by self-repair. This is consistent with the observations previously cited indicating that erythrocyte lipid and hemoglobin are lost during the cell's life span and that the cell becomes smaller and less deformable. If fragmentation progresses through the formation of vesicles, most are probably cleared locally as soon as they are formed. Evidence suggests that the spleen and bone marrow also can function as effective clearance sites for those vesicles which gain access to the circulation (Bocci *et al.*, 1980). The final event in the process could be either whole cell lysis without repair or progressive fragmentation to a hemoglobin-containing dust.

The progressive entry of water into erythrocytes results in swelling until a critical volume is reached, causing membrane defects and loss of cell constituents (Seeman, 1967), and is actually followed by a membrane repair. The tendency to undergo osmotic lysis, i.e., osmotic fragility, is greatly enhanced by loss of glycolytic intermediates and ATP which results from cell aging.

Erythrophagocytosis, the ingestion of intact erythrocytes by phagocytic cells, has been viewed principally as a pathological mechanism of erythrocyte destruction. However, as discussed, it has been proposed that erythrophagocytosis is also a physiological process (Marton, 1970, 1975). The appearance of a neoantigen with surface-bound IgG may be the consequence of multiple metabolic and mechanical injuries and be the signal for clearance of normal cells at the end of their life span.

It is likely that in reality each of the processes merges, and that all are important in erythrocyte aging and destruction. Circulating erythrocytes either freshly released or nearing the end of their life span undergo a series of changes. There is an early loss of lipid followed later by loss of hemoglobin, increased membrane permeability, changes in cation content, decreased deformability, and increased fragility and rearrangement of membrane structure. Eventually, irreversible changes take place, and there is destruction of erythrocytes in various microenvironments.

Following destruction of erythrocytes in various sites, the globin chains of hemoglobin are hydrolyzed into their component amino acids which eventually enter the metabolic pool (Ehrenreich and Cohn, 1968), and iron is reutilized.

Heme is catabolized to bilirubin which is conjugated in the liver and excreted in the bile where it is further reduced in the colon to urobilinogen. The RES plays a critical role in many of these steps.

2. HEMOGLOBIN AND HEME METABOLISM

2.1. REGULATION OF HEME METABOLISM

All normal mammalian hemoglobin molecules consist of tetramers of four globin chains with each chain bound to a heme prosthetic group. In the normal adult, each hemoglobin molecule contains two α chains and two non-α chains. It is the non-α chains which determine the identity of the hemoglobin. In the normal adult human, hemoglobin A ($\alpha_2\beta_2$) is 97% of the hemoglobin present, while hemoglobin A_2 ($\alpha_2\delta_2$) and hemoglobin F ($\alpha_2\gamma_2$) represent less than 3 and 1%, respectively. Mature red cells contain virtually only complete hemoglobin molecules and do not synthesize hemoglobin. There are only minute amounts of unbound chains in the erythrocyte. Thus, strict control of both polypeptide chain synthesis and heme synthesis must occur throughout erythropoiesis to assure this balanced synthesis of hemoglobin. Although it is beyond the scope of this chapter to cover this prolific field in depth, we will summarize the major molecular control mechanisms in normal heme synthesis in order to understand the molecular basis of its degradation and bilirubin formation.

The study of heme turnover has been one of the major areas of research in this laboratory for the past 20 years. The biosynthesis of porphyrin and heme and the control mechanisms regulating this pathway have been extensively characterized in liver tissue and cell culture systems (Gidari and Levere, 1977; Bottomley, 1977). Less extensive investigations have characterized the pathways in erythropoietic tissue. The principal difficulties which have hampered the study of the heme synthetic pathway in the immature erythron are the association of the erythroid tissues with a variety of other marrow cells, the dynamic state of red cell precursors as a continuously developing tissue, and a lack of sensitive specific methods for enzyme assays. Murine erythroleukemia cells, transformed by Friend leukemia virus in tissue culture (Friend *et al.*, 1971), have been used to study heme and globin synthesis (Ebert and Ikawa, 1974; Granick and Sassa, 1978). In the presence of dimethylsulfoxide (DMSO), these cells begin to differentiate into early erythroid cells with an increase in globin mRNA and hemoglobin synthesis and are recognizable as erythroblasts (Nudel *et al.*, 1977). On the other hand, study of heme degradation in the erythroid cells has not been reported except by Ibraham and Levere (1980) and by Hoffman *et al.* (1980) in a human leukemic cell line and in mouse bone marrow. These cells have been used as models for examining the steps involved in heme turnover. The following scheme summarizes the enzymatic steps involved in heme metabolism and bilirubin formation.

As shown in Fig. 1, in the first enzymatic step glycine and succinyl-coenzyme A are combined within the mitochondrion to form δ-aminolevulinic acid (ALA). The synthesis of ALA is medited by ALA synthetase (ALA-S), and probably proceeds through a short-lived intermediate, α-amino-β-ketoadipic acid, which yields ALA after spontaneous decarboxylation. The cytoplasmic enzyme ALA dehydratase (ALA-D) catalyzes the condensation of two molecules of ALA forming the monopyrrole porphobilinogen (PBG). The synthesis of the first tetrapyrrole in the pathway, uroporphyrinogen (Urogen) III, requires the action of two enzymes, uroporphyrinogen I synthetase (Uro-S) and uroporphyrinogen III cosynthetase (Uro-CoS), and utilizes four molecules of PBG as the substrate. Uro-S, in absence of Uro-CoS, is able to direct the synthesis only of the Urogen I isomer, which is not utilized for the formation of heme. In the cytoplasm, the four acetic acid side chains of Urogen III are sequentially decarboxylated by the enzyme(s) uroporphyrinogen decarboxylase (Uro-D), yielding coproporphyrinogen (Coprogen) III. The enzymatic production of the vinyl groups of protoporphyrinogen (Protogen) IX occurs by the modification of two of the four propionic acid substituents of Coprogen III within the mitochondrion. This reaction is catalyzed by the enzyme coproporphyrinogen decarboxylase (Copro-D), which is also called coporphyrinogenase. In the ultimate reaction, the dehydrogenation of Protogen IX by the enzyme protoporphyrinogen oxidase (Proto-O) results in the production of protoporphyrin (Proto) IX. These latter two reactions have been combined in Fig. 1 and only the net product Proto IX is shown. In the final reaction of the pathway, the mitochondri-

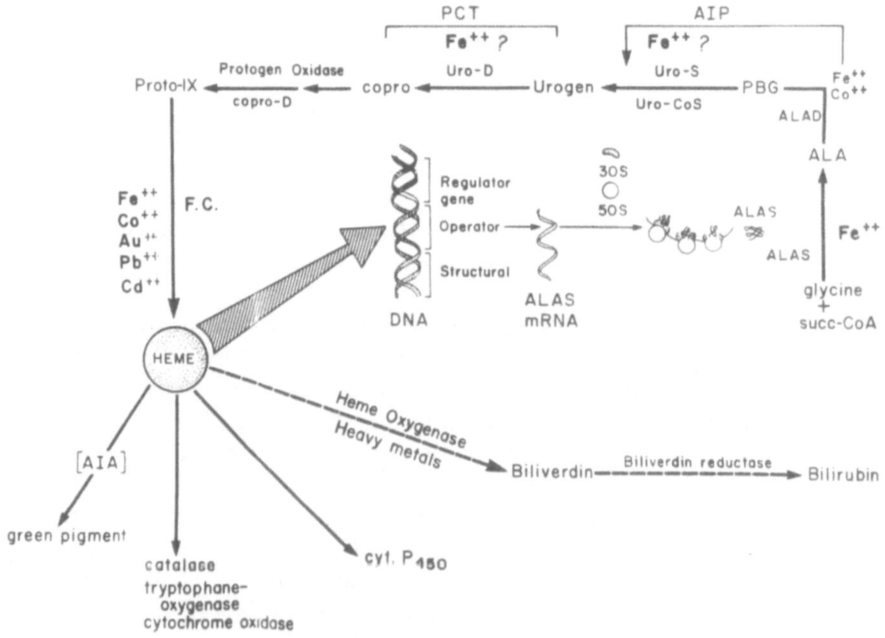

FIGURE 1. Heme biosynthetic and degradative pathways and their regulation. In the liver cells, heme regulates ALA synthetase by negative-feedback inhibition or repression. In the erythroid cells, heme may act as inducer of ALA synthetase and other heme pathway enzymes. Defective synthetic steps in acute intermittent porphyria (AIP) and porphyria cutanea tarda (PCT) are shown.

al enzyme ferrochelatase mediates the insertion of ferrous iron (Fe^{2+}) into the Proto IX ring to form heme.

Heme is the most ubiquitous tetrapyrrole in mammalian cells and constitutes the prosthetic moiety of other widespread hemoproteins in addition to hemoglobin, including myoglobin, catalase, mitochondrial cytochromes, tryptophan pyrrolase, and the microsomal hemoproteins b_5 and P-450. Excess heme is oxidatively degraded to form an open-chain tetrapyrrole, biliverdin IX. As shown in Fig. 1, this step is catalyzed by the newly purified enzyme protein called heme oxygenase with 1 mole of carbon monoxide being formed and the central metal being released. The biliverdin formed is reduced at the central methane bridge by the action of a cytosol enzyme, biliverdin reductase, to form bilirubin. Heme may also be degraded to products other than bilirubin under certain conditions. When various drugs such as the barbiturate analog allylisopropylacetamide (AIA) are administered to animals, the molar ratio of carbon monoxide produced to bilirubin recovered exceeds 1.0, indicating a conversion of heme to metabolites other than bilirubin (Landaw *et al.*, 1970). These metabolites, commonly referred to as "green pigments," were first observed by Schwartz and Ikeda (1952), DeMatteis and Unseld (1976), and DeMatteis (1978). Heme may also be degraded to other products, as in patients with congenital Heinz body hemolytic anemia, a condition in which the urine contains dark pigments. These have been tentatively identified as dipyrroles by Kreimer-Birnbaum *et al.* (1966) in normal rodents exposed to phenylhydrazine. Here, the heme moiety of the oxidatively denatured hemoglobin is degraded to metabolites exhibiting solubility and spectroscopic properties of meso-bilifuscins (Goldstein *et al.*, 1968).

There are important interrelationships between the heme synthetic and degradative pathways. If hepatic heme is generated in the mitochondria in excess of the physiological requirement of red cells and the prosthetic groups of cytochromes or enzymes, this "free heme" hypothetically would have two effects, each of which would tend to result in a net decrease of heme concentration: (1) the increased heme would repress the formation of δALA-S; (2) the increased heme would be degraded more rapidly by the induction of heme oxygenase. In contrast, when "free heme" decreases, the synthesis of δALA-S is not repressed. Actually, more of this enzyme would be generated and therefore maximum heme would be synthesized because this enzyme is the proposed rate-limiting enzyme of the heme biosynthetic chain.

However, this hypothesis applies to liver cells and not to the erythroid cells for several reasons: (1) heme synthesis is also limited by the rate of globin synthesis (Levere and Granick, 1967) and the availability of ferrochelatase (Sassa, 1976; Rutherford *et al.*, 1979), the terminal enzyme in the heme biosynthetic pathway; (2) the concentration of free heme necessary to inhibit erythroid ALA-S must be higher than 25 μM which is an unphysiological concentration (Ibraham *et al.*, 1978); (3) evidence has been presented by Ebert *et al.* (1979) and Sassa (1976) suggesting multiple sites of regulation of heme synthesis in erythroleukemia cells; (4) heme appears to stimulate both globin and nonglobin synthesis (Ross and Sautner, 1976; Dabney and Blaudet, 1977; Lodish,

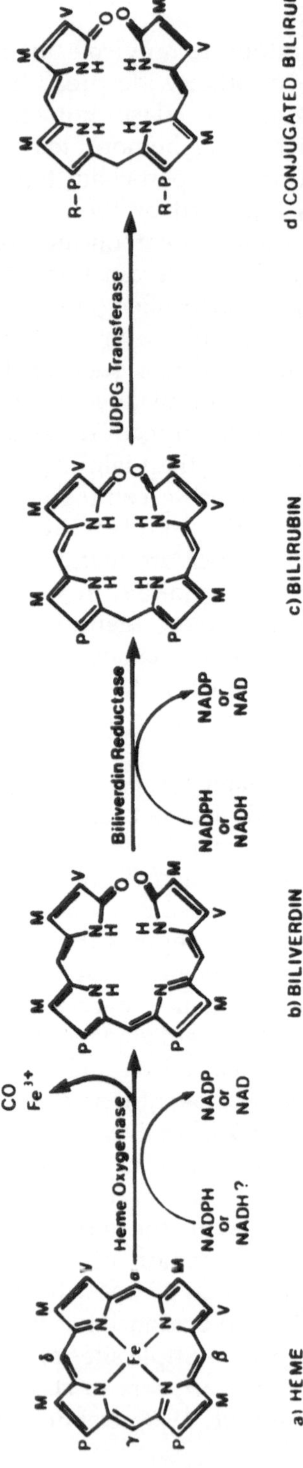

FIGURE 2. Structure of heme, heme catabolites, and conjugated bilirubin by the sequential action of enzymes (heme oxygenase, biliverdin reductase, and UDPG transferase) and the release of iron and carbon monoxide. M, Methyl; V, vinyl; P, propionic acid side chains; R, conjugated glycoside group.

1973; Hoffman and Ross, 1980; Mathews *et al.*, 1973; Granick and Sassa, 1978) and the key enzymes involved in its own synthesis, ALA-S, ALA-D, Uro-S, and ferrochelatase. There is some evidence that early heme formation is important in initiating differentiation of erythroid cells, further increasing ALA-S (Hoffman and Ross, 1980; Ibraham *et al.*, 1982a).

2.2. MECHANISMS OF HEME DEGRADATION

Several mechanisms have been proposed to explain the process of heme oxidation. Masters and Schacter (1976) have shown that heme is nonenzymatically degraded to biliverdin by proteolytically purified cytochrome C-reductase when incubated with NADPH under physiological pH and temperature. O'Carra and Colleran (1969) demonstrated that oxidation of hemoprotein with ascorbic acid at physiological pH and temperature produced all four possible isomers of biliverdin (Kendrew *et al.*, 1960) which, however, did not occur *in vivo*. Tenhunen *et al.* (1969) described an enzymatic mechanism for the degradation of heme to carbon monoxide and biliverdin. The rate-limiting enzyme in this process was characterized as a microsomal enzyme protein, heme oxygenase (Maines *et al.*, 1977). This protein has been thought to accept only NADPH as a source of electron donor in heme catabolism to bilirubin (Yoshida *et al.*, 1974). However, NADH can act as electron donor in the microsomal heme oxygenase system (Ibraham and Levere, 1980). It is not clear whether the presence of a transhydrogenase in marrow cells is responsible for transferring the reducing equivalent from NADH to NADPH in heme oxidation, whether its multiple isozymes are present in marrow cells, or whether NADH acts as electron donor in the biliverdin reductase, the second enzyme in the heme oxygenase system.

The reaction shown in Fig. 2 represents the major pathway for heme degradation in mammals whereby heme is degraded to the bile pigment biliverdin with concurrent elimination of iron and the appropriate methene-bridge carbon atom as CO. Biliverdin is subsequently converted into bilirubin by the NADPH-dependent enzyme biliverdin reductase (EC 1.3–1.24). Because of the asymmetry in the arrangement of the side chains around the porphyrin periphery, the methene-bridge carbon atoms designated α, β, γ, and δ are not equivalent and cleavage at these positions can give rise to four bile pigment isomers. In man the α isomers are present almost exclusively, indicating apparent substrate specificity. The hepatic cells are uniquely designed to metabolize those substances to more polar water-soluble derivatives that can be secreted into the bile. Although bilirubin constitutes the only identified heme-related pigment of uncontaminated mammalian bile, there is evidence that heme may be degraded to compounds other than bilirubin under certain conditions as described in Section 2.2.5.

2.2.1. Heme Oxygenase

Microsomal heme oxygenase has now been well defined as the major enzyme system responsible for heme catabolism (Pimstone *et al.*, 1971a,b; Maines and Kappas, 1974; Tenhunen *et al.*, 1970). Heme oxygenase catalyzes the oxidative degradation of iron protoporphyrin IX to form the natural biliverdin isomers IXα. The activity of heme oxygenase is found in most tissues that have been studied, including kidney, brain, lung (Tenhunen *et al.*, 1970), and gut mucosa (Raffin *et al.*, 1974), but the major sites of activity are spleen, liver, and bone marrow (Tenhunen *et al.*, 1970; Maines and Kappas, 1974; Ibraham and Levere, 1980), all of which contain a high proportion of RE cells which are important for the sequestration and breakdown of senescent red cells as detailed earlier.

In our laboratory, we examined the enzymatic activity of rodent marrow cells and the capacity for bilirubin formation. Rat bone marrow possesses more heme oxygenase activity than the liver (Ibraham and Levere, 1980). The marrow enzyme activity requires NADPH or NADH as an electron donor for heme degradation as in the case with the liver enzyme. Since bilirubin is formed by a complex enzyme system, it is possible that either heme oxygenase or biliverdin reductase or both can be electron acceptors. Although the activity of the enzyme is greater with NADPH, the oxygen incorporated into heme molecules is derived from air as no bilirubin formation is detected when the incubation is carried out in an atmosphere of N_2. This is in agreement with the result of King and Brown (1978), who have shown, using ^{18}O labeling, that both of the terminal oxygen atoms in bilirubin are derived from molecular O_2.

The α-specificity of the enzymatic heme cleavage has attracted the attention of many workers (Frydman *et al.*, 1979) who have shown that the enzyme is specific not only for iron proto heme but also for hemin XIV and hemin III. The presence of heme oxygenase activity in marrow cells has led other investigators to search further for the capacity of the bone marrow cell population for heme degradation in humans as well as rodents. We have demonstrated that heme oxygenase activity is not limited to phagocytic cells. In a human leukemic cell line, K562, activity varied reciprocally with hemoglobin synthesis (Hoffman *et al.*, 1980). We have shown, furthermore, that there is also a reciprocal relationship between heme oxygenase levels and numbers of erythroid colonies in an *in vitro* bone marrow culture system (Kochen *et al.*, 1981; Ibraham *et al.*, 1980). In addition, heme oxygenase increases concurrently with iron accumulation in the marrow as it does in liver cells (Ibraham *et al.*, 1979, 1980). The enzyme activity in marrow is also potentiated by other metals such as gold (Dresner *et al.*, 1981).

2.2.2. Iron Metabolism and Heme Oxidation

Quantitative alterations in iron have significant effects on hemoglobin metabolism. Iron deficiency is the most common clinical condition in which there is limitation in hemoglobin synthesis. Ferrous iron is the physiological metal substrate for the final enzyme of the heme biosynthetic pathway, ferrochelatase, and possibly acts as a regulator for ALA-S. Therefore, an adequate supply of iron is necessary for normal heme synthesis. In iron deficiency anemia, erythrocyte protoporphyrin levels are increased. Definitive studies of ferrochelatase levels in iron deficiency have not yet been performed. As iron depletion progresses the hemoglobin concentration falls and abnormally shaped red cells are seen in the peripheral blood. The malformed red cells of iron deficiency anemia may have a reduced life span (Layrisse *et al.*, 1965). In rats experimental iron deficiency anemia appears to be associated with ineffective erythropoiesis and early rapid destruction of newly formed cells in the peripheral blood (McKee *et al.*, 1968). Iron-deficient erythrocytes have an impaired ability to pass through Millipore filters *in vitro* and this may be associated with *in vivo* changes leading to RE sequestration and trapping (Card and Weintraub, 1971). Heme oxygenase activity in iron deficiency remains to be investigated.

In various conditions characterized by excessive bone marrow iron, heme metabolism is also altered. In hereditary or idiopathic sideroblastic anemias there is impairment of heme synthesis (Bottomley, 1977). Enhanced iron absorption and altered iron metabolism are also often found in patients exposed to benzene. There may be inhibition of RNA and DNA synthesis and therefore impaired iron incorporation into protoporphyrin IX to form heme. The pattern of iron transport to benzene toxicity and sideroblastic anemias is characterized by increased ineffective erythropoiesis, with an increased plasma iron turnover and a diminished iron utilization, as demonstrated with ferrokinetic studies. Ponka and Neuwirt (1974), from a series of studies *in vitro* with rabbit reticulocytes, have proposed that heme itself controls cellular iron uptake. With impaired heme synthesis, iron uptake is enhanced. Although intracellular iron incorporation into heme is reduced, mitochondrial iron, low-molecular-weight cytoplasmic iron, and ferritin iron pools increase. Almost all features characteristic of the sideroblastic anemias in patients can be ascribed to a primary disturbance of heme synthesis (Heilmeyer, 1966; Cartwright and Deiss, 1975; Aoki, 1980). Primary refractory anemia and thalassemia are also characterized by increased bone marrow iron, and increased bile pigment production derived from the bone marrow (White *et al.*, 1967). The increased heme oxygenase is not secondary to the hemolysis but may be to increased iron content. Dresner *et al.* (1981) have presented evidence that marrow microsomal heme oxygenase is increased in iron-overloaded rats (Table 1). They have presented evidence that heavy metals such as gold slightly increase the heme oxygenase activity. However, gold potentiates induction of heme oxygenase activity in iron-overloaded

TABLE 1. EFFECT OF GOLD ON IRON-OVERLOADED RAT BONE MARROW HEME OXYGENASE[a]

Condition	Heme oxygenase (pmoles bilirubin formed/10 mg per min)	% of control
Control	275 ± 35	100
Gold	309 ± 26	112
Iron overload	398 ± 43	144
Iron overload plus gold	467 ± 41	170

[a]Heme oxygenase was assayed as described previously (Ibraham and Levere, 1980) except that NADH and NADPH were used as electron donors at a concentration of 0.9 mM. Each value is the mean of three separate experiments performed in duplicate. Microsomal fractions were used at a concentration of 10 mg/ml and incubated for 20 min at 37°C. Iron-overloaded rats were given i.p. iron dextran over 6 months, and gold-treated rats, i.p. gold for 2 weeks.

rats. These results suggest that iron and gold have synergistic effects on induction of heme oxygenase. This demonstrates directly the capacity of the bone marrow for heme catabolism, and that iron and possibly other heavy metals may regulate bilirubin formation.

The precise effects of accumulated iron are not yet understood. The presence of excess iron may enhance cellular heme turnover and result in the induction of heme oxygenase. Thus, an increased rate of heme degradation would ensue as evidenced by a depression in benzidine-positive erythroid colonies to varying concentrations of Epo. It is possible that an increase in cellular heme in iron-overloaded marrow cells in turn accelerates terminal cell maturation and results in a reduced response to Epo and prostaglandin. In fact, it has been suggested that an increase in cellular heme and hemoglobin synthesis acts like a histone to combine with DNA and thereby act to terminate its own synthesis in the cell as it approaches maturity (Tooze and Davies, 1963; Levere and Granick, 1965). Recently, Bonanou-Tzedaki *et al.* (1981) have demonstrated that excess heme during erythroid hemolysis in turn accelerates terminal cell differentiation and a reduction in DNA synthesis.

Therefore, we assume that under steady-state conditions a more mature precursor cell pool exists in the iron-overloaded bone marrow due to an increased heme turnover. Thus, as Bonanou-Tzedaki *et al.* (1981) observed accelerated terminal differentiation and reduction in DNA synthesis with hemolysis, in a similar manner we observed no enhancement in the generation of benzidine-positive erythroid colonies. Furthermore, a second cell pool may exist and undergo a normal level of heme synthesis and differentiation. This second cell population is then able to respond to addition of exogenous hemin and Epo resulting in normal or increased numbers of erythroid colonies. In fact, the existence of an abnormal erythroid precursor clone has been implicated in hereditary sideroblastic anemia (Lee *et al.*, 1968); however, two distinct cell clones have not been proven in idiopathic sideroblastic anemia (Bottomly, 1980).

The chain of the enzymatic steps leading to heme biosynthesis or its degradation may be influenced by iron overload. For example, bone marrow ALA-S activity is equally reduced in both congenital and idiopathic sideroblastic anemia (Pasanen *et al.*, 1981; Aoki *et al.*, 1974) and this may be viewed as a sufficient mechanism to impair heme biosynthesis in these disorders. Iron excess is known to have adverse effects on the anemia of such disorders since phlebotomy improves the anemia in some instances (Frenck and Jacobs, 1976; Hines, 1976). Recently, we have demonstrated that iron is not the physiological regulator or inhibitor of ALA-S. In fact, excess iron was accompanied by an increase in ALA-S activity by about 50% (Table 2). This result may suggest that the decreased activity of ALA-S in sideroblastic anemia observed in our laboratory and others (Ibraham *et al.*, 1983; Bottomley, 1980; Pasanen *et al.*, 1981) is not due to a direct effect of iron alone on the enzyme activity, but may instead be due to other factor(s) associated with the iron overload. In contrast, ALA-D activity was decreased by about 70% in iron-overloaded rat bone marrow. In spite of this decrease in ALA-D activity, a normal level of porphyrin synthesis in these cells was observed. This result suggests that, as in hepatic cells, bone marrow ALA-D is probably present far in

TABLE 2. RELATIONSHIP BETWEEN ALA SYNTHETASE, ALA DEHYDRATASE, PORPHYRIN, HEME OXYGENASE ACTIVITIES, AND IRON STATES IN RAT BONE MARROW CELLS[a]

	ALA-S (pmoles ALA formed/ 3.2×10^6 cells per hr)	ALA-D (nmoles PBG formed/ 3.2×10^6 cells per hr)	Porphyrin (pmoles/ 5×10^6 cells)	Hemeoxygenase (pmoles bilirubin formed/ 3.2×10^6 cells per hr)
Control	357 ± 61	32.9 ± 3.2	961 ± 120	320 ± 115
Iron overload	476 ± 42	10.5 ± 2.5	880 ± 93	729 ± 126
Iron deficient	333 ± 58	53.4 ± 2.8	1484 ± 136	429 ± 128

[a]ALA synthetase activity was measured as described in the Methods section, Data for ALA. Dehydratase activity was assayed in the presence of DTT. The experimental procedures used for various measurements are described in detail in the Materials and Methods section. Data are the mean of two to four determinations (Ibraham *et al.*, 1982b).

excess of what is needed for normal heme synthesis, and that inhibition of ALA-D activity by more than 90% is required in order to achieve a detectable inhibition of heme synthesis (Tschudy *et al.*, 1981). Conversely, iron deficiency resulted in a 50% increase in ALA-D activity, which indirectly suggests that iron overload inhibits ALA-D activity. Furthermore, iron-overloaded bone marrow demonstrated increased levels of heme oxygenase activity, the rate-limiting enzyme in heme degradation. This elevated activity in heme oxygenase would result in a greater decrease in cellular heme and the observed decrease in benzidine-positive erythroid colonies as was seen in the iron-overloaded marrow cells. Maintenance of normal cellular heme is shown to inhibit the cAMP-independent protein kinase activity associated with the soluble translational inhibitor, leading to an increased formation of an initiation complex, i.e., Met-tRNA$_f$–40S ribosomal subunit. Thus, it is assumed that heme promotes not only the synthesis of globin and nonglobin protein by interfering with either the formation of the initiation complex translation of mRNA or inhibiting the activity of the translational inhibitor (Freedman and Rosman, 1976) and also functions as an essential component for erythroid cell differentiation in normal marrow cells. Alternatively, iron may cause an increase in lipid peroxidation (Wills, 1969), organelle and membrane damage, and subsequently a decrease in erythroid differentiation. These factors taken together could thus account for a major decrease in erythropoiesis *in vivo* or *in vitro*. It is of future interest to determine what effect iron chelators such as desferroxamine (Graziano and Cerami, 1977) or pharmacologic agents like 5-azacytidine (Ley *et al.*, 1982) have on iron-overloaded animals, since these compounds have been shown to be beneficial to patients with excess iron and an impairment of heme synthesis. Iron may act as inducer of ALA-S (Ibraham *et al.*, 1979; Bonkowsky *et al.*, 1981) and inhibit uroporphyrinogen decarboxylase and/or uroporphyrinogen III cosynthetase (Kushner *et al.*, 1975). However, others have reported no effect (Woods *et al.*, 1981) or activation (Blekkenhorst *et al.*, 1979) of the decarboxylase by ionic iron.

2.2.3. Factors Influencing Heme Degradation

Multiple factors influence heme catabolism. These include heme, hemoglobin, sex, heavy metals, hormones, and drugs. Hemoglobin can induce heme oxygenase. Tenhunen *et al.* (1970) showed that when rats were treated with hemoglobin, hepatic heme oxygenase activity was increased severalfold. Subsequently, Pimstone *et al.* (1971a) demonstrated that induction of heme oxygenase by hemoglobin injection resulted in *de novo* synthesis of enzyme protein. Induction of heme oxygenase by heme or hemoglobin has also been reported in macrophages, the arachnoid and choroid plexuses of the brain, kidney, and liver (Roost *et al.*, 1972; Pimstone *et al.*, 1971b; Maines *et al.*, 1977). DeSchepper and Vander Stock (1971) showed that there were pronounced sex differences in the urinary excretion of bilirubin following the injection of excess amounts of hemoglobin. Bilirubin excretion of the male animals increased nearly 10-fold higher than that of females.

The discovery by Kappas and Maines (1976) and Drummond and Kappas (1981) that heavy metals regulate synthesis of heme oxygenase has provided a major impetus for studies in this new area. They demonstrated that as a result of metal induction of heme oxygenase, there was a marked depletion of cellular free heme. This potent induction effect of metals on heme oxygenase has been confirmed by others (DeMatteis and Unseld, 1976; Correia and Schmid, 1975; Ibraham *et al.*, 1981b). Heavy metals apparently induce heme oxygenase in all organs studied, including liver, kidney, heart, lung, intestine, and bone marrow. The exact mechanism by which heavy metals induce heme oxygenase and therefore increase bilirubin production is not completely understood.

Heme oxygenase activity is also influenced by hormonal factors. Hypoglycemia increases hepatic heme oxygenase. Parenteral glucagon and epinephrine cause a five- to sixfold increase in hepatic heme oxygenase (Bakken *et al.*, 1972). Several prostaglandins, in particular prostaglandin E, have been found to decrease heme oxygenase in live cultures (Gemsa *et al.*, 1975). Several authors have reported that prostaglandins have erythropoietic properties and may act as stimulators of hemopoietic stem cell proliferation (Feher and Gidali, 1974; Fink and Fisher, 1977; Dukes *et al.*, 1973). It is unclear as yet whether this effect could be mediated through inhibition of heme oxygenase.

There may be an adaptive increase in heme oxygenase in various hemolytic anemias. Schacter *et al.* (1976) reported that total heme oxygenase activity was increased in the spleens of patients with chronic hemolytic anemia although the enzyme activity per milligram of microsomal protein was not significantly increased when compared to patients with normal spleens. Recently, Granick and Sassa (1978) have reported that in mice with severe congenital hemolytic anemia there is an adaptive increase in heme oxygenase activity. They assumed that this adaptive mechanism develops during long-term hemolysis and serves to counteract the ability of free heme to repress its own synthesis which would diminish cellular cytochrome P-450 content. In fact, the affinity of heme oxygenase for heme, like that of hemopexin (Morgan *et al.*, 1976), appears to be greater than that of α and β globin chains.

Several drugs have been studied with regard to their effects on bilirubin formation. Phenobarbital administration to rats augments the formation of early labeled bilirubin from [^{14}C]glycine but not from ALA (Levitt *et al.*, 1968; Robinson, 1972). Hepatic heme oxygenase is significantly increased in vitamin C-deficient guinea pigs (Walsch and Degkwitz, 1980). The same investigators reported that splenic heme oxygenase activity decreased to 50% of the control value after 21 days of vitamin C omission. Thus, it is clear that several apparently unrelated factors affect heme oxygenase activity and bilirubin formation.

2.2.4. Hematological Disorders and Heme Oxygenase

We have suggested that regulation of the intact heme synthetic pathway is critical to the normal proliferation and differentiation of developing erythroid cells and that heme oxygenase may play a central role in this process (vide infra). The heme degradative potential of the bone marrow in sideroblastic anemia (SA) was measured in one patient with X-linked SA and three patients with idiopathic acquired SA (IASA). In both IASA and X-linked SA, heme oxygenase activity was increased about 40–60% (Ibraham *et al.*, 1982b). In human leukemic cell line K562, heme oxygenase activity declined in the induced cell after 1.5 days of incubation, and this activity further diminished by 3.5 days, remaining low for the rest of the incubation period (Hoffman *et al.*, 1980). The decrease in erythroid heme oxygenase activity may be an important feature of the erythroid differentiation process, perhaps augmenting the conservation of heme for incorporation into hemoglobin.

Normally, up to 85% of bile pigment is derived from the breakdown of hemoglobin from senescent erythrocytes. The remaining 15% comprises the early labeled peak of bile pigments. This was first demonstrated by London and West (1950) by isotope techniques. As would be expected from our knowledge of the erythrocyte life span, most of the ^{15}N label of [^{15}N]glycine appears in bile approximately 120 days after administration, but from 10 to 20% is recovered in fecal bile pigment even before labeling of erythrocytes is completed. The early labeled bile pigment is derived from several sources including the breakdown of hepatic heme-containing compounds such as cytochrome P-450, degradation of cytoplasmic hemoglobin lost at the time of normoblast nuclear extrusion, hemoglobin derived from remodeled reticulocyte membranes, or from the actual destruction of reticulocytes.

Evidence has been presented that there are two compartments to the early labeled peak. The early component may be derived largely from hepatic heme compounds (Yamamoto *et al.*, 1965; Ibraham *et al.*, 1966; Yannoni and Robinson, 1978). [^{14}C]-δALA, a heme precursor which cannot penetrate immature red cells, selectively labels this initial component, reaching maximum labeling within only 2–3 hr. At this time, virtually no radioactivity has been incorporated into hemoglobin heme, but hepatic heme is highly labeled. The late component is related to heme synthesis in the erythropoietic system. This component is suppressed when bone marrow aplasia is induced by irradiation (Israels *et al.*, 1963) and in patients with aplastic anemia (Barrett *et al.*, 1966). This latter component is increased in various disorders where there is significant ineffective erythropoiesis with intramedullary erythrocyte destruction or in compensatory erythroid hyperplasia. Such disorders include the megaloblastic anemias (Israels *et al.*, 1963), iron deficiency anemia (Isarels *et al.*, 1963), SA (Berk *et al.*, 1972; Barrett *et al.*, 1966), congenital erythropoietic porphyria (Gray *et al.*, 1964), and thalassemia (White *et al.*, 1967). Early labeled bibirubin can also be augmented when erythropoiesis is stimulated by hypoxia or by hemorrhage (Ou and Smith, 1978). The increase in the erythropoietic component of the early labeled peak during hemorrhage is derived from rapid destruction of immature "stress' reticulocytes, presumably in the bone marrow or liver because the effect is unaltered by splenectomy.

2.2.5. Alternate Pathways of Heme Degradation

A minor proportion of heme may normally be metabolized via pathways which do not give rise to bilirubin. The final products of these pathways are dipyrroles rather than tetrapyrroles, and may arise from abnormal complexing of heme with globin side chains (Landaw *et al.*, 1970: Winterbourn and Carrel, 1974). Other heme-derived products have been discussed in Section 2.1.

2.3. TRANSPORT, CONJUGATION, AND EXCRETION OF BILIRUBIN

2.3.1. Hepatic Pathway

Once the heme moiety disassociates from its hemoprotein, it is degraded by the heme oxygenase system on the endoplasmic reticulum to biliverdin, and the latter subsequently is reduced in the cytosol to bilirubin by bilirubin reductase. Bilirubin formed in the spleen or bone marrow reaches the bloodstream where it circulates bound in a 1 : 1 molar ratio to albumin (Ostrow and Schmid, 1963) and secondarily to globulin (Cooke and Roberts, 1969). Hepatocyte uptake of unbound bilirubin which becomes dissociated from albumin may take place via a carrier-mediated transport mechanism (Reichen and Berk, 1979). In the hepatocyte cytoplasm, bilirubin is bound to two high-affinity low-molecular-weight proteins, x and ligandin. These proteins have high affinities for carcinogens, heme, steroids, bile acids, glutathione, and bilirubin (Litwack *et al.*, 1971; Ketterer *et al.*, 1975; Carne *et al.*, 1979). These proteins have also been extensively studied by Bass *et al.* (1977).

Bilirubin bound to its protein carriers is nonpolar and relatively insoluble in aqueous solution and, thus, not easily excreted. As seen in Fig. 3, bilirubin is then conjugated with one or two molecules of sugar, predominantly glucuronic acid, to form an ester, glucuronide. This conjugation is thought to disrupt the multiple intramolecular hydrogen bonds responsible for the nonpolar character of bilirubin, and thereby facilitate pigment excretion (Schmid, 1978). Bilirubin

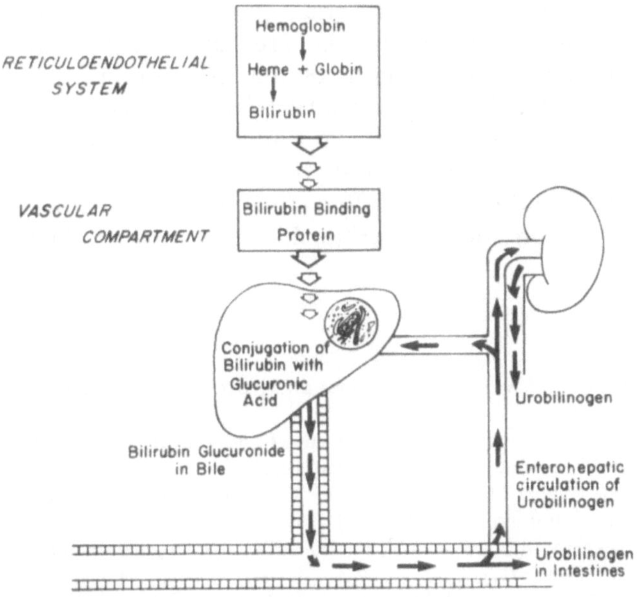

FIGURE 3. Scheme of heme degradation and bilirubin formation including its enterohepatic circulation and excretion of urobilinogen. Conjugated bilirubin in the bile is converted by intestinal bacteria to urobilinogen which is excreted mostly in the stool and partially in the urine.

monoglucuronide is formed by a microsomal enzyme, UDP-glucuronyltransferase [UDP-glucuronate β-glucuronyltransferase (Acceptor unspecific), EC 2.4.1.17], with UDP-glucuronic acid (UDPG) serving as a sugar donor. Jansen *et al.* (1977) have demonstrated that conversion of bilirubin to monoglucuronide and diglucuronide requires two separate enzyme systems. Furthermore, Chowdhury *et al.* (1978) demonstrated the formation of bilirubin diglucuronide by liver preparations from patients with Crigler–Najjar syndrome type I and from Gunn rats deficient in UDP-glucuronyltransferase activity for bilirubin. In contrast, Blankaert *et al.* (1979, 1980) have presented evidence that hepatic formation of both bilirubin mono- and diglucuronide is catalyzed by one microsomal enzyme, UDP-glucuronyltransferase. At this time it is unsettled whether there are one or two enzyme systems carrying out this reaction. By conjugation, bilirubin is converted to more polar and water-soluble derivatives which is essential for excretion into the bile canaliculi.

2.3.2. Intestinal Pathways

The transport of conjugated bilirubin from the endoplasmic reticulum to the bile involves the Golgi apparatus and canalicular membranes. In the bile, conjugated bilirubin passes from canaliculi to the intralobular bile ductules, interlobular intrahepatic bile ducts, to the gallbladder, and finally to the duodenum (Fig. 3). Bilirubin glucuronides are then hydrolyzed to unconjugated pigments by both the alkaline pH of the intestine and β-glucuronidase derived from intestinal epithelial cells and intestinal bacteria. Free bilirubin is converted through a sequence of hydrogenation reactions by the action of intestinal bacteria to various urobilinogens (Troxler *et al.*, 1968; Watson, 1969). Urobilinogens are formed when two methene bridges and two vinyl groups of bilirubin are reduced. Urobilinogens are then partially dehydrogenated to form urobilin in the gut (Elder *et al.*, 1972) and some urobilinogen is reabsorbed and undergoes an enterohepatic circulation and a small fraction is excreted in the urine.

REFERENCES

Allison, A. C., 1960, Turnovers of erythrocytes and plasma proteins in mammals, *Nature (London)* **188:**37.

Aoki, Y., 1980, Multiple enzymatic defects in mitochondria in hematological cells of patients with primary sideroblastic anemia, *J. Clin. Invest.* **66:**43.

Aoki, Y., Urata, G., Wada, O., and Takaku, F., 1974, Measurement of δ-aminolevulinic acid synthetase activity in human erythroblasts, *J. Clin. Invest.* **53:**1326.

Bakken, A. F, Thaler, M. M., and Schmid, R., 1972, Metabolic regulation of hepatic heme oxygenase activity, *J. Biol. Chem.* **51:**530.

Barrett, P. V. D., Cline, M. J., and Berlin, N. I., 1966, The association of the urobilin "early peak" and erythropoiesis in man, *J. Clin. Invest.* **45:**1657.

Bartosz, G., Soszynski, M., and Retelewska, W., 1981, Aging of the erythrocyte. X. Immunoelectrophoretic studies on the denaturation of superoxide dismutase, *Mech. Ageing Dev.* **17:**237.

Bass, N. M., Kirsch, R. E., Tuff, S. A., and Saunders, S. J., 1977, Radioimmunoassay of ligandin, *Biochim. Biophys. Acta* **494:**131.

Bentley, S. A., 1977, Red cell survival studies reinterpreted, *Clin. Haematol.* **6:**601.

Berk, P. D., Bloomer, J. R., Howe, R. B., Blaschke, T. F., and Berlin, N. I., 1972, Bilirubin production as a measure of red cell lifespan, *J. Lab. Clin. Med.* **79:**364.

Berlin, N. I., and Berk, P. D., 1975, The biological life of the red cell, in: *The Red Cell* (D. M. Surgenor, ed.), pp. 957–1019, Academic Press, New York.

Bernstein, R. E., 1959, Alterations in metabolic energetics and cation transport during aging of red cells, *J. Clin. Invest.* **38:**1572.

Bessis, M., 1972, *Cellules du sang normal et pathologique*, Masson, Paris.

Bessis, M., and Bricka, M., 1952, Aspect dynamique des cellules du sang: Son étude par la microcinematographie en contraste de phase, *Rev. Hematol.* **7:**407.

Bissell, D. M., Hammaker, L., and Schmid, R., 1972a, Liver sinusoidal cells: Identification of a subpopulation for erythrocyte catabolism, *J. Cell Biol.* **54:**107.

Bissell, D. M., Hammaker, L., and Schmid, R., 1972b, Hemoglobin and erythrocyte catabolism in rat liver: The separate roles of parenchymal and sinusoidal cells, *Blood* **40:**812.

Blankaert, N., Gollan, J., and Schmid, R., 1979, Bilirubin diglucuronide synthesis by UDP-glucuronic acid dependent enzyme system in rat liver microsomes, *Proc. Natl. Acad. Sci. USA* **76:**2037.

Blankaert, N., Gollan, J., and Schmid, R., 1980, Mechanism of bilirubin diglucuronide formation in intact rats, *J. Clin. Invest.* **65:**1332.

Blekkenhorst, G. H., Eales, L., and Pimstone, N. R., 1979, Iron and porphyria cutanea tarda: Activation of choporphyrinogen decarboxylase by ferrous iron, *S. Afr. Med. J.* **56:**918.

Bocci, V., 1981, Determinants of erythrocyte ageing: A reappraisal, *Br. J. Haematol.* **48:**515.

Bocci, V., Pessina, G. P., and Paules, L., 1980, Studies of factors regulating the ageing of human erythrocytes. III. Metabolism and fate of erythrocyte vesicles, *Int. J. Biochem.* **11:**139.

Bonanou-Tzedaki, S. A., Sohi, M., and Arnstein, H. R. V., 1981, Regulation of erythroid cell differentiation by haemin, *Cell Differ.* **10:**267.

Bonkowsky, H. C., Healey, J. F., Sinclair, P. R., Sinclair, J. F., and Pomeroy, J. S., 1981, Iron and the liver: Acute and long-term effects of iron-loading on hepatic haem metabolism, *Biochem. J.* **196**:57.

Bottomley, S. S., 1977, Porphyrin and iron metabolism in sideroblastic anemia, *Semin. Hematol.* **14**:169.

Bottomley, S. S., 1980, Sideroblastic anemia, in: *Iron in Biochemistry and Medicine II* (A. Jacobs and M. Worwood, eds.), pp. 363–392, Academic Press, New York.

Branemark, P. I., and Bagge, V., 1977, Intravascular rheology of erythrocytes in man, *Blood Cells* **3**:11.

Brok, F., Ramot, B., Zwang, E., and Danon, D., 1966, Enzyme activities in human red blood cells of different age groups, *Isr. J. Med. Sci.* **2**:291.

Bunn, H. F., 1972, Erythrocyte destruction and hemoglobin catabolism, *Semin. Hematol.* **9**:3.

Cambell, F. R., 1972, Ultrastructural studies of transmural migration of blood cells in the bone marrow of rats, mice and guinea pigs, *Am. J. Anat.* **135**:521.

Card, R. T., and Weintraub, L. R., 1971, Metabolic abnormalities of erythrocytes in severe iron deficiency, *Blood* **37**:725.

Carne, T., Tipping, E., and Ketterer, B., 1979, The binding and catalytic activities of forms of ligandin after modification of its thiol groups, *Biochem. J.* **177**:433.

Cartwright, G. E., and Deiss, A., 1975, Medical progress: Sideroblasts, siderocytes, and sideroblastic anemia, *N. Engl. J. Med.* **292**:185.

Chandra, P., Chaudhery, S. A., Rosner, F., and Kagan, M., 1975, Transient histiocytosis with striking phagocytosis of platelets, leukocytes, and erythrocytes, *Arch. Intern. Med.* **135**:989.

Chapman, R. G., and Schaumburg, L., 1967, Glycolysis and glycolytic enzyme activity of aging red cells in man, *Br. J. Haematol.* **13**:665.

Chen, L. T., and Weiss, L., 1973, The role of the sinus wall in the passage of erythrocytes through the spleen, *Blood* **41**:529.

Chien, S., 1977, Principals and techniques for assessing erythrocyte deformability, *Blood Cells* **3**:71.

Chowdhury, J. R., Jansen, P. L. M., Fischberg, E. B., Daniller, A., and Arias, I. M., 1978, Hepatic conversion of bilirubin monoglucuronide to diglucuronide in uridine diphosphate-glucuronosyltransferase-deficient man and rat by bilirubin glucuroniside glucuronosyltransferase, *J. Clin. Invest.* **62**:191.

Coburn, R. F., 1970, Enhancement by phenobarbital and diphenylhydantoin of carbon monoxide production in normal man, *N. Engl. J. Med.* **283**:512.

Cokelet, G. R., and Meiselman, H. J., 1968, Rheological comparison of hemoglobin solutions and erythrocyte suspensions, *Science* **162**:275.

Cooke, J. R., and Roberts, L. B., 1969, The binding of bilirubin to serum proteins, *Clin. Chim. Acta* **26**:425.

Correia, M. A., and Schmid, R., 1975, Effect of cobalt on microsomal cytochrome P-450: Differences between liver and intestinal mucosa, *Biochem. Biophys. Res. Commun.* **65**:1378.

Dabney, B. J., and Blaudet, A. L., 1977, Increase in globin chains and globin mRNA in erythroleukemia cells in response to hemin, *Arch. Biochem. Biophys.* **179**:106.

DeBruyn, P. P. H., Michelson, S., and Thomas, T. B., 1971, The migration of blood cells of the bone marrow through the sinusoidal wall, *J. Morphol.* **133**:417.

DeMatteis, F., 1978, Hepatic porphyrias caused by 2-allyl-2-isopropylacetamide, 3,5-diethoxycarbonyl-1,4-dihydrocollidine, griseofulvin and related compounds, in: *Heme and Hemoproteins* (F. DeMatteis and W. N. Aldridge, eds.), pp. 129–156, Springer-Verlag, Berlin.

DeMatteis, F., and Unseld, A., 1976, Increased liver haem degradation caused by foreign chemicals: A comparison of the effects of 2-allyl-2-isopropylacetamide and cobaltous chloride, *Biochem. Soc. Trans.* **4**:205.

DeSchepper, J., and Vander Stock, J., 1971, Influence of sex on the urinary bilirubin excretion at increased free plasma haemoglobin levels in whole dogs and in isolated normothermic perfused dog kidneys, *Experientia* **27**:1264.

Dresner, D., Ibraham, N. G., and Levere, R. D., 1981, Modulation of bone marrow heme and protein synthesis by trace elements, *Environ. Res.* **27**:112.

Drummond, G. S., and Kappas, A., 1981, Patent heme-degrading action of antimony and antimony containing parasiticidal agents, *J. Exp. Med.* **153**:245.

Dukes, P. P., Shore, N. A., Hammond, P., Ortega, J. A., and Data, M. C., 1973, Enhancement of erythropoiesis by prostaglandins, *J. Lab. Clin. Med.* **82**:704.

Durocher, J. R., Payne, R. C., and Conrad, M. E., 1975, Role of sialic acid in erythrocyte survival, *Blood* **45**:11.

Eadie, G. S., and Brown, I. W., Jr., 1953, Red blood cell survival studies, *Blood* **8**:1110.

Ebert, P. S., and Ikawa, Y., 1974, Induction of δ-aminolevulinic acid synthetase during erythroid differentiation of cultured leukemia cells 38155, *Proc. Soc. Exp. Biol. Med.* **146**:601.

Ebert, P. S., Bonkowsky, H. L., and Deisseroth, A., 1979, Evidence for multiple sites of regulation of heme synthesis in murine erythroleukemia cells, *J. Natl. Cancer Inst.* **62**:1247.

Ehrenreich, B. A., and Cohn, Z. A., 1968, Fate of hemoglobin pinocytosed by macrophages *in vitro*, *J. Cell Biol.* **38**:244.

Ehrenstein, G. V., and Lockner, D., 1958, Sites of the physiological breakdown of the red blood corpuscles, *Nature (London)* **181**:911.

Elder, G., Gray, L. H., and Nicholson, D. C., 1972, Bile pigment fate in gastrointestinal tract, *Semin. Hematol.* **9**:71.

Evans, E. A., and Hochmuth, R. M., 1976a, Membrane visco-elasticity, *Biophys. J.* **16**:1.

Evans, E. A., and Hochmuth, R. M., 1976b, Membrane visco-plastic flow, *Biophys. J.* **16**:13.

Evans, E. A., Waugh, R., and Melnik, L., 1976, Elastic compressibility modulus of red cell membrane, *Biophys. J.* **16**:585.

Feher, I., and Gidali, J., 1974, Prostaglandin E2 as stimulator of haemopoietic stem cell proliferation, *Nature (London)* **247**:550.

Finch, C. A., Hegsted, M., Kinney, T. D., Thomas, E. D., Roth, C. E., Haskins, D., Finch, S., and Fluharty, R. G., 1950, Iron metabolism: The pathophysiology of iron storage, *Blood* **5**:983.

Fink, G., and Fisher, J. W., 1977, Stimulation of erythropoiesis by beta adrenergic agonists. I. Characterization of activity in polycythemic mice, *J. Pharmacol. Exp. Ther.* **202**:192.

Freedman, M. L., and Rosman, J., 1976, A rabbit reticulocyte model for the role of hemin-controlled repressor in hypochronic anemia, *J. Clin. Invest.* **57**:594.

Frenck, T. J., and Jacobs, P., 1976, Sideroblastic anemia associated with iron overload treated with phlebotomy, *S. Afr. Med. J.* **50**:596.

Friend, C., Scher, W., Holland, J. G., and Sato, R., 1971, Hemoglobin synthesis in murine virus-induced leukemic cells *in vitro:* Stimulation of erythroid differentiation by dimethyl sulfoxide, *Proc. Natl. Acad. Sci. USA* **68**:378.

Frydman, R. B., Awruch, J., Tomaro, M. L., and Frydman, B., 1979, Concerning the specificity of heme oxygenase: The enzymatic oxidation of synthetic hemins, *Biochem. Biophys. Res. Commun.* **87**:928.

Ganzoni, A. M., Oakes, R., and Hillman, R. S., 1971, Red cell aging in vivo, *J. Clin. Invest.* **50**:1373.

Gemsa, D., Woo, C. H., Webb, D., Fudenberg, H. H., and Schmid, R., 1975, Erythrophagocytosis by macrophages: Suppression of heme oxygenase by cyclic AMP, *Cell. Immunol.* **15**:21.

Gidari, A. S., and Levere, R. D., 1977, Enzymatic formation and cellular regulation of heme synthesis, *Semin. Hematol.* **14**:145.

Goldstein, G. W., Hammaker, L., and Schmid, R., 1968, The catabolism of Heinz bodies: An experimental model demonstrating conversion to non-bilirubin catabolites, *Blood* **31**:388.

Granick, J. L., and Sassa, A., 1978, Hemin control of heme biosynthesis in mouse Friend virus-transformed erythroleukemia cells in culture, *J. Biol. Chem.* **253**:5402.

Gray, C. H., Kulezycka, A., Nicholson, D. C., Magnus, I. A., and Rimington, C., 1964, Isotope studies on a case of erythropoietic protoporphyria, *Clin. Sci.* **26**:7.

Graziano, J. H., and Cerami, A., 1977, Chelation therapy for the treatment of thalassemia, *Semin. Hematol.* **14**:127.

Heilmeyer, L., 1966, *Disturbances in Heme Synthesis*, pp. 103–178, Thomas, Springfield, Ill.

Hines, J. D., 1976, Effect of pyridoxine plus chronic phlebotomy on the function and morphology of bone marrow and liver in pyridoxine-responsive sideroblastic anemia, *Semin. Hematol.* **13**:133.

Hochmuth, R. M., Mohandas, N., and Blackshear, P. L., Jr., 1973, Measurement of the elastic modulus for red cell membrane using a fluid mechanical technique, *Biophys. J.* **13**:747.

Hoffman, L. M., and Ross, J., 1980, The role of heme in the maturation of erythroblasts: The effects of inhibition of pyridoxine metabolism, *Blood* **55**:762.

Hoffman, R., Ibraham, N. G., Murnane, M. J., Diamond, A., Forget, B. G., and Levere, R. D., 1980, Hemin control of heme biosynthesis and catabolism in a human leukemia cell line, *Blood* **56**:567.

Hughes-Jones, N. C., 1961, The use of ^{51}Cr and ^{59}Fe as red cell labels to determine the fate of normal erythrocytes in the rabbit, *Clin. Sci.* **20**:315.

Ibraham, G. W., Schwartz, S., and Watson, C. J., 1966, Early labeling of bilirubin from glycine and δ-aminolevulinic acid in bile fistula dogs, with special reference to stimulated versus suppressed erythropoiesis, *Metabolism* **15**:1129.

Ibraham, N. G., and Levere, R. D., 1980, Nucleotide requirements for the bone marrow heme oxygenase system, *Life Sci.* **26**:525.

Ibraham, N. G., Gruenspecht, N. R., and Freedman, M. L., 1978, Hemin feedback inhibition at reticulocyte δ-aminolevulinic acid synthetase and δ-aminolevulinic acid dehydratase, *Biochem. Biophys. Res. Commun.* **80**:722.

Ibraham, N. G., Hoffstein, S. T., and Freedman, M. L., 1979, Induction of liver cell haem oxygenase in iron-overloaded rats, *Biochem. J.* **180**:257.

Ibraham, N. G., Lutton, J., and Levere, R. D., 1980, A study of heme synthetic and degradative enzymes during *in vitro* erythroid colony development, *Clin. Res.* **28**:314A.

Ibraham, N. G., Nelson, J. C., and Levere, R. D., 1981b, Control of δ-aminolevulinate synthase and haem oxygenase in chronic iron-overloaded rats, *Biochem. J.* **200**:35.

Ibraham, N. G., Lutton, J., and Levere, R. D., 1982a, The role of heme biosynthetic and degradative enzymes in erythroid colony development: The effect of hemin, *Br. J. Haematol.* **50**:17.

Ibraham, N. G., Hoffman, R., Lutton, J. D., Kim Ritchey, A., and Levere, R. D., 1982b, Studies of heme metabolism in sideroblastic anemia, *Blood* **60**:68.

Ibraham, N. G., Hoffman, R., Lutton, J. D., Hoffman, R., Ritchey, A., and Levere, R. D., 1983, Examination of abnormalities of heme synthesis in sideroblastic erythroid cells, *Clin. Res.* **31**(2):315.

Israels, L. G., Skanderberg, J., Goyda, H., Zingg, W., and Zipursky, A., 1963, A study of the early-labelled fraction of bile pigment: The effect of altering erythropoiesis on the incorporation of [2-^{14}C]glycine into haem and bilirubin, *Br. J. Haematol.* **9**:50.

Jacob, H. S., 1974, Pathologic states of the erythrocyte membrane, *Hosp. Pract.* **9**:47.

Jansen, P. L. M., Chowdhury, J. R., Fischberg, E. B., and Arias, I. M., 1977, Enzymatic conversion of bilirubin monoglucuronide to diglucuronide by rat liver plasma membranes, *J. Biol. Chem.* **252**:2710.

Kappas, A., and Maines, M. D., 1976, Tin: A potent inducer of heme oxygenase in kidney, *Science* **192**:60.

Kay, M. M. B., 1975, Mechanism of removal of senescent cells by human macrophages *in situ*, *Proc. Natl. Acad. Sci. USA* **72**:3521.

Kay, M. M. B., 1978, Role of physiological autoantibody in the removal of senescent human red cells, *J. Supramol. Struct.* **9**:555.

Keene, W. R., and Jandl, J. H., 1965, Studies of the reticuloendothelial mass and sequestering functions of rat bone marrow, *Blood* **26**:157.

Keitt, A. S., and Bennett, D. C., 1966, Pyruvate kinase deficiency and related disorders of red cell glycolysis, *Am. J. Med.* **41**:762.

Kendrew, J. C., Dickerson, R. E., Strandberg, B. E., Hart, R. G., Davies, D. R., Phillips, D. C., and Shore, V. C., 1960, Structure of myoglobin: A three-dimensional Fourier synthesis at 2Å resolution, *Nature (London)* **185**:422.

Ketterer, B., Tipping, E., Meuweissen, J., and Beale, D., 1975, Ligandin, *Biochem. Soc. Trans.* **3**:626.

King, R. F. G. J., and Brown, S. B., 1978, The mechanism of haem catabolism: A study of haem breakdown in spleen microsomal fraction and in a model system by ^{18}O labelling and metal substitution, *Biochem. J.* **174**:103.

Kochen, J., Ibraham, N., Lutton, J., and Levere, R., 1981, Congenital sideroblastic anemia (CSA): A microenvironment defect, *Pediatr. Res.* **15**:827.

Kreimer-Birnbaum, M., Pinkerton, P. H., Bannerman, R. M., and Hutchison, H. E., 1966, Dipyrrolic urinary pigments in congenital Heinz-body anaemia due to Hb Koln and in thalassaemia, *Br. Med. J.* **2**:396.

Krögh, A., 1930, *The Anatomy and Physiology of Capillaries*, Hafner, New York.

Kushner, J. P., Steinmuller, D. P., and Lee, G. R., 1975, The role of iron in the pathogenesis of porphyria cutanea tarda. II. Inhibition of uroporphyrinogen decarboxylase, *J. Clin. Invest.* **56**:661.

La Celle, P. L., and Arkin, B., 1970, Acquired rigidity: A possible determinant of normal RBC life span, *Blood* **36**:837.

Landaw, S. A., Callahan, E. W., Jr., and Schmid, R., 1970, Catabolism of heme *in vivo:* Comparison of the simultaneous production of bilirubin and carbon monoxide, *J. Clin. Invest.* **49**:914.

Layrisse, M., Linaries, J., and Roche, M., 1965, Excess hemolysis in subjects with severe iron deficiency anemia associated and nonassociated with hookworm infection, *Blood* **25**:73.

Leblond, P. F., Lyonnais, J., and Delage, J. M., 1978a, Erythrocyte populations in pyruvate kinase deficiency anemia following splenectomy. I. Cell morphology, *Br. J. Haematol.* **39**:55.

Leblond, P. F., Coulombe, L., and Lyonnais, J., 1978b, Erythrocyte population in pyruvate kinase deficiency anemia following splenectomy. II. Cell deformability, *Br. J. Haematol.* **39**:63.

Lee, G. R., MacDiarmid, W. D., and Cartwright, G. E., 1968, Hereditary, X-linked, siderochrestic anemia: The isolation of two erythrocyte populations differing in Xg^a blood type and porphyrin content, *Blood* **32**:59.

Levere, R. D., and Granick, S., 1965, Control of hemoglobin synthesis in the cultured chick blastoderm by δ-aminolevulinic acid synthase: Increase in the rate of hemoglobin formation with δ-aminolevulinic acid, *Proc. Natl. Acad. Sci. USA* **54**:134.

Levere, R. D., and Granick, S., 1967. Control of hemoglobin synthesis in the cultured chick blastoderm, *J. Biol. Chem.* **242**:1903.

Levitt, M., Schacter, B. A., Ziporsky, A., and Israels, L. G., 1968, The nonerythropoietic component of early bilirubin, *J. Clin. Invest.* **47**:1281.

Ley, T. J., DeSimone, J., Anagnov, N. P., Keller, G. H., Humphries, R. K., Turner, P. H., Young, N. S., Heller, P., and Nienhms, A. W., 1982, 5-Azacytidine selectively increases globin synthesis in a patient with β-thalassemia, *N. Engl. J. Med.* **307**:1469.

Litwack, G., Ketterer, B., and Arias, I. M., 1971, Ligandin: A hepatic protein which binds steroids, bilirubin, carinogens and a number of exogenous organic ions, *Nature (London)* **234**:466.

Lodish, H. L., 1973, Biosynthesis of reticulocyte membrane proteins by membrane-free polyribosomes, *Proc. Natl. Acad. Sci. USA* **70**:1526.

London, I. M., and West, R., 1950, The formation of bile pigment in pernicious anemia, *J. Biol. Chem.* **184**:359.

Lorand, L., Siefring, G. E., Jr., and Lowe-Krenz, L., 1979, Enzymatic basis of membrane stiffening in human erythrocytes, *Semin. Hematol.* **16**:65.

Lux, S. E., 1979, Spectrin–actin membrane skeleton of normal and abnormal red blood cells, *Semin. Hematol.* **16**:21. M

McKee, L. C., Jr., Wasson, M., and Heyssel, R. M., 1968, Experimental iron deficiency in the rat: The use of ^{51}Cr, $DF^{32}P$ and ^{59}Fe to detect haemolysis of iron-deficient cells, *Br. J. Haematol.* **14**:87.

Maines, M. D., and Kappas, A., 1974, Cobalt induction of hepatic heme oxygenase; with evidence that cytochrome P-450 is not essential for this enzyme activity, *Proc. Natl. Acad. Sci. USA* **71**:4293.

Maines, M. D., Ibraham, N. G., and Kappas, A., 1977, Solubilization and partial purification of heme oxygenase from rat liver, *J. Biol. Chem.* **252**:5900.

Marton, P. F., 1970, Erythrophagocytosis in the human bone marrow, *Scand. J. Haematol.* **7**:177.

Marton, P. F., 1975, Erythrophagocytosis in the human bone marrow as disclosed by iliacal bone biopsies, *Scand. J. Haematol.* **14**:153.

Masters, B. S. S., and Schacter, B. A., 1976, The catalysis of heme degradation by purified NADPH-cytochrome C reductase in the absence of other microsomal proteins, *Ann. Clin. Res.* **8**(17):18.

Mathews, M. B., Hunt, T., and Brayley, A., 1973, Specificity of the control of protein synthesis of haemin, *Nature New Biol.* **243**:230.

Mieschner, P., 1956, Le méchanisme de l'erythroclasie a l'etat normal, *Rev. Hematol.* **11**:24859.

Mohandas, N., Phillips, W. M., and Bessis, M., 1979, Red blood cell deformability and hemolytic anemias, *Semin. Hematol.* **16**:95.

Morgan, W. T., Liem, H. H., Sutor, R. P., and Müller-Eberhard, U., 1976, Transfer of heme from heme-albumin to hemopexin, *Biochim. Biophys. Acta* **444**:435.

Nudel, U., Salmon, J. D., Terada, M., Bank, A., Rifkind, R. A., and Marks, P. A., 1977, Differential effects of chemical inducers on the expression of β globin genes in murine erythroleukemia cells, *Proc. Natl. Acad. Sci. USA* **74:**1100.

O'Carra, P. A., and Colleran, E., 1969, Haeme catabolism and coupled oxidation of haemoproteins, *Fed. Eur. Biol. Soc. Lett.* **5:**295.

Ostrow, J. D., and Schmid, R., 1963, The protein-binding of [C^{14}]bilirubin in human and murine serum, *J. Clin. Invest.* **42:**1286.

Ou, L., and Smith, R. P., 1978, Hemoglobinemia in rats exposed to high altitude, *Exp. Hematol.* **6:**473.

Pasanen, A. V. O., Vuopio, P., Borgstrom, G. H., and Tenhunen, R., 1981, Heme synthesis in refractory sideroblastic anemia associated with the preleukemic syndrome, *Br. J. Haematol.* **27:**35.

Pettit, J. E., 1977, Spleen function, *Clin. Haematol.* **6:**639.

Pimstone, N. R., Engel, P., Tenhunen, R., Seitz, P. T., Marver, H. S., and Schmid, R., 1971a, Inducible heme oxygenase in the kidney: A model for the homeostatic control of hemoglobin catabolism, *J. Clin. Invest.* **50:**2042.

Pimstone, N. R., Tenhunen, R., Seitz, P. T., Marver, H. S., and Schmid, R., 1971b, The enzymatic degradation of hemoglobin to bile pigments by macrophages, *J. Exp. Med.* **133:**1264.

Piomelli, S., Lurinsky, G., and Wasserman, L. R., 1967, The mechanism of red cell aging. I. Relationship between cell age and specific gravity evaluated by ultra-centrifugation in a discontinuous density gradient, *J. Lab. Clin. Med.* **69:**659.

Ponka, P., and Neuwirt, J., 1974, Haem synthesis and iron uptake by reticulocytes, *Br. J. Haematol.* **28:**1.

Raffin, S. B., Choong, H. W., Roost, K. T., Price, D. C., and Schmid, R., 1974, Intestinal absorption of hemoglobin iron-heme cleavage by mucosal heme oxygenase, *J. Clin. Invest.* **54:**1344.

Reichen, J., and Berk, P. D., 1979, Isolation of an organic anion binding protein from rat liver plasma membrane fractions by affinity chromatography, *Biochem. Biophys. Res. Commun.* **91:**484.

Rifkind, R. A., 1966, Destruction of injured red cells *in vivo*, *Am. J. Med.* **41:**711.

Robinson, S. H., 1972, Formation of bilirubin from erythroid and non-erythroid sources, *Semin. Hematol.* **9:**43.

Roost, K. T., Pimstone, N. R., Diamond, I., and Schmid, R., 1972, The formation of cerebrospinal fluid xanthochromia after subarachnoid hemorrhage: Enzymatic conversion of hemoglobin to bilirubin by the arachnoid and choroid plexus, *Neurology* **22:**973.

Ross, J., and Sautner, D., 1976, Induction of globin mRNA accumulation by hemin in cultured erythroleukemic cells, *Cell* **8:**513.

Rous, P., 1923, Destruction of the red blood corpuscle in health and disease, *Physiol. Rev.* **3:**75.

Rous, P., and Robertson, O. H., 1917, The fate of erythrocytes. I. The findings in healthy animals, *J. Exp. Med.* **25:**651.

Rutherford, T., Thompson, G. G., and Moore, M. R., 1979, Heme biosynthesis in Friend erythroleukemia cells: Control by ferrochelatase, *Proc. Natl. Acad. Sci. USA* **76:**833.

Sansone, G., 1962, Anemia emolitiche acquisite dell'infanzia, *Pediatr. Int.* **12:**77.

Sassa, S., 1976, Sequential induction of enzymes of the heme biosynthetic pathway in Friend erythroleukemia cells in culture, *J. Exp. Med.* **143:**305.

Schacter, B. A., Yoda, B., and Israels, L. G., 1976, Human spleen heme oxygenase in normal, hemolytic and other pathological states, *Ann. Clin. Res.* **8**(17)**:**28.

Schmid, R., 1978, Bilirubin metabolism: State of the art, *Gastroenterology* **74:**1307.

Schwartz, S., and Ikeda, K., 1955, Studies of porphyrin synthesis and interconversion, with special reference to certain green porphyrins in animals with experimental hepatic porphyria, in: *Porphyrin Biosynthesis and Metabolism* (G. E. W. Wolstenholme and E. C. P. Miller, eds.), pp. 209–228, Churchill, London.

Scott, J. F., and Deane, H. W., 1966, Liver and gallbladder, in: *Histology* (R. O. Greep, ed.), 2nd ed., p. 538, McGraw–Hill, New York.

Seeman, P., 1967, Transient holes in the erythrocyte membrane during hypotonic hemolysis and stable holes in the membrane after lysis by saponin and lysolecithin, *J. Cell Biol.* **32:**55.

Shattil, S. J., and Cooper, R. A., 1972, Maturation of macroreticulocyte membranes *in vivo*, *J. Lab. Clin. Med.* **79**:215.

Sheetz, M. P., and Singer, S. J., 1977, On the mechanism of ATP-induced shape changes in human erythrocyte membranes. I. The role of the spectrin complex, *J. Cell Biol.* **73**:638.

Sherton, C. C., and Kabat, D., 1976, Changes in RNA and protein metabolism preceding onset of hemoglobin synthesis in cultured Friend leukemia cells, *Dev. Biol.* **48**:118.

Shotton, D., Thompson, K., Wofsy, L., and Branton, D., 1978, Appearance and distribution of surface proteins of the human erythrocyte membrane: An electron microscope and immunochemical labeling study, *J. Cell Biol.* **76**:512.

Simpson, F., and Kling, J. M., 1967, The mechanism of denucleation in circulating erythroblasts, *J. Cell Biol.* **35**:237.

Skutelsky, E., and Danon, D., 1970, Comparative study of nuclear expulsion from the late erythroblast and cytokinesis, *Exp. Cell. Res.* **60**:427.

Tavassoli, M., 1977, Adaptation of marrow sinus wall to fluctuation in the rate of cell delivery: Studies in rabbits after bloodletting, *Br. J. Haematol.* **35**:25.

Tavassoli, M., 1978, Red cell delivery and the function of the marrow–blood barrier: A review, *Exp. Hematol.* **6**:257.

Tavassoli, M., and Crosby, W. H., 1973, Fate of the nucleus of the marrow erythroblast, *Science* **179**:912.

Teitel, P., 1977, Basic principles of the "filterability test" (FT) and analysis of erythrocyte flow behavior, *Blood Cells* **3**:55.

Tenhunen, R., Marver, H. S., and Schmid, R., 1969, Microsomal heme oxygenase: Characterization of the enzyme, *J. Biol. Chem.* **244**:6388.

Tenhunen, R., Marver, H. S., and Schmid, R., 1970, The enzymatic catabolism of hemoglobin: Stimulation of microsomal heme oxygenase by hemin, *J. Lab. Clin. Med.* **75**:410.

Tizianello, A., Pannacciuli, I., Salvidio, E., and Ajmar, F., 1961, A quantitative evaluation of the splenic and hepatic share in normal hemocromatosis, *Acta Med. Scand.* **169**:303.

Tooze, J., and Davies, H. G., 1963, The occurrence and possible significance of hemoglobin in the chromosomal regions of mature erythrocyte nuclei of the newt *Triturus cristatus cristatus*, *J. Cell Biol.* **16**:501.

Troxler, R. F., Dawber, N. H., and Lester, R., 1968, Synthesis of urobilinogen by broken cell preparation of intestinal bacteria, *Gastroenterology* **54**:568.

Tschudy, D. P., Hess, R. A., and Frykholm, B. C., 1981, Inhibition of δ-aminolevulinic acid dehydratase by 4,6-dioxoheptanoic acid, *J. Biol. Chem.* **256**:9915.

Ultmann, J. E., and Gordon, C. S., 1965, Life span and sites of sequestration of normal erythrocytes in normal and splenectomized mice and rats, *Acta Haematol.* **33**:118.

Van Gastel, C., Van der Berg, D., de Gier, J., and Van Dienen, L. L. M., 1965, Some lipid characteristics of normal red blood cells of different age, *Br. J. Haematol.* **11**:193.

Walsch, S., and Degkwitz, E., 1980, Influence of L-ascorbate deficiency on the metabolism of hepatic microsomal cytochrome P-450 in guinea pigs, *Hoppe-Seyler's Z. Physiol. Chem.* **361**:79.

Walter, H., and Selby, F., 1966, Counter-current distribution of red blood cells of slightly different ages, *Biochim. Biophys. Acta* **112**:146.

Watson, C. J., 1969, Gold from dross: The first century of the urobilinoids, *Ann. Intern. Med.* **70**:839.

Wayland, H., 1973, Photosensor methods of flow measurement in the microcirculation, *Microvasc. Res.* **5**:336.

Weed, R. I., 1970, The importance of erythrocyte deformability, *Am. J. Med.* **49**:147.

Weed, R. I., La Celle, P. L., and Merril, E. W., 1969, Metabolic dependence of erythrocyte deformability, *J. Clin. Invest.* **48**:795.

Weiss, L., 1962, The role of the spleen in the removal of normally aged red cells, Am. J. Anat. **111**:175.

Weiss, L., 1974, A scanning electron microscopic study of the spleen, *Blood* **43**:665.

Weiss, L., and Tavassoli, M., 1970, Anatomical hazards to the passage of erythrocytes through the spleen, *Semin. Hematol.* **7**:372.

Westerman, M. P., Pierce, L. E., and Jensen, W. N., 1963, Erythrocyte lipids: A comparison of normal young and old populations, *J. Clin. Lab. Med.* **62**:394.

White, P., Coburn, R. F., Williams, W. J., Goldwein, M. I., Rother, M. L., and Shafer, B. C., 1967, Carbon monoxide production associated with ineffective erythropoiesis, *J. Clin. Invest.* **46:**1986.

Williams, E. D., Szur, L., Glass, H. I., Lewis, S. M., Pettit, J. E., and Ahuja, S., 1974, Measurement of red cell destruction in the spleen, *J. Lab. Clin. Med.* **84:**134.

Wills, E. D., 1969, Lipid peroxide formation in microsomes, *Biochem. J.* **113:**315.

Winterbourn, C. C., and Batt, R. D., 1970, Lipid composition of human red cells of different ages, *Biochim. Biophys. Acta* **202:**1.

Winterbourn, C. C., and Carrel, R. W., 1974, Studies of hemoglobin denaturation and Heinz body formation in the unstable hemoglobins, *J. Clin. Invest.* **54:**678.

Wintrobe, M. M., Lee, G. R., Boggs, R., Bithell, T. C., Athens, J. W., and Foerster, J., 1974, *Clinical Hematology,* pp. 195–220, Lea & Febiger, Philadelphia.

Woods, J. S., Kardish, R., and Fowler, B. A., 1981, Studies on the action of porphyrinogenic trace metals on the activity of hepatic uroporphyrinogen decarboxylase, *Biochem. Biophys. Res. Commun.* **103:**264.

Yamamoto, T., Skanderberg, J., Zipursky, A., and Israels, L. G., 1965, The early appearing bilirubin: Evidence for two components, *J. Clin. Invest.* **44:**31.

Yannoni, C. Z., and Robinson, S. H., 1978, Early labeled heme synthesis from delta-aminolevulinic acid-4-[^{14}C] in rats: Comparison with glycine-2-[^{14}C], *Proc. Soc. Exp. Med.* **158:**466.

Yoshida, T., Takohashi, S., and Kikuchi, G., 1974, Partial purification and reconstitution of the heme oxygenase sytem from pig spleen microsomes, *J. Biochem.* **75:**1187.

Zuelzer, W. W., Robinson, A. R., and Hsu, T. H. J., 1968, Erythrocyte pyruvate kinase deficiency in non-spherocytic hemolytic anemia: A system of multiple genetic markers?, *Blood* **32:**33.

9

Hemostasis

LAURENCE A. SHERMAN and JOHN E. KAPLAN

1. INTRODUCTION

For discussion of the interaction of the RES and hemostatic system, a brief review of the latter is desirable (Murano, 1980; Hougie and Baugh, 1980). Body hemostasis is maintained by three major elements: vascular contraction, platelet aggregation, and a complex plasma protein coagulation mechanism. These interact in the following way. When a vessel is severed or injured, temporary cessation or slowing of blood loss occurs via vasoconstriction. Then platelets adhere to exposed subendothelium and aggregate to each other. Stimuli for aggregation include collagen and small amounts of thrombin. A temporary plug of platelets is formed. The plasma coagulation system is concomitantly activated, but proceeds at a slower rate. This system is initiated by Factor XII activation, probably usually by collagen in the subendothelium. A complex interaction exists between XII, high-molecular-weight kininogen, and prekallikrein. This results in further activation of the coagulation system, fibrinolytic system, and formation of bradykinin. The coagulation enzyme system cascade continues by activation of Factor XI which in turn activates Factor IX. Factor IX_a forms a complex with Factor VIII, Ca^{2+}, and phospholipid to activate Factor X. The phospholipid is furnished by platelets. X_a, in turn, complexes with V, Ca^{2+}, and phospholipid to convert Factor II (prothrombin) to thrombin. Thrombin cleaves two small pairs of polypeptides from fibrinogen. The remaining molecule is fibrin monomer, which is then capable of spontaneously polymerizing via noncovalent linkage. Fibrin polymers are relatively insoluble and at some point in polymerization they come out of solution as a gel (clot). The gel is made up of fibrin strands which trap erythrocytes and serve to link the aggregated platelets in a more permanent seal.

LAURENCE A. SHERMAN • Missouri/Illinois Regional Red Cross Blood Services, St. Louis, Missouri 63108. JOHN E. KAPLAN • Department of Physiology, Albany Medical College of Union University, Albany, New York 12208.

Thrombin has additional actions:

1. It activates Factor XIII which covalently cross-links the fibrin monomer together. This gives greater clot stability and resistance to fibrinolysis.
2. Thrombin activates Factors XIII and V making them more effective cofactors for IX_a and X_a, respectively.
3. The aforementioned aggregation of platelets by thrombin is probably one of the early stages in hemostasis.

The above cascade beginning with Factor XII is the "intrinsic system," i.e., involves only moieties found in the blood itself. An additional system is the extrinsic system. There a tissue moiety, tissue thromboplastin, activates blood coagulation Factor VII. The latter activates Factor X. Factor X_a then functions as in the intrinsic system.

2. CLEARANCE OF COAGULATION SYSTEM ACTIVATORS

The earliest studies relating the RES to hemostasis and thrombosis suggested that the RES removed altered components of the hemostatic system from the circulation. Most of these interactions were conceived of as being antithrombotic in nature and were generally felt to be significant on a systemic, as opposed to a local, level. This concept developed from early observations that freely circulating blood rarely clots whereas static blood clots quite readily. It was therefore felt that somewhere within the circulation there existed a selective blood filter which could remove activated procoagulants or activators of coagulation from the circulation. Early studies in this regard by Spaet *et al.* (1961) indicated that the RES was active in clearance of blood thromboplastin. *Blood thromboplastin* was the term used at that time to identify a fraction of activated plasma found after activated factor X (X_a) complexed with phospholipid. These studies employed radiolabeled substances and determined that intravenously injected blood thromboplastin was cleared by cells of the RES, primarily those of the liver. It was additionally determined that intraportal injection, which exposes the injected substances to the hepatic Kupffer cells prior to exposure to other vascular beds, provided significant protection against the systemic consumption of prothrombin and fibrinogen compared to intraaortic and intravenous injections. If the RES was blockaded by prior injection of colloidal carbon the capacity of intraportal injection to protect against coagulation was eliminated. This reduced protective effect was associated with reduced hepatic uptake of radiolabeled thromboplastin. This group as well as Halpern *et al.* (1953) demonstrated conversely that injection of thromboplastin delayed blood clearance of colloidal carbon. In a later study, Arakawa and Spaet (1963) showed that peritoneal macrophages possessed the capacity to phagocytize and inactivate blood thromboplastin *in vitro*.

In more recent years, it has become apparent that the RES is involved in the clearance of other activators of coagulation as well. Substances which directly or indirectly activate the blood coagulation system and which are cleared by the RES include endotoxin, bacteria, collagen, erythrocyte stroma, immune complexes, and activated platelets (Filkins, 1971; Rogers, 1960; Schneidkraut and Loegering, 1978; Haakenstad and Mannik, 1974; Kaplan, 1981). The factors influencing the clearance of these substances are discussed in greater detail elsewhere in this volume and will not be elaborated on here. These substances variously activate the intrinsic and extrinsic mechanisms of blood coagulation and by removing them from the circulation, the RES may reduce their potential thrombogenic effect. The extrinsic blood coagulation system is activated by release of tissue thromboplastin from damaged tissue (tissue factor). An additional source of thromboplastin is that which is released from the postphagocytic or activated phagocyte (Niemetz and Morrison, 1977). These cells are especially effective in releasing this substance when they have been previously exposed to bacteria or endotoxin (Niemetz and Morrison, 1977). Endotoxin, bacteria, and immune complexes may also result in the activation of the intrinsic coagulation cascade. A limited number of studies suggest that this results from direct activation of Factor XII (Rodriguez-Erdmann, 1964). Other studies, however, suggest that such activation occurs as a result of a complement-mediated, platelet-dependent action (Sandberg *et al.*, 1975). Activated platelets may contribute to the activation of Factors IX, XI, and XII, and supply membrane binding sites associated with Factor V as well as phospholipid for the later stages of blood coagulation (Walsh, 1981). Substantial evidence exists that when injection of these activators is preceded by depression of the RES, a more substantial thrombotic response occurs.

It is notable that under experimental conditions, a single injection or slow infusion of moderate doses of these agents is often innocuous. However, if the experimental animal is conditioned by depression of the RES, severe consequences may result. Studies by Good and Thomas (1952) and by Lee *et al.* (1966) observed that when phagocytic function of the RES was inhibited by prior injection of colloid, a single injection of endotoxin was capable of inducing the generalized Schwartzman reaction in association with fibrin degradation in the terminal renal vascular bed. Other studies have demonstrated that when RES depression is present, the thromboplastic effects of hemolyzed blood result in augmented consumption of Factors VI, VIII, and fibrinogen (Rabiner and Friedman, 1968).

The collagen-containing subendothelium lining of blood vessels probably comprises the most significant physiological activator of blood coagulation, and the intrinsic system in particular. In this regard, cells of the RES have been demonstrated to endocytize both particulate and soluble collagen as well as colloidal particles coated with denatured collagen (gelatin) (Saba and Jaffe, 1980). Cells of the RES also clear some nonbiological activators of intrinsic blood coagulation. The vascular sinus-lining macrophages effectively remove intravenously injected silica from the circulation and alveolar macrophages similarly engulf inhaled silica particles.

3. CLEARANCE OF PROCOAGULANTS AND ANTIPROTEASES

A more limited number of studies have indicated that the RES can clear activated blood coagulation proteases from circulating blood. The Kupffer cells of the perfused liver have been demonstrated to remove active thrombin and this removal is prolonged after colloid-induced RES depression (Gans *et al.*, 1967). *In vivo* studies have demonstrated the hepatic RES clearance of activated Factor X–phospholipid complex (Spaet *et al.*, 1961). Other studies have indicated that activated Factors IX and XI are removed by the liver but did not identify the cell type involved (Deykin, 1966; Deykin *et al.*, 1968). However, the close biochemical similarity of Factor IX with both Factors X and thrombin suggests that RE cells may be responsible for this observation. In addition, the removal of Factor X–phospholipid complex may indicate that the RES may be capable of ingesting other multimolecular procoagulant complexes, especially those which include phospholipid.

It has been further demonstrated that the RES clears complexes formed by the protease inhibitor α_2-macroglobulin and circulating proteases (Ohlsson, 1971; Debanne *et al.*, 1973). α_2-Macroglobulin is a relatively nonspecific inhibitor of serine proteases. Of interest with regard to hemostasis is its inhibition of thrombin, kallikrein, and plasmin (Ogston and Bennett, 1977). α_2-Macroglobulin–protease complexes are rapidly removed from the circulation by cells of the RES, especially those of the liver with lesser uptake by the spleen and bone marrow (Ohlsson, 1971; Debanne *et al.*, 1973). α_2-Macroglobulin–protease complexes are taken up *in vitro* by alveolar and peritoneal macrophages, but not by peripheral blood leukocytes (Debanne *et al.*, 1975). Internalization occurs following linking of the complex to specific reutilizable receptors (Kaplan, 1980; Kaplan and Nielsen, 1979a,b). The receptor is specific for complexed versus free α_2-macroglobulin but is not dependent upon which protease is bound (Debanne *et al.*, 1975). Studies have also indicated that the RES takes up proteases in complex with α_2-antitrypsin (Ohlsson, 1971; Debanne *et al.*, 1973). These α_1-antitrypsin–protease complexes are cleared more slowly and are less concentrated in the liver, relative to other RE cell-containing organs when compared to complexes containing α_2-macroglobulin. α_1-Antitrypsin is believed to have inhibitory activity for thrombin and activated Factor XI (Ogston and Bennett, 1977). It has a lower affinity for these enzymes than does α_2-macroglobulin. The RES plays a role in the synthesis and localization of antiproteases as well. Surface-bound α_2-macroglobulin has been identified upon macrophage surfaces, and cultured monocytes synthesize and secrete the protein (Hovi *et al.*, 1977). α_1-Antitrypsin has been identified in alveolar macrophages, Kupffer cells, and splenic macrophages as well as in tumors of histiocytic origin (Isaacson *et al.*, 1979; Gupta *et al.*, 1979). These cells synthesize as well as store and release this antiprotease. Antithrombin III activity has been localized in association with hepatic Kupffer cells (Zubairov *et al.*, 1970). Antithrombin III has enhanced protease binding in the presence of heparin and is probably the mediator of the prophylactic and therapeutic effects of the drug. Recent evidence suggests that RE cells play an important regulatory role with regard to the synthesis of acute-phase proteins by virtue of release of endogenous pyrogens. Thus, the RES regulates antiprotease synthesis, contains cell-associated antiproteases, and removes antiprotease–proteases complexes from the circulation as well as clears activated procoagulants. These collected activities provide the capacity to limit the extent of fibrin formation as a consequence of activation of the coagulation cascade.

4. CLEARANCE OF FIBRINOGEN DERIVATIVES

A large body of evidence supports the concept that the RES removes fibrinogen derivatives, especially fibrin, from the circulation and is active in their degradation. The Kupffer cells of the liver have been documented by immunofluorescence and electron microscopy to contain fibrin microaggregates following low-grade intravascular coagulation induced by thrombin, endotoxin, or infusion of finely dispersed fibrin (Lee and McCluskey, 1962; Prose *et al.*, 1965). Other investigators have histologically identified fibrin on the surface of Kupffer cells but have been unable to so identify fibrin within Kupffer cell vacuoles (Emeis and Lindeman, 1976). These investigators believed that fibrin was degraded on the surface of the RE cells. Other investigators, however, have identified fibrin immunofluorescently, electron microscopically, and isotopically within Kupffer cells (Gans *et al.*, 1967; Lee and McCluskey, 1962; Prose *et al.*, 1965). Radiolabeled fibrin is removed from the circulation by the RES during liver perfusion and *in vivo* and its removal is delayed during colloid-induced RES depression (Gans *et al.*, 1967; Sherman *et al.*, 1975). Similarly delayed clearance as well as reduced hepatic fibrin localization is observed during RES depression in association with traumatic shock and with thrombin-induced intravascular coagulation (Snedeker *et al.*, 1978), as well as during fibronectin depletion. While small amounts of radiolabeled fibrin are efficiently and completely cleared by the RES, larger amounts are cleared less completely and result in the formation of disseminated microemboli, especially in the renal and pulmonary circulation (Sherman *et al.*, 1977). Thus, the fibrin clearance capacity of the RES can be saturated with levels of fibrin with conversion of physiological levels of fibrinogen to fibrin. Fibrin clearance is diminished in association with chemically induced (methylpalmitate) RES depression and augmented in association with zymosan-induced RES stimulation (Snedeker and Kaplan, 1981). Evidence has been provided that the RES also clears fibrin degradation products and fibrin–fibrinogen complexes and that the clearance rates are related to the status of RES function (Gans and Lowman, 1967; Gans *et al.*, 1967; Barnhart and Cress, 1967; Sherman *et al.*, 1975). Isolated peritoneal macrophages specifically bind fibrin fibrinogen complexes on their membrane surfaces (Sherman and Lee, 1977; Colvin and Dvorak, 1975).

Recent evidence has suggested that identification and ingestion of fibrin may be mediated by plasma fibronectin, which is a high-molecular-weight dimeric glycoprotein present in plasma and other extracellular fluids. Tissue fibronectin is a chemically and immunologically similar protein associated with cell surfaces and basement membranes. Prior to the adoption of the term *fibronectin,* the plasma and tissue forms have variously been termed *cold-insoluble globulin, opsonic protein, fibroblast antigen,* and several other names. The fibronectins have been implicated in functions including cell adhesion, spreading and locomotion, structural orientation of the interstitial matrix, and opsonization of particulate matter, prior to RES phagocytosis (Mosher, 1980; Mosesson and Amrani, 1980; Saba and Jaffe, 1980; Yamada and Olden, 1978).

In the past few years, diverse areas in biology have focused on the structure and function of fibronectin. Originally described as fibrin-binding protein (Morrison *et al.*, 1948; Mosesson and Umfleet, 1970), it is now thought to have importance in other areas. Under the influence of coagulation Factor XIII (fibrin-stabilizing factor), it can be covalently bound to fibrin (Mosher, 1975) and collagen (Mosher *et al.*, 1980) *in vitro.* Additionally, noncovalent binding to the α chain of fibrin occurs at low temperatures (Mosesson, 1978). Recent evidence has documented the role of opsonic plasma fibronectin in the humoral control of RE phagocytic function. Biochemical and immunological analysis of fibronectin revealed its identity with the protein previously referred to as α_2-*surface-binding glycoprotein* or *opsonin* (Saba *et al.*, 1978). Specific opsonic depletion occurs following intravenous injection of gelatinized coloids and also in patients and experimental animals following surgery, burn injury, and traumatic shock and some models of intravascular coagulation (Blumenstock *et al.*, 1977; Kaplan and Saba, 1976; Mosher, 1976; Mosher and Williams, 1978; Scovill *et al.*, 1976; Lanser *et al.*, 1980; Saba and DiLuzio, 1969). Experimental depression of plasma fibronectin levels by injection of gelatinized colloids or monospecific antiserum to fibronectin is associated with a marked decrease of phagocytic capacities, whereas return of plasma levels to normal restores phagocytic activity (Blumenstock *et al.*, 1977; Kaplan *et al.*, 1976; Saba and DiLuzio, 1969). Limited *in vivo* data exist concerning the fibronectin-fibrin interaction. Bang *et al.* (1973) have utilized gel chromatography to study blood samples in disseminated intravascular coagulation (DIC). Fibronectin was shown to have an apparently larger molecular size after DIC, suggesting that it was now bound to fibrin. Both immunoreactive fibronectin and its opsonic activity can be absorbed by immobilized fibrin *in vitro* and depleted *in vivo,* following infusion of fibrin monomer (Stathakis and Mosesson, 1977; Stathakis *et al.*, 1978; Snedeker and Kaplan, 1980). Radiolabeled fibronectin plasma levels drop abruptly with defibrination with thrombin. This does not occur with defibrination by ancrod. Presumably, the degradation of the α chain of fibrinogen by ancrod prevents fibronectin–fibrin interaction as the α chain of fibrin contains the fibronectin–binding site (Stathakis and Mosesson, 1977; Stathakis *et al.*, 1978). The carboxy-terminal portion of the fibronectin chain contains a high-affinity fibrin- and collagen-binding site (Hormann and Seidl, 1980). Two lower-affinity fibrin-binding regions have recently been identified (Sekiguchi and Hakomori, 1980). Apparently, endogenously formed fibrin can circulate in complex with fibronectin, which may serve as a focus for formation of fibrin–fibrinogen complexes. The fibronectin–fibrin–fibrinogen interaction, which has been studied extensively with regard to formation of cold-precipitable complexes, may promote the solubility of fibrin at physiological temperature. Bang *et al.* (1973) reported that plasma fibrionectin was incorporated into soluble fibrin complexes formed *in vitro* and suggested that fibronectin helped maintain the solubility of the complex. Recent studies have demonstrated that fibronectin maintains the solubility of fibrin monomer and resolubilizes fibrin complexes in a purified *in vitro* system (Kaplan and Snedeker, 1980). Systematic administration of monospecific antiserum to fibronectin results in defective RES uptake of fibrin monomer (Snedeker and Kaplan, 1980). It has also been found that the development of hypofibrinogenemia in response to experimentally induced thrombosis is enhanced by prior fibronectin depletion and limited by prior fibronectin supplementation (Kaplan *et al.*, 1979). Fibronectin may mediate the interaction of fibrinogen with cells. Jilek and Hormann (1978) demonstrated that the capacity of macrophages to bind fibrin can be destroyed by trypsinization and restored by the presence of fibronectin. Converse results with trypsinization have been noted in one of the present authors' laboratories. Colvin *et al.* (1979) described a role for fibronectin in the adherence of fibrinogen to cultured fibroblasts. Some of the variations in reported data may be related to variation in methods of preparing both fibronectin as well as fibrin–fibrinogen complexes. The binding of fibrin to collagen is also limited by fibronectin (Kaplan and Snedeker, 1980). Fibronectin depletion has been observed in patients with intravascular coagulation in association with advanced malignancy and similar observations have been made in animal models of intravascular coagulation in association with trauma, Rocky Mountain spotted fever, spontaneous atrial thrombus formation, and *Pseudomonas* bacteremia (Mosher and Williams, 1978; Scovill *et al.*, 1976; Matsuda *et al.*, 1978; Schumacker and Saba, 1977; Kaplan and Saba, 1976; Mosher, 1976; Kaplan and Dodds, unpublished observations).

Impaired RES clearance has been correlated with low blood fibronectin levels in various shock syndromes (Saba and Jaffe, 1980; Scovill *et al.*, 1977). Acute respiratory distress syndrome (ARDS) has been posited as occuring secondary to decreased fibronectin. However, Jay *et al.* (1980) have demonstrated low fibronectin levels in acute non-ARDS pulmonary disease as well as a sizeable proportion of patients with chronic obstructive pulmonary disease. These latter findings may indicate a complicated balance of production and utilization. Elucidation of the exact pathophysiological roles of fibronectin and the RES in these disorders will require substantial further data.

We suggest that the multifaceted interactions elaborated upon above collectively comprise a physiological mechanism which subserves the function of maintaining vascular integrity through the preservation of blood fluidity. The importance of these findings is emphasized by the protective effect of RES clearance against the consequence of intravenous coagulation and the diminution of such protection by prior depression of the RES. Lee (1962) observed that when reduction of RES depression by colloids was followed by thrombin, a generalized Schwartzman reaction was consistently found associated with fibrin deposition of terminal vascular beds. Gans (1966) demonstrated that in dogs carbon blockade of the RES prior to thrombin injection, resulted in hemorrhagic necrosis of the gastrointestinal tract, as well as gross embolic and thrombotic occlusion of hepatic vessels. Busch and Saldeen (1973) have found in rats that intraportal administration of thrombin resulted in reduced pulmonary and renal deposition of fibrin as compared to that observed after intravenous or intraaortic injection of thrombin. Sherman *et al.* (1975) and later Snedeker and Kaplan (1980) found that small doses of soluble fibrin could be effectively cleared by the RES while the effusion of larger doses resulted in microthrombi with peripheral vascular events. Esnouf and Marshall (1968) reported that only following RES depression did the induction of coagulation by *Ancistrodon rhodostoma* venom in dogs lead to increased pulmonary vascular resistance, hypotension, and the deposition of clot material in the pulmonary circulation. Schumacker and Saba (1977) reported that in dogs following RES blockade thrombin infusion resulted in impaired pulmonary gas exchange as manifested by alveolar arterial oxygen gradient, increased dead space, reduced arterial PO_2, and also that RES blockade prior to thrombin infusion in rats resulted in increased pulmonary fibrin localization.

5. PRODUCTION OF HEMOSTATICALLY ACTIVE SUBSTANCES

5.1. SYNTHESIS

The interaction of the RES and hemostatic system extends beyond modulating or dampening various activated hemostatic moieties. Certain substances produced in RE cells are active in hemostasis. Two major categories of substances have been described. The first are those with coagulant activity or interactions and the second are a group of proteases, including plasminogen activator (PA), which lead to degradation of fibrin.

5.1.1. Coagulation System Moieties

For some time it has been recognized that a procoagulant material is found in leukocytes. This has been variously called *Procoagulant activity* (Muhlfelder *et al.*, 1978) or *tissue factor* (Rickles *et al.*, 1973). A variety of data have been gathered concerning this material although there are questions concerning which cell type has the bulk of this activity. The uncertainty centers on neutrophils versus monocytes as being the source of activity (Edwards and Rickles, 1978). This is, in fair part, related to the difficulties inherent in obtaining totally homogeneous preparations of each of these cells. Tissue factor appears to function via the extrinsic pathway of the coagulation system. Its activity is increased by exposure of the cells to endotoxin (lipid A) (Niemetz and Morrison, 1977), surface adherence (van Ginkel *et al.*, 1977), etc. Demonstration of contaminating endotoxin in certain agents which stimulate tissue factor activity such as Con A (Rickles *et al.*, 1979) necessitates caution in interpreting some reports in the literature. The generation of tissue factor in response to endotoxin is of the same order of sensitivity as the limulus lysate assay (Rickles *et al.*, 1979).

Biological material known to increase tissue factor in leukoytes includes platelet membrane lipoprotein (Niemetz and Marcus, 1974), Con A (Rickles *et al.*, 1979), and interaction with T cells and their products. The latter has been shown to be a requisite for phytohemagglutinin activity. The procoagulant activity does not appear to be a nonspecific release reaction, and stimulation can occur by at least two discrete mechanisms, one requiring T cells and the other not (Edwards and Rickles, 1980). The tissue factor enhancement effect of stimulants is not well correlated with other properties, such as mitogenicity.

Several interactions with the humoral immune system have also been described. Antigen–antibody complexes serves as stimulants for tissue factor (Rothberger *et al.*, 1977). Complement-activated plasma does so as well (Muhlfelder *et al.*, 1979). The latter effect has been ascribed to a fragment of C5, which can also be produced by proteolytic enzymes.

A recent study has indicated that peritoneal macrophages in culture synthesize a precoagulated material which possesses characteristics similar to Factor VIII (Maier and Ulevitch, 1981). Synthesis of this activity is enhanced by the presence of endotoxin. Osterud *et al.* (1981) have reported synthesis of Factors II, V, VII, IX, X, and tissue factor by mouse macrophages in culture. All of the observations have been made on the basis of *in vitro* cell culture studies. Correlates have been drawn with frequent occurrence of intravascular coagulation in leukemic patients (Gralnick and Abrell, 1973). However, to date, substantive direct data are not available. Additionally, the relative contribution of leukocyte procoagulant or tissue factor activity to normal hemostasis, remains to be determined.

Alitalo *et al.* (1980) and Johansson *et al.* (1979) have indicated synthesis and secretion of fibronectin by peritoneal macrophages and cultured blood monocytes, respectively. The degree to which this synthesis may be modified by "activation" or hormones is unknown.

5.1.2. Fibrinolytic Moieties

The contribution of RE cells to fibrino(geno)lytic activity may be divided into "classical" and nonclassical fibrinolysis. Classical fibrinolysis may be used to denote the plasminogen–plasmin system. Most interest in the role of the RES has focused on the secretion of PA by macrophages. Unkeless *et al.* (1974) first described this activity and also noted that it was essentially only a product of induced versus resident macrophages. Although first described after thioglycollate injections, other more pathophysiologic stimuli such as endotoxin (Gordon *et al.*, 1974), microorganisms (Gordon, 1978), and fibrin (Sherman *et al.*, 1977, 1981) have been described. Endotoxin appears to prepare or activate the cells for subsequent production and release of PA in response to phagocytosis. Without pretreatment with endotoxin, only modest amounts of PA appear in response to phagocytosis. Immunologic stimulation of macrophages by BCG or *T. cruzi* has been shown to lead to PA release on restimulation. The second challenge with these agents can be *in vivo* or *in vitro*. Purified sensitized T lymphocytes are also capable of stimulating macrophages to release PA (Newman *et al.*, 1978).

A prerequisite for production of PA in these tissue culture studies appears to be material found in serum. Customarily, the cells are in serum-containing media while adhesion occurs. A second serum-free incubation is used to demonstrate PA. Drapier *et al.* (1979) have shown that if the first incubation is in the absence of serum, then no PA results.

The control of the release of macrophage proteases has been extensively investigated by Web and co-workers (Werb and Gordon, 1975a,b). In particular, they have shown for a number of species that glucocorticoids inhibit secretion of proteases such as PA, elastase, nonspecific fibrinolytic protease, and collagenase (Werb *et al.*, 1978). Conversely, progesterone was shown to block the inhibitory effect of dexamethasone. Moreover, the basic observation that the fibrinolytic proteases are present in very small amounts in resting cells versus large amounts in elicited cells, further suggests that pathophysiologic variation may occur.

A potential added interaction with the hemostatic system is shown by data that both PA and nonplasminogen-dependent fibrinolysis are increased in response to soluble fibrin (Sherman *et al.*, 1981). The initial work on PA indicated that fibrin proteolysis required plasminogen, e.g., that PA was the major or sole enzyme causing fibrinolysis (Unkeless *et al.*, 1974). Work in our laboratory (Sherman *et al.*, 1977, 1981) and in other laboratories (Chang *et al.*, 1977) has also demonstrated nonplasminogen-dependent fibrinogenolytic activity in activated vs. resting cells. Subsequent data by Werb and her co-workers have shown that other enzymes secreted by activated macrophages (elastase, collagenase) are capable of directly degrading fibrin (Banda and Werb, 1980). A nonplasmin fibrinogenolytic enzyme has also been found in neutrophils (Plow and Edgington, 1975).

Further data have shown that small amounts (> 5 μg/ml) of soluble fibrin–fibrinogen complexes are capable of stimulating resting mouse macrophages in culture to produce both PA and other fibrinogenolytic enzymes. The amounts of fibrin involved are of a magnitude which is produced during intravascular coagulation. This stimulating effect may be dependent on a binding site for soluble fibrin found on guinea pig macrophage membranes (Sherman and Lee, 1977). Fibrin is bound in approximately 10-fold greater amounts than is fibrinogen. As previously noted, controversy exists as to the necessity for fibronectin in this process. In the original report, trypsinization, which destroys fibronectin, did not alter binding. Conversely, Jilek and Hormann (1978) found that added fibronectin enhanced binding of soluble fibrin.

The vast majority of the reports concerning macrophage proteases and hemostasis have focused on fibrino(geno)lysis as the endpoint. However, other interactions are likely to occur as well. Plasmin, resulting from PA action, has wide effects on hemostasis. Factors V, VIII, and fibronectin are among the proteins degraded by plasmin. Other neutral proteases may degrade these moieties as well. Kopec *et al.* (1980) have expanded on prior data by showing structural and functional degradation of Factor VIII by elastase and a chymotrypsin-like protease, but not collagenase or gelatinase. These enzymes were all from neutrophils.

5.2. REGULATION OF SYNTHESIS

Indirect influences of the RES on the coagulation system have not attracted wide interest. Some data are available, primarily concerning leukocytes and fibrinogen synthesis. Several investigators have proposed that a substance released from leukocytes (LEM) stimulates fibrinogen synthesis (Weidner *et al.*, 1979). Negative reports may reflect use of an inadequate number of leukocytes. Data have also been presented that stimulation of leukocytes or inflammation is required for release of this material, which does not appear to be leukocyte pyrogen. Kernoff *et al.* (1981) have recently reported that leukocytes are not necessary for inflammation-induced increases in fibrinogen synthesis. They found comparable increases in granulocytopenic animals as compared to controls.

The synthesis of coagulation moieties other than fibrinogen is known to increase in inflammation (acute-phase reactants). Factor VIII and the fibrinolytic system changes are the best examples. Whether their synthesis is controlled by factors similar to LEM is unknown.

6. INFLUENCES OF THE COAGULATION SYSTEM ON THE RES

A major direct influence of the coagulation system is RES depression resulting from intravascular coagulation. Kaplan and Saba (1978) have shown this phenomenon after thrombin-induced intravascular coagulation. Ahlgren *et al.* found that both the phagocytosis and the catabolism of microaggregates in the RES were impaired following injection of the thrombin-like enzyme difibrase. Earlier experiments using colloidal carbon particles assessed the effect of intravascular coagulation on particulate localization. These studies found that injections of thromboplastin or polybrene concomitant with carbon clearance produced a change in the tissue distribution of carbon (Halpern *et al.*, 1953; Lazar *et al.*, 1968). Localization of carbon shifted from the liver to the lung and was often associated with a delay in blood clearance. An additional indirect depression is also likely via the fibronectin system. During intravascular coagulation, fibronectin levels are reduced because of the interaction with fibrin. Presumably, low fibronectin will decrease opsonization and overall RES function. Recent studies suggest that plasminolytically derived degradation products of plasma fibronectin are inhibitors of both *in vivo* and *in vitro* phagocytosis (Ehrlich *et al.*, 1981).

Changes in RES function as the result of specific coagulation moieties have not been extensively studied. The aforementioned increases in fibrinolytic enzymes after exposure to soluble fibrin may indicate other systems are also operative. Preliminary attempts to demonstrate similar effects from enzyme inhibitor complexes which are cleared by the RES have had negative results.

Hemostatic moieties may influence the RES in a complex manner through other areas of the immune system. Various stimulating effects of fibrin degradation products on lymphocytes have been described. A vasoactive effect of fibrinopeptide B has been described chiefly in *ex vivo* experiments. It has additionally been demonstrated that the products of plasmin-digested fibrin and fibrinogen are chemotactic for monocytes (Richardson *et al.*, 1976). It is unclear whether such an effect can be obtained *in vivo*, and its hemostatic significance to DIC with shock is unknown.

7. CONCLUSION

The foregoing has outlined a complex interplay between the hemostatic system and the RES, which is only incompletely understood. The biologic significance of many tissue culture or *in vitro* observations needs to be determined. Several additional areas require investigation. The biochemical nature of various humoral and intracellular regulators of these interactions must be elucidated. Additionally, the effect on the RES of soluble fibrin raises the possibility that other biochemically activated products of thrombosis (platelet release proteins, fibrin degradation products, etc.) have regulatory action. These are but a few areas for future studies.

REFERENCES

Ahlgren, T., Berghen, L., Lagergren, H., Lahnborg, G., and Shildt, P., 1976, Phagocytic and catabolic function of the reticuloendothelial system in dogs subjected to defibrinogenation, *Thrombosis Research* **8:**819.

Alitalo, K., Hovi, T., and Vaheri, A., 1980, Fibronectin is produced by human macrophages, *J. Exp. Med.* **151:**602.

Arakawa, T., and Spaet, T. H., 1963, *In vitro* inactivation of rabbit blood thromboplastin by macrophages, *Proc. Soc. Exp. Biol. Med.* **113:**71.

Banda, M. J., and Werb, Z., 1980, The role of macrophage elastase in the proteolysis of fibrinogen, plasminogen and fibronectin, *Fed. Proc.* **39:**1756.

Bang, N. V., Hansen, M. S., Smith, G. F., and Mosesson, M. W., 1973, Properties of soluble fibrin polymers encountered in thrombotic states, *Thromb. Diath. Haemorrh. Suppl.* **56:**75.

Barnhart, M. I., and Cress, D. C., 1967, Plasma clearance of products of fibrinolysis, *Adv. Exp. Med. Biol.* **1:**492.

Blumenstock, F. A., Weber, P. B., Saba, T. M., and Laffin, R., 1977, Electroimmunoassay of alpha-2-globulin opsonic protein levels in the evaluation of reticuloendothelial function, *Am. J. Physiol.* **232:**R80.

Busch, D., and Saldeen, T., 1973, Amount of fibrin in different organs after intravenous, intraportal or intra-aortal injection of thrombin in the rat, *Thromb. Diath. Haemorrh.* **29:**87.

Chang, M. L., Bang, N. U., Truex, L., Boxer, L., Mattler, L. E., and Marks, L. A., 1977, Degration of soluble fibrin complexes, fibrinogen and fibrin by macrophage enzymes, *Thromb. Haemostasis* **38:**102.

Colvin, R. B., and Dvorak, G. F., 1975, Fibrinogen/fibrin on the surface of macrophages: Detection, distribution, binding requirements, and possible role in macrophage adherence phenomena, *J. Exp. Med.* **142:**1377.

Colvin, R. B., Gardner, P. I., Roblin, R. O., Verderber, E. L., Lanigan, J. M., and Mosesson, M. W., 1979, Cell surface fibrinogen-fibrin receptors on cultured human fibroblasts: Association with fibronectin (cold-insoluble globulin, LETS protein) and loss in SV40 transformed cells, *Lab. Invest.* **41:464.**

Debanne, M. T., Regoeczi, E., and Dolovich, J., 1973, Serum protease inhibitors in the blood clearance of subtilisin A, *Br. J. Exp. Pathol.* **54:**571.

Debanne, M. T., Bell, R., and Dolovich, J., 1975, Uptake of proteinase–α-macroglobulin complexes by macrophages, *Biochim. Biophys. Acta* **411:**295.

Deykin, D., 1966, The role of the liver in serum-induced hypercoagulability, *J. Clin. Invest.* **45:**256.

Deykin, D., Cochios, F., DeCamp, G., and Lopez, A., 1968, Hepatic removal of activated Factor X by the perfused rabbit liver, *Am. J. Physiol.* **214:**414.

Drapier, J. C., Tenu, J. P., Lemaire, G., and Petit, J. F., 1979, Regulation of plasminogen activator secretion in mouse peritoneal macrophages. I. Role of serum studied by a new spectrophotometric assay for plasminogen activators, *Biochimie* **61:**463.

Edwards, R. L., and Rickles, F. R., 1978, On the origin of leukocyte procoagulant activity, *Thromb. Res.* **13**:307.

Edwards, R. L., and Rickles, F. R., 1980, The role of human T cells (and T cell products) for monocyte tissue factor generation, *J. Immunol.* **125**:606.

Ehrlich, M. I., Krushell, J. S., Blumenstock, F. A., and Kaplan, J. E., 1981, Depression of phagocytosis by plasmin degradation products of plasma fibronectin, *J. Lab. Clin. Med.* **98**:263.

Emeis, J. J., and Lindeman, J., 1976, Rat liver macrophages will not phagocytize fibrin during disseminated intravascular coagulation, *Haemostasis* **5**:193.

Esnouf, M. P., and Marshall, R., 1968, The effect of blockade of the reticuloendothelial system and of hypotension on the response of dogs to *Ancistrodon rhodostoma* venom, *Clin. Sci.* **35**:261.

Filkins, J. P., 1971, Comparison of endotoxin detoxification by leukocytes and macrophages, *Proc. Soc. Exp. Biol. Med.* **137**:1396.

Gans, H., 1966, Preservation of vascular patency as a function of reticuloendothelial clearance, *Surgery* **60**:1216.

Gans, H., and Lowman, J., 1967, Uptake of fibrin and fibrin degradation products by the isolated perfused rat liver, *Blood* **29**:526.

Gans, H., McLeod, J., and Lowman, J. T., 1967, A new technique for the preparation of an *in vivo* labelled fibrinogen, *Blood* **29**:517.

Good, R. A., and Thomas, L., 1952, Studies on the generalized Schwartzman reaction. II. The production of bilateral cortical necrosis of the kidneys by a single injection of bacterial toxin in rabbits previously treated with thorotrast or trypan blue, *J. Exp. Med.* **96**:625.

Gordon, S., 1978, Regulation of enzyme secretion by mononuclear phagocytes: Studies with macrophage plasminogen activator and lysozyme, *Fed. Proc.* **37**:2754.

Gordon, S., Unkeless, J., and Cohn, Z. A., 1974, Induction of macrophage plasminogen activator by endotoxin stimulation and phagocytosis: Evidence for a two stage process, *J. Exp. Med.* **140**:995.

Gralnick, H. R., and Abrell, E., 1973, Studies of the procoagulant and fibrinolytic activity of promyelocytes in actue promyelocytic leukemia, *Br. J. Haematol.* **24**:89.

Gupta, P. K., Frost, J. K., Geddes, S., Aracıl, B., and Davidovski, F., 1979, Morphological identification of alpha-2-antitrypsin in pulmonary macrophages, *Hum. Pathol.* **1**:345.

Haakenstad, A. O., and Mannik, M., 1974, Saturation of the reticuloendothelial system with soluble immune complexes, *Immunology* **112**:1939.

Halpern, B. N., Benacerraf, B., and Biozzi, G., 1953, Quantitative study of the granulopectic activity of the reticuloendothelial system. I. The effect of the ingredients present in India ink and of substances affecting blood clotting *in vivo* on the fate of carbon particles administered intravenously in rats, mice and rabbits, *Br. J. Exp. Pathol.* **34**:426.

Hormann, H., and Seidl, M., 1980, Affinity chromatography on immobilized fibrin monomer. III. The fibrin affinity center of fibronectin, *Hoppe-Seyler's Z. Physiol. Chem.* **361S**:1449.

Hougie, C., and Baugh, R. F., 1980, Current views on blood coagulation and haemostatic mechanism, in: *Blood Coagulation and Haemostasis* (J. M. Thomson, ed.), p. 1, Churchill Livingstone, Edinburgh.

Hovi, T., Mosher, P., and Vaheri, A., 1977, Cultured human monocytes synthesize and secrete α_2 macroglobulin, *J. Exp. Med.* **145**:1580.

Isaacson, P., Wright, D. H., Judd, M. A., and Mepham, B. L., 1979, Primary gastrointestinal lymphoma, *Cancer* **43**:1805.

Jay, S., Bang, N., Stropes, L., Marks, C., and Campbell, S., 1980, Plasma fibronectin concentration in acute and chronic lung disease, *Clin. Res.* **28**:744A.

Jilek, F., and Hormann, H., 1978, Fibronectin (cold-insoluble globulin). V. Mediator of fibrin-monomer binding to macrophages, *Hoppe-Seyler's Z. Physiol. Chem.* **359**:1603.

Johansson, S., Rubin, K., Hook, M., Ahlgren, T., and Seljelid, R., 1979, *In vitro* biosynthesis of cold-insoluble globulin (fibronectin) by mouse peritoneal macrophages, *FEBS Lett.* **105**:313.

Kaplan, J., 1980, Evidence for reutilization of surface receptors for α-macroglobulin–protease complexes in rabbit alveolar macrophages, *Cell* **19**:197.

Kaplan, J. E., 1981, The role of the reticuloendothelial system in control of hemostatic and thrombotic mechanisms, in: *Physiology of the Reticuloendothelial System* (T. M. Saba and B. M. Altura, eds.), Raven Press, New York.

Kaplan, J., and Nielsen, M. L., 1979a, Analysis of macrophage surface receptors. I. Binding of α-macroglublin–protease complexes to rabbit alveolar macrophages, *J. Biol. Chem.* **254:**7323.

Kaplan, J., and Nielsen, M. L., 1979b, Analysis of macrophage surface receptors. II. Internalization of α-macroglobulin–trypsin complexes by rabbit alveolar macrophages, *J. Biol. Chem.* **254:**7329.

Kaplan, J. E., and Saba, T. M., 1976, Humoral deficiency and reticuloendothelial depression after traumatic shock, *Am. J. Physiol.* **230:**7.

Kaplan, J. E., and Saba, T. M., 1978, Platelet removal from the circulation by the liver and spleen, *Am. J. Physiol.* **235:**H314.

Kaplan, J. E., and Snedeker, P. W., 1980, Maintenance of fibrin solubility by plasma fibronectin, *J. Lab. Clin. Med.* **96:**1054.

Kaplan, J. E., Saba, T. M., and Cho, E., 1976, Serological modification of reticuloendothelial capacity and altered resistance to traumatic shock, *Circ. Shock* **3:**203.

Kaplan, J. E., Blumenstock, F. A., and Saba, T. M., 1979, A radial immunodiffusion method for the measurement of rat fibrinogen and fibrin degradation products, *Vox Sang.* **36:**65.

Kernoff, L. M., Colman, J., and Rawlings, E., 1981, Acute phase stimulation of fibrinogen synthesis: Evidence against a major mediator role for granulocytes, *Thromb. Haemostasis* **46:**238.

Kopec, M., Bykowska, K., Lopaciuk, S., Jelenska, M., Kaczanowska, J., Sopata, I., and Wojtecka, E., 1980, Effects of neutral proteases from human leukocytes on structure and biological properties of human Factor VIII, *Thromb. Haemostasis* **43:**211.

Lanser, M. E., Saba, T. M., and Scovill, W. A., 1980, Opsonic glycoprotein (plasma fibronectin) levels after burn, *Ann. Surg.* **192:**776.

Lazar, B., Biliczki, F., and Kavaes, K., 1968, The phagocytic function of the reticuloendothelial system in rats treated with polybrene, liquoid and compound 48/80, *Pharmacology* **1:**253.

Lee, L., 1962, Reticuloendothelial clearance of circulating fibrin in the pathogenesis of the generalized Schwartzmann reaction, *J. Exp. Med.* **115:**1065.

Lee, L., and McCluskey, R. J., 1962, Immunohistochemical demonstration of the reticuloendothelial clearance of circulating fibrin aggregates, *J. Exp. Med.* **116:**611.

Lee, L., Prose, P. H., and Cohen, M. H., 1966, Role of reticuloendothelial system in diffuse low-grade intravascular coagulation, *Thromb. Diath. Haemorrh. Suppl.* **87:**66.

Maier, R. V., and Ulevitch, R. V., 1981, Bacterial lipopolysaccharide (LPS) induces a unique procoagulant activity (PCA) in explanted rabbit hepatic macrophages, *Circ. Shock* **8:**216.

Matsuda, M., Yoshida, N., Cliki, N., and Wakalayachis, C., 1978, Distribution of cold-insoluble globulin in plasma and tissues, *Ann. N.Y. Acad. Sci.* **312:**74.

Morrison, P. R., Edsall, J. T., and Miller, S. G., 1948, Preparation and properties of serum and plasma proteins. XCIII. The separation of purified fibrinogen from fraction I of human plasma, *J. Am. Chem. Soc.* **70:**3103.

Mosesson, M. W., 1978, Structure of human plasma cold-insoluble globulin and the mechanism of its precipitation in the cold with heparin or fibrin-fibrinogen complexes, *Ann. N.Y. Acad. Sci.* **312:**11.

Mosesson, M. W., and Amrani, D. L., 1980, The structure and biological activities of plasma fibronectin, *Blood* **56:**145.

Mosesson, M. W., and Umfleet, R. A., 1970, The cold-insoluble globulin of human plasma. I. Purification, primary characterization and relationship to fibrinogen and other cold-insoluble fraction components, *J. Biol. Chem.* **245:**5728.

Mosher, D. F., 1975, Cross-linking of cold-insoluble globulin by fibrin-stabilizing factor, *J. Biol. Chem.* **250:**6614.

Mosher, D. F., 1976, Changes in plasma cold-insoluble globulin concentration during experimental Rocky Mountain spotted fever infection in rhesus monkeys, *Thromb. Res.* **9:**37.

Mosher, D. F., 1980, Fibronectin, *Prog. Haemostasis Thromb.* **5:**111.

Mosher, D. F., and Williams, E., 1978, Fibronectin concentration is decreased in plasma of severely ill patients with disseminated intravascular coagulation, *J. Lab. Clin. Med.* **91:**729.

Mosher, D. F., Schad, P. E., and Vann, J. M., 1980, Cross-linking of collagen and fibronectin by Factor XIIIa: Localization of participating glutaminyl residues to a tryptic fragment of fibronectin, *J. Biol. Chem.* **255:**1181.

Muhlfelder, T. W., Khan, I., and Niemetz, J., 1978, Factors influencing the release of procoagulant-tissue factor activity from leukocytes, *J. Lab. Clin. Med.* **92**:65.

Muhlfelder, T. W., Niemetz, J., Kreutzer, D., Beebe, D., Ward, P. A., and Rosenfeld, S. I., 1979, C_5 chemotactic fragment induces leukocyte production of tissue factor activity, *J. Clin. Invest.* **63**:147.

Murano, G., 1980, A basic outline of blood coagulation, in: *Seminars in Thrombosis and Hemostasis* (E. F. Mammen, ed.), Vol. VI, p. 140.

Newman, W., Gordon, S., Hammerling, U., Senik, A., and Bloom, B. R., 1978, Producer of migration inhibition factor (MIF) and an inducer of plasminogen activator (IPA) by subsets of T cells in MLC, *J. Immunol.* **120**:927.

Niemetz, J., and Marcus, A. J., 1974, The stimulatory effect of platelets and platelet membranes on the procoagulant activity of leukocytes, *J. Clin. Invest.* **54**:1437.

Niemetz, J., and Morrison, D. C., 1977, Lipid A as the biologically active moiety in bacterial endotoxin (LPS)-initiated generation of procoagulant activity by peripheral blood leukocytes, *Blood* **49**:947.

Ogston, D., and Bennett, B., 1977, Naturally occurring inhibitors of coagulation, in: *Haemostasis: Biochemistry, Physiology and Pathology* (D. Ogston and B. Bennett, eds.), Wiley, New York.

Ohlsson, K., 1971, Elimination of ^{125}I trypsin α-macroglobulin complexes from blood by reticuloendothelial cells in dog, *Acta Physiol. Scand.* **81**:269.

Osterud, B., Lindahl, U., Bogwald, J., and Selegelid, R., 1981, The extravascular coagulation system: The production of prothrombin, Factors V, X, IV, VII and tissue factor in macrophages, *Thromb. Haemostasis* **46**:14.

Plow, E. F., and Edgington, T. S., 1975, An alternative pathway for fibrinolysis. I. The cleavage of fibrinogen by leukocyte proteases at physiologic pH, *J. Clin. Invest.* **56**:30.

Prose, P. H., Lee, L., and Balk, S. D., 1965, Electron microscopic study of the phagocytic fibrin-clearing mechanism, *Am. J. Pathol.* **47**:403.

Rabiner, S. F., and Friedman, L. H., 1968, The role of intravascular hemolysis and the reticuloendothelial system in the production of a hypercoagulable state, *Br. J. Haematol.* **14**:105.

Richardson, D. L., Pepper, S. D., and Kay, A. B., 1976, Chemotaxis for human monocytes by fibrinogen-derived peptides, *Br. J. Haematol.* **32**:507.

Rickles, F. R., Hardin, J. A., Pitlick, F. A., Hoyer, L. W., and Conrad, M. E., 1973, Tissue factor activity in lymphocyte cultures from normal individuals and patients with hemophilia A, *J. Clin. Invest.* **52**:1427.

Rickles, F. R., Levin, J., Rosenthal, I., and Atkins, A., 1979, Functional interaction of concanavalin A and bacterial endotoxin (lipopolysaccharide): Effects on the measurement of endogenous pyrogen release, human mononuclear cell tissue factor activation, lymphocyte DNA synthesis, and gelatin of limulus amebocyte lysate, *J. Lab. Clin. Med.* **93**:128.

Rodriguez-Erdmann, F., 1964, Studies on the pathogenesis of the generalized Schwartzman reaction. III. Trigger mechanism for the activation of the prothrombin molecule, *Thromb. Diath. Haemorrh.* **12**:471.

Rogers, D. E., 1960, Host mechanisms which act to remove bacteria from the blood stream, *Bacteriol. Rev.* **24**:50.

Rothberger, H., Zimmerman, T. S., Spiegelberg, H. L., and Vaughan, J. H., 1977, Leukocyte procoagulant activity: Enhancement of production *in vitro* by IgG and antigen-antibody complexes, *J. Clin. Invest.* **59**:549.

Saba, T. M., and DiLuzio, N. R., 1969, Reticuloendothelial blockade and recovery as a function of opsonic activity, *Am. J. Physiol.* **216**:197.

Saba, T. M., and Jaffe, E., 1980, Plasma fibronectin (opsonic glycoprotein): Its synthesis by vascular endothelial cells and role in cardiopulmonary integrity as related to reticuloendothelial function, *Am. J. Med.* **68**:577.

Saba, T. M., Blumenstock, F. A., Weber, P., and Kaplan, J. E., 1978, Physiologic role of cold-insoluble globulin in systemic host defense: Implications of its characterization of the opsonic α_2-surface binding glycoprotein, *Ann. N.Y. Acad. Sci.* **312**:43.

Sandberg, A. L., Siraganian, R. P., and Mergenhagen, S. E., 1975, Biological consequences of endotoxin interaction with complement, in: *Gram-negative Bacterial Infections and Mode of Endotoxin Actions—Pathophysiological, Immunological and Clinical Aspects* (B. Urbaschek, R. Urbaschek, and E. Neter, eds.), pp. 329–334, Springer-Verlag, Berlin.

Schneidkraut, M. J., and Loegering, D. J., 1978, Effect of hemolyzed blood on reticuloendothelial function and susceptibility to hemorrhagic shock, *Proc. Soc. Exp. Biol. Med.* **159**:418.

Schumacker, P. T., and Saba, T. M., 1977, Augmentation of pulmonary insufficiency during low-grade intravascular coagulation by prior impairment of hepatic RE clearance activity, *Physiologist* **20**:85.

Scovill, W. A., Saba, T. M., Kaplan, J. E., Bernard, H. R., and Powers, S. R., Jr., 1976, Deficits in reticuloendothelial humoral mechanisms after trauma, *J. Trauma* **16**:898.

Scovill, W. A., Saba, T. M., Kaplan, J. E., Bernard, H. R., and Powers, S. R., Jr., 1977, Disturbances in circulating opsonic activity in man after operative and blunt trauma, *J. Surg. Res.* **22**:709.

Sekiguchi, K., and Hakomori, S., 1980, Identification of two fibrin-binding domains in plasma fibronectin and unequal distribution of these domains in two different subunits: A preliminary note, *Biochem. Biophys. Res. Commun.* **97**:709.

Sherman, L. A., and Lee, J., 1977, Specific binding of soluble fibrin to macrophages, *J. Exp. Med.* **145**:76.

Sherman, L. A., Harwig, S., and Lee, J., 1975, *In vitro* formation and *in vivo* clearance of fibrinogen-fibrin complexes, *J. Lab. Clin. Med.* **86**:100.

Sherman, L. A., Lee, J., and Jacobson, A., 1977, Quantitation of the reticuloendothelial system clearance of soluble fibrin, *Br. J. Haematol.* **37**:231.

Sherman, L. A., Lee, J. L., and Stewart, C. C., 1981, Release of fibrinolytic enzymes by macrophages in response to soluble fibrin, *J. Reticuloendothelial Soc.* **30**(5):317–329.

Snedeker, P. W., and Kaplan, J. E., 1980, Reticuloendothelial clearance and vascular localization of soluble fibrin monomer, *Circ. Shock* **7**:207.

Snedeker, P. W., and Kaplan, J. E., 1980, Reticuloendothelial clearance and vascular localization of soluble fibrin monomers, *Arc. Shock.* **7**:207.

Snedeker, P. W., Kaplan, J. E., and Saba, T. M., 1978, Effect of traumatic shock and alteration of reticuloendothelial function on the vascular clearance of soluble fibrin, *Physiologist* **21**:113.

Spaet, T. H., Horowitz, H. T., Zucker-Franklin, D., Clintron, J., and Biezenski, J. J., 1961, Reticuloendothelial clearance of blood thromboplastin by rats, *Blood* **17**:196.

Stathakis, N. E., and Mosesson, M. W., 1977, Interactions among heparin, cold-insoluble globulin and fibrinogen in the formation of the heparin precipitable fraction of plasma, *J. Clin. Invest.* **60**:855.

Stathakis, N. E., Mosesson, M. W., Chen, A. B., and Galankis, D. K., 1978, Cryoprecipitate of fibrin-fibrinogen complexes induced by cold-insoluble globulin of plasma, *Blood* **51**:1211.

Unkeless, J., Gordon, S., and Reich, E., 1974, Secretion of plasminogen activator by stimulated macrophages, *J. Exp. Med.* **139**:834.

van Ginkel, C. J. W., van Aken, W. G., Oh, J. I. H., and Vrecken, J., 1977, Stimulation of monocyte procoagulant activity by adherence to different surfaces, *Br. J. Haematol.* **37**:35.

Walsh, P. N., 1981, Platelets and coagulation proteins, *Fed. Proc.* **40**:2086.

Weidner, N., Itteryah, T. R., Wochner, R. D., and Sherman, L. A., 1979, Investigation of an inflammatory humoral factor as a stimulator of fibrinogen synthesis, *Thromb. Res.* **15**:651.

Werb, Z., and Gordon, S., 1975a, Secretion of a specific collagenase by stimulated macrophages, *J. Exp. Med.* **142**:346.

Werb, Z., and Gordon, S., 1975b, Elastase secretion by stimulated macrophages: Characterization and regulation, *J. Exp. Med.* **142**:361.

Werb, Z., Foley, R., and Munck, A., 1978, Glucocorticoid receptors and glucocorticoid-sensitive secretion of neutral proteases in a macrophage line, *J. Immunol.* **121**:115.

Yamada, K., and Olden, K., 1978, Fibronectins—Adhesive glycoproteins of cell surface and blood, *Blood* **6**:195.

Zubairov, D. M., Andrushko, I. A., and Davydov, V. S., 1970, Hemocoagulatory properties of Kupffer cells, *Bull. Exp. Biol. Med.* **70**:1370.

10

Platelets

JOHN E. KAPLAN and DUDLEY G. MOON

1. INTRODUCTION

The importance of the RES in regulating blood platelet dynamics is well established. Upon review of the pertinent literature, however, much of the evidence supporting the interaction between platelets and the RES is found to be indirect. There are relatively few comprehensive studies of the physiological mechanisms underlying the removal of senescent, damaged, and altered platelets. The need to understand these mechanisms has been most apparent in the clinical management of thrombocytopenic and hypersplenic syndromes in which excess platelet destruction has been implicated. Thus, a large number of studies have been performed concerning the survival of platelets in these syndromes. Most of the current knowledge of platelet recognition and removal by the RES derives from such studies. This review will begin with a discussion of phagocytosis of platelets in the hypersplenic/thrombocytopenic, normal, and thrombotic states. The mechanisms of recognition of platelets by the RES will be considered particularly in the context of alterations which enhance platelet removal. Next, the literature indicating that platelets interact directly with circulating microparticulates will be reviewed. Evidence will be presented that platelets may be phagocytic in their own right and may augment RES phagocytosis by virtue of their ability to agglutinate foreign particulates. Following this will be a consideration of RES clearance of proaggregatory agents. Finally, the most recent advances in our understanding of potential platelet–macrophage interactions will be reviewed. Both platelets and macrophages participate in inflammation and tissue repair. Evidence now suggests that platelets are also involved in immune reactions. Since platelets and macrophages react with and release a number of common mediators, it seems quite reasonable that they might modulate one another's function. Before proceeding, a brief discussion of platelet morphology and phys-

JOHN E. KAPLAN and DUDLEY G. MOON • Department of Physiology, Albany Medical College of Union University, Albany, New York 12208.

iology is in order. More extensive reviews of this aspect are available (Zucker, 1980; Holmsen *et al.*, 1977; Day *et al.*, 1978; de Gaetano and Garattini, 1978).

2. PHYSIOLOGY AND ANATOMY OF PLATELETS

The mammalian blood platelet is a nonnucleated cell fragment irregularly discoid in shape with a diameter of 1–6 μm. The human platelet is usually 1–2 μm in diameter. They originate in the bone marrow by budding from megakaryocytes. Though sparce in organelles, platelets contain two readily identifiable populations of granules. Dense granules contain small molecules including amines, adenine nucleotides, and calcium. The α granules contain proteins including molecules which modulate and participate in the hemostatic process and conventional lysosomal constituents. It has been suggested that these granules are functionally heterogeneous, some acting as storage reservoirs for such molecules, fibrinogen, and platelet factor 4 and others representing lysosomes and peroxisomes.

The major physiological function of platelets is generally believed to relate to the maintenance of integrity and continuity of the vascular lining. This is accomplished primarily through participation in hemostasis as well as in a poorly defined relationship to the support of vascular endothelial integrity. The role of platelets in hemostasis is accomplished via sequential adhesion, aggregation, secretion and contraction at a site of vascular injury. Briefly, circulating platelets adhere upon exposure of the collagenous subendothelium. Once adherent, secretion of ADP and elaboration of thromboxane A_2 (TXA_2) lead to platelet aggregation and the formation of a hemostatic plug. Adhesion and aggregation are both stimuli for secretion so a positive feedback cycle results. ADP released from injured tissue or thrombin elaborated through activation of coagulation may initiate these reactions via stimulation of aggregation. Concomitantly with these responses, contraction of platelet actinomycin initiates retraction and organization of the platelet plug. Release of serotonin and TXA_2 leads to local vasoconstriction in the area of the plug. Deposition of areas not requiring hemostatic intervention leads to thrombosis. Detachment of a hemostatic plug or aggregation initiated without adhesion may lead to thromboembolization. Because high flow velocities limit the role of coagulation in achieving hemostasis on the arterial side of the circulation, platelet events assume special significance in these areas. Platelets are the primary components of arterial thrombi and thromboemboli. Under most conditions these events are precluded by the uninterrupted continuity of the nonreactive luminal face of the vascular endothelium and by elaboration of the platelet-inhibiting prostaglandin, prostacyclin (PGI_2).

The reason for elaborating upon these processes is to emphasize that in the course of executing normal function, platelets undergo shape changes and membrane alterations. These transitions may thus characterize the activated platelet as "altered self" or "effete self," characteristics often associated with recognition and removal by the RES. In addition, since platelets are blood components which retain the capacity to impair vascular patency, the conception of the RES as a selective filter active in preventing vascular occlusion arises.

3. PHAGOCYTOSIS OF PLATELETS

3.1. HYPERSPLENIC AND THROMBOCYTOPENIC STATES

The early observations which led to the concept of phagocytosis of platelets by the cells of the RES derived primarily from studies of thrombocytopenia and hypersplenism. While a number of facts obtained from both clinical studies and animal models of the syndromes coalesced to implicate the RES–platelet relationship, our modern concepts can be traced to two specific developments. The first was a series of animal studies by Tocantins and his colleagues (Tocantins, 1936a,b, 1938; Tocantins and Stewart, 1939), as well as others (Ledingham and Bedson, 1915; Bedson, 1924; Elliott and Whipple, 1940; Leonard and Falconer, 1941) which demonstrated that heterologous antiserum to platelets could be developed. This antiserum was capable of producing fulminant thrombocytopenia, a syndrome resembling thrombocytopenia purpura in humans, and splenectomy ameliorated the effects of the antiserum. Thus, these experiments produced the concept that thrombocytopenia was related to the existence of a humoral factor which could sensitize platelets to splenic destruction. The applicability of these animal data to the human disease was suggested by the apparent correlation between the presence of circulating platelet antibodies and favorable response of a given patient to splenectomy.

A hypothesis more directly implicating the RES function of the spleen was put forth by Doan (1949) to explain hematopenias in general and thrombocytopenia in particular, associated with splenomegaly or hypersplenism. His hypothesis, based in part on the comprehensive anatomical studies by Bjorkman (1947), was that hypersplenism was associated anatomically with both an enlarged splenic reservoir of formed elements among the splenic vascular sinuses and pulp cords and a hyperplasia of RE cell elements. Functionally, the increased reservoir resulted in increased sequestration of formed elements. This hypersequestration provided a partial explanation of the hematopenias. The hypersequestration was also believed to result in increased vascular stasis and cellular contact within the spleen. This was believed to lead to increased mechanical injury, humoral sensitization, and potential for contact-phagocytosis, which, in conjunction with the RES hypertrophy, resulted in augmented destruction of formed elements by the RES.

The further development of this area of research awaited two related technological developments. The first was the evolution of techniques to radioactively label platelets so their survival in the circulation could be evaluated. The relative merits of different techniques and the interpretation of the shape of the platelet survival curve have been reviewed in detail (Harker and Ross, 1978; Harker, 1977; Mustard, 1978; Mustard *et al.*, 1966; George and Morgan, 1978; Davey, 1966). Since investigations in this area primarily dealt with human subjects, the determination of sites of platelet removal from the circulation awaited the development of external scintigraphic techniques.

Early studies combining these methods were reported by Castaldi and Firkin (1963). They reported that ^{51}Cr-labeled platelets in normal subjects could be detected by external scanning over the spleen. Their reported data are also suggestive of hepatic localization. Patients with idiopathic thrombocytopenia purpura (ITP) were characterized by shortened platelet life span and augmented splenic localization of platelets. Similar findings were obtained in patients with congestive splenomegaly in association with portal cirrhosis. Patients with congestive splenomegaly of other origin and splenomegaly in association with hereditary spherocytosis had excessive splenic platelet localization but normal platelet survival. Although suggestive, early studies by these investigators and others (Najean *et al.*, 1963; Davey and Lander, 1964) provided equivocal data due to the early scanning techniques and nonoptimal methods of platelet labeling. Najean *et al.* (1963), although suggesting a rough correlation between splenic platelet localization and response to splenectomy in ITP patients, found external scanning "beset with difficulties." Aster and Jandl (1964), using improved techniques, studied a smaller group of similar patients with much the same results. They additionally related these findings to platelet clearance in patients with platelet isoimmunities due to multiple platelet transfusion and to the effects of injected platelet isoantibody-containing serum into normal subjects. Both isoimmune subjects had reduced platelet survival. The subject with moderately reduced survival showed augmented surface radioactivity over the spleen. The subject with extremely short platelet survival had surface radioactivity over the liver and spleen with an estimated 80% hepatic sequestration. In subsequent studies, these investigators injected incremental quantities of serum containing platelet isoantibodies into normal subjects. They found that small quantities of antibody reproduced the platelet survival and distribution characteristics of the former subject while large amounts simulated the latter subject. Intermediate doses produced platelet survival and distribution characteristics graded between the two subjects. Similar observations were made on patients with ITP. In chronic ITP, transfused platelets were slowly destroyed in the spleen, while in acute severe ITP, transfused platelets were rapidly destroyed in the liver. These findings led to the development of the general hypothesis that platelets which were mildly damaged or sensitized by immunologic or other mechanisms are destroyed in the spleen, while severely damaged platelets are removed from the circulation by the liver. They additionally reported that administration of adrenocorticosteroids which often ameliorate the symptoms of ITP led to prolonged platelet survival.

Shulmand and co-workers presented a series of studies which largely supported these conclusions (Shulman *et al.*, 1965a,b). They determined that injection of platelet isoantibodies or plasma from ITP patients resulted in loss of cirulating platelet label, decreased platelet count, and enhanced splenic sequestration of label. Splenectomized or corticosteroid-treated patients were resistant to the thrombocytopenic effects of these plasmas. Much larger quantities of the plasma, however, led to thrombocytopenia in these subjects, with the highly sensitized platelets now localized in the liver. They also found a relative shift of platelet destruction site from spleen to liver with increased sensitization. Their data indicated that excessive levels of ITP factor leading to hepatic sequestration explained the resistance of certain patients to splenectomy and steroid therapy. They found that patients with hereditary spherocytosis were resistant to sensitization by ITP factor. They suggested that since RES destruction of erythrocytes was ongoing in these patients, that the RES was blockaded with respect to uptake of other substances. These investigators found they could reproduce this resistance with injection of erythrocyte stroma and suggested RE blockade as a potential therapeutic modality in resistant ITP. Corroborative findings with respect to sensitization were reported in animal studies by Baldini (1966a,b,c). Dogs were immunized with weekly injections of homologous platelets. After repeated infusions, platelet survival became progressively reduced. The increased rate of removal was associated with shift in primary site from spleen to liver.

A series of patient studies in subsequent years dealt with the relationship of liver versus spleen uptake, the nature of the disease process, and the response to therapy (Aster, 1972). Solomon and Clatanoff (1967) reporting upon 10 ITP patients and Kinlough *et al.* (1966) reporting results from patients with ITP as well as thrombocytopenia in association with lymphosarcoma, Hodgkin's disease, myelofibrosis, and cirrhosis reported rough correlation between splenic localization of labeled platelets and the response to splenectomy. Najean *et al.* (1963, 1967) and Najean and Ardaillon (1971) reported the most comprehensive studies in this regard. In several hundred patients, they clearly describe the full spectrum between splenic, hepatic, and mixed sequestration. Splenic sequestration was more common in children and hepatic appeared most frequently in adults. They found that the distribution of platelets in a given subject remained the same upon repeated determinations even over intervals of months to years. They found a high correlation between the site of sequestration and the ability to predict both successful and unsuccessful response to splenectomy in patients with splenic versus hepatic platelet sequestration. They did not agree that the site of sequestration correlated with the severity of the disease. These were at variance with the reports of Aster and Keene (1969) and Baldini (1966a,b,c). These authors reported a correlation between severity in terms of platelet count and the site of sequestration. They reported, however, that patients with hepatic platelet destruction sometimes respond favorably to splenectomy.

Despite this information it is worthy to note that over this interval there was little histological confirmation of platelet uptake by splenic macrophages in ITP patients. Several groups (Summerell and Gibbs, 1972; Hill *et al.*, 1963; Landing *et al.*, 1961; Saltzstein, 1961; Chandler and Hand, 1961) reported the existence of lipid-laden macrophages or "foamy histiocytes" during ITP. They suggested that the source of this lipid was phagocytized platelets. Koepke *et al.* (1968) using immunofluorescent techniques identified sequestered and partially destroyed platelets in "foamy histiocytes." These findings were confirmed electron microscopically by Firkin *et al.* (1969) and later by other investigators. Tavassoli and McMillan (1972) reported electron microscopic evidence of platelet destruction in the cordal compartment of the red pulp, particularly in areas adjacent to the marginal zone. Platelet destruction was apparent both within macrophages and by a "self-perpetuating extracellular mechanism" with ensuing phagocytosis of platelet "debris." Zucker-Franklin and Karpatkin (1977) have provided histological evidence of platelet phagocytosis and degradation by circulating monocytes. Walsh and Barnhart (1969) reported immunofluorescent evidence of hepatic macrophage platelet phagocytosis during thrombocytopenia after injection of zymosan particles which are inducers of RES hyperactivity. The injection of several other types of colloidal particles has also been associated with thrombocytopenia. Phagocytized platelets have been identified in the spleen and/or liver following injection of carbon, thorium dioxide, ferritin, or emulsified lipid (David-Ferreira, 1961; French, 1968; French and Barcat, 1967). An additional study (Cronkite *et al.*, 1967) localized platelet radiolabel in the splenic macrophages of rats rendered thrombocytopenic by whole-body irradiation. These studies are of particular interest because of the difficulty in distinguishing between localization of platelets within phagocytic cells as opposed to reservoirs in the sinuses of the liver and spleen by such techniques as external scanning and whole tissue counting. Indeed, this was a key question in the interpretation of studies of the type presented in the preceding material.

Several approaches were utilized to differentiate pooled versus internalized platelets both in thrombocytopenic and hypersplenic patients and in rats rendered hypersplenic with methylcellulose. The pooled proportion was distinguished by platelets released during perfusion of organs removed at splenectomy (Penny *et al.*, 1966; Aster, 1969), release of splenic labeled platelets upon epinephrine-induced splenic contraction (Aster, 1966; Wright *et al.*, 1951), and in animals, splenic label loss upon exchange transfusion (Aster, 1969). Collectively, these studies have allowed the thrombocytopenic syndrome to be classified into two general categories (Cohen *et al.*, 1961). Excluded, for the purposes of this review, are those conditions in which platelet production is inadequate, or platelets are consumed in thrombosis or hemostasis. One is characterized by thrombocytopenia, splenomegaly, and normal or mildly reduced platelet survival time. This category results from excessive platelet pooling in a splenic reservoir exchangeable with the vascular compartment. The second category is characterized by thrombocytopenia, not necessarily in association with splenomegaly, and shortened platelet survival time. This category is the result of excessive platelet destruction by cells of the RES due to either platelet sensitization or RES hyperactivity. This latter category, which includes ITP as well as thrombocytopenia in association with reticuloendotheliosis, provided the basis of the initial concept of platelet phagocytosis by the RES.

A limited number of *in vitro* studies have provided additional documentation of the capacity of phagocytic cells to ingest platelets. Platelet sensitization with serum from ITP patients or specifically-induced platelet antibody, has been demonstrated to support phagocytosis by splenic macrophages and blood monocytes as well as granulocytes (Handin and Stossel, 1974, 1978; McMillan *et al.*, 1974; Verp and Karpatkin, 1975). Splenic macrophages and granulocytes from ITP patients avidly ingest antibody-coated platelets. This is especially true in untreated patients or patients resistant to steroid therapy. Cells from patients whose disease is amenable to steroid control demonstrate reduced *in vitro* platelet phagocytosis (Handin and Stossel, 1978).

3.2. NORMAL AND THROMBOTIC STATES

A scan of the literature regarding the role of the RES in the kinetics of circulating blood platelets reveals that platelet survival in the circulation has received considerably more attention than have the sites and mechanisms involved in platelet removal. The labeling methods, shapes of survival curves, and the implications of these shapes have been reviewed (Harker and Ross, 1978; Harker, 1977; Mustard, 1978; Mustard *et al.*, 1966; George and Morgan, 1978; Davey, 1966). The temporal pattern of circulating platelet radioactivity is complex and provides a distinctly different impression depending upon the phase observed. The pattern of circulating platelet radioactivity after injection of population labeled (labeled platelets representing a sampling of all circulating platelets) can be separated into three phases. Over a period of minutes up to 1 to 2 hr, most label disappears, a varying portion returns to the circulation over the next several hours, and then diminishes linearly over a period of days (Davey and Lander, 1964; Davey, 1966). The initial rapid clearance reflects sequestration in the liver, spleen, and lung and may result from alterations or damage to platelets in the course of isolation and labeling. A portion of these platelets reenter the circulation, presumably "repaired." The remainder of the label persists in the organs and are destroyed by the RES elements in these organs. Depending upon the damage to platelets which occurs prior to injection, either of these fates may predominate in a relative or nearly complete manner. The slow linear loss reflects age-dependent or senescent destruction. Under certain circumstances, the latter phase may be more rapid and appear exponential in nature. This reflects random or age-independent destruction which may result from platelet damage or consumption. When cohort labeling (labeling of an age-related group of platelets) is utilized, the shape of the survival curve is altered to support the suggestion of senescent destruction in association with a finite life span (Harker and Ross, 1978).

Several studies have addressed the role of the RES in the removal of "normal" and "damaged" or "altered" platelets. The quotes derive from the inability to characterize what is a normal platelet once isolated and labeled. The rapid sequestration phase observed after platelet injection creates doubt that any investigators have dealt with a truly normal platelet. The reference for normal platelets is therefore the least damaged platelet preparation. Platelets with reduced survival or increased avidity for RES uptake result from specific and intentional alteration or damage or from some undefined alteration.

An example of this is provided in early studies comparing platelets collected in plasma anticoagulated with citrate or EDTA (Kinlough *et al.*, 1966; Aster and Jandl, 1964). A greater proportion of "EDTA platelets" were initially removed from the circulation and a smaller proportion returned to the circulation compared with "citrate platelets." Also, those "EDTA platelets" which returned to the circulation survived a shorter period than "citrate platelets." In addition, the relative hepatic to splenic distribution of the more rapidly cleared "EDTA platelets" was elevated although both preparations localized primarily in the liver. Thus, the subtle difference between anticoagulation with EDTA versus citrate results in shortened survival and augmented hepatic destruction consistent with damaged platelets. Since these studies, "citrate platelets," as prepared in these original studies or with slight modification, have provided the standard for normalcy and it is these preparations which will hereafter be referred to as normal.

Heyssel *et al.* (1967) reported on the survival of normal platelets in rats. The ability to interpret their findings is limited by their use of [^{14}C]serotonin, a redistributable and excretable label, to follow platelets. They reported that greatest organ radioactivity was found in the spleen and liver, respectively. Although this never amounted to more than 20–30% of injected ^{14}C activity, the total recoverable activity was only 10–20% within 4 days postinjection. Aster (1969) utilized ^{51}Cr for similar studies. He reported temporally increasing localization of platelet label in the RES organs with greater than 80% found in the liver, spleen, and marrow 4 to 5 days postinjection. The survival and RES uptake were essentially unchanged after splenectomy. The percentage of platelets which normally went to the spleen was then found in the liver. This provided support for the senescent theory of survival in that such major alteration of RES capacity did not affect the survival of these normal platelets. These findings were corroborated in normal and asplenic man by external scanning. These investigators further determined that the nonsoluble portion of osmotically lysed platelets was rapidly sequestered with predominantly hepatic localization. Studies have recently reported (Heyns *et al.*, 1980) the kinetics and distribution of "indium-labeled platelets" with computer-assisted scintigraphy in normal humans. They reported an early distribution of 40% to the liver and spleen with predominantly splenic localization. At the end of 9 days, over 70% of the platelets were localized to the liver and spleen with the liver containing slightly more platelets. These findings have recently been supported by others (Peters *et al.*, 1980; Klonizakis *et al.*, 1980).

Han and Baker (1964) reported platelets with splenic phagocytes of normal rats. Similar observations were reported by French (1967) and French and Barcat (1968). Gans *et al.* (1966) reported the uptake of "intact" and "disintegrated" (not further defined) rat platelets in a perfused liver system. Both preparations were rapidly removed from the perfusate and localized to the Kupffer cells by radioautography. Prior perfusion with unlabeled platelets blocked the uptake of labeled platelets. Curiously, the intact and damaged preparations were phagocytized at the same rate.

Kaplan and Saba (1978) analyzed the short-term blood clearance and organ localization of mildly *in vitro* injured platelets in rats. Platelet disappearance conformed to a two-compartment exponential with quantifiable rapid and slow compartments. The rate and proportion of platelet-cleared rapid compartment decreased as platelet load increased, suggesting a high-affinity finite-capacity saturable removal mechanism. The platelets cleared by the rapid system correlated directly and quantitatively with those located in the liver. Colloid-induced RES blockade resulted in reduced rate of clearance and reduction of hepatic platelet uptake. Splenic sequestration of platelets was increased after RES blockade. Additional studies with this platelet preparation (Snedeker and Kaplan, unpublished observation) demonstrated augmented hepatic platelet uptake during RES hyperactivity induced by zymosan. During depression of RES phagocytosis induced with methylpalmitate, the blood clearance as well as hepatic and splenic localization of labeled platelets was reduced. These studies employed kinetic analysis to assess the capacity of the removal system. These data indicated that the high-affinity mechanisms approached saturation when challenged with quantities of platelets below circulating levels.

Several factors have been documented to shorten platelet survival, presumably as a result of RES uptake. Additional mechanisms of platelet destruction as a result of coagulation-mediated and surface-related processes will not be discussed. Several parameters related to collection and storage of platelets influence platelet survival. As discussed above, the survival of platelets collected from EDTA-anticoagulated blood is signficantly reduced. Platelets stored at 4°C for 24 hr are rapidly removed from the circulation in contrast to platelets stored at 22°C (Murphy and Gardner, 1969; Slichter and Harker, 1976). This may be due to irreversible change in the normal discord configuration. Exposure to pH below 6.0 during storage also results in rapidly cleared platelets (Slichter and Harker, 1976; Murphy, 1976). Exposure to either elevated or reduced oxygen tensions *in vivo* leads to shortened platelet survival through an intrinsic modification which is not well characterized (Harker and Ross, 1978). Platelet survival time is reduced by *in vitro* exposure to a number of proteolytic enzymes including trypsin, chymotrypsin, and plasmin (Packham *et al.*, 1977). Alterations of polysaccharide side chains of glycoproteins have a distinct effect on platelet survival. These may be of special significance by simulating the normal senescent mechanism of platelet recognition as will be elaborated on in the next section. Enzymatic removal of sialic acid residue *in vitro* with neuraminidase (Greenberg *et al.*, 1975) or by intravenous injection of neuraminidase enhances the rate of platelet clearance (Grottum and Jeremic, 1973), as does modification of sialic acid with sodium periodate (Cazenave *et al.*, 1976). Exposure of platelets to viruses which remove surface sialic acid similarly shortens platelet survival (Terada *et al.*, 1966; Scott *et al.*, 1978). These data collectively suggest that events which result in alteration of the platelet surface, especially with respect to membrane glycoproteins, and/or loss of discoid shape, result in removal of platelets from the circulation.

The effect of thrombotic stimuli on RES clearance of platelets is a question upon which general agreement has not been reached. Reimers *et al.* (1976) have demonstrated that platelets treated with thrombin *in vitro* circulate normally after dissolution of aggregates with plasmin and prostaglandin E. Platelets infused after ADP treatment and spontaneous disaggregation survived normally (Packham *et al.*, 1980). Mustard *et al.* (1966) have noted that platelet survival is normal after intravenous infusion of thrombin or ADP. In these studies, there were transiently reduced platelet counts over about an hour or less. These studies, however, minimized the role of fibrin in the response. The fibrin in the *in vitro* thrombin-aggregated platelets was removed enzymatically. These results are complicated by the report by this same group that plasmin treatment shortened platelet survival (Packham *et al.*, 1977). Survival of both thrombin-treated and plasmin-treated controls was short, on the order of 4 days. By intravenous infusion, the platelet emboli would be primarily directed to the fibrinolytically active and generally nonphagocytic pulmonary vasculature. The generally reversible nature of *in vivo* platelet aggregation suggested by these studies is in question and does not appear consistent with prolonged thrombocytopenia often observed during intravascular coagulation. Indeed, Walsh and Barnhart (1969) reported immunofluorescent evidence of platelet uptake by hepatic RE cells after thrombin-induced intravascular coagulation, although none was observed following endotoxin injection. Hand and Chandler (1962) reported electron microscopic evidence of phagocytosis of platelets during the course of atherosclerotic metamorphosis of *in vitro*-formed autologous rabbit pulmonary thromboemboli. Similar observations were made upon *in vitro* incubation of thrombi formed from human plasma (Chandler and Hand, 1961). Poole (1966) reported phagocytosis of platelets by monocytes in organizing mural arterial thrombi forming on fabric grafts in the aorta. Jorgensen *et al.* (1967) identified macrophage uptake of platelets in platelet-rich mural thrombi in swine carotid arteries.

Platelet uptake has recently been studied during the thrombotic sequelae of injury. Berman (1975) reported enhancement of splenic platelet localization upon burn injury in animals with prelabeled platelet pools. They believed this reflected phagocytosis of platelets involved in thrombus formation. Kaplan and Saba (1977) studied the effect of trauma on hepatic and splenic localization of *in vitro*-injured platelets. They reported diminished platelet uptake in this model which has been associated with RES depression.

Several studies have demonstrated the capacity of phagocytic cells to ingest platelets *in vitro*. In general, these represented portions of studies reported in the previous section with regard to phagocytosis of antibody-sensitized platelets. These studies demonstrated that nonsensitized platelets are also taken up to some extent, albeit less than that observed for sensitized platelets, by splenic macrophages, blood monocytes and granulocytes (Handin and Stossel, 1974, 1978; McMillan *et al.*, 1974; Verp and Karpatkin, 1975).

An additional phagocytic function of the RES with regard to platelet economy is the uptake of megakaryocytic nuclei by the RES elements of bone marrow and lungs (Kaufmann *et al.*, 1965; Tinggaard-Pedersen, 1974).

4. RECOGNITION OF PLATELETS BY THE RES

4.1. IMMUNE RECOGNITION

It is now well accepted that excessive platelet destruction in ITP is associated with autoimmune platelet recognition. Several lines of evidence suggested an immune nature to this platelet disorder, now often referred to as *autoimmune thrombocytopenia purpura*. The earliest evidence was that of Tocantins and others which determined that the syndrome developing following injection of heterologous platelet antiserum closely resembles that observed with ITP (Harrington *et al.*, 1956). Two clinical observations contributed to the concept of an immunological basis for ITP. One was the observation that infants born to mothers with ITP were frequently purpuric themselves (Epstein *et al.*, 1950; Grandjean, 1948). Thus, an agent that could cross the placenta, such as an immunoglobulin, was suspect. Second, patients with acquired idiopathic hemolytic anemia often had episodes of idiopathic thrombocytopenia as well (Fisher, 1947; Evans and Duane, 1949; Evans *et al.*, 1951). Acquired diopathic hemolytic anemia was already demonstrated to be autoimmune in nature. Based upon these observations, several different groups of investigators reported results which formed the scientific basis of an immune origin for ITP. Evans *et al.* (1951) detected platelet agglutinins in some ITP patients. Harrington *et al.* (1951) reported that the plasma of many ITP patients induced the entire syndrome of ITP when infused into normal recipients. Hirsch and Gardner (1951) made the observation that normal platelets had a brief survival when injected into ITP patients. Thus, it was determined that a circulating antiplatelet factor with characteristics attributable to immunoglobulins was circulating in ITP patients.

The factor has now been characterized in appreciable detail (Karpatkin, 1980). The factor is apparently IgG in that it is heat labile at 56°C and can be neutralized by rabbit antiserum to human IgG but not by antiserum to IgM, IgA, or IgD (Karpatkin *et al.*, 1972a; Steffen, 1960; Karpatkin and Siskind, 1969). The activity binds to platelets and the eluted activity is inhibited by anti-IgG (Karpatkin *et al.*, 1972a; Shulman *et al.*, 1965b; Tullis, 1956; Steffen, 1960; Karpatkin and Siskind, 1969). The molecular weight and sedimentation coefficient are appropriate for IgG (Steffen, 1960; Karpatkin and Siskind, 1969). The heavy chain has been characterized as that of the IgG3 subclass and the light chains may be either kappa or lambda (Karpatkin and Siskind, 1974). The factor from any given patient possesses a degree of specificity to the platelet of that patient (Karpatkin and Siskind, 1974). It is notable that severity of the disease correlates well with quantity of antiplatelet antibody bound to platelets, but not with the quantity circulating (Karpatkin *et al.*, 1972b). The nature of the platelet antigen has not been identified. The source of antibody appears to be the spleen (Karpatkin *et al.*, 1972b; McMillan *et al.*, 1972). Thus, splenectomy for ITP removes the source of immunoglobulin as well as a major phagocytic organ. Evidence cited earlier has demonstrated the capacity of the antiplatelet factor of ITP to sensitize platelets to phagocytosis both *in vivo* and *in vitro*.

There is no evidence that immune recognition plays a role in normal senescent recognition and turnover. It is notable, however, that platelet-bound IgG is frequently found subsequent to septicemia and may thus play a role in the development of the thrombocytopenia which often accompanies septicemia (Kelton and Neame, 1971).

4.2. SENESCENT RECOGNITION—ROLE OF SIALIC ACID

As indicated earlier, a large body of evidence supports the concept that platelets undergo a process of senescence which results in their loss from the circulation. These senescent changes have been related to both structural and functional parameters. Older platelets are less adherent to collagen (Hirsch *et al.*, 1968), less reactive to aggregating and releasing stimuli (Karpatkin, 1969a), less metabolically active (Karpatkin, 1969b), and less capable of shortening bleeding time *in vivo* (Harker and Slichter, 1972; Shulman *et al.*, 1968). Older platelets are smaller and less dense, possibly due to loss of dense granules. Platelets lose sialic acid with aging, presumably from surface glycoproteins and/or gangliosides.

Grottum and Jeremic (1973) reported the gradual appearance of marked thrombocytopenia after injection of *Clostridium perfringens* neuraminidase into rabbits. They suggested that removal of sialic acid rendered platelets more susceptible to removal by the RES. Greenberg *et al.* (1975) made similar observations when as little as 8–10% of platelet sialic acid was removed from platelets by neuraminidase treatment *in vitro* and subsequent reinjection of platelets. They similarily suggested that sialic adic was necessary for recognition as "self" and its removal led to recognition by the RES. Support of these studies derived from the finding that treatment with influenza virus which has neuraminidase activity removes sialic acid and leads to rapid platelet clearance (Terada *et al.*, 1966; Scott *et al.*, 1978; Packham *et al.*, 1977). They also reported that removal of sialic acid-containing glycopeptides by proteolytic enzymes such as plasmin, chymotrypsin, and trypsin decreased platelet survival (Packham *et al.*, 1977). Cazenave *et al.* (1976) have demonstrated that alteration of sialic acid with sodium periodate also results in rapid platelet removal. These findings, in conjunction with the observation that platelets lose sialic acid with age, strongly implicate this mechanism in senescent recognition.

It is not clear, however, by what mechanism the RES might recognize desialized platelets. These data might be clarified in the context of a large body of data related to the RES clearance of desialized erythrocytes. The early theories suggested such cells were recognized by their decreased negative charge. Other data indicated that galactose, the penultimate residue of polysaccharide side chains of both glycoproteins and gangliosides, might be recognized by a membrane receptor. Such a galactose receptor has recently been identified on Kupffer cells (Kolb *et al.*, 1980). Recent findings have suggested that clearance of desialized erythrocytes is dependent upon a plasma factor which possesses many characteristics of an immunoglobulin (Kay, 1978). This may recognize a cryptic antigen exposed by desialization. The available evidence does not permit elucidation of which, if any, of these recognition mechanisms may apply to platelets. The fact that desialization, modification of sialic acid, and removal of glycopeptides all produced the same result argues against the galactose receptor theory. All of these manipulations reduce negative surface charge and could potentially expose cryptic antigens.

4.3. RECOGNITION DURING THROMBOSIS

As previously indicated, phagocytosis of platelets by both the fixed and the mobile cellular elements of the RES in association with thrombosis has been documented. There are little definitive data regarding the recognition of such platelets, but three possibilities justify valid speculation.

Several studies suggest that an opsonic role may be subserved by fibronectin of either plasma or platelet origin. Plasma fibronectin is a high-molecular-weight dimeric glycoprotein present in plasma and other extracellular fluids (Mosher, 1980; Mosesson and Amarani, 1980; Saba and Jaffe, 1980). Tissue fibronectin is a chemically and immunologically similar protein associated with many cell surfaces, especially fibroblasts and vascular endothelial cells, in connective tissues and most basement membranes (Yamada and Olden, 1978; Pearlstein *et al.*, 1980). Prior to adoption of the term *fibronectin*, the plasma and tissue forms of the protein have been variously termed *cold-insoluble globulin* (Mosesson and Amarani, 1980), *opsonic protein* (Saba and Jaffe, 1980), *fibroblast antigen* (Ruoslahti and Vaheri, 1975), and several other names. Collectively, the plasma and tissue forms of the protein have been implicated in a variety of functions including cell adhesion, spreading and locomotion, structural orientation of the interstitial matrix, and the opsonization of particulate matter prior to RES phagocytosis (Mosher, 1980; Mosesson and Amarani, 1980; Saba and Jaffe, 1980; Yamada and Olden, 1978; Pearlstein *et al.*, 1980). With regard to the latter, plasma fibronectin has been identified as the opsonic factor responsible for recognition of gelatinized colloids. Fibronectin depletion is a primary factor in the development of colloid-induced RES blockade and appears to be an important determinant of RES function in a variety of pathological and experimental states (Saba and DiLuzio, 1969; Blumenstock *et al.*, 1977; Kaplan and Saba, 1976).

Recent evidence suggests a potential role for fibronectin in the opsonization and recognition of platelets. Fibronectin is present in platelet α secretory granules and partial release occurs when platelets are exposed to thrombin or collagen (Zucker *et al.*, 1979). Fibronectin is normally absent from the platelet surface but has been reported to be present after reaction with ADP, collagen, and especially thrombin (Plow *et al.*, 1979; Hynes *et al.*, 1978; Ginsberg *et al.*, 1980). Plow *et al.* (1979) have suggested that upon activation, interiorized fibronectin is expressed at the platelet surface and, in part, mediates adhesive and aggregative reaction. These investigators have also demonstrated that activated platelets can take up fibronectin from the medium (Ginsberg *et al.*, 1980). This is most notable with thrombin-activated platelets but also occurs to a lesser extent after exposure to ADP or collagen. Bensusan *et al.* (1978) have suggested that fibronectin is the platelet receptor for collagen although this has been questioned by other investigators. Regardless of the validity of this speculation, experimental evidence has clearly indicated that preincubation of collagen with fibronectin greatly reduces reactivity with platelets in terms of adhesion, aggregation, and release (Grinnell *et al.*, 1979; Moon and Kaplan, 1981; Bensusan *et al.*, 1978), while preincubation of platelets with gelatin of the Fab fragment of purified antibody to fibronectin reduces platelet binding to collagen (Santoro and Cunningham, 1979). Data from this laboratory have demonstrated that specific immunologically induced fibronectin depletion in rats results in decreased resistance to thrombosis upon injection of ADP or thrombin compared with non-fibronectin-depleted control rats (Kaplan *et al.*, 1980). The documented opsonic role of plasma fibronectin in conjunction with the observations that activated platelets may be phagocytized and that these platelets show surface-bound fibronectin provides strong circumstantial evidence that fibronectin may be an opsonin for activated platelets.

A second alternative is that recognition of activated platelets is achieved through the expression of surface-bound fibrinogen or fibrin. During platelet activation, fibrinogen derived from either α granules or the medium is absorbed to the platelet surface. In the presence of thrombin, this fibrinogen is converted to fibrin. The capacity of macrophages to bind fibrinogen derivatives, especially fibrin, is well documented (Sherman and Lee, 1977; Colvin and Dvorak, 1975; Jilek and Hormann, 1978). Cellular elements of the RES have been demonstrated to internalize fibrin antigen and the radiolabel of isotopically labeled fibrin (Lee and McClusky, 1962; Gans *et al.*, 1967; Sherman *et al.*, 1975; Snedeker *et al.*, 1979). Thus, this might serve as a platelet recognition mechanism. It is notable, however, that fibrin–fibrinogen recognition may be humorally mediated rather than direct. Fibronectin has recently been implicated in this phenomenon (Snedeker *et al.*, 1979).

As indicated above, plasmin treatment results in clearance of platelets, possibly as a result of removal of sialoglycopeptides. Since plasmin is often activated in association with events which lead to platelet activation, this may serve as a mechanism of platelet recognition in such situations. Such a mechanism implies that the breaking up of platelet thrombi as a result of plasminolysis of cross-linking fibrin(ogen) would produce a phagocytizable platelet as a product.

5. PLATELET–PARTICLE INTERACTION

5.1. PHAGOCYTOSIS BY PLATELETS

A considerable amount of information has accumulated with regard to interaction of platelets and foreign particulates. Much of this information indicates that platelets interiorize particles and it is a matter of controversy whether this interiorization represents phagocytosis.

Tait (1918) first reported engulfment of carbon and carmine particles by platelets. Forty years later, Danon *et al.* (1959) reported incorporation of influenza viruses into platelets after *in vitro* incubation. Shortly thereafter, David-Ferreira (1961) and Schulz (1961) described interiorization of thorium dioxide and silicon dioxide, respectively, and characterized this interiorization as phagocytosis. Subsequent studies reported that platelets from a variety of species could engulf ferritin, colloidal fat, polystyrene latex, carbon, collagen particles, quartz particles, influenza virus and murine and antigen–antibody complexes both *in vivo* and *in vitro* (David-Ferreira, 1961; Schulz and Wedell, 1962; Schulz, 1961; Hagenau *et al.*, 1963; Movat *et al.*, 1965a,b; Grottum *et al.*, 1972; Mant and Firkin, 1972; Behnke, 1967; White, 1972). The engulfment of bacteria by platelets has recently been reported, although earlier investigators reported that this interaction was minor or absent.

Mustard and Packham (1968) provided a detailed description of the sequelae associated with platelet phagocytosis of polystyrene latex. When platelets in citrated plasma are incubated with polystyrene latex, the particles adhere and local membrane invaginations occur at points of adherence. Within a few minutes, particles are found in vascular structures within platelets. Subsequently, platelets undergo changes consistent with aggregation and release reactions.

The precise relationship of platelet uptake of particles to that generally characterized as phagocytosis has been questioned. Behnke (1967) identified two distinct platelet cytoplasmic membrane systems. The surface-connected system consists of vacuoles and invaginations which are contiguous with the plasma membrane. He reported that particulate uptake resulted from adsorption to and interiorization with the surface-connected system, which remained contiguous with the surface. Since no evidence was found for segregation and digestion of interiorized material, it was concluded that the process did not represent phagocytosis. These findings were supported by White (1972). He found that after platelet uptake of latex, an electron-dense tracer was able to penetrate the channels and outline the ingested latex particles. On this basis, he similarly concluded that particles were lodged in channels of the open canicular system and that these channels remained open rather than pinched off to form sealed phagocytic vacuoles. Lewis *et al.* (1976), however, performed an almost identical experiment over a longer time period with very different results. Their early observations were consistent with the above observations of accumulation of particles in the open-channel system. Eventually, the latex was localized to electron-opaque channel system and surface membranes. They stained positively for acid phosphatase as is characteristic for phagosomes. They concluded on this basis that platelet interiorization reflects true phagocytosis. It does appear that platelets are capable of phagocytosis or some process closely resembling phagocytosis.

The process nevertheless differs in several regards from that well characterized for macrophages and granulocytes, especially with regard to metabolic requirements. The uptake of latex by platelets is inhibited by glycolytic inhibition but not by inhibition of aerobic metabolism. In addition, it is associated with increased lactate and CO_2 production (Movat *et al.*, 1965a). In contrast, the interiorization of thorotrast is not influenced by metabolic inhibitors nor is it associated with increased metabolism (Mant and Firkin, 1972). Neither latex nor thorotrast uptake requires intact microtubules in that interiorization takes place in the presence of colchicine (Mant and Firkin, 1972). Platelet uptake of several particles requires divalent cations as indicated by inhibition in the presence of EDTA (Movat *et al.*, 1965; Glynn *et al.*, 1966). Platelet particle uptake is associated with augmented synthesis of membrane lipids (Nishizawa and Haldor, 1967), There is considerable evidence that platelet phagocytosis is associated with an aggregation and release reaction although aggregation and release are not dependent upon particle uptake (Movat *et al.*, 1965a,b). Platelet uptake of particles is dependent to a degree on plasma factors, and immunoglobulins and fibrinogen have been implicated (Mustard *et al.*, 1967; Pfueller and Firkin, 1978).

It is notable that thrombocytes, circulating cellular elements in nonmammalian vertebrates which are probably homologous with platelets, play important, if not primary, phagocytic roles. This has been documented for fish, amphibians, reptiles, and birds (Chang and Hamilton, 1979; Dawson, 1933; Hartmans, 1925; Tait and Elvidge, 1926). Cellular elements have been identified in invertebrates which share both phagocytic functions and a role in coagulation. The relationship of these cells to phagocytic platelets can only be speculated upon.

5.2. PLATELETS AND COLLOID CLEARANCE

Numerous studies have demonstrated other interactions of platelets with both colloids and natural particulates including bacteria, lipopolysaccharides, viruses, and other microorganisms. This has been suggested to contribute to the removal of particles from the circulation both by embolization of platelet–particle clumps (Copley, 1978; Copley and Witte, 1976) and by supporting colloid clearance (van Aken and Vreeken, 1969; van Aken *et al.*, 1968; Dobson *et al.*, 1971). Govaerts (1921) demonstrated agglutination of platelets with intravenously injected bacteria as with carbon particles and foreign erythrocytes. These clumps were consequently localized in capillaries. These studies were supported by the observations of Tait and Elvidge (1926), who demonstrated a simultaneous drop in the number of platelets and intravenously injected quartz particles in circulating blood. Similar observations to these studies were reported by several other investigators. Heparin did not inhibit the platelet clumping with bacteria or foreign particulates.

Several observations consistent with these early studies have been made more recently. The interaction of platelets with artificial colloid particles and with bacteria and bacterial endotoxin has been variously documented both *in vitro* and *in vivo* (van Aken and Vreeken, 1968; Copley, 1978; David-Ferreira, 1964; Glynn *et al.*, 1965; Clawson, 1973; Houlihan, 1947; Salvidio and Crosby, 1960; Cohen *et al.*, 1965; Sutton, 1975; Mandell and Hook, 1969; Herring *et al.*, 1963; Walsh and Barnhard, 1969). In addition, many particles have been demonstrated to induce thrombocytopenia upon their intravenous injection (Dineen *et al.*, 1968; Herring *et al.*, 1963; Salvidio and Crosby, 1960; Dobson *et al.*, 1971). This has been interpreted by various investigators to reflect either depletion resulting in embolization of platelet–particle complexes or RES removal of these complexes (Copley, 1978; van Aken and Vreeken, 1968; Dobson *et al.*, 1971). Thrombocytopenia after injection of zymosan particles in particular has been suggested to result from excessive phagocytosis of platelets by a hyperactive RES (Walsh and Barnhart, 1969; Dineen *et al.*, 1968). A direct platelet interaction with zymosan particles has also been documented.

Copley (1978) and Copley and Witte (1976) have advanced the position that a primary function of platelets is clumping with microscopic invaders and preventing their dissemination by microembolization. This function has been termed *conglutination* or *congregation*, depending upon whether the interaction is irreversible or reversible, respectively. This function has been observed with respect to bacteria, viruses, and intravascular parasites as well as foreign colloids. Copley (1978) and Copley and Witte (1976) have described a sequence of events upon injection of myobacteria beginning with the nearly instantaneous production of embolizing platelet thrombi. This was followed by development of microvascular tortuosity and the appearance of numerous microaneurysms. The conglutanates left the circulation upon rupture of the microaneurysms and were ultimately phagocytized by leukocytes.

van Aken and colleagues (van Aken *et al.*, 1968; van Aken and Vreeken, 1969) and later Dobson *et al.* (1971) each proposed a role for platelets in mediating colloid clearance. van Aken *et al.* (1968) found that all adenosine nucleotides tested resulted in reduction in the rate of carbon clearance while infusion of platelet concentrates resulted in augmentation of such clearance. In a later study (van Aken and Vreeken, 1969), they reported that fibrinolytic activation reduced carbon clearance rate and that isolated fibrinogen degradation products produced similar results and could disaggregate platelet–particle complexes *in vitro*. Inhibition of fibrinolysis resulted in rapid particle accumulation in non-RES organs. They developed a model in which platelet–particle aggregates lodged in various microcirculatory beds. If this occurred in the sinuses of RES cell-containing organs, phagocytosis ensued whereas aggregates lodging in non-RES beds disaggregated in a fibrinolytically dependent manner. Dobson *et al.* (1971) found that carbon clearance was accelerated by antiplatelet serum and that successive injections of carbon induced more rapid clearance by a mechanism which could be blocked by thrombocytopenia. They suggested that platelet–carbon aggregates lodge or are formed in sinuses of RES organs and trap subsequently injected carbon. These models are supported by the evidence, cited earlier, documenting platelet–particle complexes within phagocytes. This concept is also supported by *in vitro* data which indicate enhanced phagocytosis of *Salmonella* by peritoneal macrophages in the presence of platelets (Mandell and Hook, 1969). Such mechanisms have potential significance with respect to clearance of miscroscopic invaders, thrombotic material, and lipoidal material.

6. RES UPTAKE OF PLATELET ACTIVATORS

In addition to the direct interactions of platelets and elements of the RES, the RES influences platelets indirectly via its effects on activators of the platelet response. These effects will not be considered in detail here, since they are dealt with elsewhere in this volume. Both bacterial products and immune complexes are activators of the platelet response and are cleared by the RES (Filkins, 1971; Haakenstad and Mannik, 1974). Collagen is also an important platelet activator and particles with surfaces of denatured or native collagen are cleared by the RES (Saba and Jaffe, 1980). Thrombin is a late component of the blood coagulation mechanism which activates platelets. The RES contributes by several mechanisms to the control of circulating thrombin activity by the modulation of the activation of blood coagulation and inhibition of active protease procoagulants (Kaplan, 1981; Sherman and Kaplan, Chapter 9, this volume). The platelet reaction can also be initiated by exposure to catecholamines. While the physiological significance is not clear, it has been demonstrated that macrophages can take up catecholamines in a manner analogous to that seen in autonomic nerve endings

(Balter and Schwartz, 1977). The RES also clears some nonbiological activators of platelets. The vascular sinus-lining macrophages effectively remove intravenously injected silica from the circulation and alveolar macrophages similarly engulf inhaled silica particles. As previously indicated, the RES clears a number of colloids which initiate platelet aggregation and release subsequent to platelet–particle aggregation and platelet phagocytosis of particles.

7. PLATELET–MACROPHAGE INTERACTIONS

It is only relatively recently that the mechanisms by which platelets may play a role in the immune response as well as the general inflammatory response have been appreciated. Both platelets and macrophages react with a common group of immune and inflammatory mediators (some putative) including IgG, complement, prostaglandins, leukotrienes, and platelet-activating factor (PAF). Furthermore, both platelets and macrophages release some of these mediators when stimulated. Therefore, it appears quite reasonable that platelets and macrophages may mutually affect one another's function. Unfortunately, there are few direct experiments addressing this hypothesis. The inflammatory response is a redundant process with a number of interacting pathways producing the overall response. This complicates the detection and study of primary platelet interactions. Species specificity of platelet reactions must also be taken into account (Dodds, 1978; MacMillian and Sim, 1970; Sinakos and Caen, 1967). This is particularly important with regard to the response to and release of mediators such as TXA_2 (Leach and Thorburn, 1982).

The rest of this section will review the literature of the platelet responses to immunoglobulin, complement, prostaglandins, leukotrienes, and PAF. The macrophage responses to these agents are described elsewhere in this series.

7.1. IMMUNE INTERACTIONS

The significance of direct effects of IgG complexes and complement on platelets resides in the implication of simultaneous activation of both platelets and macrophages. The temporal pattern of activation of these cell types and their release of mediators may play an important regulatory role in inflammation. The special relevance in the context of macrophage–platelet interactions is the secretion by activated platelets of materials which may modulate macrophage function and vice versa.

In vitro, platelets from a number of mammalian species bind immune complexes. There appear to be two different mechanisms of binding depending upon species. Primate (human and baboon) and artiodactyl (sheep, cattle, pigs, and goats) platelets bind via an Fc receptor while lagomorph (rabbit), rodent (mouse, rat, and guinea pig), carnivore (dog and cat), and perissodactyl (horse) platelets bind via a C3b receptor (Henson and Ginsberg, 1981; Becker and Henson, 1973).

Most of the experimental support for a platelet Fc receptor comes from studies of human platelets. Binding of antigen–antibody complexes triggers activation of platelets as evidenced by the release of serotonin (Humphrey and Jaques, 1955). Aggregated IgG molecules have been demonstrated to activate human platelets and this response can be inhibited by monomeric IgG (Mueller-Eckhardt and Luscher, 1968; Pfueller *et al.*, 1977). Monomers of IgG alone, however, do not activate platelets. Aggregates of at least dimer to trimer size with a minimum of 40–70 aggregates per cell bound are required for activation (Pfueller *et al.*, 1977; Ginsberg and Henson, 1978). Henson and Spiegelberg (1973) found that IgG of all four subclasses can activate platelets while the other classes of immunoglobulin do not.

This same paper also reported that only the aggregated Fc portion of IgG triggered activation. The most direct evidence for a platelet Fc receptor comes from Cheng and Hawiger (1979). These authors demonstrated isolation and partial purification of a protein from a platelet homogenate using Fc-Sepharose affinity chromatography. Elution of the affinity column with KBr yielded a glycoprotein with an apparent molecular weight of 255,000 (nonreduced). The isolated Fc fragment-binding glycoprotein formed complexes *in vitro* with aggregated IgG. The glycoprotein could also be isolated from thrombin-treated platelets suggesting that it is not a secreted product.

As mentioned previously, the other major mechanism of platelet activation by immune complexes in some species is via binding C3b and other complement components. Rabbit platelets are the best studied example of this group. The effects of complement on rabbit platelets are complex. Activation of the alternate pathway by zymosan and other compounds results in so-called "innocent bystander" lysis of the platelets most probably by deposition of the membrane attack complex on the platelet membrane. Simultaneously, direct, nonlytic activation of rabbit platelets may occur by C3B binding (Henson and Ginsberg, 1981; Henson, 1970). Both of these mechanisms result in release of serotonin, histamine, and other proinflammatory compounds. Complement activation may also play an important role in human platelet function, despite the lack of a C3b receptor. Polley and Nachman (1979) have demonstrated an enhancement of thrombin-induced aggregation in human platelets. This enhancement requires thrombin, C3, C5–C9, and can be inhibited by cyclooxygenase inhibitors suggesting prostaglandin involvement. In support of this concept, Wautier *et al.* (1979) reported a patient deficient in the C6 component of complement with a significantly reduced response to thrombin *in vitro*. Thus, regardless of the exact mechanism (Fc receptor or C3b receptor mediated), platelets can interact with immune complexes and release humoral factors which may modulate macrophage-monocyte participation in the immune and inflammatory responses.

7.2. ROLE OF PROSTAGLANDINS

There is no direct evidence at present for a prostaglandin-mediated interaction of platelets and macrophages. The available information, however, strongly suggests that such an interaction should exist in a mutually interdependent manner. Platelet aggregation is directly influenced by several prostaglandins, most notably TXA_2 and PGI_2. TXA_2 is a potent platelet-aggregating agent while PGI_2 strongly inhibits platelet aggregation. Platelets are generally considered the primary source of TXA_2 and vascular endothelial cells the main source of PGI_2. These compounds are produced via cyclooxygenase metabolism of arachidonic acid (Nalbandian and Henry, 1978). Recent evidence has indicated, however, that macrophages may serve as sources for both of these prostaglandins as well as prostanglandins of the E and F series (Scott *et al.*, 1982; Ford-Hutchinson and Doig, 1979; Gemsa *et al.*, 1979). Of particular interest is the observation of Scott *et al.* (1982) that with activation, murine macrophages decrease PGI_2 production and increase release of TXA_2. It should be noted that prostaglandins are considered to act on a local level as autocoids. Therefore, it is reasonable to suppose that in events which cause close association of platelets and macrophages (i.e., inflammation) that macrophage-derived prostaglandins might influence platelet events.

Macrophage function is also influenced by prostaglandins. Prostaglandins such as PGE and PGI_2 have a generally inhibitory effect on activities such as adherence and phagocytosis (Kaplan and Weinberg, 1982; Ford-Hutchinson and Doig, 1979). TXA_2, the primary platelet prostaglandin, does not appear to have been studied in macrophage systems. It has been reported, however, that TXA_2 mediates an augmentation of polymorphonuclear leukocyte adhesiveness (Spagnuolo *et al.*, 1980). Thus, platelet activation may modulate macrophage function. As obvious as these interactions appear to be, it is important to remember that the *in vivo* significance of prostaglandins with regard to macrophage function has not yet been established.

7.3. ROLE OF LEUKOTRIENES

Another major class of metabolites formed from arachidonic acid are the leukotrienes which are produced via the enzyme lipoxygenase. Lipoxygenase metabolites include not only the leukotrienes but also various eicosanoids such as 12-hydroxy-6,8,11,14-eicosatetraenoic acid (12-HETE). Both macrophages and thrombin-stimulated platelets release 12-HETE (Riguad *et al.*, 1979) which has been shown to be chemotactic (Goetzl and Sun, 1979) and trigger neutrophil degranulation (Stenson and Parker, 1980). Macrophages also produce a number of leukotrienes which together appear to make up the slow-reacting substance of anaphylaxis (SRS-A). Platelet leukotriene release has not been reported. With the availability of radioimmunoassays for the various lipoxygenase metabolites, research in this area is expanding rapidly (Corey, 1982).

7.4. PLATELET-ACTIVATING FACTOR AND OTHER INTERACTIONS

PAF is a phospholipid mediator which was originally recognized for its ability to aggregate platelets. It has recently been determined that the PAF of most species is 1-*O*-hexadecyl-2-acetyl-*sn*-glycero-3-phosphorylcholine or AGEPC (Pinckard *et al.*, 1980). At 10^{-10} to 10^{-8} M., PAF induces *in vitro* calcium-dependent platelet aggregation and release of dense and α-granule contents. *In vivo* PAF (purified AGEPC) causes a wheal and flare inflammatory response at 10^{-15} M (Pinckard *et al.*, 1980). IgE-sensitized basophils were the first cells identified as releasing PAF (Benveniste, 1974). It is now known that many cells including alveolar and peritoneal macrophages, platelets, and PMN produce PAF (Benveniste, 1979; Mencia-Huerta and Benveniste, 1979). Furthermore, PAF is chemotactic for PMN and promotes a variety of leukocyte responses such as enhanced respiratory burst, leukocyte aggregation and adhesion (Ingraham *et al.*, 1982). Recently, human platelets have been shown to exhibit two distinct types of binding sites for PAF (AGEPC). One platelet site was saturable, had a high affinity (K_D = 37 ± 13 nM, mean ± S.D.), and bound about 1400 AGEPC per platelet (Valone *et al.*, 1982). The other binding "site" demonstrated nearly infinite binding and was considered to represent AGEPC incorporation into cell membranes.

Other substances with PAF activity seem to exist as evidenced by Blumenthal *et al.* (1980), who reported a factor which activates platelets and is released from cultured peritoneal macrophages. This substance is similar to but distinct from PAF/AGEPC.

Several other isolated reports have provided evidence of platelet–macrophage interactions, although the overall significance of these is not possible to determine. Morland (1979) has reported that the reactivity of platelets to collagen is reduced when the collagen is preincubated with media fractions of peritoneal macrophages or lysates of the cultured cells. Some, but not all, of the inhibition was attributed to collagenase-induced modification of the collagen. Other contributing factors were not identified. Several groups of investigators, as identified earlier, have proposed that platelets enhance macrophage uptake of microorganisms and of inert colloids. Musson and Henson (1979) have reported that the presence of platelets counteracts an inhibitor of monocyte adherence *in vitro* present in plasma and serum. Johnson *et al.* (1977) have reported macrophage and platelet "cooperation" in the expression of antibody-mediated suppression of tumor growth. Sandler *et al.* (1975) have reported that serotonin, a primary secretory product of platelets, increases monocyte content of cGMP. Increased cGMP has been related to augmentation of monocyte motility and migration.

Thus, macrophages and platelets probably interact to regulate one another's participation in the reactions of immunity and inflammation.

8. SUMMARY AND CONCLUDING STATEMENT

A large amount of evidence exists which indicates a relationship between the function and bodily economy of platelets and the activation of cells of the RES. The RES is an important effector of platelet destruction during both senescent turnover and in disease. Platelets may play an important assisting role to the RES with respect to defense against microbial invaders and other particulate material. The RES removes several activators of platelets from the circulation and may therefore modulate the platelet response to thrombotic stimuli. A new spectrum of platelet–macrophage interactions seems to be emerging. The role of these interactions in immunity, inflammation, and other host defense responses may become clear in the future.

Despite a large body of information, however, comprehensive studies of the interaction of platelets and the RES have not been forthcoming in areas outside the realm of autoimmune platelet disease. Such studies would do much to clarify the relationship between these two important effectors of primary host defense.

ACKNOWLEDGMENTS. John E. Kaplan is a recipient of a Research Career Development Award (GM-00488) and is a Scholar of the Alexandrine and Alexander R. Sinsheimer Fund. Unpublished research cited in this review was supported by Grants GM-25946 and GM-21447 from the National Institute of General Medical Sciences and by the Sinsheimer Fund. Dudley G. Moon is a recipient of a National Research Service Award (HL-07194). The authors thank Mrs. Maureen Davis and Miss Susan Churan for assistance in the preparation of this manuscript.

REFERENCES

Aster, R. H., 1966, Pooling of platelets in the spleen: Role in the pathogenesis of "hypersplenic" thrombocytopenia, *J. Clin. Invest.* **45**:645.

Aster, R. H., 1969, Studies of the fate of platelets in rats and man, *Blood* **34**:117.

Aster, R. H., 1972, Platelet sequestration studies in man, *Br. J. Haematol.* **22**:259.

Aster, R. H., and Jandl, J. H., 1964, Platelet sequestration in man. I. Methods, *J. Clin. Invest.* **43**:843.

Aster, R. H., and Keene, W. R., 1969, Site of platelet destruction in idiopathic thrombocytopenic purpura, *Br. J. Haematol.* **16**:61.

Baldini, M., 1966a, Idiopathic thrombocytopenic purpura, *N. Engl. J. Med.* **274**:1245.

Baldini, M., 1966b, Idiopathic thrombocytopenic purpura, *N. Engl. J. Med.* **274**:1301.

Baldini, M., 1966c, Idiopathic thrombocytopenic purpura, *N. Engl. J. Med.* **274**:1360.

Balter, N. J., and Schwartz, S. L., 1977, Accumulation of norepinephrine by macrophages and relationships to known uptake processes, *J. Pharmacol. Exp. Ther.* **201**:636.

Becker, E. L., and Henson, P. M., 1973, *In vitro* studies of immunologically induced secretion of mediators from cells and related phenomena, *Adv. Immunol.* **17**:93.

Bedson, S. P., 1924, The effect of splenectomy on the production of experimental purpura, *Lancet* **2**:1117.

Behnke, O., 1967, Electron microscopic observations on the membrane systems of the rat blood platelet, *Anat. Rec.* **158**:121.

Bensusan, H. B., Koh, T. L., Henry, K. G., Murray, B. A., and Culp, L. A., 1978, Evidence that fibronectin is the collagen receptor on platelet membranes, *Proc. Natl. Acad. Sci. USA* **75**:5864.

Benveniste, J., 1974, Platelet-activating factor, a new mediator of anaphylaxis and immune complex deposition from rabbit and human basophils, *Nature (London)* **249**:581.

Benveniste, J., 1979, Release of platelet-activating factor by peritoneal and alveolar macrophages, *Monogr. Allergy* **14**:138.

Berman, I. R., 1975, Intravascular microaggregation and the respiratory distress syndromes, *Pediatr. Clin. North Am.* **22**:275.

Bjorkman, S. E., 1947, The splenic circulation, with special reference to the junction of the spleen sinus wall, *Acta Med. Scand.* **128**(Suppl.):191.

Blumenstock, F. A., Weber, P. B., Saba, T. M., and Taffin, R., 1977, Electroimmunoassay of alpha-2-globulin opsonic protein levels in the evaluation of reticuloendothelial function, *Am. J. Physiol.* **232:**R80.

Blumenthal, K. M., Rourke, F. J., and Wilder, M. S., 1980, Platelet activation by cultured mouse peritoneal macrophages, *J. Reticuloendothelial Soc.* **27:**247.

Castaldi, P. A., and Firkin, B. G., 1963, Studies of the life span and fate of platelets, *Aust. Ann. Med.* **12:**333.

Cazenave, J. P., Reimers, H. J., Kinlough-Rathbone, R. L., Packham, M. A., and Mustard, J. F., 1976, Effects of sodium periodate on platelet function, *Lab. Invest.* **34:**471.

Chandler, A. B., and Hand, R. A., 1961, Phagocytized platelets: A source of lipids in human thrombi and atherosclerotic plaques, *Science* **134:**946.

Chang, C., and Hamilton, P. B., 1979, The thrombocyte as the primary circulating phagocyte in chickens, *J. Reticuloendothelial Soc.* **25:**585.

Cheng, C. M., and Hawiger, J., 1979, Affinity isolation and characterization of immunoglobulin G Fc fragment-binding glycoprotein from human blood platelets, *J. Biol. Chem.* **254:**2165.

Clawson, C. C., 1973, Platelet interaction with bacteria. III. Ultrastructure, *Am. J. Pathol.* **70:**449.

Cohen, P., Gardner, F. H., and Barnett, G. O., 1961, Reclassification of thrombocytopenias by Cr^{51}-labeling method for measuring platelet life span, *N. Engl. J. Med.* **264:**1294.

Cohen, P., Braunwald, J., and Gardner, F. H., 1965, Destruction of canine and rabbit platelets following intravenous administration of carbon particles or endotoxin, *J. Lab. Clin. Med.* **66:**263.

Colvin, R. B., and Dvorak, H. F., 1975, Fibrinogen/fibrin on the surface of macrophages: Detection, distribution, binding requirements, and possible role in macrophage adherence phenomena, *J. Exp. Med.* **142:**1377.

Copley, A. L., 1978, Platelets, and physiological defense mechanisms, in: *Platelets: A Multidisciplinary Approach* (G. de Gaetano and S. Garattini, eds.), pp. 161–197, Raven Press, New York.

Copley, A. L., and Witte, S., 1976, On physiological microthromboembolization as the primary platelet function: elimination of invaded particles from the circulation and its pathogenic significance, *Thrombos. Res.* **8:**251.

Corey, E. J., 1982, Chemical studies on slow reacting substances/leukotrienes, *Experientia* **38:**1259.

Cronkite, E. P., Bond, V. P., Robertson, J. S., and Paglia, D. E., 1967, The survival, distribution, and apparent interaction with capillary endothelium of transfused radiosulfate labeled platelets in the rat, *J. Clin. Invest.* **36:**881.

Danon, D., Jersuhahny, Z., and De Vries, A., 1959, Incorporation of influenza virus in human blood platelets *in vitro:* Electron microscopical observation, *Virology* **9:**719.

Davey, M. G., 1966, The survival and destruction of human platelets, *Bibl. Haematol. (Basel)* **22:**(fasc.).

Davey, M. G., and Lander, H., 1964, The behavior of infused human platelets during the first twenty-four hours after infusion, *Br. J. Haematol.* **10:**94.

David-Ferreira, J. F., 1961, Sur la structure et la pouvoir phagocytaire des plaquettes sanguines, *Z. Zellforsch. Mikrosk. Anat.* **55:**49.

David-Ferreira, J. F., 1964, The blood platelet: Electron microscopic studies, *Int. Rev. Cytol.* **17:**99.

Dawson, A. B., 1933, The leukocytic reaction in *Necturus maculosus* to intravascular injections of colloidal carbon, with special reference to the behavior of the basophils and thrombocytes, *Anat. Rec.* **57:**351.

Day, H. J., Moloney, B. A., Nishizawa, E. E., and Rynbrandt, R. H. (eds.), 1978, *Thrombosis: Animal and Clinical Models*, Plenum Press, New York.

de Gaetano, G., and Garattini, S. (eds.), 1978, *Platelets: A Multidisciplinary Approach*, Raven Press, New York.

Dineen, J. J., Perillie, P. E., and Finch, S. C., 1968, Zymosan-induced thrombocytopenia in the rat, *J. Reticuloendothelial Soc.* **5:**161.

Doan, C. A., 1949, Hypersplenism, *Bull. N.Y. Acad. Med.* **25:**625.

Dobson, E. L., Kelly, L. S., Linney, C. R., and Lynch, J. J., 1971, The association of blood platelets and colloidal carbon removal, *J. Reticuloendothelial Soc.* **9:**376.

Dodds, W. J., 1978, Platelet function in animals: Species specificity, in: *Platelets: A Multidisciplinary Approach* (G. de Gaetano and S. Garattini, eds.), pp. 45–59, Raven Press, New York.

Elliott, R. H. E., Jr., and Whipple, M. A., 1940, Observations on the interrelationship of capillary, platelet, and splenic factors in thrombocytopenic purpura, *J. Lab. Clin. Med.* **26**:489.

Epstein, R. D., Lozner, E. L., Coffey, T. S., and Davidson, C. S., 1950, Congenital thrombocytopenic purpura: Purpura hemorrhagica in pregnancy and in the new born, *Am. J. Med.* **9**:44.

Evans, R. S., and Duane, R. T., 1949, Acquired hemolytic anemia. I. The relation of erythrocyte antibody production to activity of the disease. II. The significance of thrombocytopenia and leucopenia, *Blood* **4**:1196.

Evans, R. S., Takahashi, K., Duane, R. T., Payne, R., and Liu, C. K., 1951, Primary thrombocytopenic purpura and acquired hemolytic anemia, *Arch. Intern. Med.* **87**:48.

Filkins, J. P., 1971, Comparison of endotoxin detoxification by leukocytes and macrophages, *Proc. Soc. Exp. Biol. Med.* **137**:1396.

Firkin, B. G., Wright, R., Miller, S., and Stokes, E., 1969, Splenic macrophages in thrombocytopenia, *Blood* **33**:240.

Fisher, J. A., 1947, Cryptogenic acquired haemolytic anemia, *Q. J. Med.* **16**:245.

Ford-Hutchinson, A. W., and Doig, M. V., 1979, Prostaglandins and macrophages, *Agents Actions Suppl.* **6**:151.

French, J. E., 1967, Blood platelets: Morphological studies on their properties and life cycle, *Br. J. Haematol.* **13**:595.

French, J. E., and Barcat, J. A., 1968, The fine structure of platelets and platelet aggregation *in vivo*, *Prog. Biochem. Pharmacol.* **4**:550.

Gans, H., Lowman, J., and Fahr, G., 1966, Uptake of intact and disintegrated rat platelets by the isolated perfused rat liver, *Fed. Proc.* **25**:703.

Gans, H. V., Subramanian, J. T., and Lowman, B. H. T., 1967, Preservation of vascular potency as a function of reticuloendothelial clearance. II. Selectivity of phagocytosis for different clotting protein molecules derived from two clotting protein of plasma, *Surgery* **62**:698.

Gemsa, D., Seitz, M., Menzel, J., Grimm, W., Kramer, W., and Till, G., 1979, Phagocytosis-induced release of prostaglandins from macrophages, *Mongr. Allergy* **14**:194.

George, J. N., and Morgan, R. K., 1978, Platelet behavior and aging in the circulation, in: *The Blood Platelet in Transfusion Therapy* (T. J. Greenwalt and G. A. Jamieson, eds.), pp. 39–64, Liss, New York.

Ginsberg, M. H., and Henson, P. M., 1978, Enhancement of platelet response to immune complexes and IgG aggregates by lipid A-rich bacterial lipopolysaccharides, *J. Exp. Med.* **147**:208.

Ginsberg, M. H., Painter, R. G., Forsyth, J., Birdwell, C., and Plow, E. F., 1980, Thrombin increases expression of fibronectin antigen on the platelet surface, *Proc. Natl. Acad. Sci. USA* **77**:1049.

Glynn, M. F., Movat, H. Z., Murphy, E. A., and Mustard, J. F., 1965, Study of platelet adhesiveness and aggregation with latex particles, *J. Lab. Clin. Med.* **65**:179.

Glynn, M. F., Herren, R., and Mustard, J. F., 1966, Adherence of latex particles to platelets, *Nature (London)* **212**:79.

Goetzl, E. J., and Sun, F. F., 1979, Generation of unique monohydroxy eicosatetraenoic acids from arachidonic acid by human neutrophils, *J. Exp. Med.* **150**:406.

Govaerts, P., 1921, Effets de l'injection des plaquettes lavées sur l'élimination des microbes circulants dans le sang, *C. R. Soc. Biol.* **16**:1.

Grandjean, L. C., 1948, A case of purpura haemorrhagica after administration of quinine with specific thrombocytolysis demonstrated *in vitro*, *Acta Med. Scand.* **131**(Suppl. 213):165.

Greenberg, J. P., Rand, M. L., and Packham, M. A., 1975, Relation of platelet age to platelet sialic acid, *Fed. Proc.* **36**:453.

Grinnell, F., Feld, M., and Snell, W., 1979, The influence of cold insoluble globulin on platelet morphological response to substrata, *Cell Biol. Int. Rep.* **3**(7):585.

Grottum, K. A., and Jeremic, M., 1973, Neuraminidase injections in rabbits: Reduced platelet surface charge, aggregation and thrombocytopenia, *Thromb. Diath. Haemorrh.* **29**:461.

Grottum, K. A., Jorgensen, L., and Jeremic, M., 1972, Decrease of platelet surface charge during phagocytosis of polystyrene latex particles or thorium dioxide, *Scand. J. Haematol.* **9**:83.

Haakenstad, A. O., and Mannik, M., 1974, Saturation of the reticuloendothelial system with soluble immune complexes, *Immunology* **112**:1939.

Hagenau, F., Hollman, K., Levy, J., and Bioron, M., 1963, Étude au microscope électronique des plaquettes sanguines dans les leucemies humaines, *J. Microsc. (Paris)* **2**:529.

Han, S. S., and Baker, B. L., 1964, The ultrastructure of megakaryocytes and blood platelets in the rat spleen, *Anat. Rec.* **149**:251.

Hand, R. A., and Chandler, A. B., 1962, Atherosclerotic metamorphosis of autologous pulmonary thromboemboli in the rabbit, *Am. J. Pathol.* **40**:469.

Handin, R. I., and Stossel, T. P., 1974, Phagocytosis of antibody-coated platelets by human granulocytes, *N. Engl. J. Med.* **290**:989.

Handin, R. I., and Stossel, T. P., 1978, Effect of corticosteroid therapy on the phagocytosis of antibody-coated platelets of human leukocytes, *Blood* **51**:771.

Harker, L. A., 1977, The kinetics of platelet production and destruction in man, *Clin. Haematol.* **6**:671.

Harker, L. A., and Ross, R., 1978, Vessel injury, thrombosis, and platelet survival, in: *Thrombosis: Animal and Clinical Models* (H. J. Day, B. A. Moloney, E. E. Nishizawa, and R. H. Rynbrandt, eds.), Plenum Press, New York.

Harker, L. A., and Slichter, S. J., 1972, Platelet and fibrinogen consumption in man, *N. Engl. J. Med.* **287**:999.

Harrington, W. J., Minnich, V., Hollingsworth, J. W., and Moore, C. V., 1951, Demonstration of a thrombocytopenic factor in the blood of patients with thrombocytopenic purpura, *J. Lab. Clin. Med.* **38**:1.

Harrington, W. J., Minnich, V., and Arimura, G., 1956, The autoimmune thrombocytopenias, *Prog. Hematol.* **1**:166.

Hartmans, E., 1925, Beitrage zur Thrombozytengenese bei niederen Vertebraten, sowie zur Frage ihrer Stellung zum Megakaryozyten der Sänger, *Folia Haematolica* **32**:1.

Henson, P. M., 1970, Mechanisms of release of constituents from rabbit platelets by antigen–antibody complexes and complement. I. Lytic and nonlytic reactions, *J. Immunol.* **105**:476.

Henson, P. M., and Ginsberg, M. H., 1981, Immunological reactions of platelets, in: *Platelets in Biology and Pathology—2* (J. L. Gordon, ed.), pp. 265–308, Elsevier, Amsterdam.

Henson, P. M., and Spiegelberg, H. L., 1973, Release of serotonin from human platelets induced by aggregated immunoglobulins of different classes and subclasses, *J. Clin. Invest.* **52**:1282.

Herring, W. B., Herion, J. C., Walker, R. I., and Palmer, J. G., 1963, Distribution and clearance of circulating endotoxin, *J. Clin. Invest.* **42**:79.

Heyns, A. D., Lotter, M. G., Badenhorst, P. N., van Reenen, O. R., Pieters, H., Minnaar, P. C., and Retief, F. P., 1980, Kinetics distribution and sites of destruction of ^{111}indium-labelled human platelets, *Br. J. Haematol.* **44**:269.

Heyssel, R. M., Silver, L. J., Wasson, M., and Brill, A. B., 1967, The relation of blood platelet survival and distribution to ^{14}C-serotonin distribution and excretion, *Blood* **29**:341.

Hill, J. M., Speer, R. J., and Gedikogln, H., 1963, Secondary lipoidosis of spleen associated with thrombocytopenia and other blood dyscrasias treated with steroids, *Am. J. Clin. Pathol.* **39**:607.

Hirsch, E. O., and Gardner, F. H., 1951, The lifespan of transfused human platelets, *J. Clin. Invest.* **30**:649.

Hirsch, J., Glynn, M. F., and Mustard, J. F., 1968, The effect of platelet age on platelet adherence to collagen, *J. Clin. Invest.* **47**:466.

Holmsen, H., Salganicoff, L., and Fukami, M. H., 1977, Platelet behaviour and biochemistry, in: *Haemostasis: Biochemistry, Physiology and Pathology* (D. Ogston and B. Bennett, eds.), pp. 239–319, Wiley, New York.

Houlihan, R. B., 1947, Studies of the adhesion of human blood platelets and bacteria, *Blood Suppl.* **1**:142.

Humphrey, J. H., and Jaques, R., 1955, The release of histamine and 5-hydroxytryptamine (serotonin) from platelets by antigen–antibody reactions (*in vitro*), *J. Physiol. (London)* **128**:9.

Hynes, R. O., Ali, I. U., Destree, A. T., Mautner, V., Perkins, M. E., Senger, D. R., Wagner, D. D., and Smith, K. K., 1978, A large glycoprotein lost from the surfaces of transformed cells, *Ann. N.Y. Acad. Sci.* **312**:317.

Ingraham, L. M., Coates, T. D., Allen, J. M., Higgins, C. P., Baehner, R. L., and Boxer, L. A., 1982, Metabolic, membrane and functional responses of human polymorphonuclear leukocytes to platelet-activating factor, *Blood* **59**:1259.

Jilek, F., and Hormann, H., 1978, Fibronectin (cold insoluble globulin). VI. Mediation of fibrin monomer binding to macrophages, *Hoppe-Seyler's Z. Physiol. Chem.* **359**:1603.

Johnson, R. J., Pasternack, G. R., and Shin, H. S., 1977, Antibody-mediated suppression of tumor growth. II. Macrophage and platelet cooperation with murine IgGl isolated from alloantiserum, *J. Immunol.* **118:**494.

Jorgensen, L., Rowsell, H. C., Hovig, T., Glenn, M. F., and Mustard, J. F., 1967, Adenosine diphosphate-induced platelet aggregation and myocardial infarction in swine, *Lab. Invest.* **17:**616.

Kaplan, J. E., 1981, The role of the reticuloendothelial system in control of hemostatic and thrombotic mechanisms, in: *Physiology of the Reticuloendothelial System* (T. M. Saba and B. M. Altura, eds.), Raven Press, New York.

Kaplan, J. E., and Saba, T. M., 1976, Humoral deficiency and reticuloendothelial depression after traumatic shock, *Am. J. Physiol.* **230:**7.

Kaplan, J. E., and Saba, T. M., 1977, Defective reticuloendothelial clearance of platelets after whole-body trauma, *Thromb. Haemostasis* **38:**141.

Kaplan, J. E., and Saba, T. M., 1978, Platelet removal from the circulation by the liver and spleen, *Am. J. Physiol.* **235:**H314.

Kaplan, J. E., and Weinberg, D. A., 1982, Depression of reticuloendothelial clearance by prostacyclin, *Fed. Proc.* **41:**1691.

Kaplan, J. E., Baum, S. H., and Snedeker, P. W., 1980, Relationship of plasma fibronectin to thrombotic resistance, *Physiologist* **23:**173.

Karpatkin, S., 1969a, Heterogeneity of human platelets. I. Metabolic and kinetic evidence suggestive of young and old platelets, *J. Clin. Invest.* **48:**1073.

Karpatkin, S., 1969b, Heterogeneity of human platelets. II. Functional evidence suggestive of young and old platelets, *J. Clin. Invest.* **48:**1083.

Karpatkin, S., 1980, Autoimmune thrombocytopenic purpura, *Blood* **56:**329.

Karpatkin, S., and Siskind, G. W., 1969, *In vitro* detection of platelet antibody in patients with idiopathic thrombocytopenic purpura and systemic lupus erythematosus, *Blood* **33:**795.

Karpatkin, S., and Siskind, G. W., 1974, Studies on the specificity of antiplatelet autoantibodies, *Proc. Soc. Exp. Biol. Med.* **147:**715.

Karpatkin, S., Strick, N., Karpatkin, M. H., and Siskind, G. W., 1972a, Cumulative experience on the detection of anti-platelet antibody in 234 patients with idiopathic thrombocytopenic purpura, systemic lupus erythematosus and other clinical disorders, *Am. J. Med.* **52:**776.

Karpatkin, S., Strick, N., and Siskind, G. W., 1972b, Detection of splenic anti-platelet antibody synthesis in idiopathic autoimmune thrombocytopenic purpura (ATP), *Br. J. Haematol.* **23:**167.

Kaufmann, R. M., Airo, R., Pollack, S., and Crosby, W. H., 1965, Circulating megakarocytes and platelet release in the lung, *Blood* **26:**720.

Kay, M. M. B., 1978, Role of physiologic autoantibody in the removal of senescent human red cells, *J. Supramol. Struct.* **9:**555.

Kelton, J. G., and Neame, P. B., 1971, Elevated platelet associated IgG in the thrombocytopenia of septicemia, *N. Engl. J. Med.* **300:**760.

Kinlough, R. L., Bennett, R. C., and Lander, H., 1966, The place of splenectomy in haematological disorders: The value of ^{51}Cr techniques, *Med. J. Aust.* **1966:**1022.

Klonizakis, I., Peters, A. M., Fitzpatrick, M. L., Kensett, M. J., Lewis, S. M., and Lavender, J. P., 1980, Radionuclide distribution following injection of ^{111}indium-labelled platelets, *Br. J. Haematol.* **46:**595.

Koepke, J. A., Jobe, M. G., and Braunstein, H., 1968, Platelet histiocytosis of the spleen, *J. Reticuloendothelial Soc.* **5:**378.

Kolb, H., Schlepper-Schafer, J., Nagamura, Y., Osburg, M., and Kolb-Bachofen, V., 1980, Analysis of a D-galactose specific lectin on rat Kupffer cells, in: *The Reticuloendothelial System and the Pathogenesis of Liver Disease* (H. Liehr and M. Grün, eds.), pp. 117–122, Elsevier/North-Holland, Amsterdam.

Landing, B. N., Strauss, L., Crocker, A. C., Braunstein, H., Henley, W. L., Will, J. R., and Sanders, M., 1961, Thrombocytopenic purpura with histiocytosis of the spleen, *N. Engl. J. Med.* **265:**572.

Leach, C. M., and Thorburn, G. D., 1982, A comparative study of collagen induced thromboxane release from platelets of different species: Implications for human atherosclerosis models, *Prostaglandins* **24:**47.

Ledingham, J. C. C., and Bedson, S. P., 1915, Experimental purpura, *Lancet* **1:**311.

Lee, L., and McClusky, R. J., 1962, Immunohistochemical demonstration of the reticuloendothelial clearance of circulating fibrin aggregates, *J. Exp. Med.* **116:**611.

Leonard, M. E., and Falconer, E. H., 1941, Experimental thrombocytopenic purpura in the guinea pig, *J. Lab. Clin. Med.* **26:**648.

Lewis, J. C., Maldonado, J. E., and Mann, K. G., 1976, Phagocytosis in human platelets: Localization of acid phosphatase-positive phagosomes following latex uptake, *Blood* **47:**833.

MacMillan, D. C., and Sim, A. K., 1970, A comparative study of platelet aggregation in man and laboratory animals, *Thromb. Diath. Haemorrh.* **24:**385.

McMillan, R., Longmire, R. L., Yelenosky, R., Smith, R. S., and Craddock, C. G., 1972, Immunoglobulin synthesis *in vitro* by splenic tissue in idiopathic thrombocytopenic purpura, *N. Engl. J. Med.* **286:**681.

McMillan, R., Longmire, R. L., Tavassoli, M., Armstrong, S., and Yelenosky, R., 1974, *In vitro* platelet phagocytosis by splenic leukocytes in idiopathic thrombocytopenic purpura, *N. Engl. J. Med.* **290:**249.

Mandell, G. L., and Hook, E. W., 1969, The interaction of platelets, salmonella, and mouse peritoneal macrophages, *Proc. Soc. Exp. Biol. Med.* **132:**757.

Mant, M. J., and Firkin, B. G., 1972, Uptake of latex and thorotrast by human platelets *in vitro:* Effect of various chemicals demonstrating differing mechanisms and metabolic requirements, *Br. J. Haematol.* **22:**383.

Mencia-Huerta, J. M., and Benveniste, J., 1979, Platelet-activating factor and macrophages. I. Evidence for the release from rat and mouse peritoneal macrophages and not from mastocytes, *Eur. J. Immunol.* **9:**409.

Moon, D. G., and Kaplan, J. E., 1981, Plasma fibronectin inhibition of collagen and ADP-induced platelet aggregation, *Fed. Proc.* **40:**768.

Morland, B., 1979, Macrophage inhibition of collagen-induced platelet aggregation, *Thromb. Haemostasis* **42:**1207.

Mosesson, M. W., and Amarani, D. L., 1980, The structure and biological activities of plasma fibronectin, *Blood* **56:**145.

Mosher, D. F., 1980, Fibronectin, *Prog. Hemostasis Thromb.* **5:**111.

Movat, H. Z., Mustard, J. F., Taichman, N. S., and Uriuhara, T., 1965a, Platelet aggregation and release of ADP, serotonin, and histamine associated with phagocytosis of antigen–antibody complexes, *Proc. Soc. Exp. Biol. Med.* **120:**232.

Movat, H. Z., Weiser, W. J., Glynn, M. F., and Mustard, J. F., 1965b, Platelet phagocytosis and aggregation, *J. Cell Biol.* **27:**531.

Mueller-Eckhardt, C., and Luscher, E. F., 1968, Immune reactions of blood platelets. I. A comparative study of the effects on platelets of heterologous antiplatelet serum, antigen–antibody complexes, aggregated gamma globulin, and thrombin, *Thromb. Diath. Haemorrh.* **20:**155.

Murphy, S., 1976, Platelet transfusion, in: *Progress in Hemostasis and Thrombosis,* Vol. 3 (T. H. Spaet, ed.), Grune & Stratton, New York.

Murphy, S., and Gardner, F. H., 1969, Platelet preservation: Effect of storage temperature on maintenance of platelet viability—Deleterious effect of refrigerated storage, *N. Engl. J. Med.* **280:**1094.

Musson, R. A., and Henson, P. M., 1979, Humoral and formed elements of blood modulate the response of peripheral blood monocytes. I. Plasma and serum inhibit and platelets enhance monocyte adherence, *J. Immunol.* **122:**2026.

Mustard, J. F., 1978, Platelet survival, *Thromb. Haemostasis* **40:**154.

Mustard, J. F., and Packham, M. A., 1968, Platelet phagocytosis, *Ser. Haematol.* **1:**168.

Mustard, J. F., Rowsell, H. C., and Murphy, E. A., 1966, Platelet economy (platelet survival and turnover), *Br. J. Haematol.* **12:**1.

Mustard, J. F., Glynn, M. F., Nishizawa, E. E., and Packham, M. A., 1967, Platelet surface interactions: Relationship to thrombosis and hemostasis, *Fed. Proc.* **26:**106.

Najean, Y., and Ardaillon, N., 1971, The sequestration site of platelets in idiopathic thrombocytopenic purpura: Its correlation with the results of splenectomy, *Br. J. Haematol.* **21:**153.

Najean, Y., Ardaillon, N., Caen, J., Laurien, M., and Bernard, J., 1963, Survival of radiochromium-labeled platelets in thrombocytopenias, *Blood* **22:**718.

Najean, Y., Ardaillon, N., Dresch, C., and Bernard, J., 1967, The platelet destruction site in thrombocytopenic purpuras, *Br. J. Haematol.* **13**:409.

Nalbandian, R. M., and Henry, R. L., 1978, Platelet–endothelial cell interactions, *Semin. Thromb. Hemostasis* **5**:87.

Nishizawa, E. E., and Haldor, J., 1967, Phospholipid synthesis during platelet aggregation, *Fed. Proc.* **26**:760.

Packham, M. A., Greenberg, J. P., Scott, S., Reimers, H. J., and Mustard, J. F., 1977, Modifications of rabbit platelets that shorten their survival, *Blood* **50**(Suppl.):249.

Packham, M. A., Guccione, M. A., Kinlough-Rathbone, R. L., and Mustard, J. F., 1980, Platelet sialic acid and platelet survival, after aggregation by ADP, *Blood* **56**:876.

Pearlstein, E., Gold, L., and Garcia-Parch, A., 1980, Fibronectin: A review of its structure and biological activity, *Mol. Cell. Biochem.* **29**:103.

Penny, R., Rozenberg, M. C., and Firkin, B. G., 1966, The splenic platelet pool, *Blood* **27**:1.

Peters, A. M., Klonizakes, I., Lavender, J. P., and Lewis, S. M., 1980, Use of ^{111}indium-labelled platelets to measure spleen function, *Br. J. Haematol.* **46**:587.

Pfueller, S. L., and Firkin, B. G., 1978, Role of plasma proteins in the interaction of human platelets with particles, *Thromb. Res.* **12**:979.

Pfueller, S. L., Weber, S., and Luscher, E. F., 1977, Studies of the mechanism of human platelet release reaction induced by immunologic stimuli. III. Relationship between the binding of soluble IgG aggregates to the Fc receptor and cell response in the presence and absence of plasma, *J. Immunol.* **118**:514.

Pinckard, R. N., McManus, L. M., Demopoulos, C. A., Halonen, M., Clark, P. O., Shaw, J. O., Kniker, W. T., and Hanahan, D. J., 1980, Ether phosphorylcholine: Evidence for the structural and functional identity with platelet-activating factor, *J. Reticuloendothelial Soc.* **28**(Suppl.):95S.

Plow, E. F., Birdwell, C., Ginsberg, M. H., Byers, V., Taylor, L., and Braisier, A., 1979, Identification and quantitation of platelet-associated fibronectin antigen, *J. Clin. Invest.* **63**:540.

Polley, M. J., and Nachman, R. L., 1979, Human complement in thrombin-mediated platelet function: Uptake of C5B-9 complex, *J. Exp. Med.* **150**:633.

Poole, J. C. F., 1966, Phagocytosis of platelets by monocytes in organizing arterial thrombi: An electron microscopical study, *Q. J. Exp. Physiol.* **51**:54.

Reimers, H., Kinlough-Rathbone, L., Cazenave, J., Senyi, A. F., Hirsh, J., Packham, M. A., and Mustard, J. F., 1976, *In vitro* and *in vivo* functions of thrombin-treated platelets, *Thromb. Haemostasis* **35**:151.

Rigaud, M., Durane, J., and Breton, J. C., 1979, Transformation of arachidonic acid into 12-hydroxy-5,8,10,14-eicosatetraenoic acid by mouse peritoneal macrophages, *Biochim. Biophys. Acta* **573**:408.

Ruoslahti, E., and Vaheri, A., 1975, Interaction of soluble fibroblast surface antigen with fibrinogen and fibrin: Identity with cold-insoluble globulin of human plasma, *J. Exp. Med.* **141**:497.

Saba, T. M., and DiLuzio, N. R., 1969, Reticuloendothelial blockade and recovery as a function of opsonic activity, *Am. J. Physiol.* **216**:197.

Saba, T. M., and Jaffe, E., 1980, Plasma fibronectin (opsonic glycoprotein): Its synthesis by vascular endothelial cells and role in cardiopulmonary integrity as related to reticuloendothelial function, *Am. J. Med.* **68**:577.

Saltzstein, S., 1961, Phospholipid accumulation in histiocytes of splenic pulp associated with thrombocytopenic purpura, *Blood* **18**:73.

Salvidio, E., and Crosby, W. H., 1960, Thrombocytopenia after intravenous injection of India ink, *J. Lab. Clin. Med.* **56**:711.

Sandler, J. A., Clyman, R. I., Manganiello, V. C., and Vaughan, M., 1975, The effect of serotonin (5-hydroxytryptamine) and derivatives on guanosine 3',5'-monophosphate in human monocytes, *J. Clin. Invest.* **55**:431.

Santoro, S. A., and Cunningham, L. W., 1979, Fibronectin and the multiple interaction model for platelet–collagen adhesion, *Proc. Natl. Acad. Sci. USA* **76**:2644.

Schulz, H., 1961, Uber die phagozytose von kolloidalem Siliziumdioxyd durch Thrombozyten mit Bermerkungen zur submikroskopischen Struktur der Thrombozytenmembran, *Folia Haematolica* **5**:195.

Schulz, H., and Wedell, J., 1962, Elektronenmikroskopische Untersuchungen zur Frage der Fettpagozytose und des Fettransportes durch Thrombozyten, *Klin. Wochenschr.* **40:**1114.

Scott, S., Reimers, H. J., Chernesky, M., Greenberg, J. P., Kinlough-Rathbone, R., Packham, M., and Mustard, J. F., 1978, The effect of viruses on platelet aggregation and platelet survival in rabbits, *Blood* **52:**47.

Scott, W. A., Pawlowski, N. A., Murray, H. W., Andreach, M., Zrike, J., and Cohn, Z. A., 1982, Regulation of arachidonic acid metabolism by macrophage activation, *J. Exp. Med.* **155:**1148.

Sherman, L. A., and Lee, J., 1977, Specific binding of soluble fibrin to macrophages, *J. Exp. Med.* **145:**76.

Sherman, L. A., Harwig, S., and Lee, J., 1975, *In vitro* formation and *in vivo* clearance of fibrinogen-fibrin complexes, *J. Lab. Clin. Med.* **86:**100.

Shulman, N. R., Weinrach, R. S., Libre, E. P., Andrews, H. L., and Shannon, J. A., 1965a, The role of the reticuloendothelial system in the pathogenesis of idiopathic thrombocytopenic purpura, *Trans. Assoc. Am. Physicians* **78:**374.

Shulman, N. R., Marder, V. J., and Weinrach, R. S., 1965b, Similarities between known antiplatelet antibodies and the factor responsible for thrombocytopenia in idiopathic purpura: Physiologic, serologic and isotopic studies, *Ann. N.Y. Acad. Sci.* **124:**499.

Shulman, N. R., Watkins, S. P., Itscoitz, S. B., and Students, A. B., 1968, Evidence that the spleen retains the youngest and hemostatically most effective platelets, *Trans. Assoc. Am. Physicians* **81:**302.

Sinakos, Z., and Caen, J. P., 1967, Platelet aggregation in mammalians (human, rat, rabbit, guinea-pig, horse, dog): A comparative study, *Thromb. Diath. Haemorrh.* **17:**99.

Slichter, S. J., and Harker, L. A., 1976, Preparation and storage of platelet concentrates. II. Storage variables influencing platelet viability and function, *Br. J. Haematol.* **34:**403.

Snedeker, P. W., Kaplan, J. E., and Saba, T. M., 1979, Effect of intravascular coagulation on the vascular clearance of soluble fibrin monomer, *Circ. Shock* **6:**197.

Solomon, R. B., and Clatanoff, D. V., 1967, Platelet survival studies and body scanning in idiopathic thrombocytopenic purpura, *Am. J. Med. Sci.* **254:**777.

Spagnuolo, P. J., Ellner, J. J., Hassid, A., and Dunn, M. J., 1980, Thromboxane A_2 mediates augmented polymorphonuclear leukocyte adhesiveness, *J. Clin. Invest.* **66:**406.

Steffen, C., 1960, Results obtained with the antiglobulin consumption test and investigations of auto-antibody eluates in immunohematology, *J. Lab. Clin. Med.* **55:**9.

Stenson, W. F., and Parker, C. W., 1980, Monohydroxyeicosatetraenoic acids (HETEs) induce degranulation of human neutrophilis, *J. Immunol.* **124:**210O.

Summerell, J. M., and Gibbs, W. N., 1972, Splenic histiocytosis associated with thrombocytopenia, *Acta Haematol.* **48:**34.

Sutton, R. H., 1975, Carbon clearance studies in sheep: A possible interaction with fibrinogen, platelets and leukocytes, *J. Reticuloendothelial Soc.* **17:**313.

Tait, J., 1918, Capillary phenomena observed in blood cells, thigmocytes phagocytosis ameloid movement differential adhesiveness of corpuscles emigration of leukocytes, *Q. J. Exp. Physiol.* **12:**1.

Tait, J., and Elvidge, A. R., 1926, Effect upon platelets and on blood coagulation of injecting foreign particles into the blood stream, *J. Physiol. (London)* **62:**129.

Tavassoli, M., and McMillan, R., 1972, Structure of the spleen in idiopathic thrombocytopenic purpura, *Am. J. Clin. Pathol.* **64:**180.

Terada, H., Baldini, M., Ebbe, S., and Madoff, M. A., 1966, Interaction of influenza virus with blood platelets, *Blood* **28:**213.

Tinggaard-Pedersen, N., 1974, The effect of splenectomy of the megakaryocyte and platelet count in the blood of rats, *Scand. J. Haematol.* **12:**291.

Tocantins, L. M., 1936a, Experimental thrombopenic purpura in the dog, *Arch. Pathol.* **21:**69.

Tocantins, L. M., 1936b, Experimental thrombopenic purpura: Cytological and physical changes in the blood, *Ann. Intern. Med.* **9:**838.

Tocantins, L. M., 1938, The mammalian blood platelet in health and disease, *Medicine* **17:**155.

Tocantins, L. M., and Stewart, H. L., 1939, Pathological anatomy of experimental thrombopenic purpura in the dog, *Am. J. Pathol.* **15:**1.

Tullis, J. L., 1956, Identification and significance of platelet antibodies, *N. Engl. J. Med.* **255**:541.

Valone, F. H., Coles, E., Reinhold, V. R., and Goetzl, E. J., 1982, Specific binding of phospholid platelet-activating factor by human platelets, *J. Immunol.* **129**:1637.

van Aken, W. G., and Vreeken, J., 1969, Accumulation of macromolecular particles in the reticuloendothelial system (RES) mediated by platelet aggregation and disaggregation, *Thromb. Diath. Haemorrh.* **22**:496.

van Aken, W. G., Coste, T. M., and Vreeken, J., 1968, Platelet aggregation: An intermediary mechanism in carbon clearance, *Scand. J. Haematol.* **5**:333.

Verp, M., and Karpatkin, S., 1975, Effect of plasma, steroids, or steroid products on the adhesion of human opsonized thrombocytes to human leukocytes, *J. Lab. Clin. Med.* **85**:478.

Walsh, R. T., and Barnhart, M. I., 1969, Clearance of coagulation and fibrinolysis products by the reticuloendothelial system, *Thromb. Diath. Haemorrh.* **36**(Suppl.):83.

Wautier, J. L., Peltier, A. P., and Caen, J. P., 1979, Complement (C6) and platelet activation by thrombin in humans, *Thromb. Res.* **15**:589.

White, J. G., 1972, Uptake of latex particles by blood platelets, phagocytosis or sequestration?, *Am. J. Pathol.* **69**:439.

Wright, C., Doan, C. A., Bouroncle, B. A., and Zollinger, R. M., 1951, Direct splenic arterial and venous blood studies in the hypersplenic syndromes before and after epinephrine, *Blood* **6**:195.

Yamada, K., and Olden, K., 1978, Fibronectins—Adhesive glycoproteins of cell surface and blood, *Nature (London)* **275**:179.

Zucker, M. B., 1980, The functioning of blood platelets, *Sci. Am.* **242**:86.

Zucker, M. B., Mosesson, M. W., Broekman, M. J., and Kaplan, K. L., 1979, Release of platelet fibronectin (cold-insoluble globulin) from alpha granules induced by thrombin or collagen; lack of requirement for plasma fibronectin in ADP-induced platelet aggregation, *Blood* **54**:8.

Zucker-Franklin, D., and Karpatkin, S., 1977, Red-cell and platelet fragmentation in idiopathic autoimmune thrombocytopenic purpura, *N. Engl. J. Med.* **297**:517.

11

RES–Leukocyte Interactions

MARC FELDMANN, DAVID R. KATZ, and GEOFFREY H. SUNSHINE

1. INTRODUCTION

The physiology of RES–leukocyte interactions is a difficult topic to discuss in a single chapter, as this brief title encompasses an incredible diversity of cells which influence each other in a bidirectional manner. Our first premise in this discussion is that RES cells and leukocytes are the basic constituents of the immune system, and that this should be the chief focus of our attention.

We have chosen to interpret RES to include the various macrophage-like but nonlymphocyte cells which influence the immune response and which have been termed *accessory cells, adherent cells, A cells,* etc. There is no doubt that these cells have a profound effect in all *in vivo* immune responses, and possibly in other defense mechanisms, but more precise definition of their role is often not possible because the techniques for their identification in and isolation from tissues are still relatively imperfect.

2. TYPES OF INTERACTION BETWEEN RES CELLS AND LYMPHOCYTES

The conventional scheme for RES–lymphocyte envisages a phagocytic effector cell which digests material and "presents" the resultant degraded antigen to a lymphocyte (Waldron *et al.*, 1974). The cell responsible for this presentation is classically described as ruffled, highly phagocytic with Fc and C3 receptors and capable of secreting a range of mediators as well as lysosomal enzymes which play their part in the intracellular "processing" phase, before presentation to the lymphocyte (van Furth *et al.*, 1972). The peritoneal macrophage (Fig.

MARC FELDMANN and GEOFFREY H. SUNSHINE • ICRF Tumour Immunology Unit, Department of Zoology, University College London, London, England. *Present address for Sunshine:* Department of Surgery, Tufts University Veterinary School, Boston, Massachusetts. DAVID R. KATZ • Department of Pathology, The Middlesex Hospital Medical School, London, England.

1) is the common *in vitro* prototype of these cells; in addition, the cells which adhere to glass/plastic from disaggregated tissue are regarded as analogs.

There are several pieces of evidence which suggest that these cells represent only one facet of RES-lymphocyte interaction in the immune response. For example, during studies tracing the fate of radiolabeled antigen in rodent tissues, Nossal, Ada, and their colleagues (1971 review) identified a cell in the lymphoid follicles and germinal centers which retained antigen for long periods of time. This cell is not a macrophage by standard criteria: it is nonphagocytic and has long processes extending out through the tissues; it bears both Fc and C3 receptors. Further analysis of this cell has not been pursued *in vitro* as there are no effective methods to isolate them; however, Klaus *et al.* (1980) have accumulated much evidence to suggest a unique role in B memory cell function.

Another example is a second cell with long processes, the dendritic cell of Steinman and Cohn (1973), which can be separated from standard macrophages by its physical characteristics as it is poorly adherent and lacks Fc receptors. This cell may be the *in vitro* analog of the *in vivo* interdigitating cell. The function of this cell has been extensively analyzed in our laboratory and will be discussed in detail below. In this context the most significant feature of these cells is that they are able to "present" antigen to T cells but are nonphagocytic.

A third cell is the marginal zone macrophage which Humphrey and Grennan (1981) have separated from a collagenase splenic digest on the basis that they will take up uncharged polysaccharides (such as Ficoll and hydroxyethyl starch), unlike the other macrophages found in the red pulp. This cell may be responsible for B-cell stimulation (see below).

Taking these three cells as a group, they share certain common features. Each has some conventional macrophage markers, yet none shows the complete range of the classic cell. Whether this reflects a common derivation with modification for interaction with the lymphocyte or whether it reflects different cell lineages is not known; we are currently investigating this in our laboratory.

Second, there is a clear *in vivo* relationship between the different RES cell types described and the lymphocytes which they influence. Follicular dendritic cells are surrounded by large B cells, interdigitating cells by T cells, and marginal zone cells by small B cells. It is thus not unreasonable to postulate that unique microenvironments play a central role in RES–lymphocyte interactions.

Third, our recent *in vitro* studies suggest that there are in fact considerable differences in stimulatory capacity between the different cell types mentioned. This applies both to antigen-specific genetically restricted responses and to alloantigen responses. This confirms that there is considerable heterogeneity within the RES–lymphocyte interaction system but that it is possible to analyze this heterogeneity in the laboratory.

There are also other types of RES cells which play a role in interaction with lymphocytes and which are organ specific rather than "microenvironment specific." The best known is the Kupffer cell, the bone marrow-derived RES cell which localizes to the liver. van Furth *et al.* (1973) showed that these cells represent a large proportion of the total mononuclear phagocyte pool and in their anatomical location they are an important site of RES phagocytic function.

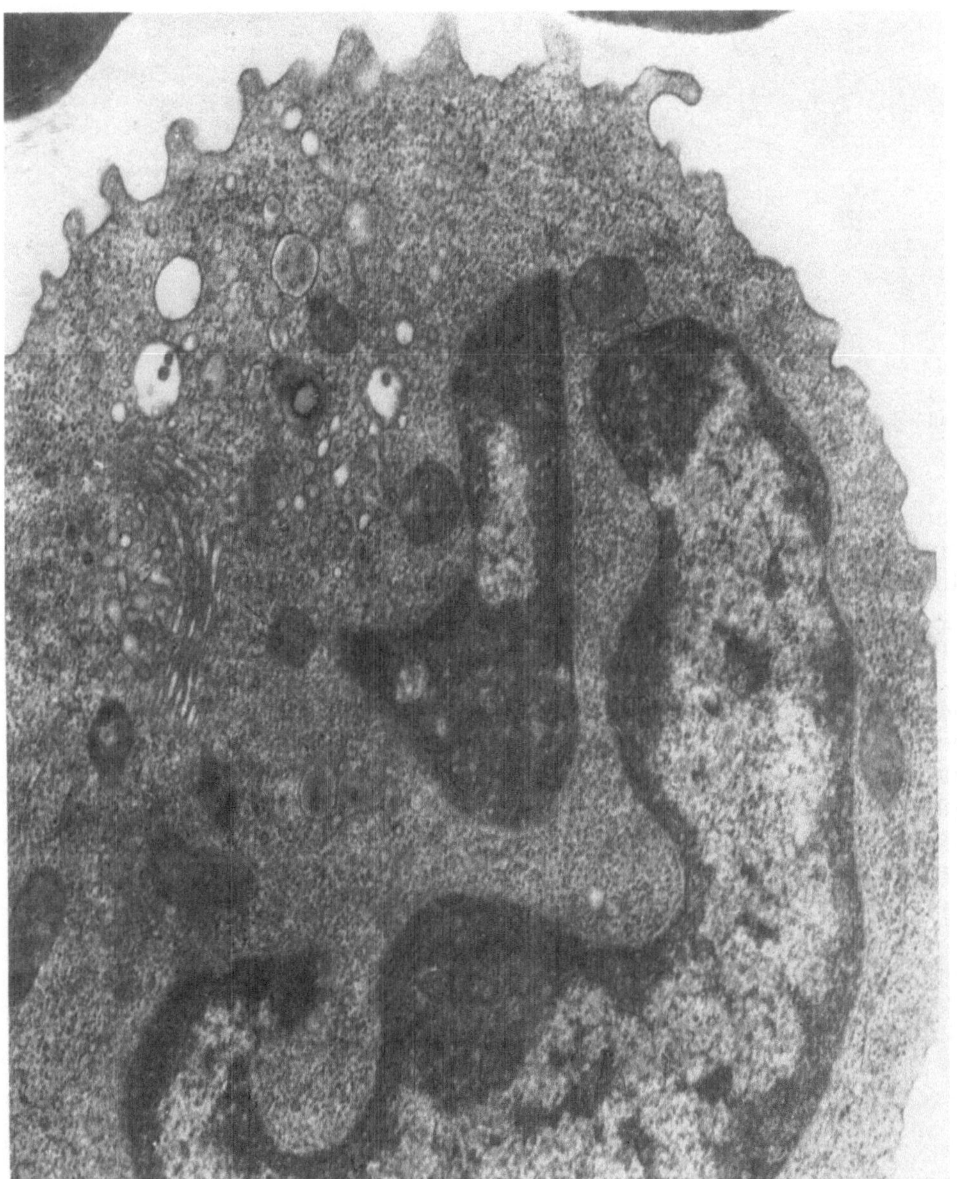

FIGURE 1. Peritoneal washout macrophage showing surface ruffles and prominent Golgi. 17,500 ×, reproduced at 80%.

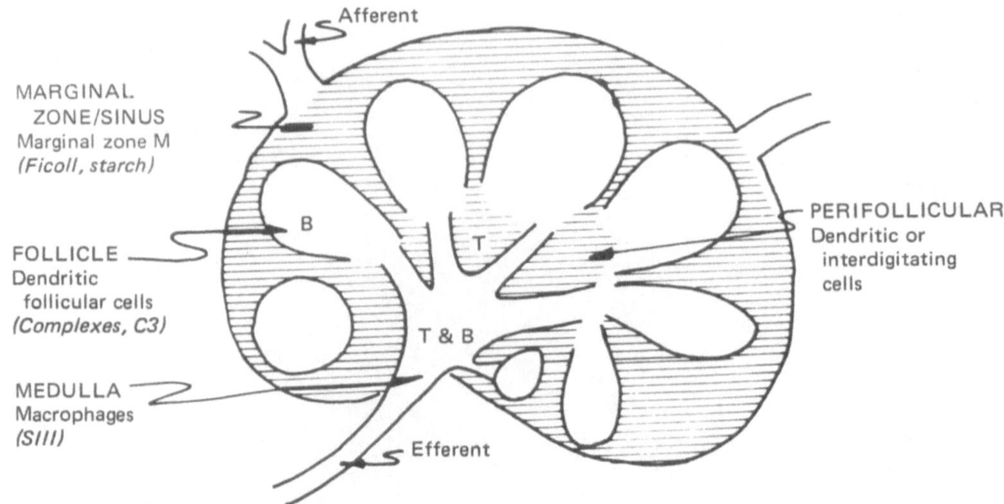

FIGURE 2. Distribution of macrophages/antigen-presenting cells in the lymph node illustrating possible relationship with lymphocytes in different microenvironmental compartments.

Direct *in vivo* interaction with leukocytes does not appear to be a significant aspect of the Kupffer cell role, but *in vitro* they are phagocytic and express Ia antigens on their surfaces and are responsible for genetic restriction in immune responses (Richman *et al.*, 1980), suggesting that they are capable of such interactions.

The current view is that the Langerhans cell is the RES representative in the epidermis. Unlike other epidermal cells, they lack desmosomes and tonofilaments, but have characteristic Birbeck granules in their cytoplasm. They are bone marrow derived and express both Ia antigens and Fc receptors. *In vivo* depletion of Langerhans cells using ultraviolet radiation alters T-cell function (Greene *et al.*, 1979), suggesting that they play a role in interaction with lymphocytes. *In vitro* guinea pig Langerhans cells can activate T cells in an antigen-specific proliferation system (Stingl *et al.*, 1980).

In summary, there are several types of RES cells involved in interaction with lymphocytes (Fig. 2). The particular organ site/microenvironment and the precise features of the cell with which each lymphocyte interacts are two essential features which must be clarified when investigating the different physiological responses within the immune system, as there is increasing evidence that this heterogeneity of cell type is a particularly important discretionary mechanism in response to antigen.

3. INTERACTION OF RES CELLS AND NONLYMPHOID LEUKOCYTES

The typical example of RES-leukocyte interactions are those between mononuclear phagocytes and their relatives and lymphoid cells in the tissues. The relative influences of these cell types on each other are the major emphasis of this review (see Sections 4–6).

However, it is important to realize that there are several other potential sites of RES–nonlymphoid leukocyte interaction and multiple potential cell types involved. For example, while peripheral blood monocytes may not interact one with the other, the tissue macrophage may act as a source of either complement components (Stecher and Thorbecke, 1967) or chemotactic substances (Unanue, 1976) which will attract other mononuclear phagocytes to a specific site. This interaction of mononuclear phagocytes may be particularly important if one accepts the notion that the microenvironment may determine the nature of the immune response to a given stimulus.

It was observed many years ago that polymorphonuclear neutrophils predominate in the acute response to injury and that monocyte-derived macrophages replace these polymorphonuclear neutrophils after a brief interval. In modern terms, this is a classic example of leukocytes releasing chemotactic substances which in turn modulate the role of other RES and leukocyte fractions. Presumably, eosinophilic neutrophils are also able to release mononuclear phagocyte chemotactic substances. Macrophages do not appear to be a source of eosinophil chemotactic factors (Czarnetski *et al.*, 1976).

Under physiological conditions there is no evidence reported that the basophil and the mast cell interact with mononuclear phagocytes. However, in the presence of degranulated mast cells, with consequential associated histamine release, there are always abundant mononuclear phagocytes which are the primary cells responsible for the removal of tissue debris (D. Lawson, personal communication).

4. INTERACTION OF RES CELLS AND T LYMPHOCYTES

4.1. INTRODUCTION

In all species studied, T-cell responses depend on RES cells functioning as accessory cells. This is most clearly demonstrated *in vitro*, but there are also examples *in vivo*, such as the delayed hypersensitivity response (Miller *et al.*, 1975). Antigen-induced proliferation (e.g., Rosenthal and Shevach, 1973), helper cell induction (Erb and Feldmann, 1975a), proliferation in response to mitogens such as Con A (Habu and Raff, 1977), the mixed lymphocyte reaction (MLR) (e.g., Greineder and Rosenthal, 1975), the induction of killer cells (Wagner *et al.*, 1972), and the release of lymphokines (Rocklin *et al.*, 1980) all require the participation of accessory cells. This participation may vary in both quantitative and qualitative aspects, and the mechanism of accessory cell–T cell interaction with soluble antigens is possibly different from that seen with cell-bound antigens as in the MLR or killer cell induction. It is unclear whether T-cell responses which have not yet been shown to require macrophages, such as suppressor cell induction (Feldmann and Kontiainen, 1976), may merely require far fewer, or perhaps qualitatively different (nonadherent?) macrophage-like cells.

Although it was originally thought that the accessory cell played a "maintenance" role in bolstering culture conditions (Alter and Bach, 1970), it has now become clear that it plays a central and critical role in immune induction, and in the following sections we discuss the systems which established their key position (reviewed by Feldmann *et al.*, 1979).

4.2. PROLIFERATIVE RESPONSES

It has been recognized for many years that when antigen is injected into the footpad of animals, a primed T-cell population can be isolated which proliferates when restimulated *in vitro* with the priming antigen. Such proliferation systems were first shown in the guinea pig.

Subsequent work indicated that the cell which proliferated was a T lymphocyte and that the system was absolutely dependent on an accessory cell whose function was the processing of antigen and its presentation to the T cells. The vast majority of these accessory cells had the characteristics of macrophages.

Analysis of the antibody response to certain synthetic antigens had previously demonstrated that the response was controlled by a gene or genes mapping to the *I* region of the major histocompatibility complex. The T proliferation assay was also found to be under the control of the same genes [immune response (Ir) genes]: if the strain of animals was not able to make an antibody response to a particular antigen, this was reflected by a lack of proliferation in the T-cell proliferation. Since the latter required only an accessory cell and T cells, it was apparent that this interaction was crucial to T-cell induction.

A classic series of experiments using high-responder × low-responder guinea pigs by Rosenthal and Shevach (1973) demonstrated that the accessory cell not only expressed *I*-region products but also strongly suggested that the cell was also a critical site of expression of the Ir gene products.

Schwartz *et al.* (1975) have developed a murine antigen-specific proliferation assay, based on the guinea pig work discussed above, also using peritoneal exudate lymphocytes. Subsequently, proliferative assays using lymph node cells have been described (e.g., Corradin *et al.*, 1977; Rosenwasser and Rosenthal, 1978).

The use of these assays in the mouse has corroborated and extended the results obtained using guinea pigs. A precise concordance between Ir gene control of proliferation and of antibody production has been found (Schwartz and Paul, 1976) which is far more precise than the concordance with helper cell induction *in vitro* or *in vivo*. Anti-Ia antisera inhibit T-cell proliferation. Using peritoneal cells, only the antibodies directed against the Ia antigens of the high-responder parent are inhibitory (Schwartz *et al.*, 1978).

Genetic restrictions of accessory cell–T cell interactions have also been found. This does not appear to be due to a suppressive effect of the MLR (Yano *et al.*, 1977). Fine mapping using *H-2* recombinant mice indicated that *I-A* compatibility between was essential (Schwartz *et al.*, 1978), concordant with the results obtained with T helper cells (Erb and Feldmann, 1975a; Singer *et al.*, 1978).

Antigens under the control of two complementing genes mapping to the *I* region have also been analyzed with the T-cell proliferation assay, and these studies have suggested that both Ir gene products must be expressed on the same antigen-presenting cell (Schwartz *et al.*, 1978).

4.3. T HELPER CELLS

The production of antibody to many antigens ("thymus dependent") requires the participation of not only B cells but also T helper cells and RES accessory cells. The requirement for RES cells in the B-cell response phase is discussed in Section 5.

In vivo systems for helper cell induction have not been analyzed for their RES cell requirement, although this could be attempted, using techniques for macrophage depletion *in vivo* such as silica or carrageenan. *In vitro*, helper cells may be induced by low doses of antigen in Marbrook-type flasks (Kontiainen and Feldmann, 1973). The process is notably dependent on the presence of adherent cells: passage of spleen, lymph node, or cortisone-resistant thymus cells through glass or polystyrene beads, nylon wool, or carbonyl iron abrogates their capacity to yield helper cells (Erb and Feldmann, 1975b). Restoration of the helper cell response with adherent, radioresistant, phagocytic peritoneal exudate cells indicates that the adherent cells involved are indeed macrophages, although it is by no means clear whether all "macrophages" can participate in this way. We are also currently testing whether the nonphagocytic dendritic cell may play a role in helper T-cell induction.

Accessory cell–T cell interaction for helper cell induction in mice was found to be restricted in the same way as in the proliferative response. The genes responsible were located in the *I* region of the *H-2* complex (Erb and Feldmann, 1975a). No helper cells were induced with macrophages differing at the *H-2* complex, regardless of the ratio of T cells or macrophages, antigen concentration, or time of assay. The genetic region which is involved in this interaction was defined by using congenic resistant lines of mice, and mice bearing recombinations within the *H-2* complex. For two antigens, KLH and (T,G)-A–L, this region was the *I-A* subregion of the *H-2* complex. It is not yet clear whether this would be true for all antigens, for example those which are also under control of Ir genes mapping in the *I-C* region.

4.4. DENDRITIC CELL–LYMPHOCYTE INTERACTION

The unique interaction between T cells and dendritic cells has been mentioned previously. Our recent experience with isolated purified dendritic cells *in vitro* indicates that this is a special type of RES–lymphocyte relationship which raises interesting questions about the mechanisms of the early induction phase of the immune response.

Dendritic cells (Fig. 3) constitute less than 1% of the spleen/lymph node cell population. They are nonphagocytic, FcR^-, Ia^+ cells which lack Mac-1 and

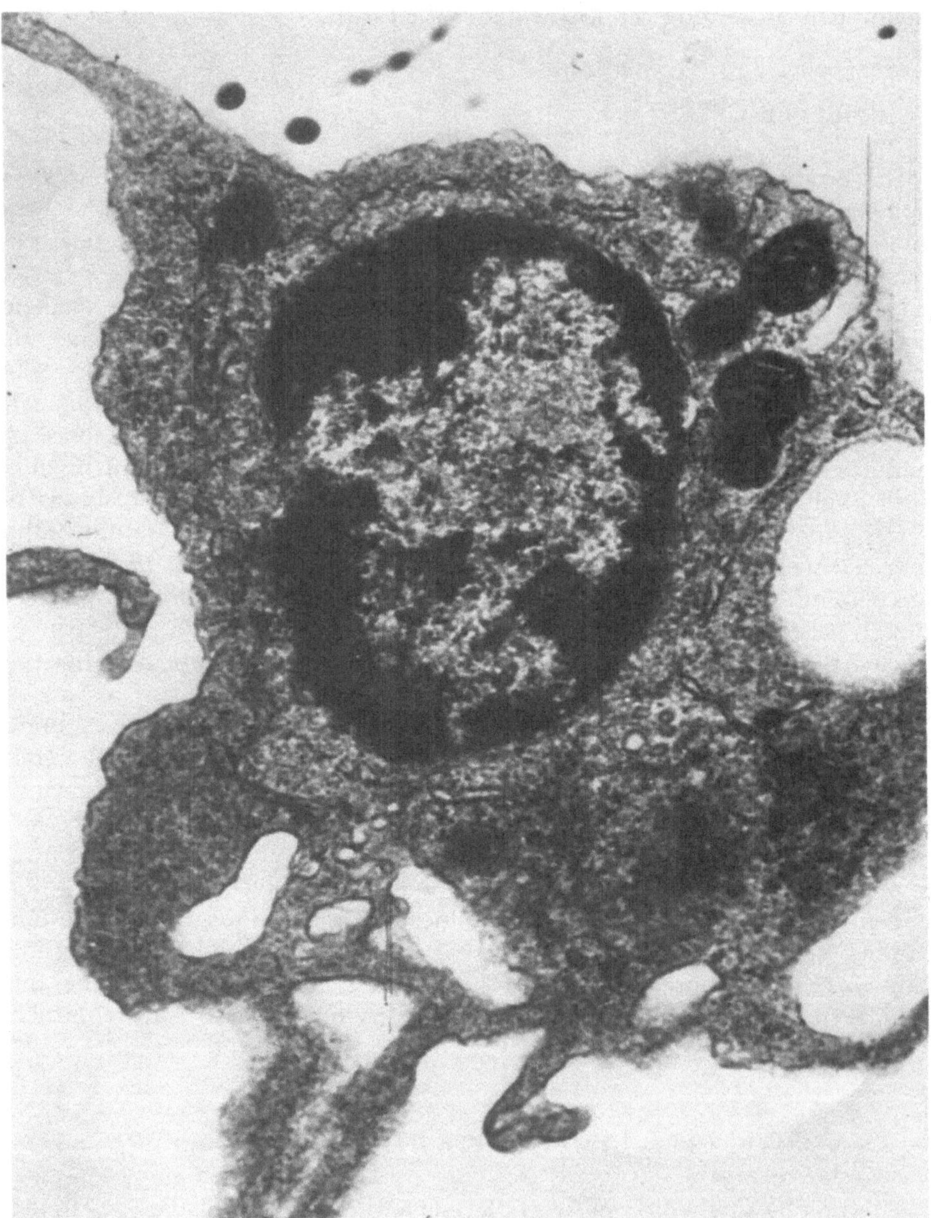

FIGURE 3. Dendritic cell showing scanty organelles and long extended processes without surface ruffles. 17,500 ×, reproduced at 80%.

F4/80 (suggested macrophage-specific) antigens defined by monoclonal antibodies (Steinman and Nussenzweig, 1980). They are the most efficient alloantigen presenters when compared with other accessory cell types; they also consistently evoke more T-cell proliferation in a syngeneic system than other accessory cells ("syngeneic" MLR). This syngeneic response could be due to retention of antigen on the cell surface from either murine or medium environment. Taken in conjunction with the allogeneic response results, however, a plausible hypothesis could also be that the dendritic cell membrane has a particular configuration or releases mediators which enable it to stimulate T cells.

Whatever the mechanism, on our evidence it must be linked to Ir gene control of the immune response since anti-Ia antisera block the interaction (Katz and Sunshine, unpublished observation). In this context, it is interesting that dendritic cells are the only FcR^- cells thus far identified which express *I*-region-encoded antigen in high concentration. Thus, our *in vitro* system raises the possibility that optimum antigen presentation requires *I*-region expression without FcR.

The role of dendritic cells in soluble antigen presentation to primary T cells has been described (Sunshine *et al.*, 1980). At present, we are investigating the relative effects of *in vitro* dendritic cell pulsing with radiolabeled antigens on T-cell responses. In view of the widespread use of peritoneal exudate cells as starter macrophage populations in this type of response, it is noteworthy that the unstimulated peritoneal exudate population is predominantly Ia^-, FcR^+, and phagocytic. This suggests at least two different mechanisms as the basis for RES–lymphocyte interaction (see below).

A further aspect of these interaction studies involves the functional identification of the T-cell subsets produced in the proliferative response. Dendritic cells undoubtedly generate alloreactive cytotoxic T cells. Their role in helper cell induction will be discussed in a later volume of this series. It is possible that they may also activate suppressor T cells as dendritic cells stimulate allogenic T-cell proliferation across the *I-J* subregion, which is believed to control the expression of products expressed on the surface of at least some T suppressor cells. It is not clear, however, whether the dendritic cell expresses I-J on its surface, as fluorescence studies have been inconclusive.

Our dendritic cell–T cell studies imply that RES–lymphocyte responses both *in vivo* and *in vitro* involve several related but distinct cell types. The importance of the lymphoid microenvironment in which these cells operate *in vivo* is also crucial to these studies and therefore unique cell surface markers for RES cells are a high priority in our further analysis of these data.

4.5. DEVELOPMENT OF THE T REPERTOIRE

In the last 5 years, chimeric mice have been used extensively to study the question of how an individual discriminates self from nonself.

The crucial role of host cells from chimeras in the development of the T-cell repertoire was first established by Zinkernagel (1978), though precisely which RES cells are involved is still moot. His experiments on T killer cells and subsequent work (Zinkernagel *et al.*, 1978; Fink and Bevan, 1978) pointed to the host thymus, and indeed to its radioresistant component, as a critical differentiation site since the capacity of T cells to recognize certain antigens depends on the haplotype of the thymus in which they develop.

The clearest experiments are those involving nude F_1 hosts grafted with irradiated parental thymus lobes. T cells from such mice recognized cells of the thymus donor, but not those of the other parent (Zinkernagel *et al.*, 1978). These results obtained with T killer cells have been confirmed with delayed hypersensitivity cells by Miller *et al.* (1979). Similar experiments have been performed with helper cells by Sprent (1978), Erb *et al.* (1979), and others.

It is as yet unclear whether this "learning process" occurs at the level of the thymic epithelial cells or at the level of a thymic antigen-presenting cell of the macrophage type. It is interesting to note, however, that both of these RES cells have been described as expressing *I*-subregion products (Jenkinson *et al.*, 1980; Beller and Unanue, 1980). Given that the expression of H-2 antigens is not uniform across the thymus structure (Rouse *et al.*, 1979), perhaps the physical location of the thymic RES cells influences differentiation of distinct cell subsets.

Subsequent experiments have indicated that the role of the thymus is not absolute (Erb *et al.*, 1979) and that other factors may also be important. In $P \rightarrow F_1$ chimeras, T cells responded to antigen presented only on the appropriate parental macrophage. The other parental macrophage was inactive. If the *only* important interaction in the development of T-cell function was in the host thymus, which is of F_1 type, both parental-type macrophages should have been effective. However, the preference for the stem cell donor macrophage indicates that a donor component is important. If the $P \rightarrow F_1$ chimeras were injected with macrophages of the other parental type and primed with antigen, the resulting T cells could interact equally with both parental-type macrophages indicating that the peripheral macrophage pool is of importance in determinining the composition of the T-cell pool, presumably by selective activation, proliferation, and hence survival of T cells.

The mechanism of the dependence of T-cell repertoire on thymus haplotype is not known. The process can be envisaged as a "learning process," presumably by selective survival and proliferation of cells with the appropriate receptors, but may be due to the generation of suppressor cells for the opposite parental haplotype (Smith and Miller, 1980). Suppressor cells have not been detected in other systems (Zinkernagel, 1978; Hodes *et al.*, 1980).

Longo and Schwartz (1980) have suggested that the bone marrow stem cell pool gradually replaces the thymic cells responsible for the intrathymic education process. By reirradiating chimeric mice 2 months after the initial bone marrow transfer, they found that the restriction specificity detected 2 months later depended on the genotype of the initial marrow inoculum, and not of the thymus.

4.6. MECHANISMS OF RES–T LYMPHOCYTE INTERACTION

In Section 2, we described how RES cells are localized at different sites such as spleen, liver, and skin, as well as in the peritoneal cavity. In *in vitro* work these RES cells have been used interchangeably and the concept of heterogeneity in this pool of cells has been a question of site of origin rather than of difference in function. Before we can analyze mechanisms of interaction we have first to establish that this interchangeable use is justified.

T-cell proliferative responses to alloantigens were demonstrated originally using whole spleen populations as stimulators. More recent reports have analyzed adherent populations extracted by a range of procedures (Minami *et al.*, 1980); Cowing *et al.* (1978) suggested that the stimulator cell may be adherent but that this was transient. Our own analysis of this spleen adherent cell pool suggests that it is heterogeneous for both phagocytic function and *I*-region surface products. An analysis of mechanisms in this type of interaction cannot exclude interactions within the adherent cell pool itself, and also cannot exclude that minor subpopulations are more important than the major components.

The standard criteria for card-carrying membership of the mononuclear phagocyte system are adherence, radioresistance, and phagocytic capacity (van Furth *et al.*, 1972). While many of the cells used in *in vitro* RES–T lymphocyte interactions fulfill these criteria, we have already shown (see Section 4.4) that the dendritic cell is neither phagocytic nor strongly adherent. This illustrates that while regarding RES–T lymphocyte interaction as purely a macrophage–T lymphocyte phenomenon may be true in some aspects of the immune response (Section 4.3), in other aspects it is an oversimplified viewpoint.

The identification of the *I*-region products on RES cells and the role of these products in immunoregulation led to the postulate that these were the markers for the RES component in stimulation of T lymphocytes. Our findings with dendritic cells would support this contention and this *I*-region recognition is probably the simplest and best understood mechanism of RES–T cell interaction. Again, however, it cannot be the only mechanism operative. Dendritic cells may bind antigens but are not phagocytic; thus, they are unlikely to be efficient processors, and this precludes them from handling some types of antigen. Having stressed the role of *I*-region products as an activating signal, it is perhaps surprising that peritoneal exudate cells are effective in several *in vitro* systems, yet few of these cells express these antigens. Finally, this theory would have to account for the presence of *I*-region products on B cells which are the one immunocompetent cell type which does not appear to have T-stimulating capacity.

Thus, our analysis of mechanism must remain incomplete partly because we believe that the interchangeable use of different accessory populations in interaction with T cells is unjustified unless the phenotype of the accessory cells has been clarified and the cells purified as far as possible.

Certain principles of the mechanism of RES–T cell interaction can, however, be inferred from the available information. First, the purification of both Fc receptor-bearing phagocytic cells and Fc^- nonphagocytic cells suggests that two RES stages may operate. Thus, one cell, such as a peritoneal macrophage, may be an efficient processor. Another cell, such as a dendritic cell, may be an efficient presenter. Sometimes both cells are present (as in a spleen adherent population) and we can derive both T proliferation and T help with this RES combination. Sometimes one or the other is present, as in the pure dendritic cell preparation or the unstimulated peritoneal washout cell, and these populations induce only one or other type of response. A similar mechanism in thymic development of the T-cell repertoire has been suggested (Beller and Unanue, 1979).

Second, there is evidence that direct cell–cell contact is at least one form of RES–T cell interaction. Antigen-dependent clusters have been observed in both mouse (Mosier, 1969) and human (Cline and Swett, 1968) systems. Rosenthal and associates have observed complex direct cell interaction between macrophages and lymphocytes in both antigen-dependent and -independent systems (Lipsky and Rosenthal, 1973, 1975). We have seen this in both syngeneic and allogeneic responses (Fig. 4).

Third, the interactive event between the cells appears to be critically dependent on *I*-region products expressed on the RES cell. Inhibition studies with anti-Ia sera in proliferative responses, genetic restriction analysis using F_1 responder cells and parental stimulators, and identification of Ia antigens on the presenting cell in the responding clusters of alloresponsive T cells are all indices of this dependence.

Fourth, direct cell–cell contact need not be the only mechanism involved in RES–T lymphocyte interaction. Erb and Feldmann (1975b) showed that soluble supernatant products of antigen-pulsed peritoneal cells are active in inducing T

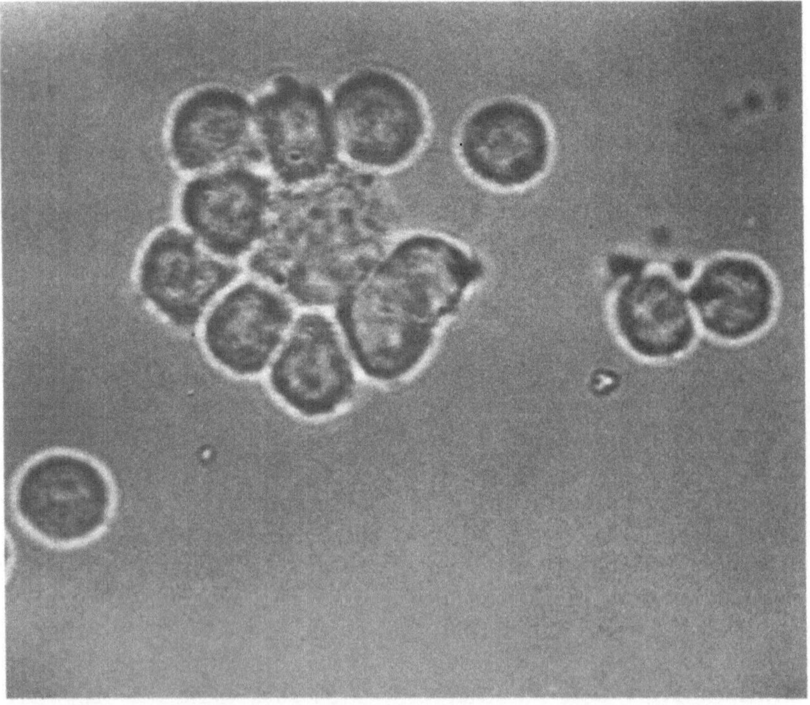

FIGURE 4. Cluster of cells in 4-day mixed leukocyte reaction showing a central dendritic cell surrounded by T cells.

helper cell differentiation. Two activities are found in these supernatants, an antigen-specific, genetically restricted factor (GRF) and a nonspecific factor (NMF) which induces helper cells in the presence of particulate antigens. Little is known about NMF although its relationship to IL-1 is likely. GRF has interesting features. It consists of two components, a fragment of immunogen and a material which binds to anti-Ia antisera. This has led to the postulate that GRF may represent the activating signal for T helper lymphocytes when it is shed from the surface of the accessory cell.

5. INTERACTION OF RES CELLS AND B LYMPHOCYTES

We have discussed the role of RES cells in the induction of the T-cell phase of the immune response. These cells also play a part in the induction and response of the B cell.

The B-cell induction process is itself conventionally separated into two types. The first group are the T-dependent responses where there is an obligatory T-cell requirement for antibody production. Here Mosier (1969) noted that adherent cell depletion depleted the response to sheep erythrocytes. Subsequent studies (Feldmann, 1972) showed that T-dependent antibody responses to soluble protein antigens were also accessory cell dependent and the nature of the system with a three-cell requirement makes it extremely difficult to clarify the precise stages involved.

Much of the work on B-cell–RES interaction has centered on so-called "T-independent" responses (Lee *et al.*, 1976; Boswell *et al.*, 1980). The study of these systems suggested at first that there were several antigens which acted directly on the B cell (Diener *et al.*, 1970). It is now becoming apparent that there is an accessory cell requirement in T-independent B-cell activation. Even the classic B-cell stimulant, lipopolysaccharide, required an accessory cell (Hammarstrom and Smith, 1978), but may need fewer such cells when compared to other antibody-inducing antigens. This is perhaps analogous to the difficulties seen in demonstrating accessory cell dependence in polyclonal T-cell responses to Con A.

The most recent viewpoint expressed is that the concept of T-cell independence is only a relative one (Mond *et al.*, 1980) and that most responses to T-independent antigens may be facilitated by inclusion of T cells within the system. This complicates further experiments with RES–B-cell responses, as we do not know whether there is a direct RES effect or whether in T-cell-depeleted samples the RES cell must act to augment the subminimal T component.

Assuming that RES cells act on B cells directly, there remain several uncertain aspects. Possibly there is a feeder effect similar to that seen in the maintenance of antigen-specific T-cell clones. Chiller (personal communication) has shown this type of feeder effect in this study of genetic restriction of B cell–macrophage interaction. Alternatively, the typical macrophage may release a factor which augments a B-cell response (e.g., Schrader, 1973). This factor may be IL-1. Macrophage monolayers are useful in permitting the growth of immature B cells (Metcalf *et al.*, 1975) and in the cloning of B-cell hybridoma lines.

Another uncertainty relates to the type of RES cell involved. The majority of these studies use either adherent cells or peritoneal cells as their macrophage source. These may not be the most relevant cell to RES–B cell interaction; for example, the follicular dendritic cell seems to be the most important activator of B memory cells to a range of antigens (Klaus *et al.*, 1980). The marginal zone macrophage has recently been shown to induce B-cell responses in our laboratory (Sunshine *et al.*, unpublished observation); however, it is not clear how precise this analysis is as the marginal zone cells are themselves still relatively impure. The mechanism of these types of RES–B-cell response is even less understood than the usual macrophage–B cell relationship.

From a different viewpoint again, antigen-specific helper factors (HF) as described by Feldmann and Basten (1972) do not induce antibody production unless there is an accessory cell present. This response is genetically restricted since only responder macrophages will stimulate F_1 (responder × nonresponder) phenotypically responder B cells (Howie and Feldmann, 1978). This demonstrated that even in the final stages of B-cell triggering where only the soluble T-cell mediator is required, an accessory RES cell remains an essential part of the B-cell induction process.

6. LYMPHOCYTE ACTIVATION OF RES CELLS

6.1. INTRODUCTION

The importance of macrophage-type cells in regulating the interaction of antigen with T lymphocytes has been extensively discussed in Section 4. It is apparent that the genetic specificity of the macrophage–T cell recognition governs the fate of the T lymphocyte: "fruitful" interaction results in the activation of the T cell. Once the T cell is activated, however, it can in its turn trigger antigen-specific and -nonspecific responses in nonimmune cells, and one of the most important effects of activated T lymphocytes is on the cells of the RES. These events have been observed both *in vivo* and *in vitro* and the characteristics of the responses are described in the following sections.

6.2. *IN VIVO* PHENOMENA

6.2.1. Antimicrobial Responses

The importance of lymphocyte–RES cell interactions has long been recognized in the control of antimicrobial immunity. Classical studies on tuberculosis demonstrated the bactericidal activity of phagocytes provided they came from immune animals (Lorie, 1942). More recent work demonstrated that lymphocytes cooperated with macrophages so that the latter could kill bacteria (Mackaness, 1969). It is now apparent that the effector arm of the responses to several viruses, intracellular bacteria, and fungal parasites involves the T lymphocyte-mediated recruitment and activation of blood monocytes and/or tissue macrophages (Blanden *et al.*, 1976). These requirements have also been shown to apply after infection by other bacteria such as *Brucella* and *Salmonella*, and viruses such as mouse pox and bebaru. Depletion of T cells by specific antisera destroys the ability of the RES population to combat the microbial infection (North, 1973).

The mechanism by which the T cells and macrophages interact *in vivo* is still incompletely understood but is believed to occur via the release of soluble mediators as *in vitro* (see below). Genetic restrictions may also govern this interaction of T cell and macrophage as it has been shown that in the mouse *H-2 I-A* subregion, identity is required between the interacting cells so that the infecting bacteria will be killed by the phagocytes (Farr *et al.*, 1979). Interestingly, in an *in vivo* graft-versus-host system, where *I-A* differences between host and donor trigger macrophage listericidal activity, disparity at the *I-J* subregion results in *reduced* macrophage killing (Zinkernagel, 1978). Since this subregion is believed to control the expression of products on suppressor T cells (Tada *et al.*, 1976; Murphy *et al.*, 1976), these findings suggest that T suppressor cells may exert a negative control over macrophage function.

6.2.2. Delayed-Type Hypersensitivity (DTH) Responses

When antigen is injected intradermally, normally in adjuvant, a complex series of cellular events occurs which culminates within 24 to 48 hr with the infiltration of mononuclear cells to the site of infection. In this infiltrate is a small population of sensitized T lymphocytes and a high proportion of recruited, nonspecifically activated macrophages. As in the case of antimicrobial immunity described above, these sensitized T lymphocytes alone can initiate DTH reactions when transferred into nonimmune animals, even though the predominant cell in the infiltrate is the macrophage (McCluskey *et al.*, 1963). This represents another example of T lymphocyte–RES cell interaction. Although the mechanisms involved in the induction of inflammatory responses remained obscure for several years, attempts to derive *in vitro* models for the DTH response ultimately had far-reaching consequences for the whole field of lymphocyte–macrophage interactions.

6.3. *IN VITRO* INTERACTIONS BETWEEN LYMPHOCYTES AND MACROPHAGES

Early attempts to derive an *in vitro* model for DTH had demonstrated that antigen inhibited the migration of cells from tissue explants of sensitized animals (Rich and Lewis, 1932). Several years later this technique was modified to study the migration of peritoneal exudate cells out of capillary tubes. It was found that a soluble, nonimmunoglobulin product replaced by activated T cells would inhibit the migration of macrophages out of the capillary tube and was thus named *migration inhibition factor* (MIF) (Bloom and Bennett, 1966; David, 1966).

The finding of this soluble mediator of T–macrophage interaction spawned a rash of *in vitro* studies in several systems which suggested that a plethora of soluble products of activated T cells could activate nonimmune cells, macrophages included (see Rocklin *et al.*, 1980, for a review). A fuller description of these mediators (named *lymphokines*) is given elsewhere in this treatise, and we will briefly describe only those which have direct relevance to the regulation of lymphocyte–macrophage interactions. Biochemical evidence on the nature of these mediators is still largely incomplete and the wide variety of lymphokines may be substantially reduced when techniques such as the production of monoclonal antibodies and the use of T hybridoma lines are effectively employed. The field has also been bedevilled by the use of bioassays to measure the small amounts of mediator released into culture supernatants which contain a vast number of soluble proteins.

Nonetheless, it is evident that soluble products from antigen- or mitogen-stimulated lymphocytes (B or T and indeed activated nonlymphoid cells) (Mackler *et al.*, 1974; Papageorgiou *et al.*, 1972) can activate macrophages, although it is not clear if one single lymphokine, MIF, is sufficient to arrest the movement *and* activate macrophages, or whether an additional activatory moiety [macrophage activation factor (MAF)] is also required (Nathan *et al.*, 1971). The function and structure of macrophages incubated in lymphokine-rich supernatant undergo changes associated with an activated state, such as increased spreading and the release of specific enzymes (Rocklin *et al.*, 1980). These functional and structural characteristics of the macrophage activated *in vitro* with MIF and those of macrophages involved in DTH responses are similar (Hammond *et al.*, 1975) and suggest that MIF plays a role *in vivo* in T–macrophage interaction. Further evidence for an *in vivo* role for soluble mediators has come from studies which show that explanted tissues produce MIF actively without further stimulation *in vitro*, and more compellingly that injected supernatants reproduce the pattern of DTH response which follows exposure to antigen (Pick *et al.*, 1979).

Another activity which has also been identified in the supernatants of activated T cells is one which is chemotactic for monocytes, that is, it enhances the directional movement of monocytes along an increasing concentration gradient of the chemotactic agent (Ward *et al.*, 1969). It may therefore be responsible for attracting monocytes into the site of an inflammatory lesion. A recently described factor has also been claimed to specifically attract Ia^+ macrophages into the peritoneum (Scher *et al.*, 1980).

Among the host of lymphokines released by T cells, two further activities must be noted which have profound effects on macrophages and the immune system. The first has been described as an "arming" factor and it activates macrophages to destroy tumor cells (which serve as a source of antigenic stimulation for the T cells) (Evans and Alexander, 1972) although the specificity of this activation is unclear (macrophage activity has also been reported to cause these changes) (Piessens *et al.*, 1975).

The second is interferon, which has reported antiviral and antitumor activities. It can enhance the phagocytic ability of monocytic cells (Donohue and Huang, 1973) and has also been reported to retard the differentiation of human monocytes to macrophages *in vitro* (Epstein, 1979). Interferon has also been described to be immunosuppressive in many responses which require macrophages (Gresser, 1977), but it is doubtful if it exerts its physiological effects exclusively through the macrophage-monocyte series of cells.

It is interesting to note that most, if not all, lymphokines generated *n vitro* by T cells require the presence of accessory cells, in common with most T-cell functions (Rocklin *et al.*, 1980). The nature of the T cell which produces the activated product is not completely resolved; different T cells may in fact produce different mediators. We may perhaps also speculate that given our understanding of the increasing complexity of the accessory cell populations, different accessory cells may be involved in activating the different T cells.

Two other types of interaction of soluble T-cell products with macrophage-monocyte cells are worthy of note. The first is the production by activated T cells of an activity known as colony-stimulating factor (CSF) which is necessary for the growth and differentiation of myeloid bone marrow precursors *in vitro* (Ruscetti and Chervenick, 1975). In the presence of the material, committed stem cells undergo several rounds of division and can differentiate along the granulocytic or monocytic pathways to mature states. The twin hematopoietic functions of this material appear to be the product of a single molecular species since CSF produced by a T hybridoma line induces both activities. It is not clear whether this *in vitro* interaction reflects an *in vivo* situation whereby T-cell products induce the differentiation of stem cells of different lineages. Nonetheless, an intriguing observation which suggests that such a connection may indeed occur *in vivo* has been described: cells from neonatally thymectomized mice have a reduced or deficient capacity to present antigens *in vitro* (Michael Feldman, personal communication). Whether this is the result of the endocrine function of the thymus or a feedback mechanism from thymocytes themselves remains an open question, but may have important ramifications for the future study of T–macrophage interaction.

7. CONCLUSION

The interaction of RES cells and leukocytes involves so many different types of interactions (since both cell types are highly heterogeneous) that no overall conclusions can be reached that encompass the spectrum of phenomena. However, it is worth pointing out that this is a major cellular crossroad—an interaction of relatively fixed tissue cells (RES) admittedly with migratory and mobile forms, with highly motile cells (leukocytes). It is to be expected that their functions are influenced and regulated by each other, and the full extent of this regulation has not yet been fully explored or defined.

The most important form of RES–leukocyte interaction is in the initiation of the immune responses. We have proposed elsewhere that all immune induction processes share this basic property; lymphocytes, phylogenetically more recent than phagocytic RES cells, have learnt to interact with antigen which has previously reacted with RES cells. It is an easy step to envisage that lymphocytes learn to recognize their other regulatory signals after they have also bound to "presenting cells." Lymphocyte traffic may be viewed as a process of scanning the appropriate "postboxes" (antigen-presenting cells of the RES) for the signals that would activate them. Homing of lymphocytes would be a reflection of this basic process. In order to regulate more efficiently the process of lymphocyte activation, each lymphocyte class has its own "postbox," and we propose that each set of lymphocytes has a different type of presenting cell.

Yet lymphocyte activation is not the only major event of the RES–leukocyte interaction. The RES has many functions, and many of these are under the control of leukocytes (Section 6). Regrettably, the permutations of these interactions cannot be fully explored until the heterogeneity of accessory cells of the RES is more fully understood. With improvements in technology—better culture conditions, selective growth factors, fusion methods, monoclonal antibodies—the prospects for unraveling the heterogeneity of the RES are indeed promising.

REFERENCES

Alter, B. J., and Bach, F. H., 1970, Lymphocyte reactivity *in vitro.* I. Cellular reconstitution of purified lymphocyte response, *Cell. Immunol.* **1:**207.

Beller, D. K., and Unanue, E. R., 1979, Evidence that thymocytes require at least two distinct signals to proliferate, *J. Immunol.* **123:**2890.

Beller, D. I., and Unanue, E. R., 1980, IA antigens and antigen presenting function of thymic macrophges, *J. Immunol.* **124:**1433.

Blanden, R. V., Hapel, A. J., Doherty, P. C., and Zinkernagel, R. M., 1976, Lymphocyte–macrophage interactions and macrophage activation in the expression of antimicrobial immunity *in vivo,* in: *Immunobiology of the Macrophage* (D. S. Nelson, ed.), pp. 367–400, Academic Press, New York.

Bloom, B. R., and Bennett, B., 1966, Mechanism of a reaction *in vitro* associated with delayed-type hypersensitivity, *Science* **153:**80.

Boswell, H. S., Ahmed, A., Scher, I., and Singer, A., 1980, Role of accessory cells in B cell activation. II. The interaction of B cells with accessory cells results in the exclusive activation of an Lyb5$^+$ B cell subpopulation, *J. Immunol.* **125:**1340.

Cline, M. J., and Swett, V. C., 1968, The interaction of human monocytes and lymphocytes, *J. Exp. Med.* **128:**1309.

Corradin, G., Etlinger, H. M., and Chiller, J. M., 1977, Lymphocyte specificity to protein antigens. I. Characterisation of the antigen-induced *in vitro* T cell dependent proliferative response with lymph node cells from primed mice, *J. Immunol.* **119:**1048.

Cowing, C., Schwartz, B. D., and Dickler, H. B., 1978, Macrophage Ia antigens. I. Macrophage populations differ in their expression of Ia antigens, *J. Immunol.* **120:**378.

Czarnetski, B., Konig, W., and Lichtenstestin, L. M., 1976, Eosinophil chemotactic factor. I. Release from polymorphonuclear leucocytes by the calcium A23187 ionophore, *J. Immunol.* **117:**229.

David, J. R., 1966, Delayed hypersensitivity *in vitro:* Its mediation by cell-free substances formed by lymphoid cell–antigen interaction, *Proc. Natl. Acad. Sci. USA* **56:**72.

Diener, E., Shortman, K., and Russell, P., 1970, Induction of immunity and tolerance *in vitro* in the absence of phagocytic cells, *Nature (London)* **225:**731.

Donohue, R. M., and Huang, K. Y., 1973, Neutralization of the phagocytosis-enhancing activity of interferon preparation by anti-interferon serum, *Infect. Immun.* **7:**501.

Epstein, L. B., 1979. The comparative biology of immune and classical interferons, *in: Biology of the Lymphokines* (S. Cohen, E. Pick, and J. J. Oppenheim, eds.), pp. 443–514, Academic Press, New York.

Erb, P., and Feldmann, M., 1975a, The role of macrophages in the generation of T-helper cells. II. The genetic control of the macrophage–T-cell interaction for helper cell induction with soluble antigen, *J. Exp. Med.* **142**:460.

Erb, P., and Feldmann, M., 1975b, The role of macrophages in the generation of T helper cells. III. Influence of macrophage derived factors in helper cell induction, *Eur. J. Immunol.* **5**:754.

Erb, P., Maier, B., Matsunaga, T., and Feldmann, M., 1979, Nature of T-cell–macrophage interaction in helper cell induction *in vitro*. II. Two stages of T helper cell differentiation analysed in irradiation and allophenic chimeras, *J. Exp. Med.* **149**:686.

Evans, R., and Alexander, P., 1972, Mechanism of immunologically specific killing of tumour cells by macrophages, *Nature (London)* **236**:168.

Farr, A. G., Kiely, J. M., and Unanue, E. R., 1979, Induction of cytocidal macrophages after *in vitro* interactions between *Listeria*-immune T cells and macrophage—Role of H-2, *J. Immunol.* **122**:2405.

Feldmann, M., 1972, Cell interaction in the immune response *in vitro*. II. The requirement for macrophages in lymphoid cell collaboration, *J. Exp. Med.* **135**:1049.

Feldmann, M., and Basten, A., 1972, Cell interactions in the immune response *in vitro*. III. Specific collaboration across a cell impermeable membrane, *J. Exp. Med.* **136**:49.

Feldmann, M., and Kontiainen, S., 1976, Suppressor cell induction *in vitro*. II. Cellular requirements of suppressor cell induction, *Eur. J. Immunol.* **6**:302.

Feldmann, M.. Rosenthal, A. S., and Erb, P., 1979, Macrophage–lymphocyte interaction in immune induction, *Int. Rev. Cytol.* **60**:149.

Fink, P. J., and Bevan, M. J., 1978, H-2 antigens of the thymus determine lymphocyte specificity, *J. Exp. Med.* **148**:766.

Greene, M. I., Sy, M. S., Kripke, M., and Benacerraf, B., 1979, Impairment of antigen-presenting cell function by ultra violet radiation, *Proc. Natl. Acad. Sci. USA* **76**:6591.

Greineder, D. K., and Rosenthal, A. S., 1975, Macrophage activation of allogeneic lymphocyte proliferation in the guinea pig mixed leukocyte culture, *J. Immunol.* **114**:154.

Gresser, I., 1977, On the varied biologic effects of interferon, *Cell. Immunol.* **34**:406.

Habu, S., and Raff, M. C., 1977, Accessory cell dependence of lectin-induced proliferation of mouse T lymphocytes, *Eur. J. Immunol.* **7**:451.

Hammarstrom, L., and Smith, C. I., 1978, Lanatoside C, a new polyclonal B cell activator, *Cell. Immunol.* **30**:377.

Hammond, M. F., Selvaggio, S. S., and Dvorak, H. F., 1975, Antigen enhanced glucosamine incorporation by peritoneal macrophages in cell-mediated hypersensitivity. I. Studies on biology and mechanism. *J. Immunol.* **115**:914.

Hodes, R. J., Hathcock, K. S., and Singer, A., 1980, Cellular and genetic control of antibody responses. VII. Absence of detectable suppression maintaining the H-2 restricted recognition of F_1–parent helper T cells, *J. Immunol.* **124**:134.

Howie, S., and Feldmann, M., 1978, Immune response (Ir) genes supressed at macrophage-β lymphocyte intereactions, *Nature* **273**:664–666.

Humphrey, J. H., and Grennan, D., 1981, Different macrophage populations distinguished by means of fluorescent polysaccharides: Recognition and properties of marginal zone macrophages, *Eur. J. Immunol.* **11**:221.

Jenkinson, E. J., Owen, J. J. T., and Aspinall, R., 1980, Lymphocyte differentiation and major histocompatibility complex antigen expression in the embryonic thymus, *Nature (London)* **284**:177.

Katz, D. R., Feldmann, M., Sunshine, G. H., and P. Erb, 1984, Role of reticuloendothelial system in T helper cell induction, in: *The Reticuloendothelial System: A Comprehensive Treatise*, Volume 6: *Immunology* (J. A. Bellanti and H. B. Herscowitz, eds.), pp. 123–140, Plenum Press, New York.

Klaus, G. G. B., Humphrey, J. H., Kunkl, A., and Dongworth, D. W., 1980, The follicular dendritic cell: Its role in antigen presentation in the generation of immunological memory, *Immunol. Rev.* **53**:3.

Kontiainen, S., and Feldmann, M., 1973, Induction of specific helper cells *in vitro, Nature (London)* **245:**285.

Lee, K. C., Shiozawa, C., Shaw, A., and Diener, E., 1976, Requirement for accessory cells in the antibody response to T cell independent antigens *in vitro, Eur. J. Immunol.* **6:**63.

Lipsky, P., and Rosenthal, A. S., 1973, Macrophage–lymphocyte interactions. I. Characteristics of the antigen independent binding of guinea pig thymocytes and lymphocytes to syngeneic macrophages, *J. Exp. Med.* **138:**900.

Lipsky, P., and Rosenthal, A. S., 1975, Macrophage–lymphocyte interactions. II. Antigen mediated physical interactions between guinea pig lymphocytes and syngeneic macrophages, *J. Exp. Med.* **141:**138.

Longo, D. L., and Schwartz, R. H., 1980, T cell specificity for H-2 and Ir gene phenotype correlates with the phenotype of thymic antigen-presenting cells, *Nature (London)* **287:**44.

Lurie, M. B., 1942, Studies on the mechanism of immunity in tuberculosis: The fate of tubercle bacilli ingested by mononuclear phagocytes derived from normal and immunised animals, *J. Exp. Med.* **75:**247.

McCluskey, R. T., Benacerraf, B., and McCluskey, J. W., 1963, Studies on the specificity of the cellular infilitrate of delayed hypersensitivity reactions, *J. Immunol.* **90:**466.

Mackaness, G. B., 1969, The influence of immunologically committed lymphoid cells on macrophage activity *in vivo, J. Exp. Med.* **129:**973.

Mackler, B. F., Altman, L. C., Rosenstreich, D. L., and Oppenheim, J. J., 1974, Induction of lymphokine production by EAC and of blastogenesis by soluble mitogens during human B-cell activation, *Nature (London)* **249:**834.

Metcalf, D., Nossal, G. J. V., Warner, N. L., Miller, J. F. A. P., Mandel, T. F., Layton, J. E., and Gutman, G. A., 1975, Growth of B lymphocyte colonies *in vitro, J. Exp. Med* **142:**15.

Miller, J. F. A. P., Vadas, M. A., Whitelaw, A., and Gamble, J., 1975, H-2 gene complex restricts transfer of delayed type hypersensitivity in mice, *Proc. Natl. Acad. Sci. USA* **72:**5095.

Miller, J. F. A. P., Gamble, J., Mottram, P., and Smith, F. I., 1979, Influence of thymic genotype on acquisition of responsiveness in delayed-type hypersensitivity, *Scand. J. Immunol.* **9:**29.

Minami, M., Shreffler, D. C., and Cowing, C., 1980, Characterisation of the stimulator cells in the murine primary mixed leucocyte response, *J. Immunol.* **124:**1314.

Mond, J. J., Mongini, P. K. A., Sieckmann, D., and Paul, W. E., 1980, Role of T lymphocytes in the response to TNP–AECM-Ficoll, *J. Immunol.* **125:**1066.

Mosier, D. E., 1969, Cell interactions in the primary immune response *in vitro:* A requirement for specific cell clusters, *J. Exp. Med.* **129:**351.

Murphy, D. B., Herzenberg, L. A., Okumura, K., and McDevitt, H. D., 1976, A new I subregion (I-J) marked by a locus (Ia-4) controlling surface determinants on suppressor T lymphocytes, *J. Exp. Med.* **144:**699.

Nathan, C. F., Karnovsky, M. L., and David, J. R., 1971, Alterations of macrophage functions by mediators from lymphocytes, *J. Exp. Med.* **133:**1356.

North, R. J., 1973, Importance of thymus derived lymphocytes in cell mediated immunity to infection. *Cell. Immunol.* **11:**166.

Nossal, G. J. V., and Ada, G. L., 1971, *Antigens, Lymphoid Cells, and the Immune Response,* Academic Press, New York.

Papageorgiou, P. S., Henley, W. L., and Glade, P. R., 1972, Production and characterisation of migration inhibitory factor(s) (MIF) of established lymphoid and non-lymphoid cell lines, *J. Immunol.* **108:**494.

Pick, E., Cohen, S., and Oppenheim, J. J., 1979, The lymphokine concept, in: *Biology of the Lymphokines* (S. Cohen, E. Pick, and J. J. Oppenheim, eds.) pp. 1–12, Academic Press, New York.

Piessens, W. F., Churchill, W. H., and David, J. R., 1975, Macrophages activated *in vitro* with lymphocyte mediators kill neoplastic but not normal cells, *J. Immunol.* **114:**293.

Rich, A. R., and Lewis, M. R., 1932, The nature of allergy in tuberculosis as revealed by tissue culture studies, *Bull. Johns Hopkins Hosp.* **50:**115.

Richman, L. R., Strober, W., and Berzofsky, J. A., 1980, Genetic control of the immune response to myogloblin. III. Determinant specific two Ir gene phenotype is regulated by the genotype of reconstituting Kupffer cells, *J. Immunol.* **124:**619.

Rocklin, R. E., Bendtzen, K., and Greineder, D., 1980, Mediators of immunity: Lymphokines and monokines, *Adv. Immunol.* **29**:55.

Rosenthal, A. S., and Shevach, E. M., 1973, Function of macrophages in antigen recognition by guinea pig T lymphocytes. I. Requirement for histocompatible macrophages and lymphocytes, *J. Exp. Med.* **138**:1194.

Rosenwasser, L. J., and Rosenthal, A. S., 1978, Adherent cell function in murine T lymphocyte recognition. I. Macrophage-dependent T cell proliferative assay in the mouse, *J. Immunol.* **120**:1991.

Rouse, R. V., van Ewijk, W., Jones, P. P., and Weissman, I. L., 1979, Expression of MHC antigens by mouse thymic dendritic cells, *J. Immunol.* **122**:2508.

Ruscetti, F. W., and Chervenick, P. A., 1975, Regulation of the release of colony-stimulating activity from mitogen-stimulated lymphocytes, *J. Immunol.* **114**:1513.

Scher, M. G., Beller, D. I., and Unanue, E. R., 1980, Demonstration of a soluble mediator that induces exudates rich in Ia-positive macrophages, *J. Exp. Med.* **152**:1684.

Schrader, J. W., 1973, The mechanism of activation of the bone marrow derived lymphocyte. III. A distinction between a macrophage derived triggering signal and the amplifying effect on triggered B lymphocytes of allogeneic interaction, *J. Exp. Med.* **138**:1466.

Schwartz, R. H., and Paul, W. E., 1976, T lymphocyte-enriched peritoneal exudate cells. II. Genetic control of antigen-induced T lymphocytes proliferation, *J. Exp. Med.* **143**:529.

Schwartz, R. H., Jackson, L., and Paul, W. E., 1975, T lymphocyte-enriched murine peritoneal exudate cells. I. A reliable assay for antigen-induced T lymphocyte proliferation, *J. Immunol.* **115**:1330.

Schwartz, R. H., Yano, A., and Paul, W. E., 1978, Interaction between antigen presenting cells and primed T lymphocytes, *Immunol. Rev.* **40**:153.

Singer, A., Cowing, C., Hathcock, K. S., Dickler, H. B., and Hodes, R. J., 1978, Cellular and genetic control of antibody responses *in vitro*. II. Immune response gene regulation of accessory cell function, *J. Exp. Med.* **147**:1611.

Smith, F. I., and Miller, J. F. A. P., 1980, Suppression of T cell specific for the non-thymic parental H-2 haplotype in thymus grafted chimeras, *J. Exp. Med.* **151**:246.

Sprent, J., 1978, Restricted helper function of F_1–parent bone marrow chimeras controlled by K and of H-2 complex, *J. Exp. Med.* **147**:1838.

Stecher, V. J., and Thorbecke, G. J., 1967, Sites of synthesis of serum proteins. I., Serum proteins produced by macrophages *in vitro*, *J. Immunol.* **99**:643.

Steinman, R. M., and Cohn, Z. A., 1973, Identification of a novel cell type in the peripheral lymphoid organs of mice, *J. Exp. Med.* **137**:1142.

Steinman, R. M., and Nussenzweig, M. C., 1980, Dendritic cells: Features and functions, *Immunol. Rev.* **53**:127.

Stingl, G., Tamaki, K., and Katz, S. I., 1980, Origin and function of epidermal Langerhans cells, *Immunol. Rev.* **53**:149.

Sunshine, G. H., Katz, D. R., and Feldmann, M., 1980, Dendritic cells induce T cell proliferation to synthetic antigens under Ir gene control, *J. Exp. Med.* **152**:1817.

Tada, T., Taniguchi, M., and David, C. S., 1976, Properties of the antigen-specific suppressive T-cell factor in the regulation of antibody response of the mouse. IV. Special subregion assignment of the genes that code for the suppressive T-cell factor in the H-2 histocompatibility complex, *J. Exp. Med.* **144**:713.

Unanue, E. R., 1976, Secretory function of mononuclear phagocytes, *Am. J. Pathol.* **83**:415.

van Furth, R., Cohn, Z. A., Hirsch, J. G., Humphrey, J. G., Spector, W. G., and Langervoort, H. L., 1972, The mononuclear phagocyte system: A new classification of macrophages, monocytes and their precursor cells, *Bull. W.H.O.* **46**:845.

van Furth, R., Disselhoff chen Dulk, M. M. C., and Mattie, H., 1973, Quantitative study on the production and kinetics of mononuclear phagocytes during an acute inflammatory reaction, *J. Exp. Med.* **138**:1314.

Wagner, H., Feldmann, M., Boyle, W., and Schrader, J. W., 1972, Cell mediated immune responses *in vitro*. III. The requirement for macrophages in cytotoxic reactions against cell bound and subcellular alloantigens, *J. Exp. Med.* **136**:33.

Waldron, J. A., Horn, R. G., and Rosenthal, A. S., 1974, Antigen-induced proliferation of guinea pig lymphocytes *in vitro:* Functional aspects of antigen handling by macrophages, *J. Immunol.* **112:**746.

Ward, P. A., Renold, H. G., and David, J. R., 1969, Leukotactic factor produced by sensitized lymphocytes, *Science* **163:**1079.

Yano, A., Schwartz, R. H., and Paul, W. E., 1977, Antigen presentation in the murine T-lymphocyte proliferative response. I. Requirement for genetic identity at the major histocompatibility complex, *J. Exp. Med.* **146:**828.

Zinkernagel, R. M., 1978, Thymus and lymphohemopoietic cells: Their role in T cell maturation in selection of T cells' H-2 restriction-specificity and in H-2 linked Ir gene control, *Immunol. Rev.* **42:**224.

Zinkernagel, R. M., Callahan, T. N., Althage, A., Cooper, S., Streilein, J. W., and Klein, J., 1978, The lymphoreticular system in triggering virus plus self-specific cytotoxic T cells: Evidence for T cell help, *J. Exp. Med.* **147:**897.

Regulatory Interactions with Blood Metabolites and Constituents

12

Glucose Regulation and the RES

JAMES P. FILKINS

. . . our conception [is] of a system of reticuloendothelial cells endowed with the faculty of exercising a multiplicity of functions [which] points strongly to its participation in the general metabolic functions.
—ASCHOFF, 1924

1. INTRODUCTION

Ludwig Aschoff is duly recognized for introduction of a unifying terminology for the confederation of body macrophages—i.e., his *reticuloendothelial system*. Less well known is that he also drew attention to numerous physiological facets of the RES which impacted on the general metabolic functions of the body and thus he offered an alternate unifying term for the collection of local and wandering histiocytes—the *histiocytic metabolic apparatus* (Aschoff, 1924). Over the ensuing years the notion of a direct role for macrophages in body metabolism has faded along with the term *histiocytic metabolic apparatus*. Only in recent years has expanding knowledge of macrophage physiology and especially their secretion of regulatory molecules or monokines with metabolic capabilities, once again drawn attention to metabolic interrelationships which involve the RES.

This chapter will update one aspect of the function of the "histiocytic metabolic apparatus," namely, the status of knowledge regarding the RES in glucoregulation. Since the primary focus of investigation for a linkage between the

JAMES P. FILKINS • Department of Physiology, Stritch School of Medicine, Loyola University of Chicago, Maywood, Illinois 60153.

RES and carbohydrate regulation has been in host energy metabolism during endotoxicosis, the first section will summarize the effects of endotoxin on glucose homeostasis, with special attention to insulin regulation. Second, the effects of perturbations of the RES by foreign particulates and colloids on glucose and insulin regulation will be reviewed. Third, since current research indicates that RES involvement in metabolism involves endocytic mediator production, the role of macrophage secretory products with major effects on glucose metabolism, i.e., "glucoregulatory monokines," will be reviewed. Lastly, an integration of the interactions of RES endocytic function to exocytic function with glucose and insulin regulation will be presented.

2. GLUCOREGULATORY ALTERATIONS IN ENDOTOXICOSIS

2.1. PHASES OF BLOOD GLUCOSE RESPONSES

Menten and Manning (1923) are generally credited with the primary experimental description of blood glucose changes in endotoxicosis. They reported an early hyperglycemic response and an ensuing severe hypoglycemia in rabbits given i.v. heat-killed organisms of the enteritidis–paratyphoid B group. Zeckwer and Goodell (1925a,b) subsequently confirmed the biphasic response of blood glucose levels in rabbits challenged with a variety of heat-killed gram-negative bacteria. Kun and Miller (1948) using crude meningococcal or *Salmonella* endotoxin in rabbits also reported a transient hyperglycemia even in fasted rabbits, which then progressed to profound hypoglycemia with blood glucose levels as low as 10 mg/100 ml.

Over the ensuing years an extensive literature has corroborated that disturbances in blood glucose regulation, i.e., glucose dyshomeostasis, are among the most characteristic and consistently observed alterations in endotoxicosis. Indeed the pattern of early hyperglycemia—a metabolic feature of compensated metabolism—which progresses to late, profound hypoglycemia—a metabolic hallmark of decompensated metabolism—has been documented in a wide range of species—chicks, mice, rabbits, rats, guinea pigs, cats, cows, ponies, horses, baboons, monkeys, and man (Agarwal, 1975; Berry, 1971; Filkins, 1979a; Hinshaw, 1976).

2.2. CHANGES IN INPUTS AND OUTPUTS OF THE GLUCOSE POOL

In overview, blood glucose levels are simply the net result of regulated inputs to vis-à-vis outputs from the circulating glucose pool. Inputs are of three primary types: (1) alimentation—either via the intestinal route or parenterally as in the clinical setting; (2) glycogen mobilization—this results in direct glucose release in the liver and in lactate release via glycolysis in muscle; and (3) gluconeogenesis—the provision of glucose due to metabolic conversion in the liver of amino acids, lactate, and glycerol. Outputs from the circulating pool are also of three primary types: (1) renal excretion—this occurs if the renal tubular transport maximum for glucose is exceeded or if renal damage occurs; (2) synthetic outflow—this refers to the use of glucose for the synthesis of glycogen, lipid, or protein; and (3) peripheral utilization—this encompasses both aerobic oxidation to CO_2 and H_2O as well as anaerobic production of lactate.

Endotoxicosis affects inputs to the glucose pool in many ways. Alimentation is decreased due to loss of appetite, an impairment of gastrointestinal motor function, and restriction of the splanchnic circulation (Berry, 1971; Fine, 1965). Massive glycogen depletion, especially in the liver, and reduction of total body carbohydrate content are constant findings in endotoxin shock (Berry, 1971, 1975). Both enhanced glycogenolysis (Hamosh and Shapiro, 1960; Zwadyk and Snyder, 1973) and reduced glycogen synthesis (Giger and McCallum, 1976; McCallum and Berry, 1973) have been implicated. Lastly, failure of gluconeogenesis is recognized as a major determinant of endotoxic hypoglycemia (Agarwal, 1975; Agarwal and Lazar, 1977; Berry, 1971, 1975, 1977; Filkins and Cornell, 1974; Filkins, 1978).

Endotoxicosis also affects outputs from the glucose pool. While neither renal excretion nor enhanced synthetic outflows are involved to a major extent, peripheral utilization and hypercatabolism of glucose are dominant factors in endotoxic hypoglycemia (Hinshaw, 1976; Holtzman *et al.*, 1974; Leach and Spitzer, 1981; Raymond *et al.*, 1981; Romanosky *et al.*, 1980).

2.3. INSULIN CHANGES IN ENDOTOXICOSIS

Circulating levels of insulin also undergo biphasic changes during endotoxicosis and sepsis. Early and/or low-dose endotoxicosis generally result in a transient hyperinsulinemia (Adeleye *et al.*, 1981; Archer *et al.*, 1978; Blackard *et al.*, 1976; Buchanan and Filkins, 1976a; Cornell, 1981; Kober *et al.*, 1981; Rayfield *et al.*, 1977; Spitzer *et al.*, 1976). The hyperinsulinemia accompanies the early hyperglycemic phase of endotoxicosis but it is generally disproportionately large to the normal glycemic stimulus (Spitzer *et al.*, 1976; Yelich and Filkins, 1980a). The early hyperinsulinemia is blunted when the hyperglycemic response to endotoxin is altered by α-adrenergic blockade with phentolamine (Filkins, 1982). In addition, enhanced autonomic sympathetic activity may negate the insulin elevation (Adeleye *et al.*, 1981; Manny *et al.*, 1977).

Hepatic uptake and degradation of insulin are not appreciably altered in endotoxicosis; in contrast, the endotoxic pancreas as evaluated in the isolated perfused model hypersecretes insulin in response to glucose, leucine, arginine, or Ca^{2+} (Goldfarb *et al.*, 1981; Yelich and Filkins, 1980a,b; Yelich and Filkins, 1981).

The late or hypoglycemic phase of endotoxin shock is generally associated with hypoinsulinemia due to lack of a glycemic stimulus and accompanying enhanced sympathetic inhibition (Adeleye *et al.*, 1981; Filkins, 1982; Leach and Spitzer, 1981). Thus, while glucoregulatory alterations are suggestive of hyperinsulinism i.e., depressed gluconeogenesis and enhanced peripheral glucose use, the circulating pancreatic insulin levels are depressed. Since nonsuppressible insulin-like activity is elevated in the later stages of endotoxicosis, it may underwrite the metabolic manifestations of hyperinsulinism (Filkins, 1981a,b; Filkins and Viray, 1982).

3. GLUCOREGULATORY ALTERATIONS AFTER RES PERTURBATIONS

It is somewhat surprising that since a great deal of evidence had linked the RES to a host of endotoxic reactions, a role for the RES in the endotoxic alterations in glucoregulation was initially overlooked. In the opinion of this author, two major types of studies converged on the concept that the alteration in glucose metabolism incident to endotoxin treatment involved altered RES function. These studies dealt with RES function in regard to hyposensitivity or tolerance to endotoxin on the one hand, and with hypersensitivity to endotoxin on the other.

In an extensive series of studies, Berry and his colleagues have evaluated the role of the RES in the interactions of corticosteroid action in the protection of carbohydrate homeostasis in endotoxin shock and tolerance (reviews, Berry, 1971, 1975, 1977). As will be discussed below, this branch of study contributed the concept that a macrophage secretory product modulates glucocorticoid action on hepatic gluconeogenesis. Conversely, Shands and his colleagues investigated the mechanism of the exaggerated hypoglycemic activity of endotoxin in the BCG-infected mouse (Shands *et al.*, 1969a,b, 1971; Senterfitt and Shands, 1978). They also concluded that the profound carbohydrate disturbances involved RES functional alterations. It was subsequently concluded that the marked sensitivity to endotoxin incident to lead salt treatment also pivoted around a RES linkup to glucose homeostasis (Filkins, 1979c). A relation between RES function and glucoregulation also arose from findings that hypoglycemia depressed RES phagocytic activity for both endotoxin and colloidal carbon (see Section 5.2). Thus, Buchanan and Filkins (1976c) reasoned that since the status of the glucoregulatory system affected RES function, could the status of the RES influence glucose homeostasis? Evidence for the latter relation is summarized below.

3.1. CHANGES IN INPUTS TO THE GLUCOSE POOL

Evaluation of RES function *in vivo* has always been hampered by a lack of adequate experimental models. At best the investigator can either depress the RES using colloid-overload or "blockade" models or stimulate the RES using a variety of microbiological products. The classic blockade model employs the colloidal carbon technique of Biozzi *et al.* (1953) to saturate the vascular lining macrophages with an endocytized ink preparation. Using this model of RES depression, Filkins and Buchanan (1977) initially explored for gross deficits in glucoregulation by evaluating changes in insulin tolerance and glucose tolerance. Carbon blockade resulted in (1) enhanced sensitivity to lethal insulin hypoglycemia, i.e., decreased insulin tolerance, and (2) an enhancement of glucose disappearance half-time, i.e., an increased glucose tolerance. Similar changes in glucose and insulin tolerance were obtained after lead-salt depression of the RES (Filkins, 1977). Cornell (1980) also reported decreased glucose disappearance times after a variety of RES perturbations. In addition, depression of the RES plus minute doses of endotoxin sensitized rats to insulin lethality and hypoglycemia (Filkins, 1977).

The status of the two major hepatic inputs to the glucose pool—glycogenolysis and gluconeogenesis—was evaluated after carbon blockade and lead-salt depression of the RES by Filkins (1979a). While glycogenolysis as reflected in depletion of glycogen levels was enhanced, gluconeogenesis was depressed. In these studies gluconeogenesis was evaluated *in vivo* using the quantitative analysis of [^{14}C]alanine conversion to [^{14}C]glucose as the index. In addition, isolated parenchymal cells from RES-depressed liver donors also manifested depressed gluconeogenic conversion of alanine, pyruvate, or lactate to glucose (Filkins and Yelich, 1980).

3.2. CHANGES IN OUTPUTS FROM THE GLUCOSE POOL

Both colloidal carbon blockade and lead-salt depression of the RES were also found to cause increased glucose oxidation. *In vivo* conversion of [^{14}C]glucose to $^{14}CO_2$ as evaluated in a rat metabolimeter was elevated twofold. In addition, the isolated epididymal fat pads from RES-depressed rats manifested enhanced oxidation of [^{14}C]glucose to $^{14}CO_2$ (Filkins and Buchanan, 1977).

3.3. INSULIN CHANGES

The changes in glucose input and glucose output after RES perturbations are consistent with hyperinsulinism—especially depressed gluconeogenesis and enhanced peripheral utilization. In initial studies serum insulin—both basal and glucose-stimulated immunoreactive insulin—levels were elevated after carbon blockade and lead treatment (Hadbavny *et al.*, 1978). Cornell (1980) also found elevation in serum insulin levels after perturbation of the RES with lipid emulsions, iron particles, and dead bacteria. Thus, in accord with studies in endotoxicosis and sepsis, RES perturbations by endocytized material resulted in elevations of circulating pancreatic insulin. Two mechanisms for the hyperinsulinism have been evaluated: (1) decreased hepatic uptake of insulin and (2) enhanced pancreatic secretion. While Brodal *et al.* (1971) and Cornell (1980) obtained evidence for a defect in insulin removal by the RES-depressed liver, Yelich and Filkins (1980a) were unable to confirm this observation. Rather a marked increase in insulin secretion by the isolated perfused pancreas from either carbon blockade or lead-treated donor rats was observed (Filkins *et al.*, 1980; Filkins and Yelich, 1982). Thus, hyperinsulinism can account for the glucoregulatory alterations after perturbation of the RES and is primarily due to enhanced pancreatic insulin secretion. The next section will review the mechanisms by which alterations in macrophage function affects glucoregulatory and insulin-regulatory systems.

4. GLUCOREGULATORY MONOKINES

The most viable hypothesis for a mechanistic link between the RES and glucose homeostasis is that macrophages secrete mediators with potent effects on carbohydrate metabolism and pancreatic endocrine function. These mediators, which may be related and/or identical biomolecules, have been termed *glucoregulatory monokines*. The four major ones are discussed in sequence below in chronological order of their discovery.

4.1. LEUKOCYTIC ENDOGENOUS MEDIATOR (LEM)

As discussed in detail in Volume 7B of this treatise, LEM is a multipotent secretion of leukocytes which among its host of functions in the infectious process may also mediate metabolic interactions (Beisel, 1975; Powanda *et al.*, 1980). Crude preparations of LEM from rabbit polymorphonuclear leukocytes elevated both insulin and glucagon plasma levels in rats (George *et al.*, 1977). While hepatic glycogen was mobilized, plasma glucose concentrations were transiently depressed. It was concluded that LEM induces the release of pancreatic hormones via a neurologic stimulus, since intracerebroventricular injection of LEM also caused insulin and glucagon elevations. Thus, among the many aspects of LEM action is an effect on the pancreas which could in turn mediate a RES role in glucoregulation.

4.2. GLUCOCORTICOID-ANTAGONIZING FACTOR (GAF)

As discussed in detail in Volume 7B of this treatise, GAF is a factor which modulates the action of glucocorticoids on hepatic gluconeogenesis. In an extensive series of studies, Berry and his colleagues (Berry, 1971, 1975, 1977) have amassed considerable evidence that the RES secretes a monokine which alters the induction of the hepatic gluconeogenic enzymes by glucocorticoids. Berry has related endotoxic hypoglycemia to the antagonism of cortisone action by GAF—presumably overproduced due to the endotoxic stimulus to the RES. McCallum (1980, 1981) has reported that the Kupffer cell is the source of a mediator—either a monokine with GAF properties or a modified endotoxin—which directly depressed gluconeogenesis in isolated mouse parenchymal cell cultures. Agarwal *et al.* (1969) and Lowitt *et al.* (1981) also showed that macrophages modulate the induction of the hepatic enzyme trytophan oxygenase. Thus, good evidence exists to support the role of a glucoregulatory monokine which acts via modification of hepatic enzyme induction by glucocorticoids (Moore *et al.*, 1976).

4.3. MACROPHAGE INSULIN-LIKE ACTIVITY (MILA)

Since hyperinsulinism occurred incident to RES perturbation, Filkins (1979b) evaluated the hypothesis that macrophages release a factor(s) with insulin-like activity. Using rat peritoneal exudate macrophages and the standard fat pad assay for insulin-like activity (ILA), a mediator with ILA was demonstrated in three different protocols: (1) injection of macrophage culture supernatants *in vivo* and removal of the fat pads for ILA assays; (2) treatment of the fat pads *in vitro* with macrophage culture supernatants; and (3) direct coincubation of the macrophages with the fat pads. The acronym MILA for macrophage insulin-like activity was proposed. Filkins (1980) also demonstrated that endotoxin increased secretion of MILA either (1) when the endotoxin was added directly to coincubations of macrophages and fat pads or (2) when endotoxin was added to macrophage cultures and the supernatant was subsequently assayed for ILA. Since ILA is elevated in terminal endotoxic hypoglycemia and septic shock, a role for MILA in carbohydrate dyshomeostasis is suggested (Filkins and Viray, 1982). It is probable that MILA is related to the growth factor(s) recently reported by DeLustro *et al.* (1980) and Glenn and Ross (1981). Thus, MILA may represent a local mediator of cell growth as well as subserving an endocrine-like role on glucose homeostasis.

4.4. MACROPHAGE INSULIN-RELEASING ACTIVITY (MIRA)

In order to explore the role of the RES in postendocytic hyperinsulinemia, i.e., elevation of circulating pancreatic immunoreactive insulin, Filkins and Yelich (1982) evaluated the effects of carbon blockade on both the hepatic removal of insulin from the circulating pool and the pancreatic secretion of insulin. In accord with past studies on the mechanism of hyperinsulinemia in endotoxicosis, colloidal carbon depression of the RES did not alter hepatic insulin removal while it did result in marked enhancement of pancreatic insulin secretion.

To provide a mechanistic link between the RES and the endocrine pancreas, the hypothesis that macrophages produce a factor which alters insulin release in the perfused pancreas was evaluated. Similar to results with MILA assays, pancreases from rats treated with macrophage culture supernatant manifested enhanced insulin secretion to a provocative glucose stimulus. The acronym MIRA for macrophage insulin-releasing activity was proposed for the monokine. The relation of MIRA to LEM has not been evaluated; however, the similarity of effects suggests that indeed macrophages of the RES may subserve an endocrine-like regulatory role on the pancreatic islets, especially in the metabolic sequelae of endotoxicosis and sepsis.

5. GLUCOSE REGULATION AND RES ENDOCYTIC FUNCTIONS

Up to this point, primary emphasis has been with how the RES affects glucoregulation. This section will deal with the converse relation, namely how altered glucoregulation (hypoglycemia) affects RES function. The effects of hyperglycemia on phagocyte function have recently been summarized (Gross and Newberne, 1980).

5.1. INTRAVASCULAR CLEARANCE DEFECTS DURING HYPOGLYCEMIA

Interest in the effects of hypoglycemia per se on intravascular phagocytosis by the RES arose from attempts to separate the diverse effects of endotoxicosis. Endotoxin is well known to cause hypoglycemia as well as to depress intravascular phagocytosis. The singular effect of hypoglycemia on RES clearance of colloidal carbon and endotoxin was evaluated by Buchanan and Filkins (1976c). There was a significant negative correlation of carbon clearance half-time to blood glucose levels—both in fed and fasted control rats as well as in rats rendered hypoglycemic to varying degrees with insulin. The maximal depression occurred at blood glucose values below 30 mg/dl. In addition, hypoglycemia severely impaired the intravascular removal of endotoxin as evaluated by lethality bioassay in lead-sensitized rats (Buchanan and Filkins, 1976c). The effects of hypoglycemia were not immediate and probably reflects the need to deplete the considerable glycogen stores of the macrophage (Gudewicz and Filkins, 1974, 1976). Further studies of the effects of hypoglycemia on RES function in the perfused liver indicated a direct effect of hypoglycemia on carbon phagocytosis; no effect of flow alterations or opsonic deficiences were indicated (Kober and Filkins, 1981; Filkins and Kober, 1981). The ability of glucose to augment the RES clearance of *E. coli* in a septicemic model has been linked to the efficacy of glucose therapy to ameliorate gram-negative septicemia and endotoxemia (Heggers *et al.*, 1976; Hinshaw *et al.*, 1974).

5.2. UNIFYING SCHEMA

An attempt to summarize the interactions of RES endocytic and exocytic functions with glucoregulation is depicted in Fig. 1. The stimulus, be it endotoxin or inert colloids, is endocytized by the macrophages of the RES. Subsequent to the endocytic act, exocytosis of a variety of glucoregulatory monokines occurs. These include GAF, MILA, LEM, and MIRA. The sum total of the effect of the release of these glucoregulatory monokines is to induce a tendency to hyperinsulinism. The hyperinsulinic state in turn mediates relative depressions in hepatic inputs to the plasma glucose pool coupled to relative increases in outputs from the peripheral glucose pool. The net tendency therefore is a reduction in the circulating plasma glucose pool, i.e., hypoglycemia. Hypoglycemia in

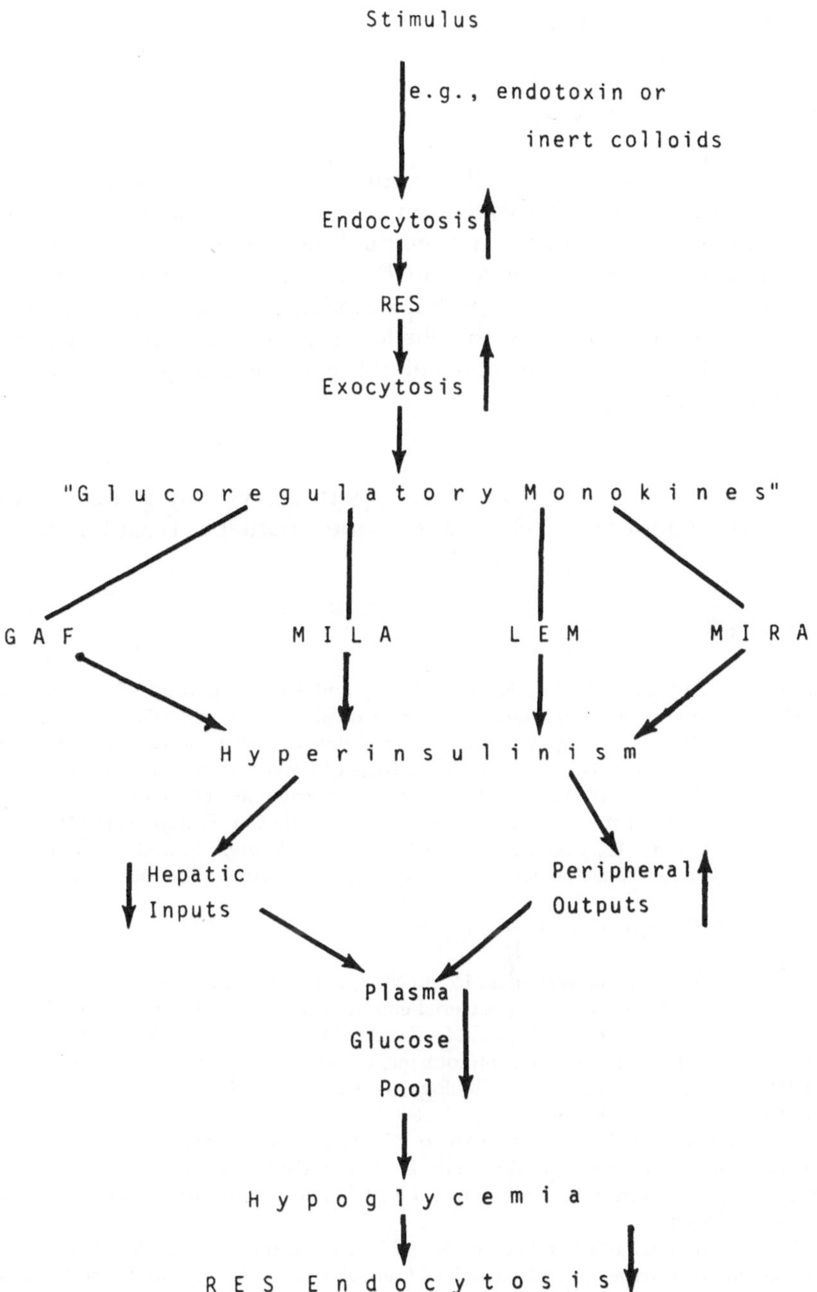

FIGURE 1. Schema of RES endocytic and exocytic functions in glucoregulation.

turn can result in defects in RES endocytosis which forms a feedback signal to remove the stimulus for the production of the exocytic release of the glucoregulatory monokines.

6. CONCLUDING REMARKS

This review attempted to reflect with accuracy the current state of knowledge regarding the role of the RES in one aspect of metabolic physiology—glucose regulation. In addition, it attempted to integrate and extend with some reasonable speculation to a role for the RES in insulin and glucose homeostasis. It is hoped that further studies, perhaps motivated by the shortcomings of this review, will continue to accept the challenge of Aschoff which introduced this chapter, namely, to pursue the concept of the RES as a major participant in host metabolic functions.

ACKNOWLEDGMENTS. This work was supported in part by United States Public Health Service Grant HL-08682 and the Bane Charitable Trust Fund.

REFERENCES

Adeleye, G. A., Al-Jibouri, L. M., Furman, B. L., and Parratt, J. R., 1981, Endotoxin-induced metabolic changes in the conscious, unrestrained rat, *Circ. Shock* **8**:543.

Agarwal, M. K., 1975, An integrative analysis of endotoxic reactions, *Naturwissenschaften* **62**:167.

Agarwal, M. K., and Lazar, G., 1977, Metabolic basis of endotoxicosis, *Microbios* **20**:183.

Agarwal, M. K., Hoffman, W. W., and Rosen, F., 1969, The effect of endotoxin and thorotrast on inducible enzymes in the isolated, perfused rat liver, *Biochim. Biophys. Acta* **177**:250.

Archer, L. T., White, G. L., Coalson, J. J., Beller, B. K., Elmore, O., and Hinshaw, L. B., 1978, Preserved liver function and leukocyte response in superlethal endotoxic shock, *Circ. Shock* **5**:279.

Aschoff, L., 1924, The reticulo-endothelial system, in: *Lectures on Pathology*, pp. 1–33, Hoeber, New York.

Beisel, W. R., 1975, Metabolic responses to infection, *Annu. Rev. Med.* **26**:9.

Berry, L. J., 1971, Metabolic effects of bacterial endotoxins, in: *Bacterial Toxins*, Vol. 5 (S. Kadis, G. Weinbaum, and S. J. Ajl, eds.), pp. 165–208, Academic Press, New York and London.

Berry, L. J., 1975, Metabolic effects of endotoxins, in: *Microbiology 1975* (D. Schlessinger, ed.), pp. 315–325, American Society of Microbiology, Washington, D.C.

Berry, L. J., 1977, Bacterial toxins, *CRC Crit. Rev. Toxicol.* 239.

Biozzi, G. B., Benacerraf, B., and Halpern, B. N., 1953, Quantitative study of the granulopectic activity of the reticuloendothelial system, *Br. J. Exp. Pathol.* **34**:441.

Blackard, W. G., Anderson, J. H., and Spitzer, J. J., 1976, Hyperinsulinism in endotoxin shock dogs, *Metabolism* **25**:676.

Brodal, B. P., Assev, S., and Eeg-Larsen, N., 1971, Degradation studies of insulin in the isolated perfused rat liver, with and without blocked reticuloendothelial system, *Horm. Metab. Res.* **3**:217.

Buchanan, B. J., and Filkins, J. P., 1976a, Insulin secretion and sensitization to endotoxin shock, *Circ. Shock* **3**: 223.

Buchanan, B. J., and Filkins, J. P., 1976b, Insulin secretion and the carbohydrate metabolic alterations of endotoxemia, *Circ. Shock* **3**:267.

Buchanan, B. J., and Filkins, J. P., 1976c, Hypoglycemic depression of RES function, *Am. J. Physiol.* **231**:265.

Cornell, R. P., 1980, Mechanisms of acute hyperinsulinemia after Kupffer cell phagocytosis, *Am. J. Physiol.* **238**:E276.

Cornell, R. P., 1981, Basal insulin and glucagon responses to portal versus systemic venous administration of endotoxin in fasted rats, *Physiologist* **24**:117.

DeLustro, F., Sherer, G. K., and LeRoy, E. C., 1980, Human monocyte stimulation of fibroblast growth by a soluble mediator(s), *J. Reticuloendothelial Soc.* **6**:519.

Filkins, J. P., 1977, Depression of RES function and sensitization to endotoxin shock and hypoglycemia, *J. Reticuloendothelial Soc.* **20**:461.

Filkins, J. P., 1978, Phases of glucose dyshomeostasis in endotoxicosis, *Circ. Shock* **5**:347.

Filkins, J. P., 1979a, RES function and glucose homeostasis, in: *Macrophages and Lymphocytes: Nature, Functions, and Interaction,* Part A (M. R. Escobar and H. Friedman, eds.), pp. 391–402, Plenum Press, New York and London.

Filkins, J. P., 1979b, Insulin-like activity (ILA) of a macrophage mediator on adipose tissue glucose oxidation, *J. Reticuloendothelial Soc.* **25**:595.

Filkins, J. P., 1979c, Role of the RES in lead salt sensitization to endotoxin shock, in: *Macrophages and Lymphocytes: Nature, Functions, and Interaction,* Part A (M. R. Escobar and H. Friedman, eds.), pp. 21–27, Plenum Press, New York and London.

Filkins, J. P., 1980, Endotoxin-enhanced secretion of macrophage insulin-like activity, *J. Reticuloendothelial Soc.* **27**:507.

Filkins, J. P., 1981a, The reticuloendothelial system and metabolic homeostasis, in: *Pathophysiology of the Reticuloendothelial System* (B. M. Altura and T. M. Saba, eds.), pp. 93–110, Raven Press, New York.

Filkins, J. P., 1981b, Serum immunoreactive (IRI) and non-suppressible insulin-like activity (NSILA) in endotoxin shock, *Physiologist* **23**:37.

Filkins, J. P., 1982, Effect of phentolamine on insulin levels and insulin responsiveness in endotoxicosis, *Adv. Shock Res.* **7**:185.

Filkins, J. P., and Buchanan, B. J., 1977, Depression of RES function and altered glucoregulation, *J. Reticuloendothelial Soc.* **21**:389.

Filkins, J. P., and Cornell, R. P., 1974, Depression of hepatic gluconeogenesis and the hypoglycemia of endotoxin shock, *Am. J. Physiol.* **227**:778.

Filkins, J. P., and Kober, P. M., 1981, Interrelations of RES endocytic and exocytic functions with glucose homeostasis in the pathogenesis of endotoxin shock, in: *Homeostasis in Injury and Shock* (A. G. B. Kovach, H. B. Stoner and J. J. Spitzer, eds.), pp. 171–180, Pergamon Press, Elmford Park, New York.

Filkins, J. P., and Viray, R. E., 1982, Comparison of immunoreactive (IRI) and non-suppressible insulin-like activity (NSILA) in endotoxic versus septic shock, *Fed. Proc.* **41**:1606.

Filkins, J. P., and Yelich, M. R., 1980, RES function and glucoregulation in endotoxicosis, in: *The Reticuloendothelial System and the Pathogenesis of Liver Disease* (H. Liehr and M. Grün, eds.), pp. 89–98, Elsevier/North Holland Biomedical Press, Amsterdam. glucoregulation

Filkins, J. P., and Yelich, M. R., 1982, Mechanisms of hyperinsulinemia after RES phagocytosis, *Am. J. Physiol.* **242**:E115.

Filkins, J. P., Janusek, L. W., and Yelich, M. R., 1980, The role of insulin and insulin-like activity in the hypoglycemic response to endotoxin, in: *Bacterial Endotoxins and Host Response* (M. K. Agarwal, ed.), pp. 361–379, Elsevier/North Holland Biomedical Press, Amsterdam.

Fine, J., 1965, Shock and peripheral circulatory insufficiency, in: *Handbook of Physiology, Circulation,* Vol. III, pp. 2037–2069, American Physiological Society, Washington, D.C.

George, D. T., Abeles, F. B., Mapes, C. A., Sobocinski, P. Z., Zenser, T. V., and Powanda, M. C., 1977, Effect of leukocytic endogenous mediators on endocrine pancreas secretory responses, *Am. J. Physiol.* **233**:E240.

Giger, O., and McCallum, R. E., 1976, Response of mouse liver glycogen cycle enzymes to endotoxin treatment, *Am. J. Physiol.* **231**:1285.

Glenn, K. C., and Ross, R., 1981, Human monocyte-derived growth factor(s) for mesenchymal cells: Activation of secretion by endotoxin and concanavalin A, *Cell* **25**:603.

Goldfarb, R. D., Weber, P., and Eisenman, J., 1981, The effect of leucine on plasma insulin following endotoxin shock in the rat, *Circ. Shock* **8**:343.

Gross, R. L., and Newberne, P. M., 1980, Role of nutrition in immunologic function, *Physiol. Rev.* **60**:188.

Gudewicz, P. W., and Filkins, J. P., 1974, Glycogen metabolism in macrophages, *J. Reticuloendothelial Soc.* **16**:1.

Gudewicz, P. W., and Filkins, J. P., 1976, Glycogen metabolism in inflammatory macrophages, *J. Reticuloendothelial Soc.* **19**:147.

Hadbavny, A. M., Buchanan, B. J., and Filkins, J. P., 1978, Insulin and the glucoregulatory alterations of RES depression, *J. Reticuloendothelial Soc.* **24**:57.

Hamosh, M., and Shapiro, B., 1960, The mechanism of glycogenolytic action of endotoxin, *Br. J. Exp. Pathol.* **41**:372.

Heggers, J. P., Robson, M. C., Jennings, P. B., and Fariss, B. L., 1976, Effects of glucose therapy on experimental *Escherichia coli* septicemia, *J. Surg. Res.* **20**:33.

Hinshaw, L. B., 1976, Concise review: The role of glucose in endotoxic shock, *Circ. Shock* **3**:1.

Hinshaw, L. B., Peyton, M. D., Archer, L. T., Black, M. B., Coalson, J. J., and Greenfield, L. J., 1974, Prevention of death in endotoxin shock by glucose administration, *Surg. Gynecol. Obstet.* **139**:851.

Holtzman, S., Schuler, J. J., Earnst, W., Erve, P. R., and Schumer, W., 1974, Carbohydrate metabolism in endotoxemia, *Circ. Shock* **1**:99.

Kober, P. M., and Filkins, J. P., 1981, Hypoglycemic depression of hepatic phagocytosis *in vivo* and in the *in situ* perfused rat liver, *Adv. Shock Res.* **5**:47.

Kober, P. M., Thomas, J. X., and Filkins, J. P., 1981, Correlation of hemodynamics to glucose and insulin changes in endotoxicosis, *Fed. Proc.* **40**:572.

Kun, E., and Miller, C. P., 1948, Effect of bacterial endotoxins on carbohydrate metabolism in rabbits, *Proc. Soc. Exp. Biol. Med.* **67**:221.

Leach, G. J., and Spitzer, J. A., 1981, Endotoxin-induced alterations in glucose transport in isolated adipocytes, *Biochim. Biophys. Acta* **643**:71.

Lowitt, S., Szentivanyi, A., and Williams, J. F., 1981, Endotoxin inhibition of dexamethasone induction of tryptophan oxygenase in suspension culture of isolated rat parenchymal cells, *Biochem. Pharmacol.* **30**:1999.

McCallum, R. E., 1980, Mediated inhibition of hepatic gluconeogenesis by endotoxin, in: *Microbiology 1980* (D. Schlessinger, ed.), pp. 87–90, American Society of Microbiology, Washington, D.C.

McCallum, R. E., 1981, Hepatocyte–Kupffer cell interactions in the inhibition of hepatic gluconeogenesis by bacterial endotoxin, in: *Pathophysiological Effects of Endotoxins at the Cellular Level* (J. A. Majde and R. J. Person, eds.), pp. 99–113, Alan R. Liss, Inc., New York.

McCallum, R. E., and Berry, L. J., 1973, Effects of endotoxin on gluconeogenesis, glycogen synthesis, and liver glycogen synthase in mice, *Infect. Immun.* **7**:642.

Manny, J., Rabinovici, N., and Schiller, M., 1977, Insulin response to continuous glucose load in endotoxin shock in the dog, *Surg. Gynecol. Obstet.* **145**:198.

Menten, M. L., and Manning, H. M., 1923, Blood sugar studies on rabbits infected with organisms of the enteritidis-paratyphoid B group, *J. Med. Res.* **44**:675.

Moore, R. N., Goodrum, K. J., and Berry, L. J., 1976, Mediation of endotoxin effects by macrophages, *J. Reticuloendothelial Soc.* **19**:187.

Powanda, M. C., Bostian, K. A., Dinterman, R. E., Fee, W. G., Fowler, J. P., Hauer, E. C., and White, J. P., 1980, Phagocytosis and the metabolic sequelae of infection, *J. Reticuloendothelial Soc.* **27**:67.

Rayfield, E. J., Curnow, R. T., Reinhard, D., and Kochicheril, N. M., 1977, Effects of acute endotoxemia on glucoregulation in normal and diabetic subjects, *J. Clin. Endocrinol. Metab.* **45**:513.

Raymond, R. M., Harkema, J. M., and Emerson, T. E., Jr., 1981, Direct effects of gram-negative endotoxin on skeletal muscle glucose uptake, *Am. J. Physiol.* **240**:H342.

Romanosky, A. J., Bagby, B. J., Bockman, E. L., and Spitzer, J. J., 1980, Increased muscle glucose uptake and lactate release after endotoxin administration, *Am. J. Physiol.* **239**:E311.

Senterfitt, V. C., and Shands, J. W., Jr., 1978, Endotoxin induced metabolic alterations in BCG infected (hyperreactive) mice, *Proc. Soc. Exp. Biol. Med.* **159**:69.

Shands, J. W., Jr., Miller, V., and Marten, H., 1969a, The hypoglycemic activity of endotoxin. I. Occurrence in animals hyperreactive to endotoxin, *Proc. Soc. Exp. Biol. Med.* **130**:413.

Shands, J. W., Jr., Miller, V., and Senterfitt, V., 1969b, Hypoglycemic activity of endotoxin. II. Mechanism of the phenomenon in BCG-infected mice, *J. Bacteriol.* **98**:494.

Shands, J. W., Jr., Senterfitt, V. C., and Miller, V., 1971, Endotoxin-induced insulin sensitivity in BCG infected mice, *Proc. Soc. Exp. Biol. Med.* **136**:983.

Spitzer, J. J., Wagner, G. G., and Blackard, W. G., 1976, The effect of glucose infusion on selected hemodynamic variables and plasma insulin concentration in dogs after *E. coli* endotoxin administration, *Circ. Shock* **3**:31.

Yelich, M. R., and Filkins, J. P., 1980a, Mechanisms of hyperinsulinemia in the endotoxic rat, *Am. J. Physiol.* **239**:E156.

Yelich, M. R., and Filkins, J. P., 1980b, Effect of arginine on lethality and insulin secretion in endotoxicosis, *Circ. Shock* **7**:200.

Yelich, M. R., and Filkins, J. P., 1981, Role of Ca^{++} in the insulin hypersecretion of the endotoxic pancreas, *Circ. Shock* **8**:224.

Zeckwer, I. T., and Goodell, H., 1925a, Blood sugar studies. I. Rapid alterations in the blood sugar levels of rabbits as result of intravenous injections of killed bacteria of various types, *J. Exp. Med.* **42**:43.

Zeckwer, I. T., and Goodell, H., 1925b, Blood sugar studies. II. Blood sugar changes in fatal bacterial anaphylaxis in the rabbit, *J. Exp. Med.* **42**:57.

Zwadyk, P., Jr., and Snyder, I. S., 1973, Effects of endotoxin on hepatic glycogen metabolism *in vitro*, *Proc. Soc. Exp. Biol. Med.* **142**:299.

13

Interaction of the Reticuloendothelial System with Blood Lipid and Lipoprotein Metabolism

T. J. C. VAN BERKEL, P. H. E. GROOT, and A. VAN TOL

1. INTRODUCTION

The RES, as defined by Aschoff (1924), comprises a wide variety of cell types with different origins and functions (van Furth *et al.*, 1972). For example, the clearance of substances from the bloodstream is mainly exerted by macrophages of the liver (Kupffer cells) and macrophages of the spleen (Stiffel *et al.*, 1970) but not, or to a low extent, by endothelial or reticular cells. In this review the attention will be focused on lipid and lipoprotein metabolism by Kupffer cells and macrophages of the spleen. *In vivo* the fate of an intravenously injected radioactive compound can be followed by autoradiography as well as by subsequent isolation of the cells. For *in vitro* studies, however, the isolation of the aforementioned cell types in an unmodified form is absolutely necessary. As this is still rather difficult at present, such *in vitro* studies have only been done with isolated (peritoneal) macrophages. These studies gave information on the mechanism of interaction, and metabolic consequences of the interaction of lipids and lipoproteins with macrophages and are also relevant for the tissue macrophages from liver and spleen (Goldstein *et al.*, 1979). For this reason we first deal with information concerning the interaction of lipids with these free macrophages (Section 2) while Section 3 deals with the evidence obtained with tissue macrophages.

T. J. C. VAN BERKEL, P. H. E. GROOT, and A. VAN TOL • Department of Biochemistry I, Medical Faculty, Erasmus University Rotterdam, 3000 DR Rotterdam, The Netherlands.

2. INTERACTION OF LIPIDS WITH FREE MACROPHAGES

2.1. ARTIFICIAL LIPID SUBSTRATES

2.1.1. Test Substances

The use of test substances to study the mechanism of the interaction of particles with macrophages is widespread. The metabolic consequences of the interaction of a particle with phagocytic cells are manifold (Roos, 1977). The role of the lipid composition of the applied particles on the handling and subsequent degradation has been reviewed (Elsbach, 1973).

Mason *et al.* (1973) and Stossel *et al.* (1971, 1972) used droplets of paraffin oil colored with oil red and followed the ingestion of these particles by cells by measuring the absorbance at 524 nm. The uptake of these lipid particles is stimulated by prior incubation of the particles with serum and requires both Mg^{2+} and Ca^{2+} (about 1 mM) for maximal uptake. In the presence of 1 mM *N*-ethylmaleimide the uptake is abolished, indicating that thiol groups are involved in this process. The uptake is inhibited by colchicine indicating the involvement of microtubules. Studies by Mason *et al.* (1973) indicate that the complement system may be involved in the uptake of lipids. After internalization of the lipids, degradation may occur, after which the degradation products are used as constitutive components by the macrophages (Patriarca *et al.*, 1972). Studies with several microorganisms, using [^{14}C]fatty acid-labeled lipids, indicate that the extent of microbial lipid degradation depends on the accessibility of the lipids. Phospholipase A activity with an acid pH optimum was detected in macrophages (Franson *et al.*, 1973), which was, at least in part, associated with the lysosomes. It is suggested that the lysosomal phospholipases exert the lipid-splitting activity and that the accumulating fatty acids and lysophospholipids are incorporated into the phagosome membrane, before being lost to the extracellular environment (Patriarca *et al.*, 1972).

2.1.2. Liposomes

In recent years liposomes have gained recognition as promising carriers for enzyme replacement therapy and drug transport (Papahadjopoulos, 1978; Finkelstein and Weissmann, 1978; Gregoriadis, 1978). Unfortunately, most of the liposomes are cleared from the circulation by cells of the RES (Papahadjopoulos, 1978; Finkelstein and Weissmann, 1978; Gregoriadis, 1978; De Barsy *et al.*, 1976). Several attempts have been made to decrease this uptake, for example by applying empty liposomes to saturate the RES (Gregoriadis and Neerunjun, 1974) before the drug- or enzyme-containing liposomes are applied. Model studies with macrophages to study possible inhibition of liposomal uptake have only been done by Torchilin *et al.* (1980). They found that coating of liposomes with albumin or γ-globulin decreases their capture by macrophages. However, the effect was rather small (~ 20%) and further studies are necessary.

2.2. NATURAL LIPID SUBSTRATES

2.2.1. Blood Lipids

Cholesterol and triglyceride suspensions are readily taken up by macrophages. Electron microscopic studies have shown that emulsions of triglyceride are taken up by peritoneal macrophages in large phagocytic vesicles. Morphological evidence suggests that the triglyceride is then hydrolyzed in these vesicles and the resulting fatty acids are metabolized (Casley-Smith and Day, 1966; Day, 1967). Lipolytic enzymes, responsible for this metabolism, have been studied in homogenates of peritoneal macrophages, and both triglyceride lipase and cholesterolesterase activities have been detected. These enzymatic activities possess an acid pH optimum which suggests that they are mainly of lysosomal origin (Day, 1967; Koster *et al.*, 1980). The free fatty acids are partly incorporated to form phospholipids, triglycerides, and cholesterolesters (Day, 1967; Day and Fidge, 1962). The hydrolysis of cholesterolesters inside the lysosomes leads to an increase of cholesterol in the cytoplasm, where it is reesterified and stored as cholesterolester (Werb and Cohn, 1972). Werb and Cohn (1972) further suggested that the macrophages excrete their excess of cholesterol, with the aid of serum lipoproteins. As discussed in Section 2.2.2, more detailed studies of Ho *et al.* (1980) indeed indicate that such a mechanism is operative in macrophages.

2.2.2. Lipoproteins

2.2.2a. Introduction. Studies with cultured human fibroblasts have led to a model for the catabolism of lipoproteins by cells. This model includes binding of lipoprotein to membrane receptors, active endocytosis and lysosomal degradation of the proteins and cholesterolesters by the action of cathepsins and acid cholesterolesterase, respectively (Goldstein and Brown, 1977). As mentioned in the previous section, this is the same model as presented by Werb and Cohn (1972), where they followed the fate of cholesterol and cholesterolester complexes in macrophages. Recently, Brown, Goldstein, and co-workers also used peritoneal macrophages (Goldstein *et al.*, 1979, 1980; Brown *et al.*, 1979; Basu *et al.*, 1979) and confirmed the model described by Werb and Cohn. In addition, they showed that the cellular cholesterol derived from receptor-mediated uptake regulates a variety of cellular functions.

2.2.2b. Nature of the Recognition Site. The interaction of lipoproteins with fibroblasts involves the recognition of the lipoprotein by a specific membrane receptor (Goldstein and Brown, 1977). On fibroblasts this receptor recognize the B apolipoprotein which is the only apolipoprotein present on human low-density lipoprotein (LDL). Goldstein *et al.* (1979) showed that mouse peritoneal macrophages also interact with the native LDL particle, although to a much lower extent than fibroblasts. The nature of the binding by macrophages is similar to the binding by human fibroblasts. In addition to the native LDL particle, mouse peritoneal macrophages also interact with a modified LDL particle (acetyl-LDL). This particle is obtained by a chemical modification of the native LDL with acetic anhydride, which leads to acetylation, resulting in an increased negative charge (Basu *et al.*, 1976).

The surface binding site that recognizes acetyl-LDL is trypsin sensitive and possesses a high affinity for this modified LDL particle (half-maximal binding at an acetyl-LDL concentration of about 25 μg/ml). Native LDL shows no interaction with this binding site in contrast to maleylated LDL, maleylated albumin, and the polysaccharides fucoidin and dextran sulfate. These latter interactions indicate that negative charges are important in the binding mechanism. When human LDL is complexed with dextran sulfate, the uptake and degradation of LDL is also markedly stimulated. This uptake is not inhibited by fucoidin, suggesting (Basu *et al.*, 1979) that it is recognized by a surface binding site different from the site necessary for acetyl-LDL uptake. However, direct binding kinetics with both modified particles or direct competition experiments were not reported. The different inhibitory effects of fucoidin could also be exerted on the LDL itself and not necessarily on the membrane receptor. A fourth recognition site on mouse peritoneal macrophages is claimed for β-VLDL isolated from the plasma of cholesterol-fed dogs (Goldstein *et al.*, 1980). Half-maximal saturation of this recognition site is obtained at 40 μg apoprotein/ml. Again no interaction of this receptor site with human LDL was observed, while the addition of fucoidin had no effect on the binding of β-VLDL. Goldstein *et al.* (1980) therefore concluded that β-VLDL from cholesterol-fed dogs binds to a receptor which is different from the receptor for acetyl-LDL.

These results lead to the conclusion that mouse peritoneal macrophages contain membrane receptors, recognizing specifically modified lipoprotein particles, normally not circulating in mouse blood, in addition to a receptor for native LDL. However, direct competition experiments between β-VLDL from cholesterol-fed dogs, human acetyl-LDL, and dextran sulfate-treated LDL have not been performed and are necessary to draw a conclusion about the specificity of the receptor(s).

2.2.2c. Handling of Lipoprotein Constituents. Whatever the nature of the receptor(s) may be, it is clear that the protein moiety of acetyl-LDL is rapidly degraded in lysosomes of peritoneal macrophages and that this degradation is much higher than for native LDL. Also the protein moiety of the LDL–dextran sulfate complex is rapidly degraded as is the β-VLDL from cholesterol-fed dogs. It appears that all these lipoproteins follow the same lysosomal route. The rapid uptake of these lipoproteins can lead to a deposition of cholesterolesters in the macrophages. This deposition appears to be mainly in the form of cholesterolesters (Brown *et al.*, 1979). The mechanism of cholesterolester deposition involves a hydrolysis–reesterification mechanism with hydrolysis of the incoming cholesterolesters in the lysosomes and reesterification in the cytosol where the newly formed esters are stored as lipid droplets. These newly formed esters can be hydrolyzed again and released from the cells in the form of cholesterol (Brown *et al.*, 1979), a similar mechanism as has been proposed by Werb and Cohn (1972) (see 2.2.1). As acceptors for this cholesterol, HDL, whole serum, erythrocytes, casein, or thyroglobulin can be used (Ho *et al.*, 1980). This indicates that the *in* *vivo* deposition of cholesterolesters in the arterial wall or elsewhere, in cholesterol-fed dogs or rabbits can be caused either by an increased uptake (for example of β-VLDL) or by a decreased excretion as a consequence of the lack of an adequate extracellular receptor for cholesterol. According to Brown *et al.* (1980) the cholesterolesters, present in the cytoplasm of isolated peritoneal macrophages, undergo a continual cycle of hydrolysis and reesterification. Hydrolysis appears to be mediated by a nonlysosomal cholesterylester hydrolase because its activity is resistant to lysosomal inhibitors like chloroquine and ammonium chloride. Reesterification is mediated by an acyl-CoA:cholesterol acyltransferase. As the reesterification reaction uses acyl-CoA derivatives that require ATP for their synthesis, the continual hydrolysis and reesterification constitutes a futile cycle. It is, however, unclear how far this represents a physiological phenomenon and occurs also *in vivo*. Further data, correlating *in vivo* and *in vitro* behavior, are therefore necessary to identify the mechanism for cholesterolester deposition in macrophages.

3. INTERACTION OF LIPIDS WITH TISSUE MACROPHAGES

3.1. ARTIFICIAL LIPID SUBSTRATES

3.1.1. Test Substances

Although cells of the RES are located in many regions of the body, the major part consists of macrophages of liver, spleen, and bone marrow which are in direct contact with the vascular compartment. Their strategic anatomic localization provides them with the opportunity to "monitor" the blood and remove material by way of recognition by specific receptors. Foreign material may be coated by opsonic glycoproteins before recognition can be established. In addition to classical test substances for the RES like colloidal carbon, silver, gold, and particulate material, artificial lipid emulsions can be used which possess an intravascular behavior and organ distribution comparable to the classical test substances (DiLuzio and Riggi, 1964). The lipid test emulsion can be used in the evaluation of the RES both in experimental animals and in human subjects. This test emulsion consists of corn oil, glycerol, and phospholipid in the ratio 10 : 10 : 1 (DiLuzio and Riggi, 1964). The emulsion is labeled with [^{131}I]triolein. Following intravenous injection into rats, the emulsion is rapidly cleared from the circulation ($t_{\frac{1}{2}}$ about 13 min) and mainly recovered in the liver (about 80% of the injected dose). The $t_{\frac{1}{2}}$ is 10-fold decreased by stimulation of the RES with glucan, the active component in the bacterial wall constituent zymosan (Riggi and DiLuzio, 1961). After uptake, deiodination of the particle occurs (Cornell and Saba, 1975) and *in vitro* the degradation capacity is localized in the lysosomal fraction. Cornell and Saba (1975) concluded that the particle is taken up intact and subsequently degraded in the lysosomes.

The ^{131}I-labeled lipid test emulsion can also be used to evaluate RE function under a variety of conditions. For instance, Holper *et al.* (1974) used the lipid test emulsion to correlate RE function with the effect of hepatic ischemia on survival. They showed that impaired phagocytosis or other functional alterations of Kupffer cells play a significant role in the pathophysiology of acute hepatic failure since derangement of RE function was most indicative of the final outcome of hepatic ischemia.

The lipid test emulsion as developed by DiLuzio and Riggi (1964) has been used to evaluate the effect of different lipids on RE function. Already in 1960, Stuart and co-workers described the striking variety in sensitivity of the RES to different lipids. Particles made from olive oil and glycerol trioleate produce a marked stimulation of phagocytosis (up to 10-fold), whereas glycerol monooleate has no action and both ethyl oleate and ethyl stearate cause a severe depression (up to 6-fold). Also Conning and Heppleston (1966) noticed a depressed phagocytic activity for carbon in rats, previously injected with ethyl palmitate, whereas triolein increased the phagocytosis. Tricaprin had an effect which was closely similar to triolein. They noticed that the stimulating effect of triolein was not equally effective in all rats. Mouton *et al.* (1975) found that the phenotypic character "responsiveness of macrophages to triolein" presents large individual variations in a population of random-bred albino mice, and is submitted to polygenic regulation. Starting from a foundation population of males and females, they could select a high- and a low-responder line. The heredity of the triolein response, calculated from the interline divergence, is 12 $\pm$ 1% which indicates that the response is determined by the cumulative effect of a group of about 27 independently segregating loci. Besides ethyl and methyl palmitate, ethyl oleate or stearate, cholesterol oleate and oleic acid (Spratt and Kratzing, 1975) also depress macrophage phagocytosis. It is relevant to note that the different triglycerides and fatty acids are normal constituents of the diet and are present in the circulation in the form of lipoproteins. The role played by the various food lipids is discussed in Section 3.2.1, but it is interesting to conclude here that the use of test substances has evaluated the genetic regulation of macrophage responsiveness to lipids. It seems interesting to compare this genetic factor with familiar manifestations of altered metabolism of lipids and cholesterol, associated with certain types of cardiovascular disease.

Lipid particles consisting of methyl methacrylate, marked with fluorescent green, have been used by Juhlin (1959) to test the phagocytic activity in alloxan-diabetic rabbits. He found that the particles disappeared more slowly from the blood in alloxan-diabetic animals than in controls. If there is a correlation between the decreased activity of the RES in diabetes and the well-known increased occurrence of atherosclerosis in these patients has not been investigated until now.

Lipid test emulsions have also been used to investigate the binding, uptake, and degradation mechanism of isolated Kupffer cells (Pisano *et al.*, 1968, 1970; van Berkel *et al.*, 1980b). Pisano *et al.* (1968) isolated RE cells from the liver with the aid of collagenase and trypsin. The cell association of the test lipid emulsion was about two- to threefold stimulated by the simultaneous presence of heparin (25 U) plus serum (50–66%) in the incubation medium. When compared to liver parenchymal cells, the cell association of the lipid emulsion to RE (nonparenchymal) cells was six times higher (expressed per cell). When this is expressed per milligram cell protein, this value is about 60 (Munthe-Kaas *et al.*, 1976). In contrast to the cell association of carbon, which increased with increasing incubation time, the cell association of the lipid emulsion is saturated within 15 min. Readdition of lipid emulsion leads to an additional cell association. This suggests that only a minor subfraction of the lipid emulsion associates with the cells. Addition of lipid emulsion did not result in a stimulation of glucose oxidation, while the acetate oxidation to CO_2 was inhibited. Only very high concentrations of cyanide or iodoacetate (10^{-1} M) inhibit the phagocytosis by Kupffer cells. To what extent the differences in metabolism of Kupffer cells and free macrophages as described by Pisano *et al.* (1970) are caused by the differences in isolation procedure, or are caused by a real difference is still not clear at the moment. Munthe-Kaas (1976) found that freshly isolated Kupffer cells were seriously modified by the isolation procedure so that a recovery time was necessary before phagocytosis could be studied. After recovery the uptake of gold was greatly inhibited by 10^{-3} M KCN and it was found that the sensitivities of Kupffer cells to 2,4-dinitrophenol, NaF, and KCN were similar to those of peritoneal macrophages. So it seems that after isolation of nonparenchymal liver with the aid of collagenase plus trypsin (Pisano *et al.*, 1968) or Pronase (Mills and Zucker-Franklin, 1969) the cells are modified. The recently described isolation procedure which uses only collagenase may circumvent this problem (van Berkel and van Tol, 1978).

The nonparenchymal liver cells (including Kupffer cells) contain, relative to parenchymal cells, a high activity of lysosomal enzymes (van Berkel *et al.*, 1975; Munthe-Kaas *et al.*, 1976; Knook and Sleyster, 1976). From the lysosomal enzymes tested, especially cathepsin D and acid lipase are highly active in nonparenchymal cells (van Berkel *et al.*, 1975). Table 1 shows the acid lipase and acid cholesterolesterase activity in nonparenchymal liver cells, as compared to parenchymal cells. The substrates for acid lipase (4-methylumbelliferyl oleate) and acid cholesterolesterase (cholesterol[1-^{14}C]oleate) were incorporated in egg yolk lecithin vesicles and assayed in total cell homogenates. As shown in Table 1, acid lipase activity (expressed per milligram cell protein) is 2.5 times more active in nonparenchymal cells than in parenchymal cells. The pH optimum for the hy-

TABLE 1. ACID LIPASE AND ACID CHOLESTEROLESTERASE ACTIVITIES IN PARENCHYMAL AND NONPARENCHYMAL LIVER CELLS

	Substrate	
Source of homogenate	4-Methylumbelliferyl oleate (nmoles/min per mg protein)	Cholesteryl[1-^{14}C]oleate (pmoles/min per mg protein)
Parenchymal cells (PC)	25 ± 3[a]	4.4 ± 0.3
Nonparenchymal cells (NPC)	63 ± 12	50.4 ± 7.0
Activity ratio NPC/PC	2.5	11.4

[a]Values are mean ± S.E.

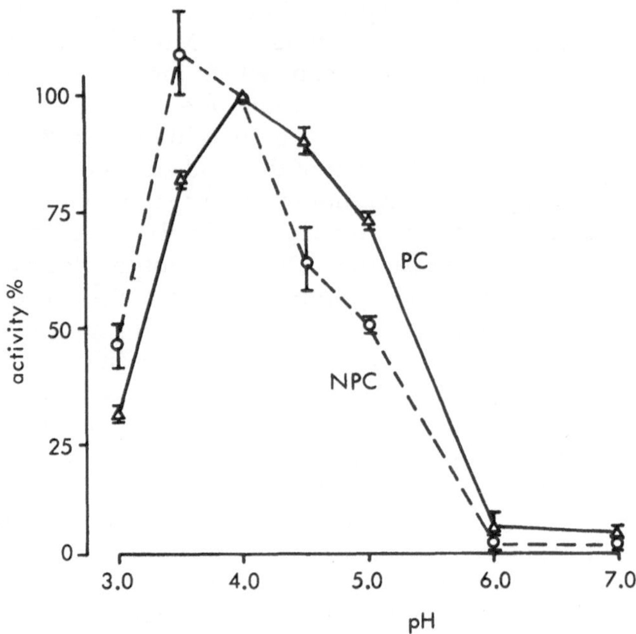

FIGURE 1. The effect of pH upon the relative activity of nonparenchymal and parenchymal cell homogenates for cholesteryl[1-^{14}C]oleate hydrolysis. ○, Nonparenchymal cell homogenates; △, parenchymal cell homogenates.

drolysis of 4-methylumbelliferyl oleate is 4.0. Because such a low pH is only found in the lysosomes, this is suggestive for a lysosomal localization. Although the acid lipase activity is highest per milligram nonparenchymal cell protein, this is even more so for the acid cholesterolesterase activity. The cholesterolesterase activity has an 11.4-fold higher specific activity in nonparenchymal cells, as compared to parenchymal cells and also exerts maximal activity at an acid pH (Fig. 1). These results indicate that the hydrolysis of cholesterolesters probably occurs in the lysosomes, where lipid particles or lipoproteins are transported to after binding and uptake.

3.1.2. Liposomes

The use of liposomes as enzyme or drug carriers has gained much interest during the past years (Desnick *et al.*, 1976). Because of their capacity to retain water-soluble components in their aqueous phase, hidden from immunological recognition, it has been argued that liposomes could be useful to deliver enzymes at an intracellular site. The application for drugs is related to the possibility that liposomes can remain in the circulation for a relatively long period and thus can maintain a certain concentration in the blood compartment. However, most types of liposomes behave as a lipid test emulsion for the RES, i.e., they are rapidly removed from the circulation and mainly recovered in the liver and spleen.

Consequently, other "target" tissues which are low in endocytotic activity such as heart, kidney, or skeletal muscle are not reached. Furthermore, the existence of morphological barriers such as capillary walls and the blood–brain barrier prevents access of injected liposomes to most tissues other than liver. In liver the occurrence of fenestrations with a diameter of 100 nm in the endothelial cell lining (Wisse, 1970) limits the maximal size of the liposomes. Indeed, most types of liposomes are not small enough to reach the liver parenchymal cells simply by the presence of this anatomical sieve. The relative involvement of parenchymal and nonparenchymal liver cells in the uptake of liposomes has given rise to a great deal of controversy (Gregoriadis and Ryman, 1972; Gregoriadis and Neerunjun, 1974; Segal *et al.*, 1974; Rahman and Wright, 1975; Tanaka *et al.*, 1975; Wisse *et al.*, 1976; De Barsy *et al.*, 1976; Roerdink *et al.*, 1977; Roerdink, 1978). Different methods were used to locate the injected material and it seems that the techniques used all have their own limitations which were not fully recognized. Gregoriadis and Ryman (1972) showed by autoradiography of rat liver tissue that 3 min after injection of [^{3}H]cholesterol-liposomes, the radioactivity was present both in hepatocytes and in Kupffer cells. However, cholesterol rapidly exchanges between lipid particles and plasma membranes (Drevon *et al.*, 1977) and it is not surprising therefore that the same group of authors (Segal *et al.*, 1974), now using nitroblue tetrazolium (NBT) as liposomal marker, found that liposomes were almost exclusively found within nonparenchymal cells. However, for microscopy the reduction of NBT to formazan is necessary and this process may be more active in nonparenchymal cells (van Berkel and Kruijt, 1977), which could lead to an underestimation of the contribution by the parenchymal cells. Rahman and Wright (1975) followed the fate of liposomes by recognizing liposomal structures within cells. According to Roerdink (1978), this must be considered with caution since glutaraldehyde-induced artifacts closely resembling liposomes are regularly observed in electron micrographs. Tanaka *et al.* (1975) isolated Kupffer cells and a parenchymal cell fraction and found, using [^{14}C]inulin entrapped in liposomes, that 70% of the total uptake was present in the Kupffer cell fraction and 30% in the parenchymal cell fraction. However, no criteria for cell purity, integrity, and recovery of label during cell isolation were given. The same authors reported a 70% inhibition of the hepatic uptake of liposomes by methylpalmitate.

Wisse *et al.* (1976) followed the fate of horseradish peroxidase entrapped in liposomes, and found the same cellular distribution (parenchymal cells as well as nonparenchymal cells) as for the free horseradish peroxidase. This observation could be explained by a serum-induced disintegration of the liposomes in the circulation. Such a serum-induced destruction of liposomes with transfer of phospholipids to serum high-density lipoproteins has been noticed by Scherphof *et al.* (1978).

From these data it is obvious that the clearance of liposomal lipids or intraliposomal substances from the blood and the uptake by the different liver cell types not necessarily represents the fate of intact liposomes. Therefore, the clearing from the blood and the cellular uptake of both the phospholipid membrane and the entrapped protein should be followed simultaneously. Also, a good recovery of the injected material in the isolated cell populations, as well as criteria showing purity and integrity of the isolated cells are necessary in order to draw reliable conclusions.

All liposomes used until now do not seem to escape from uptake by the RE cells. Blockade or saturation of RE cells either by carbon (Gregoriadis and Neerunjun, 1974) or by preinjection of empty liposomes (Caride *et al.*, 1976) inhibits the clearance rate and may result in an increased uptake by target cells. Also, the use of small unilamellar liposomes, which are cleared by RE cells at a much lower rate than large multilamellar liposomes (Roerdink, 1978; Juliano and Stamp, 1975), may lead to a lower RE contribution to the total uptake. *In vitro* studies with isolated RE cells could be useful in investigating the interaction between liposomes and these cell types. This may lead to tools which make it possible to direct liposomes away from the RE pathway, to their desired target cells (see Godfredson *et al.*, 1983).

3.2. NATURAL LIPID SUBSTRATES

3.2.1. Blood Lipids

3.2.1a. Role of RE Cells in the Uptake of Lipids Derived from the Diet. Although it is generally agreed that intravenously injected artificial test substances interact heavily with RE cells (see Section 3.1), no such agreement exists for lipids circulating in normal serum. When rats were fed cholesterol and the liver subsequently stained with sudan IV, Friedman *et al.* (1954) detected sudanophilic granules 6 hr after ingestion in Kupffer cells while parenchymal cells of the liver did not show this reaction. At 24 hr more "cholesterol" was found in Kupffer cells and also parenchymal cells reacted positively. The peripheral or portal intravenous injection of colloidal suspensions, which interfere with Kupffer cells, led to a reduction in the deposition of hepatic cholesterol, the persistence of chylomicrons in the blood, and the occurrence of hypercholesterolemia. If the fate of dietary [4-^{14}C]cholesterol is followed, both isolated parenchymal and nonparenchymal cells contain label (Friedman *et al.*, 1956). However, the integrity and purity of the isolated cells were not reported. DiLuzio (1960) and Riggi and DiLuzio (1962) showed that activation of the RES in cholesterol-fed rats, by zymosan or glucan injection, results in a lowering of the serum cholesterolester concentration (by about 60%). Also, in cholesterol-fed rabbits Lee and Ho (1975) observed an important participation of Kupffer cells in the hepatic fat accumulation. In humans, Berken and Sherman (1972) showed that olive oil ingestion leads to depression of the RES. These data suggest that the ingestion of high amounts of cholesterol and/or triglycerides leads to a quantitatively important involvement of the RES in the metabolism of blood lipids.

3.2.1b. Role of RE Cells in the Uptake of Injected Blood Lipids. The intravenous injection of chylomicron-free hypercholesteremic rat serum into normal rats does not lead to the appearance of sudanophilic granules in hepatic RE cells (Friedman and Byers, 1954). In their experiments, blocking of the RE cells with India ink did not lead to a change in the amount of sudanophilic granules in the liver or to an increase in serum cholesterol. However when total hypercholesteremic serum (including chylomicrons) was given, the simultaneous injection of India ink induced higher levels of serum cholesterol as compared to controls. Similar results were obtained by Neveu *et al.* (1956). In contrast to Friedman and Byers (1954), however, these authors showed that aggregates of denatured serum proteins inhibited the clearance of cholesterol which was present in chylomicron-free serum. According to Neveu *et al.* (1956), this difference with Friedman and Byers (1954) is due to the higher effectiveness of aggregated denatured serum proteins as compared to India ink, to block the RES. More recently, Nilsson and Zilversmit (1972) showed that intravenously injected cholesterol, either dissolved in ethanol-saline or as a colloidal suspension, is rapidly cleared from the circulation. The majority of the radioactivity (70–80%) is recovered in the subsequently isolated Kupffer cells. However, the interpretation of studies using labeled, unesterified cholesterol is difficult, because no discrimination can be made between isotope exchange and mass transfer. DiLuzio and Saba (1971) showed that stimulation of RE cells by glucan leads to an increased vascular clearance of fatty acids by the liver. Also *in vitro,* palmitic acid was readily oxidized by isolated RE cells (five times the quantity per milligram cell protein as did the parenchymal cells). Further studies concerning the utilization of lipids by the liver RE cells are necessary to find out if these cell types derive a significant portion of their energy requirements from lipids.

3.2.2. Lipoproteins

3.2.2a. Introduction. As stated in Section 2.2.2a, a model has been proposed for the interaction of a certain cell type with lipoproteins, which implies binding, uptake, and degradation of the lipoproteins on the plasma membrane, endocytotic vesicles, and lysosomes, respectively (Goldstein and Brown, 1977). In addition to intracellular catabolism of lipoproteins in the lysosomes, lipoprotein metabolism may occur extracellularly on the plasma membrane (Fielding *et al.*, 1979). This seems especially true for cells which bind lipoprotein lipase or hepatic lipase at the extracellular membrane surface. In this section the involvement of the RES in both metabolic compartments will be discussed.

3.2.2b. *In Vivo* Uptake or Metabolism of Lipoproteins. The *in vivo* uptake and metabolism of lipoproteins has been followed by using autoradiography of tissues after the injection of labeled lipoproteins, by modification of the RE function and subsequently following the half-life of labeled serum lipoproteins, or by isolation of RE cells at different times after the intravenous injection of labeled lipoproteins.

Autoradiographic localization studies have been reported with chylomicrons (Stein *et al.*, 1969), VLDL (Stein *et al.*, 1974), and HDL (Rachmilewitz *et al.*, 1972). These studies led to the conclusion that the lipoproteins initially accumulate at the hepatic cell boundary and distribute more evenly, later on. It is suggested that cholesterolesters are hydrolyzed at the cell boundary before uptake (Stein *et al.*, 1969). In the liver the autoradiographic reaction was seen predominantly over the parenchymal cells although Kupffer and endothelial cells were also labeled (Rachmilewitz *et al.*, 1972; Stein *et al.*, 1974). From the grain densities reported by Stein *et al.* (1974) it can be noticed that, especially at the short time intervals, the grain density over Kupffer and endothelial cells was higher than over parenchymal cells. However, localization studies on the basis of autoradiography are very difficult to interpret quantitatively.

Modification of RE function has often been used to identify the involvement of RE cells in the clearance of particles (see Sections 2.1.1, 3.1.1, and 3.2.1). However, this tool is used relatively little in investigations exploring the fate of lipoproteins. DiLuzio and Riggi (1967) reported no effect of colloidal carbon upon the removal of tripalmitin-labeled chylomicrons in rats; glucan was also ineffective in the removal of cholesterol-labeled chyle. However, it can be argued that inhibition or stimulation of the phagocytic function of RE cells only will influence lipoprotein binding, uptake, and degradation, if these lipoproteins follow the phagocytic route. This implies that ineffectiveness of a certain treatment on the fate of a certain lipoprotein, only indicates that this lipoprotein and the test substance do not share a common removal pathway.

A more direct and quantitative approach is the intravascular injection of labeled lipoproteins and the subsequent determination of label in isolated RE cells. We (van Berkel and van Tol, 1978) developed a protocol for such a procedure which is shown in Fig. 2. By this method, the initial total hepatic uptake can be determined and the amount of label found in the isolated cells can be compared with the initial total hepatic uptake. The isolation procedure is biochemically controlled by measuring the specific activities and distribution of the isoenzymes of pyruvate kinase (van Berkel and van Tol, 1978). This method is further established by the use of multiple biochemical markers (Table 2). The described method has recently been used to determine the relative importance of RE liver cells in the hepatic uptake of the protein moieties of rat chylomicron remnants, VLDL, VLDL remnants, LDL, and HDL (Table 3). The same approach was used for human VLDL, LDL, and HDL (Table 4). As a general conclusion it can be stated that, if expressed per milligram cell protein, the nonparenchymal (RE) liver cells accumulate 5–6 times more rat lipoproteins than parenchymal liver cells. For human LDL this value is higher (van Tol and van Berkel, 1980). This indicates that, irrespective of the kind of rat lipoproteins, nonparenchymal liver cells are about fivefold more active per milligram cell protein in their binding and/or uptake than parenchymal liver cells. Studies of Blouin *et al.* (1977) have shown that 78.0 and 6.3% of the total rat liver volume is occupied by parenchymal and nonparenchymal cells, respectively. The remaining volume (15.7%) is occupied by extracellular spaces. As the protein concentration (in mg/ml cell volume) in both populations is similar (Munthe-Kaas *et al.*, 1976), about 92.5 and 7.5% of the hepatic protein content is of parenchymal and nonparenchymal origin, respectively. Using these data, it can be calculated that about 30% of the rat lipoprotein uptake, found in isolated liver cells, is associated with the nonparenchymal cells. A similar value for rat LDL was reported by Ose *et al.* (1979). These data were obtained with rat lipoproteins labeled in the protein moiety with radioactive iodine. However, application of this label has the disadvantage that, once the lipoproteins are degraded (which, with intact

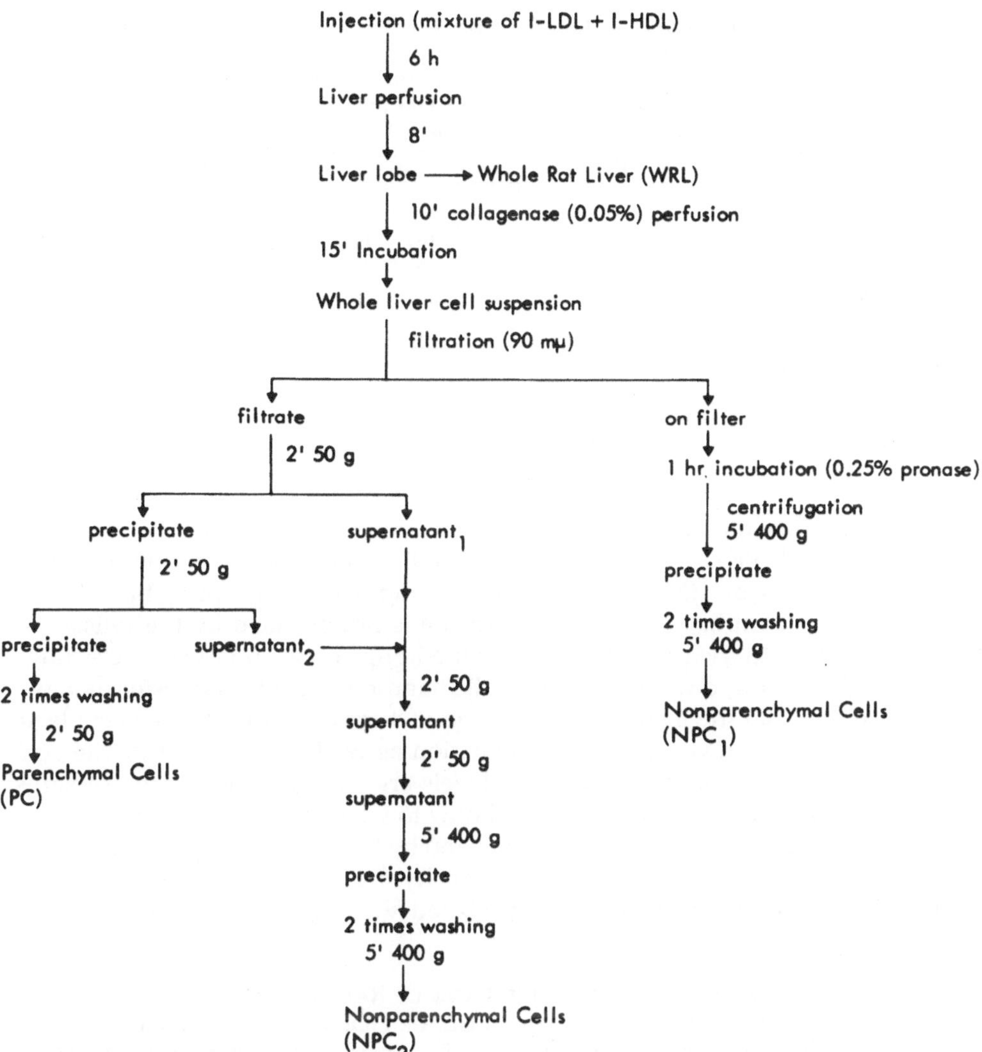

FIGURE 2. Isolation procedure for parenchymal and nonparenchymal cells from rat liver. Liver perfusion and incubations were performed in Hank's balanced salt solution at 37°C. Filtration and centrifugation steps were performed at 0°C.

cells, may occur even during the isolation procedure), the labeled iodotyrosine will leave the cell and will be recovered extracellularly. This explains the loss of iodine label during cell isolation, which leads to a relatively low recovery of the total amount of label, initially present in the whole rat liver sample. This low recovery probably will not influence the interpretation of the data. This conclusion is derived from the data in the first column of Table 3, showing the chylomicron remnant uptake. The protein moiety of the chylomicron remnants was labeled with [^{3}H]leucine and it can be seen that a similar distribution of radioac-

TABLE 2. PURITY AND INTEGRITY TEST ON THE ISOLATED CELL PREPARATIONS[a]

Source	Pyruvate kinase at 1 mM phosphoenolpyruvate	Pyruvate kinase at 1 mM phosphoenolpyruvate + 0.5 mM Fru-1,6-P_2	Cathepsin D	Peroxidase
Whole rat liver	27 ± 5	120 ± 5	11.2 ± 1.2	7.0 ± 0.5
Parenchymal cells	11 ± 1	130 ± 10	5.5 ± 1.0	2.9 ± 0.3
Nonparenchymal cells (Pronase method, NPC_1)	85 ± 8	86 ± 7	41.6 ± 3.9	80.8 ± 8.4
Nonparenchymal cells (diffential centrifugation, NPC_2)	48 ± 5	110 ± 15	18.1 ± 2.1	19.4 ± 1.4

[a]The activities are expressed as nmoles/min per mg protein ± S.E.

tivity between parenchymal and nonparenchymal liver cells is obtained as with the iodinated lipoproteins. The use of [^{3}H]leucine results in a much higher recovery (93% of the liver-associated radioactivity), probably because [^{3}H]leucine in contrast to iodinated tyrosine is not excreted by the cells.

In addition to the protein moiety of the lipoproteins, it is also interesting to see how the liver and the hepatic RE cells handle the cholesterolester moiety. It is well known that the liver is the only organ where cholesterol and/or cholesterolesters can be removed from the circulation as well as degraded to bile acids. Because high serum cholesterol(ester) levels are correlated with atherosclerosis (Dawber *et al.*, 1962), it is important to know to what extent RE cells are involved in uptake and handling of cholesterol(esters) by the liver. Van Berkel and van Tol (1979) determined the relative role of parenchymal and nonparenchymal rat liver cells in the uptake of cholesterolester-labeled VLDL remnants, LDL, and HDL.

TABLE 3. DISTRIBUTION OF [^{3}H]LEUCINE LABEL OF RAT CHYLOMICRON REMNANTS OR IODINE LABEL OF RAT VLDL REMNANTS, LDL, OR HDL BETWEEN PARENCHYMAL AND NONPARENCHYMAL LIVER CELLS 30 MIN AFTER INTRAVENOUS INJECTION INTO RATS[a]

	Rat lipoproteins			
	Chylomicron remnants	VLDL remnants	LDL	HDL
Whole rat liver (WRL)	310	105	23	20
Parenchymal cells (PC)	200	21	8	7
Nonparenchymal cells (NPC_1)	1380	106	42	46
Nonparenchymal cells (NPC_2)	1270	115	42	64
Ratio NPC_1/PC	6.7	4.9	6.1	5.3
Recovery of radioactivity in PC and NPC_1, as percent of WRL	93%	26%	41%	53%

[a]The values are expressed as percent × 10^4 of the injected dose per milligram cell protein. The recovery is expressed as percent.

TABLE 4. DISTRIBUTION OF IODINE LABEL OF HUMAN VLDL, LDL, OR HDL BETWEEN PARENCHYMAL AND NONPARENCHYMAL LIVER CELLS 30 MIN AFTER INTRAVENOUS INJECTION INTO RATS[a]

	Human lipoproteins		
	VLDL	LDL	HDL
Whole rat liver (WRL)	61	14	7
Parenchymal cells (PC)	28	3	5
Nonparenchymal cells (NPC_1)	105	33	29
Nonparenchymal cells (NPC_2)	68	31	25
Ratio NPC_1/PC	5.1	12.0	5.9
Recovery of radioactivity in PC and NPC_1, as percent of WRL	52%	39%	83%

[a]The values are expressed as percent × 10^4 of the injected dose per milligram cell protein. The recovery is expressed as percent.

The results are summarized in Table 4. It is found that nonparenchymal liver cells contain about 4 times more radioactivity (per mg cell protein) than parenchymal cells. No loss of label occurs during cell isolation because cholesterolesters, once taken up, remain associated with the liver cells. From these figures it can be calculated that about 25% of the liver-associated cholesterolesters taken up from or with lipoproteins are present in the nonparenchymal cells. Nilsson and Zilversmit (1971) arrived at a value of about 6% for chylomicron cholesterolester, while Redgrave (1970) found a somewhat higher value for total chylomicron cholesterol. Drevon *et al.* (1977) prepared cholesterolester-labeled lipoproteins *in vitro* by the action of lecithin:cholesterol acyltransferase, and determined the uptake by the different liver cell types. From their data it can be calculated, using the protein contribution assumptions mentioned above, that the contribution of nonparenchymal liver cells to the total hepatic uptake will be 25–30%, a value very close to that shown in Table 5. As a general conclusion it can be stated that the recent data, obtained with isolated nonparenchymal liver

TABLE 5. DISTRIBUTION OF [^{3}H]CHOLESTEROLESTER RADIOACTIVITY FROM VLDL REMNANTS, LDL, AND HDL BETWEEN PARENCHYMAL AND NONPARENCHYMAL LIVER CELLS[a]

	Rat lipoproteins		
	VLDL remnants	LDL	HDL
Parenchymal cells (PC)	460	120	110
Nonparenchymal cells (NPC_1)	1480	480	450
Nonparenchymal cells (NPC_2)	2020	890	600
Ratio NPC_1/PC	3.1	4.3	4.1

[a]The values are expressed as percent × 10^4 of the injected dose per milligram cell protein. The samples were obtained 30 min (VLDL remnants) or 6 hr (LDL and HDL) after injection. The recovery of the radioactivity in PC and NPC_1 as percent of WRL was around 100%.

cells with well-defined purity and integrity, indicate that a significant *in vivo* uptake of lipoproteins or lipoprotein constituents occurs in the RE liver cells. The reason for the apparent discrepancy with autoradiographic data is unclear at the moment but might be related to exchange and/or washout phenomena. As indicated in the next section, the mechanism of the involvement of RE cells in the metabolism of serum lipoproteins is further substantiated by the active *in vitro* interaction of isolated nonparenchymal cells with lipoproteins.

An indirect involvement of the nonparenchymal liver cells in the liver metabolism of large chylomicrons is suggested by Naito and Wisse (1978) and Fraser *et al.* (1978). They suggested that the fenestrations in the endothelial lining of sinusoids can filter chylomicrons, remnants, and other lipoproteins. This filter function is thought to be most effective in blocking the passage of larger chylomicrons. Further studies are required to determine the physiological importance of such a mechanism in lipoprotein catabolism.

3.2.2c. Nature of the Recognition Site. Before lipoproteins are taken up by cells, the circulating particle is likely to be recognized and bound to a cell receptor (Goldstein and Brown, 1977). The availability of procedures to isolate intact RE liver cells opens the possibility to study the mechanism and nature of the initial interaction of lipoprotein and cell. The most common procedure to isolate nonparenchymal liver cells is the method resulting in the NPC_1 preparation (Fig. 2). With this method, Pronase (0.25%) is used to destroy specifically the parenchymal liver cells. However, Pronase contains a mixture of highly active proteolytic enzymes which are expected to destroy receptors, located at the outer surface of the cell membrane. To avoid the use of Pronase we can apply a method in which only collagenase (which contains a low amount of proteolytic enzymes) is used. Subsequently, the nonparenchymal liver cells can be purified

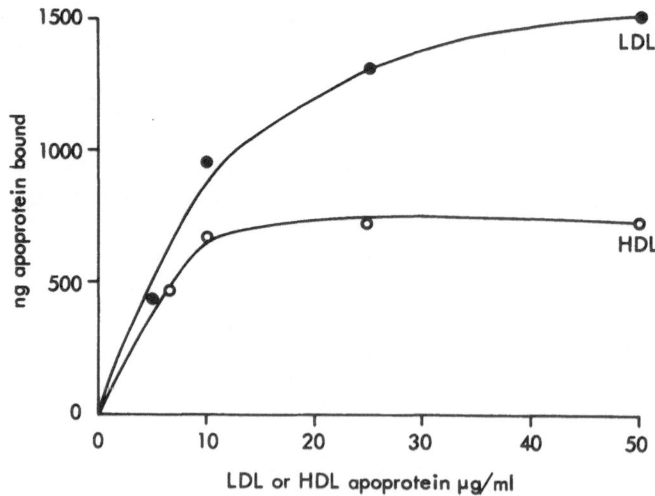

FIGURE 3. Relation of the concentration of rat [^{125}I]-LDL or [^{125}I]-HDL to the extent of binding to nonparenchymal liver cells.

from the cell mixture by differential centrifugation. Such a preparation of nonparenchymal cells is not completely pure and contains variable amounts of parenchymal cell protein (Table 2). However, because the binding of lipoproteins per milligram cell protein to such nonparenchymal cell preparations is much higher than to parenchymal cells, it can be assumed that the contribution of the parenchymal cell protein to the total binding will be relatively low. Figure 3 indicates that the binding of both rat LDL and HDL to RE liver cells follows saturation kinetics, which is indicative for the presence of a saturable high-affinity receptor for these lipoproteins. Besides LDL and HDL, VLDL remnants, prepared by extrahepatic circulation of VLDL, also bind with a high affinity (Fig. 4). From Lineweaver–Burke plots, K_m and V values for these bindings can be obtained (Table 6) which indicate that the receptor(s) present on nonparenchy-

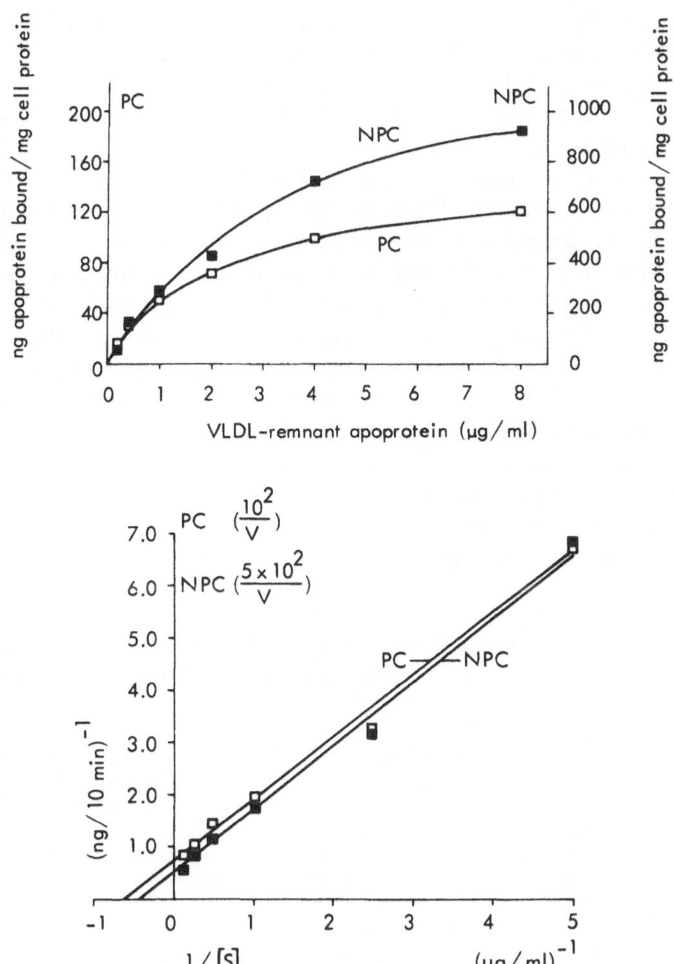

FIGURE 4. Relation of the concentration of [^{125}I]-VLDL remnant to the extent of binding to nonparenchymal (NPC) and parenchymal (PC) liver cells.

mal liver cells possesses by far the highest affinity for VLDL remnants. To determine the relation between the binding of the different lipoproteins to nonparenchymal liver cells, direct competition experiments were performed (Fig. 5). It can be noticed that the binding of labeled VLDL remnants to RE cells is inhibited most efficiently by unlabeled VLDL remnants. However, unlabeled rat LDL and HDL are also efficient inhibitors in contrast to human LDL. Similar results are obtained if the binding of labeled LDL or HDL is measured. These results indicate that rat VLDL remnants, rat LDL, and rat HDL compete for the same surface receptor with the highest apparent affinity for VLDL remnants. A comparison of the apoprotein composition of these lipoproteins (van Berkel *et al.*, 1981) indicates that the common recognition mark could be apolipoprotein E. The high apparent affinity of the receptor for VLDL remnants can be explained by multiple receptor occupancy of this lipoprotein.

Goldstein *et al.* (1979) reported the participation of the liver in the rapid uptake (within 2 min) of ^{125}I-labeled acetyl-LDL. Mahley *et al.* (1979) found a similar rapid uptake, probably in Kupffer cells, of acetoacetylated LDL. By using cultured guinea pig Kupffer cells it was shown that [^{125}I]acetyl-LDL is degraded 8 times as fast as native human [^{125}I]-LDL. Goldstein and co-workers suggest that the RE liver cells contain a similar acetyl-LDL recognition site as described for peritoneal macrophages. With freshly isolated nonparenchymal liver cells (NPC_2, Fig. 2), we found that human acetyl-LDL was degraded 50 times faster than native LDL, which indicates the presence of a recognition site for acetyl-LDL on the nonparenchymal liver cells. The question arises if the receptor described for rat lipoproteins is similar to the receptor which recognizes the modified human LDL. Direct competition experiments indicate that competition for binding occurs neither between human acetyl-LDL and the rat lipoproteins nor between acetyl-LDL and native human LDL. This indicates that one receptor is present on RE liver cells for the rat lipoproteins (probably recognizing apo E) and another one recognizing human acetyl-LDL. Native human LDL competes with neither rat lipoproteins nor acetyl-LDL, suggesting the presence of a third receptor for native human LDL.

TABLE 6. COMPARISON OF THE APPARENT K_m AND V VALUES OF PARENCHYMAL AND NONPARENCHYMAL CELLS FOR VLDL REMNANTS, LDL, AND HDL BINDING[a]

	Apparent K_m (μg apoprotein/ml)			Apparent V (ng apoprotein/mg cell protein per 10 min)		
	VLDL remnant	LDL	HDL	VLDL remnant	LDL	HDL
Nonparenchymal cells	1.88 ± 0.41	21.3 ± 5.9	11.9 ± 3.0	713 ± 155	1452 ± 144	1171 ± 159

[a]The apparent K_m and V values were obtained from Lineweaver–Burk plots like those shown in Fig. 4. The values are means of three different experiments with three different lipoprotein preparations (± S.E.).

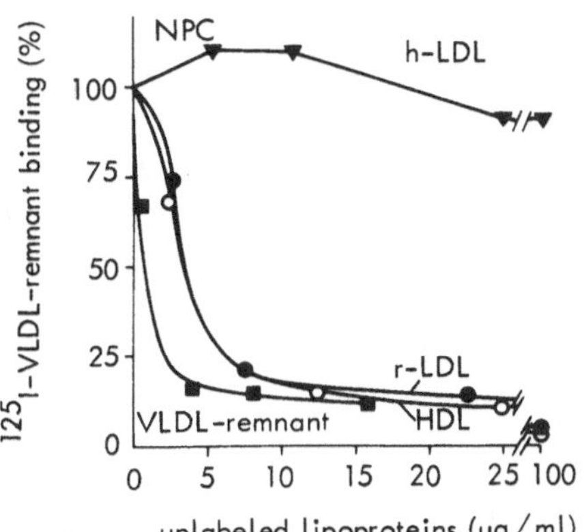

FIGURE 5. Comparison of the ability of unlabeled lipoproteins to compete with the binding of [^{125}I]-VLDL remnants. Nonparenchymal cells were incubated for 10 min with 0.53 µg/ml [^{125}I]-VLDL remnants and with the indicated amounts of unlabeled rat LDL (●), rat HDL (○), human LDL (▼), or VLDL remnant (■).

3.2.2d. Handling of Lipoprotein Constituents. The presence of specific receptors for rat lipoproteins and acetyl-LDL suggests that the RE cells of rat liver are involved in their metabolism. Figure 6 shows that in addition to binding, degradation of apoprotein also occurs. During 3 hr of incubation 4% of the total cell-associated VLDL-remnant protein is degraded. This percentage is 11% for rat LDL, 49% for rat HDL, 10% for human LDL, and about 100% for acetyl-LDL. This wide variety in degradation rate might be caused by the nature of the different receptors involved in the binding of these lipoproteins. However, the different rat lipoproteins are bound to the same receptor and still different degradation rates (4%–49%) are noticed. We therefore also compared the capacity of the RE cells to degrade the different lipoproteins in a cell-free system at the optimal pH of 4.2. The values obtained are shown in Table 7. It is remarkable that the relative differences in degradation capacities between the different lipoproteins are similar to the relative differences found with intact cells, although the intracellular capacity is actually 400 to 1000-fold higher in homogenates than used by the intact cells.

In vitro data on the role of RE cells in triglyceride and cholesterolester metabolism are greatly lacking. Studies from our laboratory (Jansen *et al.*, 1978, 1980a) have indicated that the nonparenchymal liver cells very efficiently bind the heparin-releasable liver lipase, an enzyme probably involved in the metabolism of lipoprotein phospholipids and cholesterol(esters) (Jansen *et al.*, 1980; van Tol *et al.*, 1980). A localization of this liver lipase on nonparenchymal liver cells *in vivo* was indicated by immunofluorescence and immuno-electron microscopy (Kuusi *et al.*, 1979). Together with the efficient *in vivo* uptake of cholesterolesters from rat lipoproteins (van Berkel and van Tol, 1979) by nonparenchymal cells, this suggests that these cell types can participate actively in lipoprotein cholesterolester metabolism at the plasma membrane. Drevon *et al.* (1977) showed

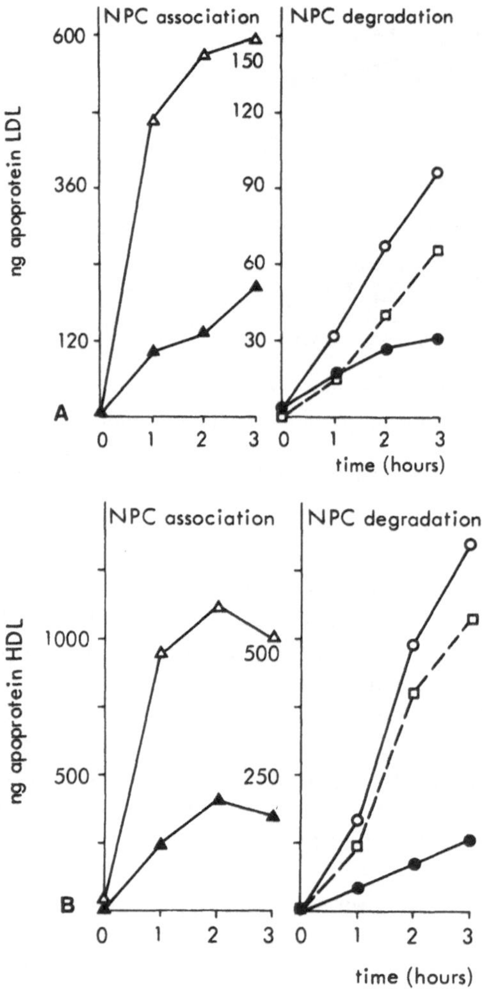

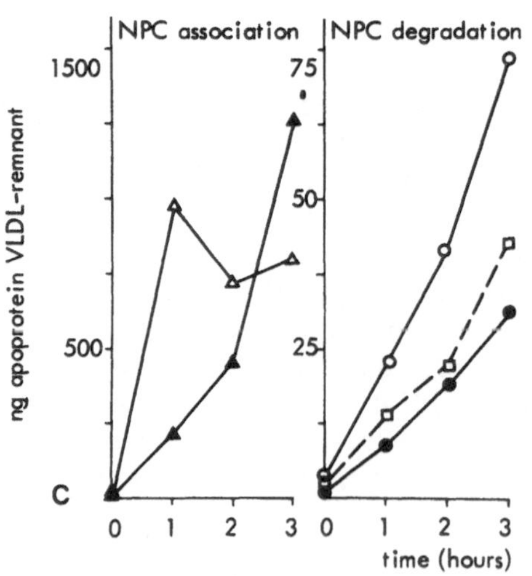

FIGURE 6. Time course of the cell association and degradation of lipoproteins by isolated nonparenchymal liver cells in the absence and presence of an excess of unlabeled lipoproteins. Nonparenchymal cells were incubated with (A) 4.0 μg/ml [^{125}I]-LDL (△,○) or 4.0 μg/ml [^{125}I]-LDL + 49.3 μg/ml unlabeled LDL (▲,●); (B) 4.6 μg/ml [^{125}I]-HDL (△,○) or 4.6 μg/ml [^{125}I]-HDL + 104.7 μg/ml unlabeled HDL (▲,●); (C) 3.3 μg/ml [^{125}I]-VLDL remnants (△,○) or 3.3 μg/ml VLDL remnants + 109.9 μg/ml unlabeled HDL (▲,●). At the indicated time, samples were withdrawn and the amount of cell-associated radioactivity (after three washings), as well as the radioactivity present in the acid-soluble water phase (degradation) were determined. Both cell association and degradation are expressed as ng apolipoprotein/mg cell protein. □- - -□ can be defined as high-affinity degradation (difference ○ versus ●).

TABLE 7. DEGRADATION OF IODINE-LABELED RAT VLDL REMNANTS, LDL, AND HDL BY NONPARENCHYMAL RAT LIVER CELL HOMOGENATES[a]

	ng apolipoprotein hydrolyzed/min/mg protein		
	VLDL remnant	LDL	HDL
Nonparenchymal cells	54	390	1750

[a]Apolipoprotein degradation was determined in the sonicated cell homogenates (sonicated twice for 30 sec at 21 kHz) at pH 4.2 and values represent the radioactivity in the acid-soluble water phase.

that, with increasing concentrations of rat HDL, nonparenchymal liver cells took up increasing amounts of cholesterolesters although no clear saturation level was reached. Nonparenchymal liver cells possess the enzymatic equipment for an efficient, intracellular degradation of both triglycerides and cholesterolesters (van Berkel *et al.*, 1980b). The hydrolytic activities are maximal at pH 4.0 for triglycerides and pH 3.5 for cholesterolesters and the enzyme(s) possess(es) a high apparent affinity for the substrates. The acid pH optimum suggests that the major proportion of the triglyceride- or cholesterolester-hydrolyzing activity is located in the lysosomes. Further studies are required to see if a cholesterolester route, as described in Section 2.2.2 for macrophages, exists also in the RE liver cells. In circulating macrophages, HDL may serve as a recipient for an excess of cholesterol(esters). In liver the adjacent liver parenchymal cells may serve as recipients for cholesterolesters internalized by nonparenchymal cells. Such a process would introduce a metabolic cooperation between the different cell types in liver (van Berkel, 1979), in which the RE cells, located at the border between intravascular and extravascular spaces, not only form the anatomical, but also the metabolic link between these two compartments.

4. PERSPECTIVES IN THE USE OF MACROPHAGE ACTIVITY IN THE PREVENTION OR TREATMENT OF ATHEROSCLEROSIS

The interaction of RE cells with blood lipids and lipoproteins theoretically opens the possibility that optimal use of this potency may lead to a significant change in the levels in circulating lipoproteins. Two possibilities arise: (1) Activation of the RES may lead to an increased interaction of this system with natural substrates. (2) Modification of lipids or lipoproteins may lead to an increased interaction of these modified substances with the RES.

In principle both possibilities may lead to the same result, that is, an increased handling of the substrate by the RES. The first possibility has been investigated in numerous systems especially by DiLuzio and co-workers (Di-

Luzio and Riggi, 1964; DiLuzio and Saba, 1971). Using this method, no evidence has been obtained that activation of the RES leads to significantly decreased levels of circulating lipoproteins. As mentioned before (Section 3.2.2b) this might be the consequence of a different removal pathway for lipoproteins and the RE test substances. The second possibility is unexplored and may have advantages, because modification of one component will only lead to a modification in interaction between the RE cells and this specific component. As discussed in Section 3.3.2, acetylation of LDL leads to a 50-fold increase in degradation of this lipoprotein by the RE liver cells (van Berkel *et al.*, 1982). It is well known that high LDL levels form an important risk factor for atherosclerosis. So, theoretically, a specific chemical modification of LDL, leading to an increased interaction with the RES, has the potential for application in the prevention or treatment of atherosclerosis.

ACKNOWLEDGMENTS. Mrs. H. Vaandrager and Messrs. J. K. Kruijt and T. van Gent are gratefully thanked for their expert technical assistance. Prof. Dr. J. F. Koster is thanked for helpful discussions and Miss A. C. Hanson for preparation of the manuscript. The Foundation for Fundamental Medical Research (FUNGO) and the Dutch Heart Foundation are acknowledged for financial support.

REFERENCES

Aschoff, A., 1924, Das reticulo-endotheliale System, *Ergeb. Inn. Med. Kinderheilkd.* **26**:1.

Basu, S. K., Goldstein, J. L., Anderson, R. G. W., and Brown, M. S., 1976, Degradation of cationized low density lipoprotein and regulation of cholesterol metabolism in homozygous familial hypercholesterolemia fibroblasts, *Proc. Natl. Acad. Sci. USA* **73**:3178.

Basu, S. K., Brown, M. S., Ho, Y. K., and Goldstein, J. L., 1979, Degradation of low density lipoprotein–dextran sulfate complexes associated with deposition of cholesteryl esters in mouse macrophages, *J. Biol. Chem.* **254**:7141.

Berken, A., and Sherman, A. A., 1972, Reticulo-endothelial system depression in man after olive oil ingestion, *Proc. Soc. Exp. Biol. Med.* **141**:656.

Blouin, A., Bolender, R. P., and Weibel, E. R., 1977, Distribution of organelles and membranes between hepatocytes and nonhepatocytes in the rat liver parenchyma: A stereological study, *J. Cell Biol.* **72**:441.

Brown, M. S., Goldstein, J. L., Krieger, M., Ho, Y. K., and Anderson, R. G. W., 1979, Reversible accumulation of cholesterylesters in macrophages incubated with acetylated lipoproteins, *J. Cell Biol.* **82**:597.

Brown, M. S., Ho, Y. K., and Goldstein, J. L., 1980, The cholesteryl ester cycle in macrophage foam cells: Continual hydrolysis and re-esterification of cytoplasmic cholesterylesters, *J. Biol. Chem.* **255**:9344.

Caride, V. J., Taylor, W., Cramer, J. A., and Gottschalk, A., 1976, Evaluation of liposome-entrapped radioactive tracers as scanning agents. Part 1. Organ distribution of liposome [^{99m}TC-DTPA] in mice, *J. Nucl. Med.* **17**:1067.

Casley-Smith, J. R., and Day, A. J., 1966, The uptake of lipid and lipoprotein by macrophages *in vitro:* An electron microscopical study, *J. Exp. Physiol.* **51**:1.

Conning, D. M., and Heppleston, A. G., 1966, Reticuloendothelial activity and local particle disposal: A comparison of the influence of modifying agents, *Br. J. Exp. Pathol.* **47**:388.

Cornell, R. P., and Saba, T., 1975, Degradation of particulate lipid by the large-granule fraction of rat liver, *Proc. Soc. Exp. Biol. Med.* **148**:430.

Dawber, T. R., Kannel, W. B., Revotskie, N., and Kayan, A., 1962, The epidemiology of coronary heart disease—The Framingham enquiry, *Proc. R. Soc. Med.* **55:**265.

Day, A. J., 1967, Lipid metabolism by macrophages and its relationship to atherosclerosis, in: *Advances in Lipid Research* (R. Paoletti and D. Kritchevsky, eds.), Vol. 5, pp. 185–207, Academic Press, New York.

Day, A. J., and Fidge, N. H., 1962, The uptake and metabolism of ^{14}C-labeled fatty acids by macrophages *in vitro*, *J. Lipid Res.* **3:**333.

De Barsy, T., Devos, P., and Van Hoof, F., 1976, A morphologic and biochemical study of the fate of antibody-bearing liposomes, *Lab. Invest.* **34:**273.

Desnick, R. J., Thorpe, S. R., and Fiddler, M. D., 1976, Toward enzyme therapy for lysosomal storage diseases, *Physiol. Rev.* **56:**57.

DiLuzio, N. R., 1960, Modification of cholesterosis and lipidosis of rats maintained on a atherogenic diet, *Nature (London)* **185:**616.

DiLuzio, N. R., and Riggi, S. J., 1964, The development of a lipid emulsion for the measurement of reticuloendothelial function, *J. Reticuloendothelial Soc.* **1:**136.

DiLuzio, N. R., and Riggi, S. J., 1967, Participation of hepatic parenchymal and Kupffer cells in chylomicron and cholesterol metabolism, *Adv. Exp. Med. Biol.* **1:**382.

DiLuzio, N. R., and Saba, T. M., 1971, Liver parenchymal and Kupffer cell metabolism of ^{14}C-labeled acetate, palmitate and triglyceride, *J. Reticuloendothelial Soc.* **10:**392.

Drevon, C. A., Berg, T., and Norum, K. R., 1977, Uptake and degradation of cholesterol ester-labeled rat plasma lipoproteins in purified rat hepatocytes and non-parenchymal liver cells, *Biochim. Biophys. Acta* **487:**122.

Elsbach, P., 1973, On the interaction between phagocytes and micro-organisms, *N. Engl. J. Med.* **289:**846.

Fielding, C. J., Vlodavsky, I., Fielding, P. E., and Gospodarowicz, D., 1979, Characteristics of chylomicron binding and lipid uptake by endothelial cells in culture, *J. Biol. Chem.* **254:**8861.

Finkelstein, M., and Weissmann, G., 1978, The introduction of enzymes into cells by means of liposomes, *J. Lipid Res.* **19:**289.

Franson, R., Beckerdite, S., Wang, P., Waite, M., and Elsbach, P., 1973, Some properties of phospholipases of alveolar macrophages, *Biochim. Biophys. Acta* **296:**365.

Fraser, R., Bosanquet, A. G., and Day, W. A., 1978, Filtration of chylomicrons by the liver may influence cholesterol metabolism and atherosclerosis, *Atherosclerosis* **29:**113.

Friedman, M., and Byers, S. O., 1954, Observation concerning the production and excretion of cholesterol in mammals. XIV. The relationship of the hepatic reticuloendothelial cell (Kupffer cell) to endogenously produced cholesterol, *Circulation* **10:**491.

Friedman, M., Byers, S. O., and Rosenman, R. H., 1954, Observations concerning the production and excretion of cholesterol in mammals. XII. Demonstration of the essential role of the hepatic reticuloendothelial cell (Kupffer cell) in the normal disposition of exogenously derived cholesterol, *Am. J. Physiol.* **177:**77.

Friedman, M., Byers, S. O., and George, S. S., 1956, Detection of dietary cholesterol-4-^{14}C in the hepatic reticuloendothelial cell of the rat, *Am. J. Physiol.* **184:**141.

Godfredsen, C. F., van Berkel, Th. J. C., Kruÿt, J. K., and Goethals, A., 1983, Cellular localization of stable solid liposomes in the liver of rats, *Biochem. Pharmacol.* **32:**3389.

Goldstein, J. L., and Brown, M. S., 1977, Atherosclerosis: The low-density lipoprotein receptor hypothesis, *Metabolism* **26:**1257.

Goldstein, J. L., Ho, Y. K., Basu, S. K., and Brown, M. S., 1979, Binding site on macrophages that mediates uptake and degradation of acetylated low density lipoprotein, producing massive cholesterol deposition, *Proc. Natl. Acad. Sci. USA* **76:**333.

Goldstein, J. L., Ho, Y. K., Brown, M. S., Innerarity, T. L., and Mahley, R. W., 1980, Cholesteryl ester accumulation in macrophages resulting from receptor-mediated uptake and degradation of hypercholesterolemic canine β-very low density lipoproteins, *J. Biol. Chem.* **255:**1839.

Gregoriadis, G., 1978, Liposomes, *Nature (London)* **201:**211.

Gregoriadis, G., and Neerunjun, D. E., 1974, Control of the rate of hepatic uptake and catabolism of liposome-entrapped proteins injected into rats: Possible therapeutic applications, *Eur. J. Biochem.* **47:**179.

Gregoriadis, G., and Ryman, B. E., 1972, Fate of protein-containing liposomes injected into rats: An approach to the treatment of storage diseases, *Eur. J. Biochem.* **24**:485.

Ho, Y. K., Brown, M. S., and Goldstein, J. L., 1980, Hydrolysis and excretion of cytoplasmic cholesteryl esters by macrophages: Stimulation by high density lipoprotein and other agents, *J. Lipid Res.* **21**:391.

Holper, K., Olcay, I., Kitahama, A., Miller, R. H., Brettscheider, L., Drapanas, T., Trejo, R. A., and DiLuzio, N. R., 1974, Effect of ischemia on hepatic parenchymal and reticuloendothelial function in the baboon, *Surgery* **76**:423.

Jansen, H., van Berkel, T. J. C., and Hülsmann, W. C., 1978, Binding of liver lipase to parenchymal and non-parenchymal rat liver cells, *Biochem. Biophys. Res. Commun.* **85**:148.

Jansen, H., van Berkel, T. J. C., and Hülsmann, W. C., 1980a, Properties of binding of lipases to non-parenchymal rat liver cells, *Biochim. Biophys. Acta* **619**:119.

Jansen, H., van Tol, A., and Hülsmann, W. C., 1980b, On the metabolic function of heparin-releasable liver lipase, *Biochem. Biophys. Res. Commun.* **92**:53.

Juhlin, L., 1959, The phagocytic activity of the reticuloendothelial system in alloxan-diabetic rabbits, *Acta Physiol. Scand.* **45**:369.

Juliano, R. L., and Stamp, D., 1975, The effect of particle size and charge on the clearance rates of liposomes and liposome encapsulated drugs, *Biochem. Biophys. Res. Commun.* **63**:651.

Knook, D. L., and Sleyster, E. C., 1976, Lysosomal enzyme activities in parenchymal and non-parenchymal cells isolated from young, adult and old rats, *Mech. Ageing Dev.* **5**:389.

Koster, J. F., Vaandrager, H., and van Berkel, T. J. C., 1980, Study of the hydrolysis of 4-methylumbelliferyl oleate by acid lipase and cholesteryl oleate by acid cholesterylesterase in human leucocytes, fibroblasts and liver, *Biochim. Biophys. Acta* **618**:98.

Kuusi, T., Nikkilä, E. A., Virtanen, I., and Kinnunen, P. K. J., 1979, Localization of the heparin-releasable lipase *in situ* in the rat liver, *Biochem. J.* **181**:245.

Lee, S. S., and Ho, K. J., 1975, Cholesterol fatty liver: Morphological changes in the course of its development in rabbits, *Arch. Pathol.* **99**:301.

Mahley, R. W., Weisgraber, K. H., Innerarity, T. L., and Windmueller, H. G., 1979, Accelerated clearance of low-density and high-density lipoproteins and retarded clearance of E apoprotein-containing lipoproteins from the plasma of rats after modification of lysine residues, *Proc. Natl. Acad. Sci. USA* **76**:1746.

Mason, R. J., Stossel, T. P., and Vaughan, M., 1973, Quantitative studies of phagocytosis by alveolar macrophages, *Biochim. Biophys. Acta* **304**:864.

Mills, D. M., and Zucker-Franklin, D., 1969, Electron microscopic study of isolated Kupffer cells, *Am. J. Pathol.* **54**:147.

Mouton, D., Bouthillier, Y., Feingold, N., Feingold, J., Decreusefond, C., Stiffel, C., and Biozzi, G., 1975, Genetic control of macrophage functions, *J. Exp. Med.* **141**:306.

Munthe-Kaas, A. C., 1976, Phagocytosis in rat Kupffer cells *in vitro*, *Exp. Cell Res.* **99**:319.

Munthe-Kaas, A. C., Berg, T., and Sjeljelid, R., 1976, Distribution of lysosomal enzymes in different types of rat liver cells, *Exp. Cell Res.* **99**:146.

Naito, M., and Wisse, E., 1978, Filtration effect of endothelial fenestrations on chylomicron transport in neonatal rat liver sinusoids, *Cell Tissue Res.* **190**:371.

Neveu, T., Biozzi, G., Benacerraf, B., Stiffel, C., and Halpern, B. N., 1956, Role of reticuloendothelial system in blood clearance of cholesterol, *Am. J. Physiol.* **187**:269.

Nilsson, A., and Zilversmit, D. B., 1971, Distribution of chylomicron cholesteryl ester between parenchymal and Kupffer cells of rat liver, *Biochim. Biophys. Acta* **248**:137.

Nilsson, A., and Zilversmit, D. B., 1972, Fate of intravenously administered particulate and lipoprotein cholesterol in the rat, *J. Lipid Res.* **13**:32.

Ose, L., Ose, T., Norum, K. R., and Berg, T., 1979, Uptake and degradation of ^{125}I-labeled high density lipoproteins in rat liver cells *in vivo* and *in vitro*, *Biochim. Biophys. Acta* **574**:521.

Papahadjopoulos, D., 1978, Liposomes and their uses in biology and medicine, *Ann. N.Y. Acad. Sci.* **308**:1.

Patriarca, P., Beckerdile, S., Pettis, P., and Elsbach, P., 1972, Phospholipid metabolism by phagocytic cells. VII. The degradation and utilization of phospholipids of various microbial species by rabbit granulocytes, *Biochim. Biophys. Acta* **280**:45.

Pisano, J. C., Filkins, J. P., and DiLuzio, N. R., 1968, Phagocytic and metabolic activities of isolated rat Kupffer cells, *Proc. Soc. Exp. Biol. Med.* **128**:917.

Pisano, J. C., Filkins, J. P., and DiLuzio, N. R., 1970, Metabolic characterization of actively phagocytozing isolated rat Kupffer cells, *J. Reticuloendothelial Soc.* **8**:25.

Rachmilewitz, D., Stein, O., Roheim, P. S., and Stein, Y., 1972, Metabolism of iodinated high density lipoproteins in the rat. II. Autoradiographic localization in the liver, *Biochim. Biophys. Acta* **270**:414.

Rahman, Y. E., and Wright, B. J., 1975, Liposomes containing chelating agents: Cellular penetration and a possible mechanism of metal removal, *J. Cell Biol.* **65**:112.

Redgrave, T., 1970, Formation of cholesteryl ester-rich particulate lipid during metabolism of chylomicrons, *J. Clin. Invest.* **49**:465.

Riggi, S. J., and DiLuzio, N. R., 1961, Identification of a reticuloendothelial stimulating agent in zymosan, *Am. J. Physiol.* **200**:297.

Riggi, S. J., and DiLuzio, N. R., 1962, The influence of reticuloendothelial stimulation on the response of rats to dietary cholesterol, *J. Lipid Res.* **3**:339.

Roerdink, F. H., 1978, Liposomes as enzyme carriers, Ph.D. thesis, University of Groningen.

Roerdink, F. H., Wisse, E., Morselt, H. W. M., Van der Meulen, J., and Scherphof, G., 1977, Cellular distribution of intravenously injected protein-containing liposomes in the rat liver, in: *Kupffer Cells and Other Sinusoidal Cells* (E. Wisse and D. L. Knook, eds.), pp. 263–272, Elsevier, Amsterdam.

Roos, D., 1977, Oxidative killing of microorganisms by phagocytic cells, *Trends Biochem. Sci.* **2**:61.

Scherphof, G., Roerdink, F., Waite, M., and Parks, J., 1978, Disintegration of phosphatidylcholine liposomes in plasma as a result of interaction with high-density lipoproteins, *Biochim. Biophys. Acta* **542**:296.

Segal, A. W., Wills, E. J., Richmond, J. E., Slavin, G., Black, C. D. V., and Gregoriadis, G., 1974, Morphological observations on the cellular and subcellular destination of intravenously administered liposomes, *Br. J. Exp. Pathol.* **55**:320.

Spratt, N. G., and Kratzing, C. C., 1975, Oleic acid as a depressant of reticuloendothelial activity in rats and mice, *J. Reticuloendothelial Soc.* **17**:135.

Stein, O., Stein, Y., Goodman, D. S., and Fidge, A., 1969, The metabolism of chylomicron cholesteryl esters in rat liver: A combined radioautographic electron microscopic and biochemical study, *J. Cell Biol.* **43**:410.

Stein, O., Rachmilewitz, D., Sanger, L., Eisenberg, S., and Stein, Y., 1974, Metabolism of iodinated very low density lipoprotein in the rat: Autoradiographic localization in the liver, *Biochim. Biophys. Acta* **360**:205.

Stiffel, C., Mouton, D., and Biozzi, G., 1970, Kinetics of the phagocytic function of reticuloendothelial macrophages *in vivo*, in: *Mononuclear Phagocytes* (R. van Furth, ed.), pp. 335–381, Blackwell, Oxford.

Stossel, T. R., Pollard, T. D., Mason, R. J., and Vaughan, M., 1971, Isolation and properties of phagocytic vesicles from polymorphonuclear leucocytes, *J. Clin. Invest.* **50**:1745.

Stossel, T. R., Mason, R. J., Pollard, T. D., and Vaughan, M., 1972, Isolation and properties of phagocytic vesicles. II. Alveolar macrophages, *J. Clin. Invest.* **51**:604.

Stuart, A. E., Biozzi, G., Stiffel, C., Halpern, B. N., and Mouton, D., 1960, The stimulation and depression of reticuloendothelial phagocytic function by simple lipids, *Br. J. Exp. Pathol.* **41**:599.

Tanaka, T., Taneda, K., Kobayashi, H., Okumara, K., Muranishi, S., and Sezaki, H., 1975, Application of liposomes to the pharmaceutical modification of the distribution characteristics of drugs in the rat, *Chem. Pharm. Bull.* **23**:3069.

Torchilin, V. P., Berdichevsky, V. R., Barsukov, A. S., and Smirnov, V. N., 1980, Coating liposomes with protein decreases their capture by macrophages, *FEBS Lett.* **111**:184.

van Berkel, T. J. C., 1979, The role of non-parenchymal cells in liver metabolism, *Trends Biochem. Sci.* **4**:202.

van Berkel, T. J. C., and Kruijt, J. K., 1977, Distribution and some properties of NADPH and NADH oxidase in parenchymal and non-parenchymal liver cells, *Arch. Biochem. Biophys.* **179**:8.

van Berkel, T. J. C., and van Tol, A., 1978, *In vivo* uptake of human and rat low density and high density lipoprotein by parenchymal and non-parenchymal cells from rat liver, *Biochim. Biophys. Acta* **530**:299.

van Berkel, T. J. C., and van Tol, A., 1979, Role of parenchymal and non-parenchymal rat liver cells in the uptake of cholesterolester-labeled serum lipoproteins, *Biochem. Biophys. Res. Commun.* **89:**1097.

van Berkel, T. J. C., Kruijt, J. K., and Koster, J. F., 1975, Identity and activities of lysosomal enzymes in parenchymal and non-parenchymal cells from rat liver, *Eur. J. Biochem.* **58:**145.

van Berkel, T. J. C., Kruijt, J. K., van Gent, T., and van Tol, A., 1980a, Saturable high affinity binding of low density and high density lipoprotein by parenchymal and non-parenchymal cells from rat liver, *Biochem. Biophys. Res. Commun.* **92:**1002.

van Berkel, T. J. C., Vaandrager, H., Kruijt, J. K., and Koster, J. F., 1980b, Characteristics of acid lipase and acid cholesteryl esterase activity in parenchymal and non-parenchymal rat liver cells, *Biochim. Biophys. Acta* **617:**446.

van Berkel, T. J. C., Kruijt, J. K., van Gent, T., and van Tol, A., 1981, Saturable high affinity binding, uptake and degradation of rat plasma lipoproteins by isolated parenchymal and non-parenchymal cells from rat liver, *Biochem. Biophys. Acta* **665:**22.

van Berkel, T. J. C., Nagelkerke, J. F., and Harkes, L., 1982, Liver sinusoidal cells and lipoprotein metabolism, in: *Sinusoidal Liver Cells* (Knook, D. L. and Wisse, E., eds.), pp. 305–318, Elsevier Biomedical Press, Amsterdam.

van Furth, R., Cohn, Z. A., Hirsch, J. G., Humphrey, J. H., Spector, W. G., and Langevoort, H. L., 1972, The mononuclear phagocyte system: A new classification of macrophages, monocytes, and their precursor cells, *Bull. W.H.O.* **46:**845.

van Tol, A., and van Berkel, T. J. C., 1980, Uptake and degradation of rat and human very low density (remnant) apolipoprotein by parenchymal and non-parenchymal rat liver cells, *Biochim. Biophys. Acta* **619:**156.

van Tol, A., van Gent, T., and Jansen, H., 1980, Degradation of high density lipoprotein by heparin-releasable liver lipase, *Biochem. Biophys. Res. Commun.* **94:**101.

Werb, Z., and Cohn, Z. A., 1972, Cholesterol metabolism in the macrophage. III. Ingestion and intracellular fate of cholesterol and cholesterol esters, *J. Exp. Med.* **135:**21.

Wisse, E., 1970, An electron microscopic study of the fenestrated endothelial lining of rat liver sinusoids, *J. Ultrastruct. Res.* **31:**125.

Wisse, E., Gregoriadis, G., and Daems, W. T., 1976, Electron microscopic cytochemical localization of intravenously injected liposome-encapsulated horseradish peroxidase in rat liver cells, in: *The Reticuloendothelial System in Health and Disease: Functions and Characteristics* (S. M. Reichard, M. R. Escobar, and H. Friedman, eds.), pp. 237–245, Plenum Press, New York.

14

Selected Aspects of Protein Metabolism in Relation to Reticuloendothelial System, Lymphocyte, and Fibroblast Function

MICHAEL C. POWANDA and ELIZABETH D. MOYER

1. INTRODUCTION

For the purposes of this discussion, the reticuloendothelial or mononuclear phagocyte system will be considered to include monocytes, histiocytes, Kupffer cells, alveolar macrophages, and phagocytic cells in spleen, lymph nodes, bone marrow, the nervous system, and the pleural and peritoneal cavities (Bloom and Fawcett, 1975). Fibroblasts, which are facultative phagocytes, are not usually included in the RES, nor are lymphocytes, but both cell types interact with mononuclear cells in the performance of their functions (Pierce, 1980; Leibovich and Ross, 1975). There are a number of plasma proteins (excluding immunoglobulins) which may play roles in host defense against infection and in wound healing. Some of these are synthesized by, catabolized within, and/or adherent to RES cells, lymphocytes, and fibroblasts. The possible consequences of these protein–cell interactions are the focus of this chapter.

Plasma protein synthesis, other than that of the immunoglobulins, was initially believed to take place solely in the liver (Miller and Bale, 1954). Later studies indicated that there is considerable extrahepatic formation of many plas-

MICHAEL C. POWANDA • Biochemistry Branch, U.S. Army Institute of Surgical Research, Fort Sam Houston, Texas 78234. *Present address:* Division of Cutaneous Hazards, Letterman Army Institute of Research, Presidio of San Francisco, California 94129. ELIZABETH D. MOYER • Departments of Surgery and Biochemistry, State University of New York, Buffalo, New York 14215. *Present address:* Department of Intravenous Nutrition, Cutter Group of Miles Laboratories, Berkeley, California 94701. The views of the authors do not purport to reflect the positions of the Department of the Army or the Department of Defense.

ma proteins (Kukral *et al.*, 1963; Weissman *et al.*, 1966). Moreover, it has even been suggested that all plasma protein synthesis by the liver is in fact the product of the RES, in particular the Kupffer cell, and not the parenchymal cell (Mehta, 1977). However, there is a wealth of evidence that hepatocytes are capable of synthesizing the vast majority of plasma proteins (Le Guilly *et al.*, 1973; Jeejeebhoy *et al.*, 1975, 1977; Knowles *et al.*, 1980). These studies do not exclude the possibility that *in vivo* the propinquity of Kupffer and parenchymal cells could facilitate interactions leading to regulation of plasma protein formation. During infection, injury, or inflammation, Kupffer cells may become activated and release leukocyte endogenous mediator (LEM) which can directly stimulate the liver to enhanced acute-phase protein synthesis (Powanda, 1977; Kampschmidt, 1978) without LEM entering the general circulation. Recent data indicate that LEM is identical with, or closely related to, lymphocyte activation factor (LAF) (Murphy *et al.*, 1980). Thus, during infection or following injury, macrophages can, with one mediator, not only regulate the activity of the nonspecific host defense/repair system (Powanda, 1977) a facet of which is the hepatic production of acute-phase proteins, but can also modulate the development of the specific immune response (Powanda, 1977; Powanda and Beisel, 1982).

In the absence of injury or infection, plasma proteins constitute approximately 3.5% of total body protein and represent 10% of whole body protein turnover (Wannemacher, 1975). Only approximately 40% of plasma protein degradation takes place in the liver. The rest occurs extrahepatically, in intestines, kidneys, skin, and phagocytes (Elwyn, 1970); thus, even in a healthy individual, plasma proteins provide a way of supplying extrahepatic tissue with nitrogen. During injury, inflammation, and/or infection, the turnover/concentration of individual plasma proteins may increase 5- to 10-fold (Zeineh and Kukral, 1970; Zeineh *et al.*, 1972; Farrow and Baar, 1973; Daniels *et al.*, 1974a,b; Powanda *et al.*, 1975, 1979; Powanda, 1977; Bostian *et al.*, 1976). At the same time, amino acid incorporation into plasma proteins increases 2- to 10-fold (Powanda *et al.*, 1972; Wannemacher *et al.*, 1974). Despite the limitations of the methods used in estimating plasma protein synthesis and turnover, it should be apparent that plasma protein metabolism during severe injury and/or infection represents a major portion of overall body nitrogen metabolism. The magnitude of nitrogen turnover involved warrants examination of the interactions of some of these plasma proteins with RES cells, lymphocytes, and fibroblasts to ascertain how these proteins may affect host defense against infection and wound healing. This is especially true when one realizes that these cells are themselves capable of synthesizing and catabolizing some of these proteins.

2. INTERACTIONS OF SPECIFIC PROTEINS WITH RES, LYMPHOCYTES, AND FIBROBLASTS

2.1. α_1-ANTITRYPSIN

α_1-Antitrypsin (α_1-AT) concentration increases during infection (Bostian *et al.*, 1976; Powanda, 1977) and following injury (Minchin Clarke *et al.*, 1971). *In vitro* studies indicate that α_1-AT is capable of inhibiting pancreatic and neutrophil elastase, pancreatic trypsin and chymotrypsin, neutrophil cathepsin G, thrombin, plasmin, acrosin, tissue kallikrein, Factor Xa, Factor XIa, skin and synovial collagenase and urokinase, as well as serine proteases released by a variety of microorganisms such as *B. subtilis*, *Proteus* and *Pseudomonas aeruginosa* (Laurell and Jeppsson, 1975; Harpel, 1982; Schapira *et al.*, 1982; Travis and Salvesen, 1983). Thus it would seem that α_1-AT could function to limit tissue damage due to proteases released by injured tissue, activated phagocytes and invading microorganisms, as well as to modulate clot formation and lysis. However, estimates of the *in vivo* half-time of inhibition suggest that α_1-AT is only exceedingly effective against elastase (Travis and Salvesen, 1983). These estimates of *in vivo* inhibition do not appear to take into consideration the possibility of surface binding of the inhibitor which can increase the efficiency of inhibition (Johnson and Varani, 1981), or the conditions which constitute the microenvironment of the inflammatory lesion, all of which may enhance the ability of α_1-AT to inhibit a variety of serine proteases at sites of inflammation.

α_1-AT is synthesized by the liver (Le Guilly *et al.*, 1973), but production of α_1-AT by the monocyte-macrophage family has been documented (Isaacson *et al.*, 1979; Wilson *et al.*, 1980; White *et al.*, 1981). Moreover, there is indirect evidence of local production of α_1-AT in bronchial secretions (Szabó *et al.*, 1980). However, other data suggest that the bulk of α_1-AT associated with peritoneal macrophages is derived from plasma (Remold-O'Donnell and Lewandrowski, 1983). Remold-O'Donnell and Lewandrowski (1983) present evidence that peritoneal macrophages do synthesize an inhibitor of trypsin and elastase which is of lower molecular weight than α_1-AT and which appears not to be detectable in normal plasma. It may be that culture conditions, the method of obtaining the macrophages and the origin of the macrophages account for these disparate data. The important, but unanswered, questions are, which and how much protease inhibitor is produced at the site of inflammation. Though α_1-AT is not synthesized by lymphocytes (Wilson *et al.*, 1980), it appears to reversibly bind to lymphocytes, seemingly more so to B than to T cells (Bata *et al.*, 1981).

Production of α_1-AT appears to be essential in that α_1-AT deficiency is associated with various pathological conditions such as chronic obstructive pulmonary disease (Lieberman, 1973), an inherited form of cirrhosis (Eriksson and Larsson, 1975), and an increased risk of arthritis (Cox and Huber, 1980), as well as chronic active hepatitis and cryptogenic cirrhosis (Hodges *et al.*, 1981). Emphysema associated with heavy smoking may be at least partly caused by an induced functional deficiency of α_1-AT resulting from the local inactivation of this protease inhibitor (Janoff *et al.*, 1983); this functional deficiency occurs despite normal concentrations of α_1-AT in serum and five to ten times more α_1-AT in alveolar macrophages (Olsen *et al.*, 1975).

The balance between proteases and protease inhibitors has a conceptually obvious role in controlling the amount of tissue damage at a site of injury/infection. The fact that proteases seem to be involved in peroxide and superoxide production by polymorphonuclear leukocytes and monocytes (Goldstein *et al.*, 1979; Kitagawa *et al.*, 1980), can induce chemotactic activity (Hatcher *et al.*, 1977) and stimulate immunoglobulin synthesis (Stein-Streilein and Hart, 1980), are able to potentiate thymocyte blastogenesis (Ulrich, 1979) as well as activitate fibroblast cell division (Blumberg and Robbins, 1975) would suggest that the balance between proteases and protease inhibitors could have widespread consequences with respect to the inflammatory process. α_1-AT at an inflammatory nidus, whether as the result of diffusion from plasma or of secretion/production locally by macrophages, may therefore not only limit tissue destruction due to proteases but also regulate wound repair and modulate the development of immunity. In excess, however, α_1-AT could inhibit tissue regeneration, granulocyte infiltration, and phagocyte killing of microorganisms (Powanda and Moyer, 1981), as well as suppress the immune response (Arora *et al.*, 1978).

2.2. α_2-MACROGLOBULIN

α_2-Macroglobulin (α_2-MG) is the other major protease inhibitor found in serum (Laurell and Jeppsson, 1975). It is a zinc metalloprotein (Parisi and Vallee, 1970) and is capable of inhibiting a wide variety of proteolytic enzymes such as kallikrein, trypsin, elastase, cathepsin B, and collagenase, and to a lesser extent, plasmin and thrombin (Laurell and Jeppson, 1975). Plasma α_2-MG concentration is variously reported to be unchanged or decreased by infection and injury (Daniels *et al.*, 1974b; Bostian *et al.*, 1976; Werner and Cohnen, 1969), yet there is an increased turnover of this protein in burn patients (Farrow and Baar, 1973) and increased utilization after major surgery (Dickson and Manning, 1972). This seemingly atypical acute-phase protein response may in part be due to the relatively high molecular weight of α_2-MG (725,000). The size of α_2-MG restricts its extravasation even into inflamed tissue (Kushner and Somerville, 1971). Another factor may be related to its site of synthesis: α_2-MG is not only produced by liver (Le Guilly *et al.*, 1973), but also by monocytes (Hovi *et al.*, 1977), lymphocytes (Tunstall and James, 1974), and fibroblasts (Mosher and Wing, 1976). Thus, one may hypothesize that α_2-MG produced in the liver may only be important in clearing circulating proteases, while α_2-MG synthesized by monocytes, lymphocytes and fibroblasts would provide α_2-MG in the vicinity of injury or inflammation.

Fibroblasts not only produce α_2-MG but also bind it to specific membrane surface receptors (Willingham *et al.*, 1979). After binding, α_2-MG is internalized and degraded by these cells (Maxfield *et al.*, 1978; Mosher and Vaheri, 1980). Since α_2-MG, insulin, epidermal growth factor, and low density lipoproteins share a common path of internalization (Maxfield *et al.*, 1978; Via *et al.*, 1982), binding of one of these substances could possibly either inhibit or facilitate the rate of clearance and inactivation of the others. If α_2-MG were to facilitate insulin clearance by fibroblasts, then the increased turnover of α_2-MG during injury might also be a factor in the generation of insulin resistance such as has been observed during injury and infection (Ryan *et al.*, 1974; Wolfe *et al.*, 1979). In fact, enhanced turnover of α_2-MG at the wound with a concomitant acceleration in insulin clearance may explain why insulin resistance appears to increase in proximity to the wound (Turinsky and Patterson, 1979), though this may merely reflect the presence of dead tissue. Consistent with but not proof of this hypothesis is the fact that though traumatic injury induces insulin resistance *in vivo, in vitro* evaluation of fat pad and muscle tissue metabolism indicates that insulin resistance is not due to a decreased capacity of these tissues to respond to insulin (Nelson and Filkins, 1979). Conversely, one might also wonder if the increased serum α_2-MG levels in diabetics (James *et al.*, 1980) result from decreased α_2-MG turnover in the absence of insulin. A similar relationship might exist between α_2-MG and epidermal growth factor and low density lipoproteins whereby α_2-MG could regulate cell proliferation.

α_2-MG complexed with proteases is taken up by fibroblasts (van Leuven *et al.*, 1978), and to even a greater extent by macrophages (Debanne *et al.*, 1975; Kaplan and Nielson, 1979). The proteases may bind directly to α_2-MG or be transferred from α_1-AT (Ohlsson, 1975). α_1-AT–protease complexes apparently are not cleared by macrophages, in contrast to α_2-MG–protease complexes which are (Dolovich *et al.*, 1975). While binding of proteases to α_1-AT apparently completely inactivates them (Laurell and Jeppsson, 1975), proteolytic enzymes bound to α_2-MG still retain activity toward denatured proteins and peptide hormones (Rinderknecht and Geokas, 1973), especially in the acidic conditions which would exist at the inflammatory site (Harpel, 1982). Formation of α_2-MG–protease complexes thus not only helps to limit tissue damage by removing the proteases, but also focuses the residual proteolytic activity of the complexes on denatured proteins and peptides such as would be found at the site of injury or inflammation. In addition, the rapid clearing of α_2-MG–protease complexes (half-life < 10 min) insures that even such residual proteolytic activity will be limited in duration. If not cleared rapidly, α_2-MG–protease complexes might stimulate B cells and lead to the formation of autoantibodies (Teodorescu *et al.*, 1982). Since macrophages not only secrete proteases, but stimulate fibroblasts to do so as well, e.g., collagenase (Huybrechts-Godin *et al.*, 1979), α_2-MG production by both macrophages and fibroblasts may be a means by which these cells attempt to localize and restrict the action of the proteolytic enzymes they release and thereby modulate inflammation. However, there are data from *in vitro* studies which indicate that endocytosis of α_2-MG–protease complexes by macrophages actually leads to the release of newly formed neutral proteases (Vischer and Berger, 1980), which suggests that α_2-MG may under some circumstances perpetuate inflammation.

Hubbard (1978) has proposed that the relative amounts of free and protease-complexed α_2-MG control cell proliferation and exert immunomodulatory effects. The idea that free α_2-MG may enhance cell proliferation is derived from the observation that transformed, but not untransformed fibroblasts synthesize α_2-MG (Mosher and Wing, 1976) and that elevated concentrations of α_2-MG are present in normal individuals during periods of active tissue growth. For example, α_2-MG is present in children at much higher levels than in adults (Weeke and Krasilnikoff, 1972). Alternatively, the increased levels of α_2-MG may reflect the role α_2-MG plays as a zinc carrier, since zinc is essential to growth and development, or the increased levels may be the result, rather than the cause, of increased cell proliferation. As additional evidence for his hypothesis, Hubbard has noted that α_2-MG–protease complexes inhibit mouse L-cell division.

The evidence that Hubbard presents of a role of α_2-MG and α_2-MG–proteases in modulating the immune system is no less tenuous than for its proposed role in cell proliferation, but it is nonetheless provocative. If this hypothesis is valid, it might explain how parasites such as *Schistosoma mansoni* (which bears host-like α_2-MG antigenic determinants on its surface) persist so long. In support of Hubbard's hypothesis of the role of α_2-MG in regulating the immune response, α_2-MG has been shown to modulate lymphocyte proliferation in response to both mitogens and antigens (Goutner *et al.*, 1976). At low doses α_2-MG potentiates and at high doses markedly inhibits this response. α_2-MG in moderate doses also has the ability to restore humoral responsiveness in X-irradiated mice, while high doses of α_2-MG are inhibitory (Tunstall and James, 1975). α_2-MG and its subunits have been shown to inhibit natural killing and antibody-dependent cell-mediated cytotoxicity (Gravagna *et al.*, 1982). In addition, diabetics who have suppressed cell-mediated immunologic responses (Mahmoud *et al.*, 1976) also have increased α_2-MG levels (James *et al.*, 1980). A benefit of this proposed system would be to reduce the reactivity of the immune system to antigenic sites unmasked as a result of injury. This could prevent the development of autoimmune disease or an anaphylactic reaction upon subsequent exposure of similar sites during wound repair or reinjury (Munster, 1976). A drawback of such a system may arise in the case of a major, slow-healing injury, e.g., a severe burn, continually releasing large quantities of proteases which in turn would generate considerable amounts of α_2-MG–protease complexes. If high levels of α_2-MG protease do in fact reduce immunoreactivity, this would result in an extended period of immunosuppression and increase susceptibility to infection. One wonders whether the findings of Glaser *et al.* (1973), that an α-globulin rich fraction of human serum decreased resistance of mice to infection, may be related to α_1-AT and α_2-MG inhibition of phagocyte superoxide production and/or the effects of the protease inhibitors on immune responsiveness.

Another role for α_2-MG appears to be that of a carrier protein for macrophage activating factor produced by lymphocytes (McDaniel *et al.*, 1976), thereby providing another means of communication between cells of the RES and the immune system. It is also possible that α_2-MG/α_2-MG–protease binding by lymphocytes may, if the α_2-MG is internalized and degraded, provide some of the zinc which appears to be required for DNA replication in lymphocytes (Williams and Leob, 1973).

2.3. C-REACTIVE PROTEIN

C-reactive protein (CRP) has been shown to be synthesized by hepatocytes in response to inflammation (Kushner and Feldmann, 1978). The rapid increase in plasma concentration of CRP elicited by injury and infection is seemingly in proportion to the extent of injury or severity of disease (Werner, 1969; Fischer *et al.*, 1976; Kushner *et al.*, 1978). CRP has been reported to accumulate at the site of injury following myocardial infarction (Kushner *et al.*, 1963) or localized inflammation (Kushner and Kaplan, 1961). Plasma CRP concentration does not, however, increase in all instances of extensive tissue damage, and is not always found at the site of inflammation, for example in childhood dermatomyositis (Haas *et al.*, 1982). The exact function of this protein in the inflammatory response is not known. However, CRP can initiate agglutination reactions (Osmand *et al.*, 1975) and complement consumption (Kaplan and Volanakis, 1974; Siegel *et al.*, 1974) via the classical pathway (Mold and Gewurz, 1981) as well as enhance phagocytosis of certain pathogenic microorganisms (Hokama *et al.*, 1962; Kindmark, 1971; Mold *et al.*, 1982). CRP can also inhibit aggregation of platelets and the release of platelet enzymes (Fiedel and Gewurz, 1976). In addition, CRP has been reported to enhance the movement of polymorphonuclear leukocytes into experimentally induced wounds (Ahlstedt, 1980), and increase the local temperature when injected into the joints of rabbits with experimentally induced arthritis (Phillips, 1982). Increases in CRP would thus seem designed to facilitate the clearance of microorganisms and necrotic tissue as well as to modulate the local inflammatory response.

Another facet of CRP function appears to be the regulation of cellular and humoral immunity. In earlier studies CRP was found to bind selectively to T lymphocytes (Mortensen *et al.*, 1975), preferentialy to antigen- but not mitogen-stimulated T lymphoblasts (Croft *et al.*, 1976). More recent studies from that laboratory have failed to reproduce these results, but do demonstrate that CRP, in the presence of appropriate ligands, binds to B as well as T cells and to mononuclear cells (James *et al.*, 1982). Nonetheless, Mortensen (1982) reports that purified CRP and CRP–C-polysaccharide complexes inhibit pokeweed mitogen-induced but not T-independent generation of polyclonal immunoglobin-secreting cells. Selective inhibition of T-dependent antigen production could lead to depressed cellular immunity with normal or even enhanced humoral immunity, at least with regard to some antigens.

In point of fact, surgical and mild to moderate thermal injury produce just such a response: decreased cell-mediated immunity and enhanced generation of antibody-forming cells, seemingly in proportion to the extent of injury (Howard and Simmons, 1974; Rapaport and Bachvaroff, 1976; Kinnaert *et al.*, 1978). Even severe injury which results in a depression in both T- and B-lymphocyte populations has less of an effect on B cells than T cells (Volenec *et al.*, 1977). This increase in antibody production, however, may be due to a decrease in the clearance function of the RES resulting in an increased persistence of the antigen (Rapaport and Bachvaroff, 1976). It is thus conceivable that CRP may be one of the factors responsible for the anergy which accompanies injury. Since CRP appears to increase in proportion to the degree of injury/inflammation, it is also possible that during severe injury, CRP might increase to such a concentration as to compromise both humoral and cell-mediated immunity, paving the way for infection. In support of this hypothesis, it should be noted that the levels of CRP necessary to markedly inhibit *in vitro* both aspects of immunity (50–100 μg/ml) can be achieved *in vivo*. In fact, following severe thermal injury (60–90% total body surface), CRP may reach 300–500 μg/ml (Powanda, unpublished results). Recent studies indicate that such high levels of CRP and CRP-complexes can enhance cell-mediated cytotoxicity (Vetter *et al.*, 1983).

Advances in assay techniques demonstrate that CRP is not only present in plasma during injury and infection, but can be found in plasma from healthy individuals as well (Claus *et al.*, 1976). Osgood (1954) has estimated that there are about 6×10^{12} lymphocytes/70-kg man with 1.5×10^{10} in circulation. Normal adults have approximately 1 μg CRP/ml serum or about 2×10^{16} molecules of CRP in circulation; thus, some 10^6 molecules of CRP are available to bind to each circulating lymphocyte. This suggests that CRP may play a role in immune modulation even in healthy individuals. However, a caveat is in order: the results obtained by various investigators concerning the effects of CRP on immune function at times seem to be at odds. This may be due to the assay systems employed, the purity of biologic reagents, or simply the complexity of the system, wherein the presence of other plasma proteins could alter the results.

Finally, recent findings have confirmed that CRP in plasma is at least in part associated with very-low-density lipoproteins (Cabana *et al.*, 1982). The significance of these data is unknown, but it should be remembered that lipoproteins, particularly low- and very-low-density lipoproteins may affect the immune system (Section 2.5).

2.4. TRANSFERRIN

Transferrin is the major circulating iron-binding protein in healthy individuals. There are transient decreases in plasma transferrin concentration during inflammation (Powanda *et al.*, 1979), infection (Bostian *et al.*, 1976), and following surgery (Werner and Cohnen, 1969). Transferrin is synthesized by liver (Thorbecke *et al.*, 1973), macrophages (Stecher and Thorbecke, 1976a), and lymphoblasts (Stecher and Thorbecke, 1967b). Dermal histiocytes (Beamish *et al.*, 1971), splenic and pulmonary macrophages (O'Shea *et al.*, 1973), and fibroblasts (Octave *et al.*, 1979) are able to take up transferrin-bound iron. Rabbit alveolar macrophages bind 80% iron-saturated transferrin more readily than 10 or 50% saturated transferrin (Wyllie, 1977). Transferrin can also bind and transport zinc (Evans and Winter, 1975); zinc transferrin attaches to lymphocytes (Phillips, 1976) and enhances nucleic acid synthesis in stimulated lymphocytes (Phillips and Azari, 1974).

Uptake of highly saturated transferrin by macrophages may be one of the ways in which iron is sequestered so as to prevent, or at least limit, microorganism dissemination (Weinberg, 1978; Kluger and Rothenberg, 1979). However, lactoferrin is also accumulated by macrophages (Markowetz *et al.*, 1979), and lactoferrin rather than transferrin may play the major role in the sequestration of iron during inflammatory diseases (van Snick *et al.*, 1974). Both lactoferrin and transferrin when saturated with iron can enhance hydroxyl radical production by neutrophils and their particulate fractions (Ambruso and Johnston, 1981), and thus both compounds may be involved in the microbicidal activity of neutrophils, and perhaps other phagocytic cells as well. The iron that accumulates in the macrophages is preferentially taken up by mitochondria (Wyllie, 1979), perhaps to be incorporated into the electron transport chain, but some of it is also stored as ferritin. Uptake of iron by fibroblasts may provide the cofactor for proline hydroxylase which is involved in collagen synthesis and thus facilitate wound healing.

2.5. LIPOPROTEINS

Lipoproteins generally fall into four main categories: chylomicron, very-low-density lipoprotein (VLDL), low-density lipoprotein (LDL), and high-density lipoprotein (HDL). Chylomicrons are formed postprandially, and transport lipid absorbed from the intestine to peripheral tissue. Through the action of tissue lipoprotein lipase loosely bound to the endothelial wall, the chylomicron triglyceride is hydrolyzed. During hydrolysis, the surface materials are, for the most part, transferred to HDLs, while the remnant is removed by the liver (Redgrave, 1970). VLDLs, on the other hand, are primarily synthesized by the liver to transport hepatically produced lipids, and differ from chylomicrons in that the degradation of these particles by lipoprotein lipase is less complete: low-density lipoproteins (rather than remnants removed by the liver) result. (For review, see Cryer, 1981.)

Low density lipoproteins are in turn removed from the plasma and are metabolized by one of two routes: (1) receptor mediated pathway, and (2) "scavenger" pathway. Cultured human fibroblasts (Goldstein and Brown, 1974) and lymphocytes (Ho *et al.*, 1977) recognize, bind, and degrade LDL as a function of the apoprotein normally present in the highest concentration on LDLs, apo-B. (Mahley *et al.*, 1977). Macrophages also have a receptor-mediated pathway for LDL uptake; however, these receptors seem to recognize only apo-B protein that has been chemically modified (Goldstein *et al.*, 1979). The significance of this difference is unclear, but one attractive hypothesis is that LDLs present at the site of mural thrombosis might be reacted upon by malondialdehyde released by platelets as a function of prostaglandin synthesis (Goldstein and Brown, 1983). Since LDLs are known to damage vascular epithelia (Henriksen *et al.*, 1979), macrophage removal of these altered lipoproteins may facilitate revascularization and wound healing. The "scavenger" pathway appears to become extremely important in familial hypercholesterolemia, when LDL receptors are absent or defective (Brown and Goldstein, 1974). Normal removal of LDLs by the receptor mediated system being inhibited, most of the LDL removal appears to take place through the scavenger pathway. As a result, macrophages become engorged with LDL cholesterol and become "foam" cells, precursors to atherosclerotic plaque (Goldstein and Brown, 1983).

Cell density appears to affect the degree of LDL binding by fibroblasts, with rapidly growing cells having more LDL receptors than stationary cultures (Kruth *et al.*, 1979). In fact, lipoproteins seem to be required for dermal fibroblast proliferation, since serum from abetalipoproteinemic patients is very much less effective than normal serum in promoting the growth of these cells in culture (Layman *et al.*, 1980). Addition of LDL to lipoprotein deficient serum increases skin fibroblast replication, the cellular content of protein and cholesterol and the secretion of glycosaminoglycans (Spain *et al.*, 1979). LDL also is capable of regulating cholesterol synthesis in lymphoid cells (Kayden *et al.*, 1976; Ho *et al.*, 1977), while studies with macrophages indicate that VLDL enhances rather than depresses cholesterol synthesis and accumulation (Goldstein *et al.*, 1980).

HDLs contain more protein than any of the other lipoprotein classes, hence their relatively greater density. The major protein component of HDLs is apo-AI, which is a potent activator of lecithin cholesterol acyl transferase (Scanu *et al.*, 1982). It is not surprising, therefore, that HDL appears to play a pivotal role in the regulation of cholesterol metabolism: macrophage "foam" cells incubated in the absence of HDLs, appear to continuously esterify and de-esterify the accumulated cholesterol in an ATP-consuming futile cycle; but in the presence of HDLs, the cycle is disrupted, and free cholesterol is released (Brown *et al.*, 1980). At the same time, macrophages release apo-E, a protein which facilitates lipoprotein binding to receptors [apo-E binds even more avidly to the lipoprotein receptor than apo-B (Innerarity and Mahley, 1978)], facilitating the removal of lipoproteins enriched with the cholesterol released by macrophages. It has also been found that a subclass of HDL, HDL_3, which is relatively low in cholesterol (Durrington, 1982) stimulates fibroblast cholesterol synthesis whereas HDL_2, which has been hypothesized to be formed from HDL_3 as a result of accepting cellular lipids (Eisenberg, 1980), inhibits fibroblast cholesterol synthesis (Daerr *et al.*, 1980).

The apoprotein content of lipoproteins is essential to their recognition and metabolism, as can be seen from the discussion above. In some cases, however, for reasons which are not yet clearly understood, the apoprotein portion of lipoproteins, particularly LDL and HDL, appear to form complexes which can only be separated in the presence of a disulphide reducing agent, such as dithiothreitol. When these complexes are formed, abnormal lipoproteins result, at least some of which have been found to compromise immune function. One of these, an abnormal form of LDL, has been shown to bind to lymphocytes and attenuate sheep erythrocyte rosette formation by T lymphocytes: rosette inhibiting factor (RIF) (Chisari and Edgington, 1975). Defective T-lymphocyte erythrocyte rosette formation is observed in association with cancer, viral infection (Wybran and Fudenberg, 1973) and autoimmune diseases (Messner *et al.*, 1973), circumstances in which lipid metabolism is generally altered. Normal human sera contain a subspecies of LDL, designated low-density lipoprotein inhibitor (LDL-In), which contains another apoprotein complex, exerts its effects by inhibiting lymphocyte proliferation rather than differentiation (Curtiss *et al.*, 1980) and can affect both B and T lymphocyte populations. Mitogenic stimulation of lymphocytes has also been shown to be inhibited by autooxidized LDL (Schuh *et al.*, 1978). Apoprotein complexes have been reported in HDL as well (Tada *et al.*, 1979; Weisgraber and Mahley, 1978). While the importance of HDL apoprotein complexes to immune function has yet to be demonstrated, it is perhaps significant that an HDL subfraction has also been reported to inhibit autoerythrocyte rosette formation by mouse thymus cells (Hsu *et al.*, 1980).

LDL conceivably could play a significant role in wound healing, fostering replication of fibroblasts, supplying cholesterol for membrane synthesis, and stimulating glycosaminoglycan secretion so as to provide both cell components and the extracellular matrix for tissue regeneration. VLDL, by stimulating cholesterol synthesis and accumulation by macrophages, may alter the fluidity of the membranes of these cells (Emmelot, 1977), which may affect their movement through tissue to the sites of injury or inflammation. Both LDL and VLDL may modulate rather than abrogate immune responsiveness by acting as a dampening mechanism; VLDL cannot entirely suppress the immune response even at six times the maximal inhibitory dose (Chisari, 1980), and the effect of LDL-In is influenced by antigen dose, being more inhibitory at low antigen doses (Curtiss *et al.*, 1977).

2.6. FIBRONECTIN

Fibronectin, also known as LETS protein, cold-insoluble globulin, opsonic protein, and fibroblast surface antigen, exists in both circulating and cell surface forms which appear identical immunologically (Mosesson, 1977; Yamada and Olden, 1978). These two forms are also similar in the ability to promote both cell attachment to collagen and cell spreading, but differ in subunit molecular weight and the capacity to agglutinate erythrocytes as well as to restore the morphology of transformed cells toward normal (Yamada and Kennedy, 1979). Hepatocytes (Voss *et al.*, 1979), peritoneal macrophages (Johansson *et al.*, 1979), monocytes (Alitalo *et al.*, 1980), fibroblasts, and certain epithelial cells (Yamada and Olden, 1978) are among the many cells which secrete fibronectin.

Fibronectin appears to be required for cell attachment not only *in vitro* (Grinnell and Feld, 1979; Blaauboer and Paine, 1979) but also *in vivo* (McDonald *et al.*, 1979; Fyrand, 1979; Weiss and Reddi, 1981). Both circulating (exogenous) and endogenous (secreted by the cell) fibronectin may be incorporated into the extracellular matrix (Hayman and Ruoslahti, 1979). Fibronectin mediates the binding of fibrin to macrophages (Jilek and Hörmann, 1978). Fibronectin also reportedly acts as the collagen receptor on platelet cell membranes (Bensusan *et al.*, 1978), but antibodies against fibronectin only slightly reduce adhesion, suggesting that either fibronectin is not ordinarily involved in platelet binding to collagen, or that it is only one of several adhesive factors (Santoro and Cunningham, 1979). On the other hand, only miniscule amounts of collagen are required to stimulate platelet aggregation (Kronick and Jimenez, 1979), and thus one might anticipate that very little fibronectin would be required. Fibronectin does seem necessary for platelets to spread on fibrinogen or collagen-coated substrate and to lyse (Grinnell *et al.*, 1979). Fibronectin binding to platelets is stimulated by thrombin, but not by adenosine diphosphate or epinephrine (Plow and Ginsberg, 1981).

Cold-insoluble globulin, the circulating form of fibronectin, localizes at the site of injury (Kaplan *et al.*, 1976). Fibronectin appears to accumulate at the site of epidermal lesions in a variety of skin diseases (Fyrand, 1980). Fibronectin binds to denatured collagens more avidly than to native collagens, and amongst native collagens more readily to Type III than to Type I (Ruoslahti *et al.*, 1981). In the initial stages of wound healing, there is an increased synthesis of Type-III collagen. As repair progresses, Type-I collagen becomes predominant (Minor, 1980). The relative affinity of fibronectin for the various forms of collagen is consistent with its tendency to localize at sites of injury and may even influence the pattern of collagen synthesis and cell proliferation during wound healing. Fibronectin not only has a binding site for collagen (Balian *et al.*, 1979) [the same site binds fibrin(ogen) and actin] (Ruoslahti and Vaheri, 1975; Keski-Oja *et al.*, 1980) but also for cell surfaces (Ruoslahti and Hayman, 1979), DNA, and staphylococci (Ruoslahti *et al.*, 1981). The interaction between cold-insoluble globulin and fibrin or collagen appears to be mediated by Factor XIII (Mosher, 1976a; Mosher *et al.*, 1979). The cross-linking of fibronectin to collagen and fibrin and the binding to actin may be important in wound healing both in masking actin and collagen so as to prevent an immune response to these otherwise occult antigens as well as providing nidi for the establishment of matrices for macrophage and fibroblast interactions (Leibovich and Ross, 1975).

Cold-insoluble globulin reportedly facilitates phagocytosis by Kupffer cells (Blumenstock *et al.*, 1976) and other RE cells (Niehaus *et al.*, 1980). However, at physiological concentrations fibronectin binding to staphylococci does not promote phagocytosis of bacteria by polymophonuclear leucocytes, monocytes, or alveolar macrophages and depletion of fibronectin from serum does not result in a measurable loss of opsonic activity (Verburgh *et al.*, 1981). Cold-insoluble globulin transiently decreases in concentration during infection (Mosher, 1976b) and following injury (Lanser *et al.*, 1980). In individuals who die from injury, or who become septic following injury, the decrease persists and is exacerbated (Berghem *et al.*, 1979). Treatment of septic surgical and trauma patients with cold-insoluble globulin lessens septicemia and pulmonary insufficiency, improves renal function and shortens the duration of the recovery period (Saba *et al.*, 1978). Fibronectin and its soluble form, cold-insoluble globulin, thus could be important in maintaining resistance to infection as well as in wound healing.

3. SUMMARY AND CONCLUSIONS

Table 1 summarizes the information regarding the potential of RES cells, lymphocytes and fibroblasts to synthesize, bind, and/or degrade selected plasma proteins. Much of these data were gathered from *in vitro* studies, and thus it is difficult to assess the *in situ* capacity of these cells for synthesis, binding and/or catabolism of many of these proteins. However, fibronectin is clearly made *in vivo* by these cells, and the circumstantial evidence indicates that α_1-AT is produced locally by alveolar macrophages. On the other hand, analysis of plasma protein polymorphism of donors and recipients in the case of hepatic transplantation indicates that the liver is the prime source of circulating transferrin and α_1-AT (Alper *et al.*, 1980) in noninjured, noninfected individuals. There was, however, one reported instance when the α_1-AT did not convert to the donor type and remained that of the recipient. Moreover, one should remember that a

TABLE 1. POTENTIAL FOR RES CELLS, LYMPHOCYTES, AND FIBROBLASTS TO SYNTHESIZE, BIND, AND/OR CATABOLIZE SELECTED PLASMA PROTEINS

	Synthesis	Binding	Catabolism
RES cells	α1-Antitrypsin α2-Macroglobulin Transferrin Fibronectin	α2-Macroglobulin C-reactive protein Transferrin Fibronectin Low-density lipoprotein Very-low-density lipoprotein High-density lipoprotein	α2-Macroglobulin Transferrin Low-density lipoprotein High-density lipoprotein?
Lymphocytes	α2-Macroglobulin Transferrin	α1-Antitrypsin α2-Macroglobulin Transferrin Low-density lipoprotein C-reactive protein Rosette-inhibiting factor Low-density lipoprotein inhibitor Very-low-density lipoprotein	Low-density lipoprotein
Fibroblasts	α2-Macroglobulin Fibronectin	α2-Macroglobulin Transferrin Fibronectin Low-density lipoprotein High-density lipoprotein	α_2-Macroglobulin Low-density lipoprotein High-density lipoprotein

transplanted liver contains RE cells which may contribute to the hepatic production of certain plasma proteins.

It is conceivable that proteins produced by RE cells, lymphocytes, and fibroblasts do not enter in appreciable concentrations into circulation but are formed and degraded locally. In a microenvironment, even small amounts of some of these proteins, especially macromolecules like α_2-MG, which would have little tendency or opportunity to diffuse, could exert profound effects. It is also possible that there is little or no synthesis of these proteins by the cells in question in healthy individuals, but marked production following stimulation of these cells as during inflammation, infection, and/or injury.

Evidence has been accumulated that cells of the RES, lymphocytes, and fibroblasts synthesize, bind, and/or degrade selected plasma proteins. Moreover, these plasma proteins, whether of hepatic origin or produced by the cells under discussion, appear to have the capacity to profoundly influence the metabolism and function of these cells as well as acting as a means of communication between these cells. The extent to which these proteins actually affect these cells in a healthy individual and what value interactions might have in host defense against infection and in wound healing remain to be determined.

ACKNOWLEDGMENT. E. D. M. was supported in part by Grant GM-15768-09A1 from the National Institute of General Medical Science.

REFERENCES

Ahlstedt, S., 1980, Inflammatory potency of antigen–antibody complexes and C-reactive protein in a wound chamber model in rats, *Int. Arch. Allergy Appl. Immunol.* **62**:341.

Alitalo, K., Hovi, T., and Vaheri, A., 1980, Fibronectin is produced by human macrophages, *J. Exp. Med.* **151**:602.

Alper, C. A., Raum, D., Awdeh, Z. L., Petersen, B. H., Taylor, P. D., and Starzl, T. E., 1980, Studies of hepatic synthesis *in vivo* of plasma proteins, including orosomucoid, transferrin, α_1-antitrypsin, C8, and factor B, *Clin. Immunol. Immunopathol.* **16**:84.

Ambruso, D. R., and Johnston, R. B., Jr., 1981, Lactoferrin enhances hydroxyl radical production by human neutrophils, neutrophil particulate fractions, and an enzymatic generating system, *J. Clin. Invest.* **67**:352.

Arora, P. K., Miller, H. C., and Aronson, L. D., 1978, α_1-Antitrypsin is an effector of immunological stasis, *Nature (London)* **274**:589.

Balian, G., Click, E. M., Crouch, E., Davidson, J. M., and Bornstein, P., 1979, Isolation of a collagen-binding fragment from fibronectin and cold-insoluble globulin, *J. Biol. Chem.* **254**:1429.

Bata, J., Deviller, P., and Revillard, J. P., 1981, Binding of alpha 1 antitrypsin (α_1 protease inhibitor) to human lymphocytes, *Biochem. Biophys. Res. Commun.* **98**:709.

Beamish, M. R., Jobbins, K., and Cavill, I., 1971, The cellular distribution of transferrin-bound iron in the skin, *Br. J. Dermatol.* **85**:49.

Bensusan, H. B., Koh, T. L., Henry, K. G., Murray, B. A., and Culp, L. A., 1978, Evidence that fibronectin is the collagen receptor on platelet membranes, *Proc. Natl. Acad. Sci. USA* **75**:5864.

Berghem, L., Ahlgren, T., and Lahnborg, G., 1979, Immunoassay of macrophage stimulating α_2SB (surface binding) glycoprotein in pig plasma following missile trauma, *Acta Chir. Scand. Suppl.* **489**:239.

Blaauboer, B. J., and Paine, A. J., 1979, Attachment of rat hepatocytes to plastic substrata in the absence of serum requires protein synthesis, *Biochem. Biophys. Res. Commun.* **90**:368.

Bloom, W., and Fawcett, D. W., 1975, *A Textbook of Histology*, 10th ed., pp. 178–181, Saunders, Philadelphia.

Blumberg, P. M., and Robbins, P. W., 1975, Effect of proteases on activation of resting chick embryo fibroblasts and on cell surface proteins, *Cell* **6**:137.

Blumenstock, F., Saba, T. M., Weber, P., and Cho, E., 1976, Purification and biochemical characterization of a macrophage stimulating alpha-2-globulin opsonic protein, *J. Reticuloendothelial Soc.* **19**:157.

Bostian, K. A., Blackburn, B. S., Wannemacher, R. W., Jr., McGann, V. G., and Beisel, W. R., 1976, Sequential changes in the concentration of specific serum proteins during typhoid fever infection in man, *J. Lab. Clin. Med.* **87**:577.

Brown, M. S., and Goldstein, J. L., 1974, Familial hypercholesterolemia: Defective binding of lipoproteins to cultured fibroblasts associated with impaired regulation of 3-hydroxy-3-methylglutaryl coenzyme A reductase activity, *Proc. Natl. Acad. Sci. USA* **71**:788.

Brown, M. S., and Goldstein, J. L., 1983, Lipoprotein metabolism in the macrophage: Implications for cholesterol deposition in atherosclerosis, *Ann. Rev. Biochem.* **52**:223.

Brown, M. S., Ho, Y. K., and Goldstein, J. L., 1980, The cholesteryl ester cycle in macrophage foam cells, *J. Biol. Chem.* **255**:9344.

Cabana, V. G., Gerwurz, H., and Siegel, J. N., 1982, Interaction of very low density lipoproteins (VLDL) with rabbit C-reactive protein, *J. Immunol.* **128**:2342.

Chisari, F. V., 1980, Modulation of the *in vivo* immune response by human plasma very low-density lipoproteins, *Cell. Immunol.* **52**:223.

Chisari, F. V., and Edgington, T. S., 1975, Lymphocyte E rosette inhibitory factor: A regulatory serum lipoprotein, *J. Exp. Med.* **142**:1092.

Claus, D. R., Osmand, A. P., and Gewurz, H., 1976, Radioimmunoassay of human C-reactive protein and levels in normal sera, *J. Lab. Clin. Med.* **87**:120.

Cox, D. W., and Huber, O., 1980, Association of severe rheumatoid arthritis with heterozygosity for α_1-antitrypsin deficiency, *Clin. Genet.* **17**:153.

Croft, S. M., Mortensen, R. F., and Gewurz, H., 1976, Binding of C-reactive protein to antigen-induced but not mitogen-induced T lymphoblasts, *Science* **193**:685.

Cryer, A., 1981, Tissue lipoprotein lipase activity and action in lipoprotein metabolism, *Int. J. Biochem.* **13**:525.

Curtiss, L. K., DeHeer, D. H., and Edgington, T. S., 1977, *In vivo* suppression of the primary immune response by a species of low density serum lipoprotein, *J. Immunol.* **118**:648.

Curtiss, L. K., DeHeer, D. H., and Edgington, T. S., 1980, Influence of the immunoregulatory serum lipoprotein LDL-In on the *in vivo* proliferation and differentiation of antigen-binding and antibody-secreting lymphocytes during a primary immune response, *Cell. Immunol.* **49**:1.

Daerr, W. H., Gianturco, S. H., Patsch, J. R., Smith, L. C., and Gotto, A. M., 1980, Stimulation and suppression of 3-hydroxy-3-methyl glutaryl coenzyme A reductase in normal human fibroblasts by high density lipoprotein subclasses, *Biochim. Biophys. Acta* **619**:287.

Daniels, J. C., Larson, D. L., Abston, S., and Ritzmann, S. E., 1974a, Serum protein profiles in thermal burns. I. Serum electrophoretic patterns, immunoglobulins, and transport proteins, *J. Trauma* **14**:137.

Daniels, J. C., Larson, D. L., Abston, S., and Ritzmann, S. E., 1974b, Serum protein profiles in thermal burns. II. Protease inhibitors, complement factors, and C-reactive protein, *J. Trauma* **14**:153.

Debanne, M. T., Bell, R., and Dolovich, J., 1975, Uptake of proteinase–α–macroglobulin complexes by macrophages, *Biochim. Biophys. Acta* **411**:295.

Dickson, I. R., and Manning, C. W., 1972, Changes in serum-alpha$_2$-macroglobulin levels in children after major bone surgery, *Lancet* **2**:484.

Dolovich, J., Debanne, M. T., and Bell, R., 1975, The role of alpha$_1$-antitrypsin and alpha-macroglobulins in the uptake of proteinase by rabbit alveolar macrophages, *Am. Rev. Respir. Dis.* **112**:521.

Durrington, P. N., 1982, High-density lipoprotein cholesterol: Methods and clinical significance, *CRC Crit. Rev. Clin. Lab. Sci.* **18**:31.

Eisenberg, S., 1980, Plasma lipoprotein conversion: The origin of low-density and high-density lipoprotein, *Ann. NY Acad. Sci.* **348**:30.

Elwyn, D. H., 1970, The role of the liver in regulation of amino acid and protein metabolism, in: *Mammalian Protein Metabolism* (H. N. Munro, ed.), pp. 523–557, Academic Press, New York.

Emmelot, P., 1977, The organization of the plasma membranes of mammalian cells: Structure and relation to function, in: *Mammalian Cell Membranes* (G. A. Jamieson and D. M. Robinson, eds.), Vol. II, pp. 1–54, Butterworths, London.

Eriksson, S., and Larsson, C., 1975, Purification and partial characterization of PAS-positive inclusion bodies from the liver in alpha$_1$-antitrypsin deficiency, *N. Engl. J. Med.* **292**:176.

Evans, G. W., and Winter, T. W., 1975, Zinc transport by transferrin in rat portal blood plasma, *Biochem. Biophys. Res. Commun.* **66**:1218.

Farrow, S. P., and Baar, S., 1973, The metabolism of α_2-macroglobulin in mildly burned patients, *Clin. Chim. Acta* **46**:39.

Fiedel, B. A., and Gewurz, H., 1976, Effects of C-reactive protein on platelet function. I. Inhibition of platelet aggregation and release reactions, *J. Immunol.* **116**:1289.

Fischer, C. L., Gill, C., Forrester, M. G., and Nakamura, R., 1976, Quantitation of "acute-phase proteins" postoperatively: Value in detection and monitoring of complications, *Am. J. Clin. Pathol.* **66:**840.

Fryand, O., 1979, Studies on fibronectin in the skin. I. Indirect immunofluorescence studies in normal human skin, *Br. J. Dermatol.* **101:**261.

Fryand, O., 1980, Studies on fibronectin in the skin. VI. Intra-epidermal depositions in vulgar psoriasis, lupus erythematosus, bullous pemphigoid and dermatitis herpetiformis, *Acta Dermatovener* **60:**393.

Glaser, M., Nelken, D., Ofek, I., Bergner-Rabinowitz, S., and Ginsburg, I., 1973, Alpha globulin decreases resistance of mice to infection with group A *Streptococcus, J. Infect. Dis.* **127:**303.

Goldstein, B. D., Witz, G., Amoruso, M., and Troll, W., 1979, Protease inhibitors antagonize the activation of polymorphonuclear leukocyte oxygen consumption, *Biochem. Biophys. Res. Commun.* **88:**854.

Goldstein, J. L., and Brown, M. S., 1974, Binding and degradation of low density lipoproteins by cultured human fibroblasts: Comparison of cells from a normal subject and from a patient with homozygous familial hypercholesterolemia, *J. Biol. Chem.* **249:**5153.

Goldstein, J. L., and Brown, M. S., 1983, Familial hypercholesterolemia, in: *Metabolic Basis of Inherited Disease* (J. B. Stanbury, J. B. Wyngaarden, D. S. Fredrickson, J. L. Goldstein, and M. S. Brown, eds.), 5th edition, p. 672–712, McGraw-Hill, New York.

Goldstein, J. L., Ho, Y. K., Basu, S. K., and Brown, M. S., 1979, Binding site on macrophages that mediates uptake and degradation of acetylated low density lipoprotein, producing massive cholesterol deposition, *Proc. Nat. Acad. Sci. USA,* **76:**333.

Goldstein, J. L., Ho, Y. K., Brown, M. S., Innerarity, T. L., and Mahley, R. W., 1980, Cholesteryl ester accumulation in macrophages resulting from receptor-mediated uptake and degradation of hypercholesterolemic canine β-very low density lipoproteins, *J. Biol. Chem.* **255:**1839.

Goutner, A., Simmler, M. C., Tapon, J., and Rosenfeld, C., 1976, Modulation by α-2 macroglobulin of human lymphocyte proliferation in response to mitogens and antigen, *Differentiation* **5:**171.

Gravagna, P., Gianazza, E., Arnaud, P., Neels, M., and Ades, E. W., 1982, Modulation of the immune response by plasma protease inhibitors. II. Alpha$_2$-macroglobulin subunits inhibit natural killer cell cytotoxicity and antibody-dependent cell-mediated cytotoxicity, *Scand. J. Immunol.* **15:**115.

Grinnell, F., and Feld, M. K., 1979, Initial adhesion of human fibroblasts in serum-free medium: Possible role of secreted fibronectin, *Cell* **17:**117.

Grinnell, F., Feld, M., and Snell, W., 1979, The influence of cold insoluble globulin on platelet morphological response to substrata, *Cell Biol. Int. Rep.* **3:**585.

Haas, R. H., Dyck, R. F., Dubowitz, V., and Pepys, M. B., 1982, C-reactive protein in childhood dermatomyositis, *Ann. Rheum. Dis.* **41:**483.

Harpel, P. C., 1982, Blood proteolytic enzyme inhibitors: their role in modulating blood coagulation and fibrinlytic enzyme pathways, in: *Hemostasis and Thrombosis: Basic Principles and Clinical Practice* (R. W. Colman, J. Hirsch, V. J. Marder and E. W. Salzman, eds.), pp. 738–747, J. B. Lippincott Co., Philadelphia.

Hatcher, V. B., Lazarus, G. S., Levine, N., Burk, P. G., and Yost, F. J., Jr., 1977, Characterization of a chemotactic and cytotoxic proteinase from human skin, *Biochim. Biophys. Acta* **483:**160.

Hayman, E. G., and Ruoslahti, E., 1979, Distribution of fetal bovine serum fibronectin and endogenous rat cell fibronectin in extracellular matrix, *J. Cell Biol.* **83:**255.

Henricksen, T., Evensen, S. A., and Carlander, B., 1979, Injury to cultured endothelial cells induced by low-density lipoproteins: protection by high density lipoproteins, *Scand. J. Clin. Lab Invest.* **39:**369.

Ho, Y. K., Faust, J. R., Bilheimer, D. W., Brown, M. S., and Goldstein, J. L., 1977, Regulation of cholesterol synthesis by low density lipoprotein in isolated human lymphocytes: Comparison of cells from normal subjects and patients with homozygous familial hypercholesterolemia and abetalipoproteinemia, *J. Exp. Med.* **145:**1531.

Hodges, J. R., Millward-Sadler, G. H., Barbatis, C., and Wright, R., 1981, Heterozygous MZ alpha$_1$-antitrypsin deficiency in adults with chronic active hepatitis and cryptogenic cirrhosis, *N. Engl. J. Med.* **304:**557.

Hokama, Y., Coleman, M. K., and Riley, R. F., 1962, *In vitro* effects of C-reactive protein on phagocytosis, *J. Bacteriol.* **83:**1017.

Hovi, T., Mosher, D., and Vaheri, A., 1977, Cultured human monocytes synthesize and secrete α_2-macroglobulin, *J. Exp. Med.* **145:**1580.

Howard, R. J., and Simmons, R. L., 1974, Acquired immunologic deficiencies after trauma and surgical procedures, *Surg. Gynecol. Obstet.* **139:**771.

Hsu, K-H. L., Ghanta, V. K., Duncan, L. A., Hunt, C. E., and Hiramoto, R. N., 1980, Characterization of an auto-rosette inhibitory fraction in mouse serum, *J. Immunol.* **125:**1298.

Hubbard, W. J., 1978, Hypothesis: Alpha-2 macroglobulin–enzyme complexes as suppressors of cellular activity, *Cell. Immunol.* **39:**388.

Huybrechts-Godin, G., Hauser, P., and Vaes, G., 1979, Macrophage–fibroblast interactions in collagenase production and cartilage degradation, *Biochem. J.* **184:**643.

Innerarity, T. L., and Mahley, R. W., 1978, Enhanced binding by cultured human fibroblasts of apo-E containing lipoproteins as compared with low density lipoproteins, *Biochemistry* **17:**1440.

Isaacson, P., Jones, D. B., and Judd, M. A., 1979, α_1-Antitrypsin in human macrophages [Letter], *Lancet* **2:**964.

James, K., Merriman, J., Gray, R. S., Duncan, L. J. P., and Herd, R., 1980, Serum α_2-macroglobulin levels in diabetes, *J. Clin. Pathol.* **33:**163.

James, K., Baum, L. L., Vetter, M. L., and Gewurz, H., 1982, Interactions of C-reactive protein with lymphoid cells, *Ann. NY Acad. Sci.* **389:**274.

Janoff, A., Carp, H., Laurent, P., and Raju, L., 1983, The role of oxidative processes in emphysema, *Am. Rev. Respir. Dis.* **127:**531.

Jeejeebhoy, K. N., Ho, J., Greenberg, G. R., Phillips, M. J., Bruce-Robertson, A., and Sodtke, U., 1975, Albumin, fibrinogen and transferrin synthesis in isolated rat hepatocyte suspensions: A model for the study of plasma protein synthesis, *Biochem. J.* **146:**141.

Jeejeebhoy, K. N., Ho, J., Mehra, R., Jeejeebhoy, J., and Bruce-Robertson, A., 1977, Effects of hormones on the synthesis of α_1 (acute-phase) glycoprotein in isolated rate hepatocytes, *Biochem. J.* **168:**347.

Jilek, F., and Hörmann, H., 1978, Fibronectin (cold-insoluble globulin). V. Mediation of fibrin-monomer binding to macrophages, *Hoppe-Seylers Z. Physiol. Chem.* **359:**1603.

Johansson, S., Rubin, K., Höök, M., Ahlgren, T., and Seljelid, R., 1979, *In vitro* biosynthesis of cold insoluble globulin (fibronectin) by mouse peritoneal macrophages, *FEBS Lett.* **105:**313.

Johnson, K. J., and Varani, J., 1981, Substrate hydrolysis by immune complex-activated neutrophils: Effects of physical presentation of complexes and protease inhibitors, *J. Immunol.* **127:**1875.

Kampschmidt, R. F., 1978, Leukocytic endogenous mediator, *J. Reticuloendothelial Soc.* **23:**287.

Kaplan, J., and Nielsen, M. L., 1979, Analysis of macrophage surface receptors. II. Internalization of α-macroglobulin–trypsin complexes by rabbit alveolar macrophages, *J. Biol. Chem.* **254:**7329.

Kaplan, J. E., Molnar, J., Saba, T. M., and Allen, C., 1976, Comparative disappearance and localization of isotopically labeled opsonic protein and soluble albumin following surgical trauma, *J. Reticuloendothelial Soc.* **20:**375.

Kaplan, M. H., and Volanakis, J. E., 1974, Interaction of C-reactive protein complexes with the complement system. I. Consumption of human complement associated with the reaction of CRP with pneumococcal C-polysaccharide and with choline phosphatides, lecithin and sphingomyelin, *J. Immunol.* **112:**2135.

Kayden, H. J., Hatam, L., and Beratis, N. G., 1976, Regulation of 3-hydroxy-3-methylglutaryl coenzyme A reductase activity and the esterification of cholesterol in human long term lymphoid cell lines, *Biochemistry* **15:**521.

Keski-Oja, J., Sen, A., and Todaro, G. J., 1980, Direct association of fibronectin and actin molecules *in vitro*, *J. Cell Biol.* **85:**527.

Kindmark, C.-O., 1971, Stimulating effect of C-reactive protein on phagocytosis of various species of pathogenic bacteria, *Clin. Exp. Immunol.* **8:**941.

Kinnaert, P., Mahieu, A., and Van Geertruyden, N., 1978, Stimulation of antibody synthesis induced by surgical trauma in rats, *Clin. Exp. Immunol.* **32:**243.

Kitagawa, S., Takaku, F., and Sakamoto, S., 1980, Evidence that proteases are involved in superoxide production by human polymorphonuclear leukocytes and monocytes, *J. Clin. Invest.* **65:**74.

Kluger, M. J., and Rothenburg, B. A., 1979, Fever and reduced iron: Their interaction as a host defense response to bacterial infection, *Science* **203**:374.

Knowles, B. B., Howe, C. C., and Aden, D. P., 1980, Human hepatocellular carcinoma cell lines secrete the major plasma proteins and hepatitis B surface antigen, *Science* **209**:497.

Kronick, P. L., and Jimenez, S. A., 1979, Direct measurement of the amount of bound collagen which stimulates platelet aggregation, *Thromb. Haemostasis* **41**:498.

Kruth, H. S., Avigan, J., Gamble, W., and Vaughan, M., 1979, Effect of cell density on binding and uptake of low density lipoprotein by human fibroblasts, *J. Cell Biol.* **83**:588.

Kukral, J. C., Sporn, J., Louch, J., and Winzler, R. J., 1963, Synthesis of alpha- and beta-globulins in normal and liverless dog, *Am. J. Physiol.* **204**:262.

Kushner, I., and Feldmann, G., 1978, Control of the acute phase response: Demonstration of C-reactive protein synthesis and secretion by hepatocytes during acute inflammation in the rabbit, *J. Exp. Med.* **148**:466.

Kushner, I., and Kaplan, M. H., 1961, Studies of acute phase protein. I. An immunohistochemical method for the localization of Cx-reactive protein in rabbits: Association with necrosis in local inflammatory lesions, *J. Exp. Med.* **114**:961.

Kushner, I., and Somerville, J. A., 1971, Permeability of human synovial membrane to plasma proteins: Relationship to molecular size and inflammation, *Arthritis Rheum.* **14**:560.

Kushner, I., Rakita, L., and Kaplan, M. H., 1963, Studies of acute-phase protein. II. Localization of Cx-reactive protein in heart in induced myocardial infarction in rabbits, *J. Clin. Invest.* **42**:286.

Kushner, I., Broder, M. L., and Karp, D., 1978, Control of the acute phase response: Serum C-reactive protein kinetics after acute myocardial infarction, *J. Clin. Invest.* **61**:235.

Lanser, M. E., Saba, T. M., and Scovill, W. A., 1980, Opsonic glycoprotein (plasma fibronectin) levels after burn injury: Relationship to extent of burn and development of sepsis, *Ann. Surg.* **192**:776.

Laurell, C.-B., and Jeppsson, J.-O., 1975, Protease inhibitors in plasma, in: *The Plasma Proteins: Structure, Function and Genetic Control* (F. W. Putnam, ed.), Vol. I, 2nd ed., pp. 229–264, Academic Press, New York.

Layman, D. L., Jelen, B. J., and Illingworth, D. R., 1980, Inability of serum from abetalipoproteinemic subjects to stimulate proliferation of human smooth muscle cells and dermal fibroblasts *in vitro*, *Proc. Natl. Acad. Sci. USA* **77**:1511.

Le Guilly, Y., Launois, B., Lenoir, P., and Bourel, M., 1973, Production of serum proteins by primary cultures of adult human liver, *Biomedicine* **19**:248.

Leibovich, S. J., and Ross, R., 1975, The role of the macrophage in wound repair: A study with hydrocortisone and antimacrophage serum, *Am. J. Pathol.* **78**:71.

Lieberman, J., 1973, Involvement of leukocytic proteases in emphysema and antitrypsin deficiency, *Arch. Environ. Health* **27**:196.

McDaniel, M. C., Laudico, R., and Papermaster, B. W., 1976, Association of macrophage-activation factor from a human cultured lymphoid cell line with albumin and α_2-macroglobulin, *Clin. Immunol. Immunopathol.* **5**:91.

McDonald, J. A., Baum, B. J., Rosenberg, D. M., Kelman, J. A., Brin, S. C., and Crystal, R. G., 1979, Destruction of a major extracellular adhesive glycoprotein (fibronectin) of human fibroblasts by neutral proteases from polymorphonuclear leukocyte granules, *Lab. Invest.* **40**:350.

Mahley, R. W., Inneraity, T. L., Pitas, R. E., Weisgraber, K. H., Brown, J. H., and Gross, E., 1977, Inhibition of lipoprotein binding to cell surface receptors of fibroblasts following selective modification of arginyl residues in arginine rich and B-apoproteins, *J. Biol. Chem.* **252**:7279.

Mahmoud, A. A. F., Rodman, H. M., Mandel, M. A., and Warren, K. S., 1976, Induced and spontaneous diabetes mellitus and suppression of cell-mediated immunologic responses: Granuloma formation, delayed dermal reactivity, and allograft rejection, *J. Clin. Invest.* **57**:362.

Markowetz, B., van Snick, J. L., and Masson, P. L., 1979, Binding and ingestion of human lactoferrin by mouse alveolar macrophages, *Thorax* **34**:209.

Maxfield, F. R., Schlessinger, J., Shechter, Y., Pastan, I., and Willingham, M. C., 1978, Collection of insulin, EGF and α_2-macroglobulin in the same patches on the surface of cultured fibroblasts and common internalization, *Cell* **14**:805.

Mehta, N. G., 1977, The site of synthesis and functions of acute phase plasma proteins: Close relationship with the reticulo-endothelial system, *Med. Hypotheses* **3:**63.

Messner, R. P., Lindström, F. D., and Williams, R. C., Jr., 1973, Peripheral blood lymphocyte cell surface markers during the course of systemic lupus erythematosus, *J. Clin. Invest.* **52:**3046.

Miller, L. L., and Bale, W. F., 1954, Synthesis of all plasma protein fractions except gamma globulins by the liver: The use of zone electrophoresis and lysine-ϵ-C^{14} to define the plasma proteins synthesized by the isolated perfused liver, *J. Exp. Med.* **99:**125.

Minchin Clarke, H. G., Freeman, T., and Pryse-Phillips, W., 1971, Serum protein changes after injury, *Clin. Sci.* **40:**337.

Minor, R. R., 1980, Collagen metabolism: A comparison of diseases of collagen and diseases affecting collagen, *Am. J. Pathol.* **98:**227.

Mold, C., Rodgers, C. P., Kaplan, R. L., and Gewurz, H., 1982, Binding of human C-reactive protein to bacteria, *Infect. Immun.* **38:**392.

Mold, C., and Gerwurz, H., 1981, Inhibitory effect of C-reactive protein on alternate C pathway activation by liposomes and *Steptococcus pneumoniae*, *J. Immunol.* **127:**2089.

Mortensen, R. F., 1979, C-reactive protein (CRP)-mediated inhibition of the induction of *in vitro* antibody formation. I. T-cell dependence of the inhibition, *Cell Immunol.* **44:**270.

Mortensen, R. F., Osmand, A. P., and Gewurz, H., 1975, Effects of C-reactive protein on the lymphoid system. I. Binding to thymus-dependent lymphocytes and alteration of their functions, *J. Exp. Med.* **141:**821.

Mortensen, R. F., 1982, Inhibition of the polyclonal antibody plaque-forming cell response of human B lymphocytes by C-reactive protein (CRP) and CRP complexes, *Cell Immunol.* **66:**99.

Mosesson, M. W., 1977, Cold-insoluble globulin (CIg), a circulating cell surface protein, *Thromb. Haemostasis* **38:**742.

Mosher, D. F., 1976a, Action of fibrin-stabilizing factor on cold-insoluble globulin and α_2-macroglobulin in clotting plasma, *J. Biol. Chem.* **251:**1639.

Mosher, D. F., 1976b, Changes in plasma cold-insoluble globulin concentration during experimental Rocky Mountain spotted fever infection in rhesus monkeys, *Thromb. Res.* **9:**37.

Mosher, D. F., and Vaheri, A., 1980, Binding and degradation of α_2-macroglobulin by cultured fibroblasts, *Biochim. Biophys. Acta* **627:**113.

Mosher, D. F., and Wing, D. A., 1976, Synthesis and secretion of α_2-macroglobulin by cultured human fibroblasts, *J. Exp. Med.* **143:**462.

Mosher, D. F., Schad, P. E., and Kleinman, H. K., 1979, Cross-linking of fibronectin to collagen by blood coagulation Factor XIII_a, *J. Clin. Invest.* **64:**781.

Munster, A. M., 1976, Post-traumatic immunosuppression is due to activation of suppressor T cells, *Lancet* **1:**1329.

Murphy, P. A., Simon, P. L., and Willoughby, W. F., 1980, Endogenous pyrogens made by rabbit peritoneal exudate cells are identical with lymphocyte-activating factors made by rabbit alveolar macrophages, *J. Immunol.* **124:**2498.

Nelson, K. M., and Filkins, J. P., 1979, Effects of traumatic injury on sensitivity to insulin, *Circ. Shock* **6:**285.

Niehaus, G. D., Schumacker, P. R., and Saba, T. M., 1980, Reticuloendothelial clearance of blood-borne particulates: Relevance to experimental lung microembolization and vascular surgery, *Ann. Surg.* **191:**479.

Octave, J.-N., Schneider, Y.-J., Hoffmann, P., Trouet, A., and Crichton, R. R., 1979, Transferrin protein and iron uptake by cultured rat fibroblasts, *FEBS Lett.* **108:**127.

Ohlsson, K., 1975, α_1-antitrypsin and α_2-macroglobulin: Interactions with human neutrophil collagenase and elastase, *Ann. N.Y. Acad. Sci.* **256:**409.

Olsen, G. N., Harris, J. O., Castle, J. R., Waldman, R. H., and Karmgard, H. J., 1975, Alpha-1-antitrypsin content in the serum, alveolar macrophages, and alveolar lavage fluid of smoking and nonsmoking normal subjects, *J. Clin. Invest.* **55:**427.

Osgood, E. E., 1954, Number and distribution of human hemic cells, *Blood* **9:**1141.

O'Shea, M. J., Kershenobich, D., and Tavill, A. S., 1973, Effects of inflammation on iron and transferrin metabolism, *Br. J. Haematol.* **25:**707.

Osmand, A. P., Mortensen, R. F., Siegel, J., and Gewurz, H., 1975, Interactions of C-reactive protein with the complement system. III. Complement-dependent passive hemolysis initiated by CRP, *J. Exp. Med.* **142:**1065.

Parisi, A. F., and Vallee, B. L., 1970, Isolation of a zinc α_2-macroglobulin from human serum, *Biochemistry* **9:**2421.

Phillips, J. L., 1976, Specific binding of zinc transferrin to human lymphocytes, *Biochem. Biophys. Res. Commun.* **72:**634.

Phillips, J. L., and Azari, P., 1974, Zinc transferrin: Enhancement of nucleic acid synthesis in phytohemagglutinin-stimulated human lymphocytes, *Cell. Immunol.* **10:**31.

Phillips, N. C., 1982, Exacerbation of experimental poly-D-lysine arthritis by C-reactive protein, *Agents and Actions* **12:**344.

Pierce, C. W., 1980, Macrophages: Modulators of immunity, *Am. J. Pathol.* **98:**10.

Plow, E. F., and Ginsberg, M. H., 1981, Specific and saturable binding of plasma fibronectin. *J. Biol. Chem.* **256:**9477.

Powanda, M. C., 1977, Changes in body balances of nitrogen and other key nutrients: Description and underlying mechanisms, *Am. J. Clin. Nutr.* **30:**1254.

Powanda, M. C., and Beisel, W. R., 1982, Hypothesis: Leukocyte endogenous mediator/endogenous pyrogen/lymphocyte—activating factor modulates the development of nonspecific and specific immunity and affects nutritional status, *Am. J. Clin. Nutr.* **35:**762.

Powanda, M. C., and Moyer, E. D., 1981, Plasma proteins and wound healing, *Surg. Gynecol. Obstet.* **153:**749.

Powanda, M. C., Wannemacher, R. W., Jr., and Cockerell, G. L., 1972, Nitrogen metabolism and protein synthesis during pneumococcal sepsis in rats, *Infect. Immun.* **6:**266.

Powanda, M. C., Cockerell, G. L., Moe, J. B., Abeles, F. B., Pekarek, R. S., and Canonico, P. G., 1975, Induced metabolic sequelae of tularemia in the rat: Correlation with tissue damage, *Am. J. Physiol.* **229:**479.

Powanda, M. C., Abeles, F. B., Bostian, K. A., Fowler, J. P., and Hauer, E. C., 1979, Differential effects of clofibrate on inflammation-induced alterations in plasma proteins in the rat, *Biochem. J.* **178:**633.

Rapaport, F. T., and Bachvaroff, R. J., 1976, Kinetics of humoral responsiveness in severe thermal injury, *Ann. Surg.* **184:**51.

Redgrave, T. G., 1970, Formation of cholesteryl ester-rich particulate lipid during metabolism of chylomicrons, *J. Clin. Invest.* **49:**465.

Remold-O'Donnell, E., and Lewandrowski, K., 1983, Two proteinase inhibitors associated with peritoneal macrophages, *J. Biol. Chem.* **258:**3251.

Rinderknecht, H., and Geokas, M. C., 1973, On the physiological role of α_2-macroglobulin, *Biochim. Biophys. Acta* **295:**233.

Ruoslahti, E., and Hayman, E. G., 1979, Two active sites with different characteristics in fibronectin, *FEBS Lett.* **97:**221.

Ruoslahti, E., and Vaheri, A., 1975, Interaction of soluble fibroblast surface antigen with fibrinogen and fibrin: Identity with cold insoluble globulin of human plasma, *J. Exp. Med.* **141:**497.

Ruoslahti, E., Engvall, E., and Hayman, E. G., 1981, Fibronectin: Current concepts of its structure and functions, *Coll. Res.* **1:**95.

Ryan, N. T., Blackburn, G. I., and Clowes, G. H. A., Jr., 1974, Differential tissue sensitivity to elevated endogenous insulin levels during experimental peritonitis in rats, *Metabolism* **23:**1081.

Saba, T. M., Blumenstock, F. A., Scovill, W. A., and Bernard, H., 1978, Cryoprecipitate reversal of opsonic α_2-surface binding glycoprotein deficiency in septic surgical and trauma patients, *Science* **201:**622.

Santoro, S. A., and Cunningham, L. W., 1979, Fibronectin and the multiple interaction model for platelet–collagen adhesion, *Proc. Natl. Acad. Sci. USA* **76:**2644.

Scanu, A. M., Byrne, R. E., and Mihovilovic, M., 1982, Functional roles of plasma high density lipoproteins, *CRC Crit. Rev. Biochem.* **13:**109.

Schapira, M., Scott, C. F., and Colman, R. W., 1982, Contribution of plasma protease inhibitors to the inactivation of kallikrein in plasma, *J. Clin. Invest.* **69:**462.

Schuh, J., Novogrodsky, A., and Haschemeyer, R. H., 1978, Inhibition of lymphocyte mitogenesis by autooxidized low-density lipoprotein, *Biochem. Biophys. Res. Commun.* **84:**763.

Siegel, J., Rent, R., and Gewurz, H., 1974, Interactions of C-reactive protein with the complement system. I. Protamine-induced consumption of complement in acute phase sera, *J. Exp. Med.* **140:**631.

Spain, M. J., Wosu, L. O., and Kalant, N., 1979, Multiple effects of serum low-density lipoprotein on cultured human fibroblasts, *Can. J. Biochem.* **57:**684.

Stecher, V. J., and Thorbecke, G. J., 1967a, Sites of synthesis of serum proteins. I. Serum proteins produced by macrophages *in vitro, J. Immunol.* **99:**643.

Stecher, V. J., and Thorbecke, G. J., 1967b, Sites of synthesis of serum proteins. III. Production of β_{1C}, β_{1E} and transferrin by primate and rodent cell lines, *J. Immunol.* **99:**660.

Stein-Streilein, J., and Hart, D. A., 1980, Protease inhibitors modulate *in-vitro* antibody formation, *Fed. Proc.* **39:**807A.

Szabó, S., Barbu, Z., Lakatos, L., László, I., and Szabó, A., 1980, Local production of proteins in normal human bronchial secretion, *Respiration* **39:**172.

Tada, N., Fidge, N., and Nestel, P., 1979, Identification and characterization of mixed disulphide complexes of E apoprotein in high density lipoprotein of patients with acute alcoholic hepatitis, *Biochem. Biophys. Res. Commun.* **90:**297.

Teodorescu, M., Ganea, D., Lee, T. T., Skosey, J. L., and Rutter, G., 1982, The effect of protease inhibitors on the polyclonal B cell activator from the serum of patients with rheumatoid arthritis, *Int. J. Immunopharmac.* **4:**1.

Thorbecke, G. J., Liem, H. H., Knight, S., Cox, K., and Müller-Eberhard, U., 1973, Sites of formation of the serum proteins transferrin and hemopexin, *J. Clin. Invest.* **52:**725.

Travis, J., and Salvesen, G. S., 1983, Human plasma proteinase inhibitors, *Ann. Rev. Biochem.* **52:**655.

Tunstall, A. M., and James, K., 1974, Preliminary studies on the synthesis of alpha$_2$-macroglobulin by human lymphocytes *in vitro, Clin. Exp. Immunol.* **17:**697.

Tunstall, A. M., and James, K., 1975, The effect of human α_2-macroglobulin on the restoration of humoral responsiveness in X-irradiated mice, *Clin. Exp. Immunol.* **21:**173.

Turinsky, J., and Patterson, S. A., 1979, Proximity to a burn wound as a new factor in considerations of postburn insulin resistance, *J. Surg. Res.* **26:**171.

Ulrich, F., 1979, Protease potentiation of thymocyte blastogenesis, *Immunology* **38:**705.

van Leuven, F., Cassiman, J. J., and van den Berghe, H., 1978, Uptake and degradation of α_2-macroglobulin–protease complexes in human cells in culture, *Exp. Cell Res.* **117:**273.

van Snick, J. L., Masson, P. L., and Heremans, J. F., 1974, The involvement of lactoferrin in the hyposideremia of acute inflammation, *J. Exp. Med.* **140:**1068.

Verburgh, H. A., Peterson, P. K., Smith, D. E., Nguyen, B-Y. T., Hoidal, J. R., Wilkinson, B. J., Verhoef, J., and Furcht, L. T., 1981, Human fibronectin binding to staphylococcal surface protein and its relative inefficiency in promoting phagocytosis by human polymorphonuclear leukocytes, monocytes and alveolar macrophages, *Infec. Immun.* **33:**811.

Vetter, M. L., Gerwurz, H., Hansen, B., James, K., and Baum, L. L., 1983, Effects of C-reactive protein on human lymphocyte responsiveness, *J. Immunol.* **130:**2121.

Via, D. P., Willingham, M. C., Pastan, I., Gotto, A. M., Jr., and Smith, L. C., 1982, Co-clustering and internalization of low-density lipoproteins and α_2-macroglobulin in human skin fibroblasts, *Exp. Cell. Res.* **141:**15.

Vischer, T. L., and Berger, D., 1980, Activation of macrophages to produce neutral proteinases by endocytosis of alpha$_2$-macroglobulin–trypsin complexes, *J. Reticuloendothelial Soc.* **28:**427.

Volenec, F. J., Mani, M. M., Clark, G. M., Robinson, D. W., and Humphrey, L. J., 1977, Peripheral blood T and B lymphocytes in patients with burns. II. Sequential rosette analyses considering burn severity and pseudomonas sepsis, *Burns* **4:**7.

Voss, B., Allam, S., Rauterberg, J., Ullrich, K., Gieselmann, V., and von Figura, K., 1979, Primary cultures of rat hepatocytes synthesize fibronectin, *Biochem. Biophys. Res. Commun.* **90:**1348.

Wannemacher, R. W., Jr., 1975, Protein metabolism (applied biochemistry), in: *Total Parenteral Nutrition: Premises & Promises* (H. Ghadimi, ed.), pp. 85–153, Wiley, New York.

Wannemacher, R. W., Jr., Powanda, M. C., and Dinterman, R. E., 1974, Amino acid flux and protein synthesis after exposure of rats to either *Diplococcus pneumoniae* or *Salmonella typhimurium, Infect. Immun.* **10:**60.

Weeke, B., and Krasilnikoff, P. A., 1972, The concentration of 21 serum proteins in normal children and adults, *Acta Med. Scand.* **192:**149.

Weinberg, E. D., 1978, Iron and infection, *Microbiol. Rev.* **42:**45.

Weisgraber, K. H., and Mahley, R. W., 1978, Apoprotein (E-A-II) complex of human plasma lipoproteins. I. Characterization of this mixed disulfide and its indentification in a high density lipoprotein subfraction, *J. Biol. Chem.* **253:**6281.

Weiss, R. E., and Reddi, A. H., 1981, Role of fibronectin in collagenous matrix-induced mesenchymal cell proliferation and differentiation *in vivo, Exp. Cell Res.* **133:**247.

Weissman, S. M., Wochner, R. D., Mullins, F. X., Wynngate, A., and Waldmann, T. A., 1966, Synthesis of plasma proteins by hepatectomized dogs, *Am. J. Physiol.* **210:**128.

Werner, M., 1969, Serum protein changes during the acute phase reaction, *Clin. Chim. Acta* **25:**299.

Werner, M., and Cohnen, G., 1969, Changes in serum proteins in the immediate postoperative period, *Clin. Sci.* **36:**173.

White, R., Lee, D., Habicht, G., and Janoff, A., 1981, Secretion of alpha-1-proteinase inhibitor by cultured alveolar rat macrophages, *Am. Rev. Respir. Dis.* **123:**477.

Williams, R. O., and Loeb, L. A., 1973, Zinc requirement for DNA replication in stimulated human lymphocytes, *J. Cell Biol.* **58:**594.

Willingham, M. C., Maxfield, F. R., and Pastan, I. H., 1979, α_2Macroglobulin binding to the plasma membrane of cultured fibroblasts: Diffuse binding followed by clustering in coated regions, *J. Cell Biol.* **82:**614.

Wilson, G. B., Walker, J. H., Jr., Watkins, J. H., Jr., and Wolgroch, D., 1980, Determination of subpopulations of leukocytes involved in the synthesis of α_1-antitrypsin *in vitro, Proc. Soc. Exp. Biol. Med.* **164:**105.

Wolfe, R. R., Durkot, M. J., Allsop, J. R., and Burke, J. F., 1979, Glucose metabolism in severely burned patients, *Metabolism* **28:**1031.

Wybran, J., and Fudenberg, H. H., 1973, Thymus-derived rosette-forming cells in various human disease states: Cancer, lymphoma, bacterial and viral infections, and other diseases, *J. Clin. Invest.* **52:**1026.

Wyllie, J. C., 1977, Transferrin uptake by rabbit alveolar macrophages *in vitro, Br. J. Haematol.* **37:**17.

Wyllie, J. C., 1979, The subcellular distribution of iron in rabbit alveolar macrophages, *Biochem. Biophys. Res. Commun.* **89:**1307.

Yamada, K. M., and Kennedy, D. W., 1979, Fibroblast cellular and plasma fibronectins are similar but not identical, *J. Cell Biol.* **80:**492.

Yamada, K. M., and Olden, K., 1978, Fibronectins—Adhesive glycoproteins of cell surface and blood, *Nature (London)* **275:**179.

Zeineh, R. A., and Kukral, J. C., 1970, The turnover rate of orosomucoid in burned patients, *J. Trauma* **10:**493.

Zeineh, R. A., Barrett, B., Niemirowski, L., and Fiorella, B. J., 1972, Turnover rate of orosomucoid in the dog with sterile abscess, *Am. J. Physiol.* **222:**1326.

15

Iron Metabolism

FARID I. HAURANI and SAMIR K. BALLAS

1. INTRODUCTION

The bone marrow of an average person needs about 30 mg of iron per day. Approximately 1 to 2 mg of iron is absorbed daily from a diet that contains roughly 10 mg of iron, in order to keep up with an equal daily loss of iron. This means that the daily needs of iron by the bone marrow are met primarily by the reutilization of iron from the RES. Therefore, this system plays a very important role in iron metabolism on a daily basis. Figure 1 attempts to depict the relationship of the RES to iron metabolism in its proper perspective as it relates to the other two major systems involved in iron metabolism, namely, the bone marrow and the gastrointestinal tract. However, this chapter deals primarily with the RE aspects of iron metabolism. First, the uptake of the iron carriers, red blood cells and hemoglobin or heme, by the RES is presented. This is followed by the processing in the RES and the release of iron from the RES back to the circulation. Finally, we conclude with iron storage. Figure 2 highlights these events of iron metabolism.

2. UPTAKE OF IRON CARRIERS BY THE RES

2.1. HAPTOGLOBIN AND HEMOPEXIN

Hemoglobin (Hb) and heme are cleared from the circulation with the help of three proteins: haptoglobin (Hp), hemopexin (Hx), and albumin. These three proteins help to conserve these iron products and prevent their loss to the urine. However, they are very inefficient systems because, in intravascular hemolysis, hemoglobulinuria ensues with severe iron loss leading to iron deficiency hypochromic anemia. Their role in conservation of iron is minimal.

FARID I. HAURANI and SAMIR K. BALLAS • Cardeza Foundation for Hematologic Research, Department of Medicine, Thomas Jefferson University, Philadelphia, Pennsylvania 19107.

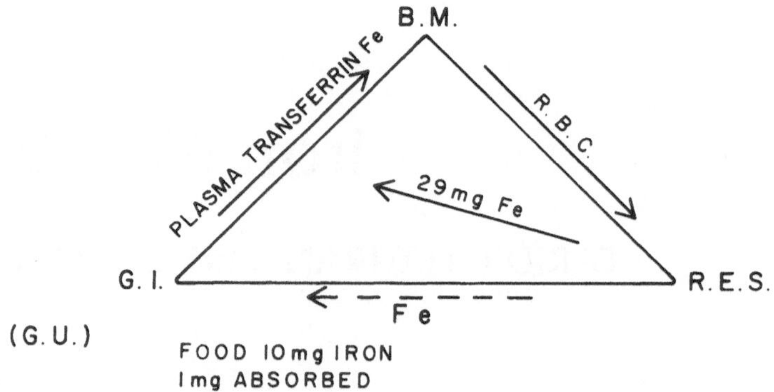

FIGURE 1. Simplifed scheme of iron metabolism.

1. Hp is an α_2-globulin, synthesized by the liver, and its plasma concentration is 50–150 mg/dl (Alper *et al.*, 1965). Under normal conditions, the half-life of this protein is 1 day and this maintains the human plasma level of Hb at a concentration of less than 0.6 mg/dl (Noyes and Garby, 1967). Hp consisting of two dimers binds with the β chains of Hb in a molar ratio of 1 : 1 (Gordon and Bearn, 1966). The Hb–Hp complex has a half-life of 9 to 30 min (Garby and Noyes, 1959). Since the production of this protein is limited, its plasma con-

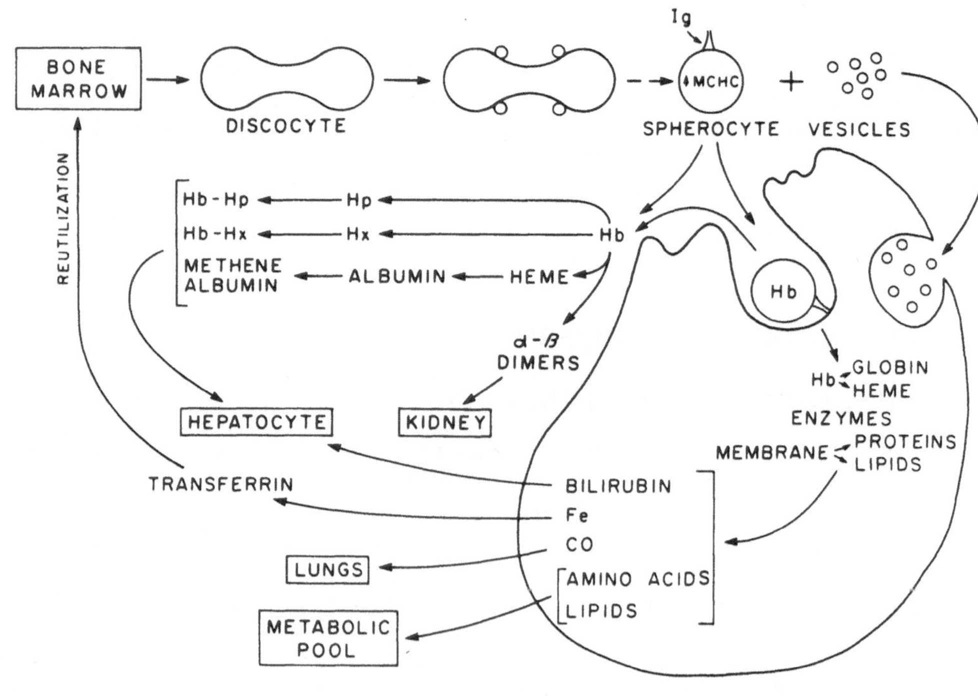

FIGURE 2. Uptake, processing, and release of iron and related compounds by the RE cell.

centration drops precipitously (less than 20 mg/dl and often is completely depleted) with any mild hemolytic process. This is so when the half-life of ^{51}Cr-tagged red blood cells is 17 days or less (normal, 28). Because Hp is one of the acute-phase reactants, its concentration rises in the presence of neoplastic or inflammatory states (Nyman, 1959). Normal values in such conditions should alert one to the coexistence of hemolysis.

2. Hx is a β globulin that binds heme (ferriprotoporphyrin IX) in an equimolar ratio (Muller-Eberhard, 1976). The complex has a half-life of 7 to 8 hr and is removed primarily by the hepatic cells (Hanstein and Muller-Eberhard, 1968). This is in contrast to the Hp–Hb which is cleared by cells of the RES, mainly the liver and bone marrow (Keene and Jandl, 1965; Jandl *et al.*, 1956). Its concentration in plasma is about 50–100 mg/dl and this seldom varies except in the presence of severe hemolytic anemia or following heme injections when its value may fall or even become depleted. It is not an acute-phase reactant. The two carcinogens 3-methylcholanthrene and benzpyrene result in its induction and increased blood concentration (Ross and Muller-Eberhard, 1970).

2.2. ERYTHROCYTE

The erythrocyte starts its lively mission as a nucleated erythroid precursor that matures into a clone of at least eight identical siblings. After a few days of sedentary residence in the bone marrow, it is denucleated and released into a turbulent world of circulation. The newly born cell, a reticulocyte, finds the new environment shocking and timidly takes refuge in the splenic habitat. There it undergoes a process of membrane remodeling (Song and Groom, 1972; Shattil and Cooper, 1971), gets rid of its intracellular ferritin by a process of exocytosis (Cartwright and Deiss, 1975), and after 1 or 2 days it is again launched into a sea of plasma to navigate a 300-mile-long journey. After 120 days the old and faltering erythrocyte, like a salmon, swims back to the RES, mostly the spleen, liver, and bone marrow, where it is phagocytized, processed, and degraded into essential metabolites.

2.2.1. Red Cell Aging

As erythrocytes age *in vivo* they are randomly removed from the circulation by the RES (Berlin and Berk, 1975). The characteristics of the senescent erythrocyte which lead to its disposal are not understood but are believed to involve changes in one or more physicochemical properties of the cell as outlined in Table 1 and discussed below.

2.2.1a. Cytoplasmic Changes in Aging Red Cells. A number of intracellular physicochemical changes occur in aging erythrocytes (Table 1). As the red cell gets older, some of its enzymes gradually lose activity with a consequent deterioration of their corresponding metabolic processes. Impairment of the glycolytic enzyme activities results in decreased availability of ATP with age (Brok *et al.*, 1966; Lichtman and Weed, 1972; Lichtman and Murphy, 1975). Loss

TABLE 1. PHYSICOCHEMICAL ALTERATIONS IN SENESCENT ERYTHROCYTES

- Cytoplasmic changes
  - Changes in glycolysis
    - Glycolytic enzyme activities decrease
    - Decreased content of ATP and 2,3-DPG
  - Changes in the hexose monophosphate shunt
    - Decreased enzyme activities
  - Changes in hemoglobin
    - Increased methemoglobin
    - Increased oxygen affinity
  - Other physicochemical changes
    - Altered cation content
      - Increased Na^+
      - Decreased K^+
      - Increased Ca^{2+}
    - Increased viscosity
- Membrane changes
  - Changes in the biophysical properties of the membrane
    - Decreased surface area
    - Decreased cell size
    - Decreased negative charge
    - Decreased deformability
    - Increased cell density
    - Increased osmotic and mechanical fragility
  - Biochemical changes
    - Decreased content of cholesterol and phospholipids
    - Decreased lipid/protein ratio
    - Decreased sialic acid
    - Decreased anionic sites
    - Decreased enzyme activities
    - Alterations in membrane proteins
    - Accumulation of hemoglobin and its subunits on the inner surface
    - Accumulation of autologous immunoglobulin on the external surface

of intracellular ATP leads to impairment of energy-dependent processes, especially the transport of sodium, potassium, and calcium. Thus, a senescent erythrocyte has increased Na^+, decreased K^+ (Bernstein, 1959), and increased Ca^{2+} content (Bocci *et al.*, 1978). Increased amounts of intracellular Ca^{2+} activate a dormant endogenous enzyme, transglutaminase, which cross-links intracellular and, more importantly, membrane proteins (Lorand *et al.*, 1976). This effect of calcium may be responsible for the decrease in red cell deformability that occurs with aging (LaCelle and Arkin, 1970). Kirpatrick *et al.* (1979), however, have questioned the role of decreasing amounts of ATP with aging and it seems unlikely that loss of cellular ATP is a major causative factor in removal of senescent red cells from the circulation.

Another phosphorylated glycolytic intermediate that decreases with red cell aging is 2,3-DPG (Haidas *et al.*, 1971). This , in turn, increases the oxygen affinity of Hb by shifting the oxygen dissociation curve to the left (Haidas *et al.*, 1971). The efficiency of oxygen transport to the tissues, therefore, becomes curtailed in older red cells. Methemoglobin, a poor transporter of oxygen, also increases with aging (Keitt *et al.*, 1966).

Activities of the enzymes in the hexose monophosphate shunt also deteriorate with red cell aging (Bonsignore *et al.*, 1964; Brok *et al.*, 1966). Glucose-6-phosphate dehydrogenase (G6PD) plays a cardinal role in the detoxification of H_2O_2 or other free radicals (Cohen and Hochstein, 1961). Gradual accumulation of these compounds leads to oxidation of reduced glutathione (GSH) and the formation of mixed disulfides (HbS–SG) by complexing GSH to the β93 cysteine residue of Hb (Allen and Jandl, 1961). Another postulated mechanism of red cell destruction with aging is that the generated H_2O_2 causes direct lipid peroxidation with resultant membrane destruction and hemolysis (Carrell *et al.*, 1975).

Williams and Morris (1980) have recently shown that the internal viscosity of the human erythrocyte may be an important determinant of its life span *in vivo*. Young erythrocytes obtained by high-speed centrifugation of a normal blood sample had a mean corpuscular Hb concentration (MCHC) of 31.7 g/dl which corresponds to an internal viscosity of 9 cP. Old erythrocytes from the same blood sample had an average MCHC of 37.5 g/dl which is equivalent to a mean internal viscosity of 54 cP. This increased internal viscosity of old cells results in a prolonged transit time through the narrow channels of the splenic microvasculature and hence a higher probability of contact with a phagocytic macrophage and subsequent elimination.

2.2.1b. Membrane Changes in Senescent Red Cells. The physicochemical alterations that occur in the membrane of aging erythrocytes have recently received greater attention. These age-dependent membrane changes are summarized in Table 1.

A number of investigators have shown that old erythrocytes have decreased fluidity and decreased deformability (Shiga *et al.*, 1979; Leonhardt *et al.*, 1978). Although this faltering resiliency and pliability of the red cell membrane may, in part, be secondary to some of the cytoplasmic alterations described above, a number of biophysical and biochemical changes in the red cell membrane itself are undoubtedly a major contributory factor. As the erythrocyte ages, it undergoes a gradual and steady discocyte–spherocyte transformation (Weed and Reed, 1966; Ganzoni *et al.*, 1971). This is manifested by the finding of decreased membrane surface area in old red cells (Greenwalt and Lau, 1978). Senescent erythrocytes lose membrane in the form of vesicles that disappear very rapidly from the circulation with half lives from 2 to 8 min (Bocci *et al.*, 1980). Membranes lipids, both cholesterol and phospholipids, are preferentially lost during aging resulting in a decreased lipid/protein membrane ratio (Shiga *et al.*, 1979). Lutz *et al.* (1977) have shown that human erythrocytes release spectrin-free vesicles during ATP depletion. This progressive membrane loss, together with the other age-dependent changes, are responsible for the demise of the cell.

Although aging red cells lose little or no membrane proteins, some investigators have demonstrated the presence of quantitative and qualitative membrane protein changes in senescent erythrocytes. Some membrane proteins are not present in newly formed red cells but gradually appear and become more prominent with age (Kadlubowski and Harris, 1974). Membrane proteins from senescent erythrocytes analyzed by polyacrylamide gel electrophoresis in SDS have less protein in band 7 and more protein in bands 2, 4, 5, and 6 (Kadlubowski, 1978). Some investigators have suggested the occurrence of progressing conformational changes among membrane proteins with aging (Bienzle and Pjura, 1977). Other reported membrane protein changes with aging include decreased density of anionic sites on the membrane surface (Gayer *et al.*, 1972) and decreased activity of several membrane-bound enzymes (Kadlubowski and Agutter, 1977). It is interesting to point out that the activity of the membrane-bound Na^+-K^+ ATPase was found not to change with red cell age (Kadlubowski and Agutter, 1977).

A number of recent studies have demonstrated some interesting and selective biochemical changes that occur on the inner and the outer surface of the red cell membrane with age. Older erythrocytes retain more Hb on the cytoplasmic surface of the membrane (Kadlubowski, 1978). Moreover, there is selective accumulation of the α-Hb chain on the inner leaflet of the membrane during *in vitro* incubation (Sears *et al.*, 1975; Ballas and Burka, 1978). This may indicate that the red cell membrane degrades the α chains less efficiently than the β polypeptides (Ballas and Burka,1979). Progressive accumulation of α chain, undoubtedly, contributes to the decreasing deformability and fluidity of an aging erythrocyte.

Whether the sialic acid content and, hence, the negative charge on the outer surface of the membrane decrease with aging is, at the present time, controversial and there are many conflicting reports in the literature. Durocher *et al.* (1975) reported decreased sialic acid concentration with aging. Desialylated erythrocytes are indeed promptly removed from the circulation (Gattegno *et al.*, 1979). The hypothesis that decreasing surface charge and sialic acid density is a mechanism of senescent red cell recognition *in vivo* is challenged by many investigators (Pessina *et al.*, 1980; Luner *et al.*, 1977). The issue seems to be resolved by the finding that human erythrocytes lose intact glycophorin together with membrane during senescence and that old cells hardly reveal desialylated surface proteins (Lutz and Fehr, 1979). Such losses would account for decreases in the level of sialic acid per cell but would not require that the concentration of sialic acid be altered in the remaining membrane. Thus, surface charge density could remain unchanged even though the total sialic acid content per cell is reduced.

Finally, a recently reported difference between young and old erythrocyte membranes which may be relevant to differential phagocytosis is the accumulation of autologous IgG exclusively on the outer surface of senescent red cell populations (Kay, 1978). Immunoglobulins eluted from the surface of old erythrocytes by thermal stripping reacted with younger red cells that had been desialylated by *Vibrio cholera* neuraminidase (Alderman *et al.*, 1980). This suggests that human red cells contain cryptic antigenic receptor sites to autologous immunoglobulins on their surface. During senescence, erythrocytes lose this membrane crypticity, expose their antigenic sites, and attach autologous immunoglobulins. This, in turn, is instrumental to the uptake of senescent erythrocytes by the RES.

2.2.2. Extracellular Destruction

Which of the alterations that occur during red cell aging are the most important factors leading to its demise is difficult to know with certainty. It could well be that all these factors operate in unison in order to dispose of an effete cell. We like, however, to stress the importance of three pathophysiologic mechanisms of aging. First, an aging erythrocyte undergoes a discocyte–spherocyte transformation. An old spherocytic cell is rigid and finds difficulty in negotiating capillary and splenic microvasculature. Second, the increased intracellular viscosity prolongs the transit time across small vessels and potentiates the ill-effects of decreased deformability. Third, accumulation of autologous immunoglobulins on the outer surface of an aging erythrocyte enables macrophages of the RES to recognize and phagocytize such a cell. Destruction of senescent erythrocytes is thus a miniature picture of hemolysis in autoimmune hemolytic anemia where cells of all ages are coated with autoantibodies and cleared by phagocytes of the RES.

Figure 2 is a scheme that summarizes the major events that occur during red cell aging and extravascular destruction. Once taken up by the macrophage, Hb, the major red cell constituent, is catabolized into globin and heme. Globin, other cytoplasmic and membrane proteins are further hydrolyzed to individual amino acids that enter the general pool of protein metabolism. Membrane lipids are degraded into cholesterol, acetate, phospholipids, and their precursors which are also reincorporated into lipid and carbohydrate metabolic pathways. Failure to catabolize these membrane lipids results in accumulation of certain intermediate metabolites in the RES that are typically found in certain lipidoses (Beutler, 1977).

The degradation of heme in the RES results in the following changes: (1) heme is converted to bilirubin and carbon monoxide is generated in the process of heme degradation and (2) iron is released and transported to the bone marrow or storage sites.

Heme oxygenase is a microsomal enzyme that degrades heme to biliverdin, CO, and Fe (Tenhunen *et al.*, 1968). This enzyme has an absolute requirement for NADPH and is inhibited by CO. It is active in spleen, liver, bone marrow, kidney, and lung. Hypoglycemia, glucagon, and epinephrine stimulate hepatic heme oxygenase activity via the action of cAMP (Bakken *et al.*, 1972). Biliverdin is further converted to bilirubin by the enzyme bilirubin reductase (Colleran and O'Carra, 1970). As the α-methene bridge of heme is cleared by heme oxygenase within the macrophages of the RES, CO is released (Coburn *et al.*, 1964). Normally, about 75% of endogenous CO originates from heme degradation, and in hemolytic anemia CO production shows close correlation with red cell survival (Coburn *et al.*, 1966). This CO is transported as carbon monoxy Hb to the lungs and excreted in expired air.

3. PROCESSING AND RELEASE OF IRON FROM THE RES

3.1. CONCEPTS AND MECHANISMS OF IRON RELEASE

3.1.1. Labile Iron Pool

As soon as iron is released from heme in the RE cell, it is ready to leave the cell for reutilization and red cell production by the bone marrow. This exit of iron is unidirectional. No transferrin iron *in vivo* seems to enter the RE cell (Finch *et al.*, 1970). The concept of early release of iron from the RE cell, based on the work of Cruz *et al.* (1942), led to the dictum "last iron to arrive at the RES is the first to leave." Several studies since then have confirmed and extended this observation, both in man and in animals.

In the original study of Cruz and co-workers it was shown that the radioiron of circulating erythrocytes in dogs, upon hemolysis was immediately reutilized in new cohorts of red blood cells. Similarly, Fillet *et al.* (1974) again in dogs, showed that intravenous injection of heat-damaged erythrocytes labeled with ^{59}Fe resulted in the release of radioiron to the circulation in two almost equal phases, early and late, with $t_{1/2}$ release of 34 min and 7 days, respectively. A 10-fold increase of the iron load to the RES did not affect the release phenomenon, neither did the level of plasma transferrin iron. Acute inflammation induced by endotoxin caused a striking decrease in iron release during both phases. In fact, the early phase of release was not detected. In other words, overactivation of the RES caused by the inflammation brought about a sharp drop in iron release from the RES to the circulation.

In man, similar observations were made. Noyes *et al.* (1960) showed that coincidental with disappearance of ^{59}Fe-tagged nonviable red blood cells from the circulation, there was an increase in surface counting over liver and spleen. The radioactivity over the splenic area decreased again within a few hours after its first appearance indicating the rapid release of iron to the circulation. When 0.01 mg red cell iron/kg was injected intravenously into a normal individual, the radioiron was detected in plasma transferrin within 30 min following the injection and rose to approximately 30% at 2 hr and 85% at 12 days. However, increasing the nonviable erythrocyte iron load to 0.1 to 1.0 mg/kg increased the $t_{1/2}$ clearance rate of these cells from the circulation but more significantly, increased the retention of iron in the RES from 15% to 50%. The same authors also found that elevation of plasma iron and storage iron by ingestion or by intravenous injections of saccharated iron oxide had no effect on retention of erythrocyte iron by the RES. These important findings suggested that the RES can process red cell iron for release at a certain maximal rate per unit time and that this iron was not diluted with any other pool of iron inside the RES. Similar observations were made again in man following the intravenous administration of ^{59}Fe Hb solution (1.5 to 2 mg/kg) well below the binding capacity of plasma Hp (Dresch and Najean, 1972). The $t_{1/2}$ for the clearance of Hb–Hp complex was 25.5 ± 4.3 min in normal individuals and without significant difference in patients with chronic disease (increased activity of the RES). In contrast, the clear-

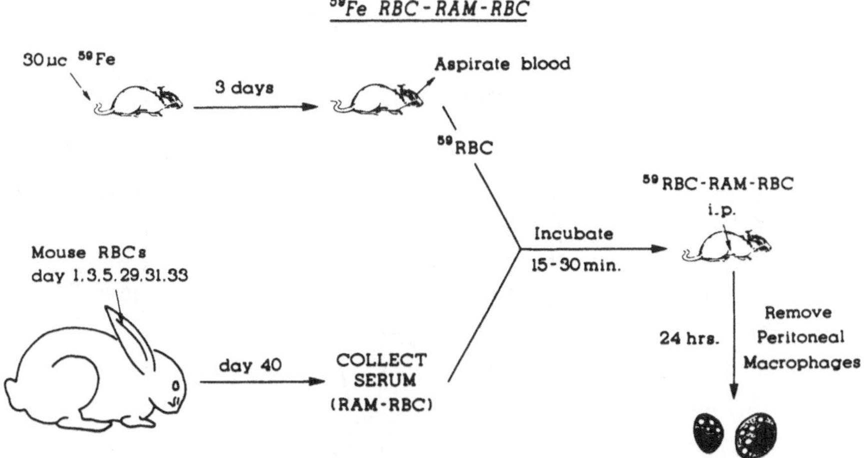

FIGURE 3. A model for the study of uptake, processing, and release of iron by macrophages.

ance of lipid emulsion or albumin aggregates was faster in patients with activated RES as a result of rheumatoid arthritis or Hodgkin's disease.

Similar *in vitro* studies confirmed the rapid release of iron by phagocytic cells and provided indirect evidence for the mechanism of such a release. Haurani and O'Brien (1972) have shown that when a lysate of murine peritoneal macrophages previously incubated with sensitized ^{59}Fe-tagged red blood cells (Fig. 3) was subjected to chromatography, the radioactivity was identified in the Hb and transferrin fractions but not in the ferritin fraction (Fig. 4). In the same experiment, it was shown that within 1 hr following phagocytosis of sensitized ^{59}Fe-tagged red blood cells, more than 50% of the radioactivity had left the macrophages and was located in the transferrin fraction of the extracellular medium (Fig. 5).

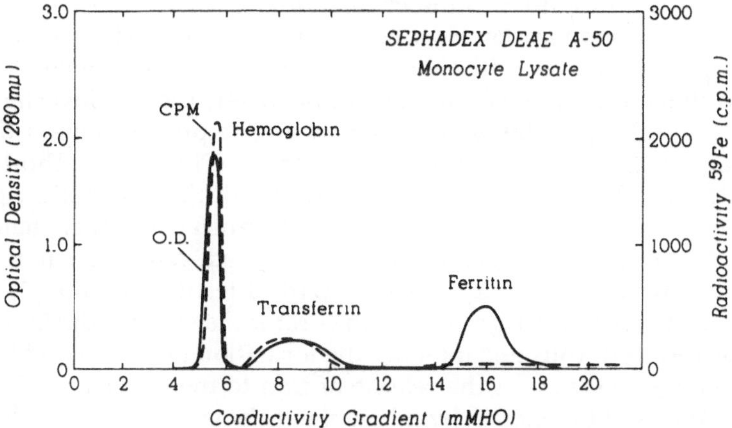

FIGURE 4. Appearance of radioiron in transferrin and hemoglobin fractions of monocyte lysate following incubation of monocytes with ^{59}Fe-sensitized red cells. Reproduced by permission of Academic Press.

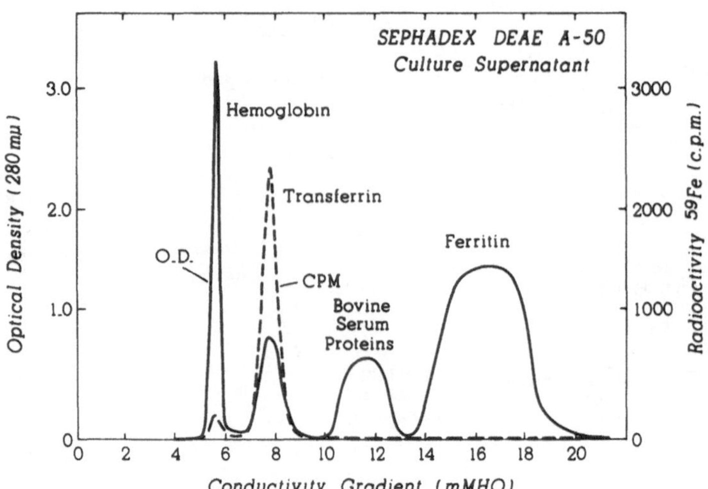

FIGURE 5. Early release of radioiron (1 hr) incubation from monocytes to transferrin of the incubation medium. Reproduced by permission of Academic Press.

Recently, Custer *et al.* (1982) using a different experimental design have confirmed and extended these observations. They too have found a transferrin-like protein in the macrophages. Within 1 hr after cultured macrophages were exposed to ^{59}Fe red cells, the iron radioactivity was observed in the Hb and ferritin fractions. In addition, there was a transient increase in ^{59}Fe associated with an intracellular protein eluting with transferrin.

The rapid release of iron from the RES immediately following its acquisition suggests the existence of a labile pool of iron within the RE cell which is separate from the stable (storage) pool of ferritin and hemosiderin. To one of the authors (F.I.H.), this labile pool is more like a canal or tract for the continuous movement of iron traffic from the RES to the circulation.

Now the question arises: How can a canal of iron exist within a framework of a stable pool in one cell without direct evidence of compartmentalization? The answer to this question is still uncertain; however, there is investigational evidence to suggest a possibility. The evidence rests primarily upon the fact that macrophages (RE cells) can synthesize transferrin (Phillips and Thorbecke, 1966; Haurani *et al.*, 1973). The iron atoms released during catabolism of Hb bind preferentially to the endogenously produced transferrin rather than to apoferritin (Fig. 6). Only excessive amounts of iron or reduced production of transferrin would allow accumulation of iron as part of ferritin and then hemosiderin. During activation of the RES as occurs in chronic disease, production of transferrin may be reduced, thus encouraging the formation of ferritin and hemosiderin and subsequent decrease in the release of iron to the circulation. The evidence for the synthesis of transferrin by macrophages was studied in our laboratory in the following fashion (Haurani *et al.*, 1973): Macrophages containing phagocytized sensitized ^{59}Fe-labeled red cells were collected from the peritoneal cavities of mice. The macrophages, about 700 million in number, were lysed,

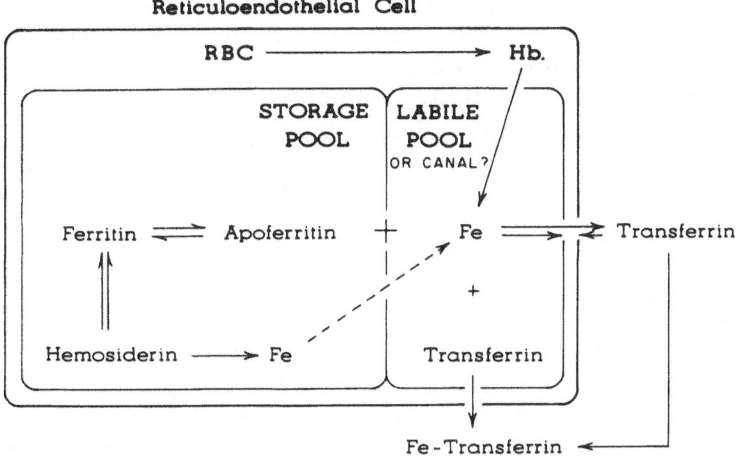

FIGURE 6. The movement of iron in the RE cells.

and the lysate without the addition of carrier transferrin was subjected to DEAE-Sephadex A-50 column chromatography. There was sufficient protein in the lysate to achieve coincident radioactivity and optical density peaks characteristic of transferrin. Macrophages were also cultured in the presence of [^{3}H]leucine in order to demonstrate that transferrin is actually synthesized by the macrophage, and not merely found there. Following culture, the lysate of the macrophage was subjected to ion-exchange chromatography. Coincident and significantly, overlapping ^{3}H and ^{59}Fe activity peaks were found (Fig. 7). After concentrating the radioactive peak, the material was subjected to immunoelectrophoresis against anti-mouse whole serum or anti-mouse transferrin. A precipitation line was obtained in the β region of transferrin.

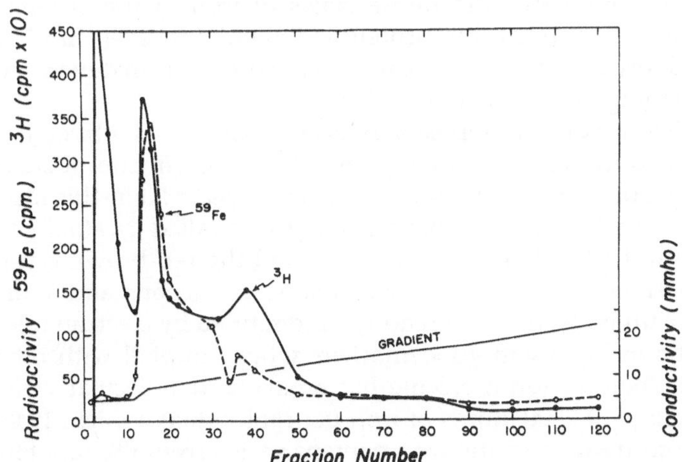

FIGURE 7. Monocytes incorporate radioiron and radioactive leucine in one peak that corresponds to transferrin. Reproduced by permission of Academic Press.

The kinetics of iron and transferrin in the circulation do not suggest that each iron atom leaving the RES has its own transferrin molecule. The $t_{1/2}$ clearance rate of plasma iron is 1 to 2 hr, whereas the $t_{1/2}$ clearance rate of plasma transferrin is at least 7 days (Katz, 1961). The plasma unsaturated transferrin molecules by adhering to the receptor sites of the macrophage are capable of extracting and binding iron.

3.1.2. Stable Iron Pool (Storage)

3.1.2a. Lactoferrin. At this point, a few remarks about lactoferrin are in order. This protein, another iron-binding protein abundant in granulocytes and widely distributed in body fluids and tissues, seems to be in search of a good physiologic role. Already, a few actions have been assigned to it: modulation of iron absorption in milk-fed infants, negative feedback regulator of colony-stimulating factor (CSF) and hence CFU-GM (Broxmeyer *et al.*, 1980), and as a bacteriostatic agent in depriving organisms of iron in infected tissues at acid pH (Masson, 1970).

It seems to have no role in iron metabolism although it shares a few characteristics with transferrin: similar molecular weight and iron-binding sites. The two proteins differ immunologically and in the pH at which they maximally bind iron (pH 7.4 for transferrin and 6.4 for lactoferrin) (Masson, 1970). The lactoferrin–iron complex is removed from the circulation by RE cells where the protein is digested and the iron is riveted to ferritin. No immediate utilization of this iron by red cells is demonstrated (van Snick *et al.*, 1977). Whether or not lactoferrin plays a role in anemia of chronic disease remains to be seen.

3.1.2b. Ferritin. Inside the RE cells, ferritin, a water-soluble protein, captures the excessive free iron ions and in so doing, acts as a detoxifying protein and also as a storage protein from which iron can be mobilized. Also, it is the protein that when lost with the macrophages and intestinal mucosa cells, in feces or urine, regulates the homeostasis of iron in the system. The rate of mobilization of iron from this protein is limited and even in idiopathic hemochromatosis, brisk bleeding may cause hypochromic iron deficiency anemia long before the storage iron is exhausted.

The existence of iron deposits in tissues was first identified by Neumann in 1868. He called the yellow-brown pigment *homosiderin* after it stained blue with potassium ferrocyanide (Prussian blue). A few years later, this insoluble protein was differentiated from a soluble form that was called *ferratin* by Schmedeberg. It was not until 1937 that Laufberger isolated the red-brown crystals of ferritin from horse spleen. (For original references, see Jacobs and Worwood, 1975.)

The ferritin molecules can readily be identified by electron microscopy. This characteristic makes them good markers when coupled with other molecules. The horse spleen ferritin molecule has an average molecular weight of 700,000 with an average of 2500 atoms of iron per molecule (Granick, 1946). This varies from tissue to tissue and the iron content varies from 0% (apoferritin) to 20%. Each molecule of ferritin has 24 subunits with a core (micelle) of hydrous ferric oxide phosphate in its center responsible for the red-brown color (Harrison *et al.*,

1974). There are several published amino acid analyses of horse and human apoferritins from liver and spleen (Harrison *et al.*, 1974). Ferritin structure is complicated by various types of heterogeneity and structural complexity. It is found in marine creatures and flowering plants. It has been studied in many animals besides horse and man. It is produced by many organs outside the RES, notably the heart, kidneys, and pancreas and in pathologic conditions by malignant cells. A single person may have more than one type of ferritin as demonstrated by gel electrophoresis and other criteria. These ferritins that differ in one or more properties are referred to as *isoferritins* (Drysdale, 1970). More acidic ferritin is found in heart, kidney, and in various neoplastic cells. Acidic ferritins are iron-rich (Powell *et al.*, 1975). Yet this concept of heterogeneity of ferritin on electrofocusing has been challenged (Bryce and Crichton, 1973). The physiologic importance of these isoferritins is not known. It might be that ferritin with less iron content serves as a platform for rapid exchange of iron intracellularly.

The uptake of iron by ferratin involves a ferrous to ferric oxidation and the release of iron involves its reduction. Apoferritin after binding with ferrous ions, acts as a ferroxidase *in vitro* since it oxidizes these ferrous ions to the ferric form. *In vitro* apoferritin is obtained from ferritin by the use of reducing agents such as sodium dithonite or thiogycollic acid at pH 4.6 to 5.4 (Harrison *et al.*, 1974). Iron is released fastest from ferritin of low iron content. Probably the iron found near the surface of the molecule is readily mobilized compared to the iron in the center. Probably, this surface iron is the last to deposit yet the first to come out.

Chelating agents such as desferrioxamine do release iron from ferritin and this agent is used therapeutically in iron overload associated with anemia (thalassemia) (Barry *et al.*, 1974). At the physiological level two mechanisms have been advanced for the release of iron from ferritin. Green and Mazur (1957) suggested that xanthine oxidase coupled with the oxidation of xanthine or hypoxanthine could bring about the reduction of iron in the ferritin molecule. Over the years, this mechanism has not gained much ground and, moreover, Crichton (1973) was unable to release ferritin iron with this system *in vitro*. The second mechanism could relate to the enzyme ferrireductase which was partially purified and which can reduce ferritin iron anaerobically in the presence of NADH and FMN.

One of the most intriguing aspects of the physiology of ferritin is the fact that administration of iron (any route or form) causes an increase in the level of ferritin in the tissues (Granick, 1946; Fineberg and Greenberg, 1955; Drysdale and Munro, 1966). Cycloheximide, which inhibits protein synthesis but not mRNA, inhibits ferritin synthesis, whereas actinomycin D, which inhibits protein synthesis through mRNA lack of formation, has no influence on ferritin synthesis (Chu and Fineberg, 1969). The studies suggest that iron induces the synthesis of ferritin at the translational level.

3.1.2c. Hemosiderin. Hemosiderin is another form of iron storage protein. However, it is insoluble and aggregates mainly in parenchymal cells. The fact that both ferritin and hemosiderin increase simultaneously and in parallel fashion after administration of iron (Shoden *et al.*, 1953; Drysdale and Munro, 1966) speaks against the suggestion advanced by Sturgeon and Shoden (1964) that hemosiderin is a degraded form of ferritin that has lost its micelles and aggregated in tissues.

3.2. METHODS OF STUDY OF RES ACTIVITY IN RELATION TO IRON

3.2.1. Initial Studies

1. Examination of peripheral blood: red cell count, Hb and hematocrit determinations, reticulocyte count, examination of peripheral blood smear.
2. Serum iron, total iron-binding capacity, and serum copper.
3. Red cell free protoporphyrin and coproporphyrin.
4. Serum Hp and ferritin.
5. Bone marrow examination and stain for iron.

3.2.2. Clearance of Colloidal Material

The clearance from the circulation of various colloidal particles has been used to reflect on the activity of the RES. Among these agents, carbon and aggregated albumin are extensively used. In overactivation of the RES, as occurs in anemia of chronic disease, there is fast clearance of the colloidal particles by the RES, whereas in RES blockade there is slow clearance of these particles from the circulation. This subject is discussed elsewhere in this volume.

3.2.3. Erythrokinetic Studies

3.2.3a. Plasma Iron Turnover and Iron Utilization by the Red Cells. The total and effective bone marrow erythropoiesis are studied by the intravenous injection of inorganic radioactive iron (^{59}Fe citrate or ^{55}Fe sulfate). The clearance of the radioactivity from the plasma transferrin is measured over a period of 2 hr, and the appearance of the same radioactivity in newly made red blood cells (i.e., iron utilization) is measured over a period of 2 weeks. The normal plasma iron turnover is approximately 30 mg/day. Iron utilization by the red blood cells is normally more than 75%.

3.2.3b. Reutilization of Iron by the Red Cells. The release of iron by the RES, transport by transferrin, and utilization by the bone marrow for Hb synthesis are studied by red cell iron *reutilization*. It is only when plasma iron turnover and red cell iron utilization, which are studied concomitantly, are normal that the study of iron reutilization by the red cells is expressive of the release of iron by the RE cells. Most of the studies on iron reutilization have been based on the injection of Hb solutions rather than on the use of intact heat-damaged red blood cells.

Five methods are available for the study of iron reutilization:

1. Reticulocyte-rich blood capable of Hb synthesis is obtained from patients with megaloblastic anemia receiving specific therapy and is incubated with ^{59}Fe for several hours. The blood is washed several times with saline to get rid of

extracellular ^{59}Fe and then is hemolyzed with water. The stroma is removed by centrifugation. The Hb solution contains ^{59}Fe (Haurani *et al.*, 1963).

2. Radioiron Hb solution was also prepared from blood obtained from a patient with polycythemia vera who received 5 to 7 mCi of ^{55}Fe (ferrous sulfate) as part of a therapeutic radiation course. At intervals of a few months, blood was collected from the donor and was hemolyzed by adding 1 part whole blood to 2.5 parts of sterile, distilled water. Hb solution was divided into small aliquots, measured, and stored at 4°C until further use (Haurani *et al.*, 1965, 1967).

3. Mice could be given 100 to 200 μCi of ^{59}Fe 2 days before being sacrificed. The blood is collected under sterile conditions from the heart and is hemolyzed and used on the same day that a patient is available for study. Dresch and Najean (1972) have used this technique successfully in more than 100 trials, and have encountered no untoward reaction. The advantages of this method are that the material could be made available at the time of the study, could be obtained at high specific activity, and could be used to do organ scanning and counting besides quantitation of the reutilization of iron.

4. Radioactive iron dextran (Yamada, 1968; Kanakakorn *et al.*, 1973) is cleared from the circulation in a way similar to the clearance of Hb; first, it has to be cleared by the RES where it is catabolized and the iron is released (Richter, 1959).

Iron utilization and reutilization can be studied simultaneously in a given patient by the use of two isotopes of iron: ^{59}Fe and ^{55}Fe. The former isotope is easy to handle and to count since it only requires a gamma scintillation counter, whereas the latter is more tedious to work with since the blood has to be processed before its radioactivity is counted in a liquid scintillation counter.

5. More recently, a colloidal suspension of hydrolyzed radioiron of high specific activity has been developed for the study of iron release from the RES (Bentley *et al.*, 1979). There are two objections to this method. First, colloidal compounds may be handled differently than damaged red cells by the RES and, therefore, are less useful as a physiologic or pathophysiologic probe (Fillet, 1977), and second, without knowing the actual size of the labile iron pool (not the storage pool) the information derived by release rates may be incomplete. The amount of iron released per day in milligrams by the RES is a function of the iron release rate and the size in milligrams of the labile iron pool. The labile iron pool is presumably decreased in the anemia of chronic disease.

3.2.3c. Iron Absorption. Iron absorption may be measured by the double-tracer technique in which ^{59}Fe is given by mouth to a fasting patient and ^{55}Fe is given by vein. Measurement of the 14-day iron utilization of both isotopes makes it possible to calculate the intestinal absorption of the oral dose. Another method for determining iron absorption is to estimate the ^{59}Fe activity lost in the stools.

3.3. FACTORS THAT DECREASE IRON RELEASE

Activation of individual cells of the RES rather than an increase in their numbers causes a sudden retention of iron in the system.

The decrease in serum iron is the first and the most rapidly occurring feature. In animals, injections of bacteria, vaccines, toxins, or leukocytic endogenous mediator (LEM) have caused a sharp fall in the serum iron and accumulation of the metal in the liver (Pekarek *et al.*, 1972). The fall in serum iron is independent of the production of pyrexia (Grieger and Kluger, 1978). Usually, the hypoferremia is established within 24 hr without significant change in the plasma iron turnover (total erythropoiesis) (Peterson, 1953; Bush *et al.*, 1956; Kampschmidt and Arredondo, 1963).

The serum TIBC fell only after the disease was well established and then the extent of the fall was moderate (Pekarek *et al.*, 1969).

These workers and others (Hershko *et al.*, 1974; Fillet, 1977) using transferrin-bound radioiron have shown that in acute inflammation, there is no significant change in the distribution of available iron (plasma iron) to tissues. In addition, the rapidity with which the fall in serum iron occurs following acute infection in man without changes in iron storage depots (Kumar *et al.*, 1978a,b) speaks against dilution of the radioactive iron in an enlarged iron storage pool and confirms the contention that the rearrangement of iron in that pool plays an important role in the sharp decrease in the flow of iron from the RES and the ensuing hypoferremia. In addition, the absorption of iron abruptly diminishes (Dubach *et al.*, 1948; Cortell and Conrad, 1967; Beresford *et al.*, 1971), and this plays a smaller role in the early production of hypoferremia.

If the state of activation of the RES continues for a few weeks, then anemia ensues. This type of anemia for no better name is called *anemia of chronic disease.* It is usually mild and not progressive, but can produce symptoms. This type of anemia complicates such chronic inflammatory states as rheumatoid arthritis and subacute bacterial endocarditis as well as malignant disorders such as Hodgkin's disease. It is occasionally seen in chronic renal disease and severe tissue injury (Cartwright, 1966). A similar type of anemia has also been described as a primary disorder (Haurani and Green, 1967; Hume *et al.*, 1973). Although its exact cause is unknown, it has been considered that the host, in its own defense, may withhold iron not only from Hb synthesis (an innocent bystander) but primarily from microbes, other invasive organisms, and possibly tumor cells (Kochan, 1973; Weinberg, 1974). The ability of the host to withhold iron in this fashion has been called *nutritional immunity* (Cartwright, 1966). Certain bacteria, for their own growth, can produce compounds (siderophores) by which they can extract iron from transferrin if this protein is at least 30% saturated with iron (for excellent reviews, see Weinberg, 1974, and Emery, 1982). The anemia is characterized by decreased plasma iron, decreased TIBC in most instances, increased serum Hp and serum ferritin, decreased bone marrow sideroblasts, and normal or increased RES iron. Of the above features, the triad of low plasma iron, decreased TIBC, and increased RES iron is distinctive and occurs in no other condition.

The studies mentioned above have shown the following results in anemia of chronic disease:

1. Peripheral blood: Mild to moderate anemia with varying degrees of hypochromia depending on the severity of anemia is usually revealed. Reticulocyte count is usually normal unless there is hemolysis.

2. Serum iron and total iron-binding capacity: Serum iron is low if not normal. In simple iron deficiency this latter parameter is usually elevated.

3. Serum copper, red cell free protoporphyrin, and red cell free coproporphyrin: These are increased (Cartwright, 1966). The increase in the red cell free protoporphyrin and red cell free coproporphyrin is similar to that of simple iron deficiency, suggesting that in both conditions, there is a lack of heme synthesis in the developing red cell. The levels of serum copper associated with infection range from 118 to 267 μg/dl (mean 191 μg/dl) compared with a normal range of 81 to 147 μg/dl (mean 144 μ/dl). The hypercupremia is due mainly to an increase in ceruloplasmin. The rise in ceruloplasmin, along with that of fibrinogen and certain other plasma proteins, is considered to be an "acute-phase reactant." [Ceruloplasmin, at least in swine, acts as a ferroxidase enzyme oxidizing ferrous to ferric ions thereby enhancing the movement of these ions from cells to plasma (Roeser *et al.*, 1970).]

4. Serum Hp: Increase in serum Hp from normal values of 50 to 150 mg/gl up to 500 mg/dl occurs in anemia of chronic disease (Nyman, 1959). In chronic disease in which serum Hp is normal, the mechanism of the anemia could be a hemolytic process as well as defective iron reutilization.

5. Serum ferritin: Ferritin and hemosiderin are the two storage proteins. Although both are intracellular proteins, ferritin with the help of a newly developed radioimmunoassay technique has been detected and quantitated in all human sera (Addison *et al.*, 1972; Jacobs *et al.*, 1972). The serum ferritin is secreted by the RE cells and is not released from dead cells (Birgegard, 1980). In one study (Lipschitz *et al.*, 1974), the geometric mean of serum ferritin in normal subjects is 59 ng/ml with a 95% confidence range of 12 to 300 ng/ml. In iron deficiency anemia, the mean serum ferritin level is 4 ng/ml (range 1–14). In 39 patients with inflammation, the mean serum ferritin is 305 with a range of 10 to 1650 ng/ml. There is no apparent relation between serum ferritin level and the duration, severity, or type of inflammatory process. There is in inflammation a significant negative correlation between serum ferritin level when matched with either the fraction of iron released from the RES or the iron reutilized by bone marrow (Bentley *et al.*, 1979). When anemia of chronic disease is complicated with iron deficiency, serum ferritin falls to a mean level of 56 ng/ml with a range of 2–270 (Reeves and Haurani, 1980). Serum ferritin is also elevated in conditions characterized by iron overload such as hemochromatosis.

6. Bone marrow: The bone marrow appears grossly "normal." When particles are stained for iron, the metal appears in the RE cells as bluish pigment. Usually, no sideroblasts are seen.

7. Results of erythrokinetic studies in anemia of chronic disease are as follows:

In malignant disease, when causes of anemia such as blood loss, hemolysis, or myelofibrosis are absent and the cardinal features of anemia of defective iron reutilization are present, erythrokinetic studies have confirmed that plasma iron turnover and iron utilization by the bone marrow are normal or almost normal but iron reutilization is diminished. This was first observed in dogs by Freireich *et al.* (1957). The degree of depression of iron red cell reutilization correlates well with the severity of anemia. In addition, iron absorption is found to be low, less than 10% (Haurani *et al.*, 1963; Haurani *et al.*, 1965b; Cartwright, 1966; Yamada, 1968; Dresch and Najean, 1972).

In patients with chronic inflammatory disease, similar findings have been obtained (Haurani *et al.*, 1965a). The values for iron red blood cell reutilization vary from 7% to 42% (Fig. 8), and iron absorption is low. In fact, more than 90% of the oral dose of radioiron is lost to the stools on the first day, indicating that there is no uptake of iron by the intestinal mucosa. Normally, food iron is acervated by intestinal mucosal cells, partly absorbed, and the rest is lost to the stools in 3 to 4 days upon desquamation of these cells (Crosby, 1963).

We performed erythrokinetic studies in one group of elderly women with hypochromic anemia, and normal or increased iron stores. These patients have poor iron absorption and defective red cell iron reutilization but have no apparent other disease. This entity is referred to as *primary defective iron reutilization* (Haurani and Green, 1967). Similar cases have been described by others (Hume *et al.*, 1973). Perhaps the slight fall of Hb concentration in the aged is the result of activation of the RES and subsequent retention of iron. This area of geriatric research has not been undertaken yet.

In attempting to explain the impaired marrow response in anemia of chronic disease, the role of erythropoietin must be considered. In patients with iron deficiency or primary hematopoietic disease, the activity of erythropoietin is directly related to the severity of the anemia in an attempt to compensate for the anemia. However, in the anemia of chronic disorder, no correlation exists between erythropoietin levels and the severity of anemia (Lukens, 1973). As a

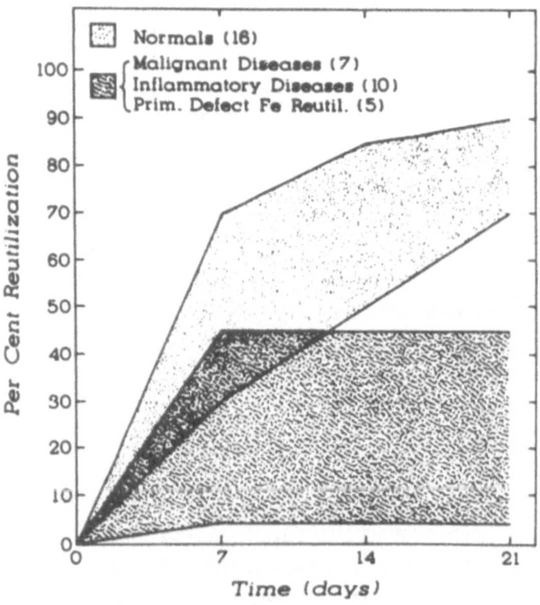

FIGURE 8. Poor iron reutilization by the red blood cells in anemia complicating malignancy or inflammation or in anemia of primary defective iron reutilization.

matter of fact, erythropoietin levels are only elevated to even a moderate degree in 11 of 31 patients with anemia of chronic disorders (Ward *et al.*, 1971). At mild levels of anemia (hematocrits > than 30%), no good correlation has been noted between erythropoietin level and degree of anemia, probably a reflection of the insensitivity of the bioassay. At severe levels of anemia, anemia of chronic disorder is usually complicated by other conditions so that one is not necessarily measuring the erythropoietin titer of only this type of anemia. It is still a possibility that an erythropoietin-blocking factor is elaborated by the chronic disease as a tightly bound antibody with a serum excess, although no erythropoietin-blocking or -inhibiting factors have been found in the sera of patients or of animals with this anemia (Ward *et al.*, 1971; Lukens, 1973). In further investigation of the marrow response to anemia, it has been found that the anemia responds normally or as expected to hypoxia or to the administration of erythropoietin (Gutnisky and Van Dyke, 1963) or cobalt (Wintrobe *et al.*, 1947), a known stimulator of erythropoietin production; indicating that the marrow is capable of a normal response to the anemia if properly stimulated.

On the other hand, other studies (Alexanian, 1972) have shown no depressed erythropoietin production when urinary rather than serum levels of erythropoietin in patients with anemia of chronic disease are assayed. The urinary erythropoietin levels are appropriate for the hematocrit. In a more recent work (Douglas and Adamson, 1975), it has been shown that patients with anemia of chronic disease have an elevated P50 and 2,3-DPG levels. Erythropoietin urinary excretion is variable and bears no relation to the hematocrit. Marrow transit times correlate inversely with the urinary or serum levels of erythropoietin, suggesting an impaired response of erythropoiesis to erythropoietin stimulation. In addition, the red cell iron utilization correlated better with plasma iron levels than with erythropoietin.

It might be that the red cell precursors in the bone marrow, in the absence of adequate amounts of iron, are less responsive to erythropoietin. Besides erythropoietin, other factors have been involved in the suppression of erythropoiesis: endotoxin (Thomas and Briscoe, 1973), neutrophilic extract (Herman *et al.*, 1978), or an inhibitor from cultured tumor cells (Zucker *et al.*, 1980).

3.4. FACTORS THAT INCREASE IRON RELEASE

It has been known for some time that repeated injections of an RE-blockading agent such as colloidal thorium dioxide cause hyperferremia. Vannotti (1957) attributed the hyperferremia to increased hemolysis by the blockaded RES. Cartwright *et al.* (1950) found no increased excretion of urobilinogen in dogs receiving repeated injections of thorium dioxide and they offered a reduced rate of uptake of iron from the circulation by the RES as an explanation for the hyperferremia. Years later, Haurani and O'Brien (1973) studying again the effect of repeated injections of thorium dioxide in dogs found that the RES-blockading agent enhances erythropoiesis. The increase in red cell production is manifested by increased reticulocyte count, increased red cell mass, and increased plasma iron turnover. There is no evidence of abnormal hemolysis when the survival of the red cells is tested by the ^{51}Cr method.

The stimulatory action of thorium dioxide on erythropoiesis is not likely mediated through the adrenal or thyroid glands since measurements of the corticosteroids and thyroxine remain the same following the administration of the blockading agent. Similarly, urinary and serum levels of erythropoietin do not change significantly and therefore, this hormone may not be held responsible for the increased red cell production following the administration of thorium dioxide. Most probably as has been shown by Hillman and Henderson (1969), provision of large amounts of iron to the bone marrow stimulates erythropoiesis and in this case thorium dioxide causes more release of iron from the RES and consequently increases the delivery of iron to the bone marrow.

However, other workers have shown that similar agents blockading the RES such as colloidal carbon (Peschle *et al.*, 1976) or crystalline silica (Rich and Kubanek, 1982) are capable of releasing erythropoietin from the macrophages. In addition to erythropoietin, other factors also released by macrophages that modulate erythropoiesis have been described (Ascensao *et al.*, 1981; Udupa and Lipschitz, 1982).

In man, testosterone enanthate was effective in correcting the anemia of patients with primary defective iron reutilization (Haurani and Green, 1967). The corrective action of testosterone was assumed to be the result of mobilization of iron from the RES by this hormone. *In vitro* studies using mouse peritoneal macrophages that have engulfed ^{59}Fe-labeled antibody-coated mouse red cells (Fig. 3) (Reeves *et al.*, 1981) have shown that hormones like testosterone, estrogen, and dexamethasone but not erythropoietin could increase the release of iron from these macrophages.

REFERENCES

Addison, G. M., Beamish, M. R., Hales, C. N., Hodgkins, M., Jacobs, A., and Llewellin, P., 1972, An immunoradiometric assay for ferritin in the serum of normal subjects and patients with iron deficiency and iron overload, *J. Clin. Pathol.* **25**:326.

Alderman, E. M., Fudenberg, H. H., and Lovins, R. E., 1980, Binding of immunoglobulin classes to subpopulations of human red blood cells separated by density-gradient centrifugation, *Blood* **55**:817.

Alexanian, R., 1972, Erythropoietin excretion in hemolytic anemia and in the hypoferremia of chronic disease, *Blood* **40**:946.

Allen, D. W., and Jandl, J. H., 1961, Oxidative hemolysis and precipitation of hemoglobin. II. Role of thiols in oxidant drug action, *J. Clin. Invest.* **40**:454.

Alper, C. A., Peters, J. H., Birtch, A. G., and Gardner, F. H., 1965, Haptoglobin synthesis. I. *In vivo* studies of the production of haptoglobin, fibrinogen and gamma-globulin by the canine liver, *J. Clin. Invest.* **44**:574.

Ascensao, J. L., Kay, N. E., Earenfight-Engler, T., Koren, H. S., and Zanjani, E. D., 1981, Production of erythroid potentiating factor(s) by a human monocytic cell line, *Blood* **57**:170.

Bakken, A. F., Thaler, M. M., and Schmid, R., 1972, Metabolic regulation of heme catabolism and bilirubin production. I. Hormonal control of hepatic heme oxygenase activity, *J. Clin. Invest.* **51**:530.

Ballas, S. K., and Burka, E. R., 1978, Catabolism of hemoglobin by human erythrocyte membranes, *J. Lab. Clin. Med.* **92**:387.

Ballas, S. K., and Burka, E. R., 1979, Protease activity in the human erythrocyte: Localization to the cell membrane, *Blood* **53**:875.

Barry, M., Flynn, D. M., Letsky, E. A., and Disdon, R. A., 1974, Long term chelation therapy in thalassemia major—Effect on liver iron concentration, liver histology and clinical progress, *Br. Med. J.* **2**:16.

Bentley, D. P., Cavill, I., Ricketts, C., and Peake, S., 1979, A method for the investigation of reticuloendothelial iron kinetics in man, *Br. J. Haematol.* **43**:619.

Beresford, C. H., Neale, R. J., and Brooks, O. G., 1971, Iron absorption and pyrexia, *Lancet* **1**:568.

Berlin, N. I., and Berk, P. D., 1975, The biological life of the red cell, in: *The Red Cell*, 2nd ed., Vol. 2 (D. M. Surgenor, ed.), pp. 957–1019, Academic Press, New York.

Bernstein, R. E., 1959, Alterations in metabolic energetics and cation transport during aging of red cells, *J. Clin. Invest.* **38**:1572.

Beutler, E., 1977, Lipid storage diseases, in: *Hematology* (W. J. Williams, E. Beutler, A. J. Erslev, and R. W. Rundles, eds.), 2nd ed., pp. 1147–1155, McGraw-Hill, New York.

Bienzle, U., and Pjura, W. J., 1977, Alteration of membrane proteins during erythrocyte aging, *Clin. Chim. Acta* **76**:183.

Birgegard, G., 1980, The source of serum ferritin during infection, *Clin. Sci.* **59**:385.

Bocci, V., Pessina, G. P., Paulesu, L., Pacini, A., and Muscettola, M., 1978, Studies of factors regulating the aging of human erythrocytes. I. The role of pH and of divalent cations, *Int. J. Biochem.* **10**:19.

Bocci, V., Pessina, G. P., and Paulesu, L., 1980, Studies of factors regulating the aging of human erythrocytes. III. Metabolism and fate of erythrocyte vesicles, *Int. J. Biochem.* **11**:139.

Bonsignore, A., Fornaini, G., Fantoni, A., Leoncini, G., and Segni, P., 1964, Relationship between age and enzymatic activities in human erythrocytes from normal and fava bean sensitive subjects, *J. Clin. Invest.* **43**:834.

Brok, F., Ramot, G. P., and Danon, D., 1966, Enzyme activities in human red blood cells of different age groups, *Isr. J. Med. Sci.* **2**:291.

Broxmeyer, H. E., DeSousa, M., Smithyman, A., Ralph, P., Hamilton, J., Kurland, J. I., and Bocknacki, J., 1980, Specificity and modulation of the action of lactoferrin, a negative feedback regulator of myelopoiesis, *Blood* **55**:324.

Bryce, C. F. A., and Crichton, R. R., 1973, Microheterogeneity in apoferritin molecules—Artifact, *Hoppe-Seylers Z. Physiol. Chem.* **354**:344.

Bush, J. A., Ashenbrucker, H., Cartwright, G. E., and Wintrobe, M. M., 1956, Anemia of infection: Kinetics of iron metabolism in anemia associated with chronic infection, *J. Clin. Invest.* **35**:89.

Carrell, R. W., Winterbourn, C. C., and Rachmilewitz, E. A., 1975, Activated oxygen and hemolysis, *Br. J. Haematol.* **30**:259.

Cartwright, G. E., 1966, The anemia of chronic disorders, *Semin. Hematol.* **3**:351.

Cartwright, G. E., and Deiss, A., 1975, Sideroblasts, siderocytes and sideroblastic anemia, *N. Engl. J. Med.* **292**:185.

Cartwright, G. E., Gubler, C. J., and Wintrobe, M. M., 1950, Anemia of infection. XII. The effect of turpentine and colloidal thorium dioxide on plasma iron and plasma copper of dogs, *J. Biol. Chem.* **184**:579.

Chu, L. L. H., and Fineberg, R. A., 1969, On the mechanism of iron-induced synthesis of apoferritin in HeLa cells, *J. Biol. Chem.* **244**:3847.

Coburn, R. F., Williams, W. F., and Forster, R. E., 1964, Effect of erythrocyte destruction on carbon monoxide production in man, *J. Clin. Invest.* **43**:1098.

Coburn, R. F., Williams, W. F., and Kahn, S. B., 1966, Endogenous carbon monoxide production in patients with hemolytic anemia, *J. Clin. Invest.* **45**:460.

Cohen, G., and Hochstein, P., 1961, Glucose-6-phosphate dehydrogenase and detoxification of hydrogen peroxide in human erythrocytes, *Science* **134**:1756.

Colleran, E., and O'Carra, P., 1970, Specificity of biliverden reductase, *Biochem. J.* **119**:16.

Cortell, S., and Conrad, M. E., 1967, Effect of endotoxin on iron absorption, *Am. J. Physiol.* **213**:43.

Crichton, R. R., 1973, Structure and function of ferritin, *Angew. Chem. Int. Ed. Engl.* **12**:57.

Crosby, W. H., 1963, The control of iron balance by the intestinal mucosa, *Blood* **22**:441.

Cruz, W. O., Hahn, P. F., and Bale, W. F., 1942, Hemoglobin radioactive iron liberated by erythrocyte destruction (acetylphenylhydrazine) promptly reutilized to form new hemoglobins, *Am. J. Physiol.* **135**:595.

Custer, G., Balcerzak, S., and Rinehart, J., 1982, Human macrophage hemoglobin-iron metabolism *in vitro*, *Am. J. Hematol.* **13**:23.

Douglas, S. W., and Adamson, J. W., 1975, The anemia of chronic disorders: Studies of marrow regulation and iron metabolism, *Blood* **45**:55.

Dresch, C., and Najean, Y., 1972, Hemoglobin iron kinetics in man, *Eur. J. Clin. Biol. Res.* **27**:930.

Drysdale, J. W., 1970, Microheterogeneity in ferritin molecules, *Biochim. Biophys. Acta* **207**:256.

Drysdale, J. W., and Munro, H. N., 1966, Regulation of synthesis and turnover of ferritin in rat liver, *J. Biol. Chem.* **241**:3630.

Dubach, R., Callender, S. T. E., and Moore, C. V., 1948, Studies in iron transportation and metabolism. VI. Absorption of radioactive iron in patients with fever and with anemias of varied etiology, *Blood* **3**:526.

Durocher, J. R., Payne, R. C., and Conrad, M. E., 1975, Role of sialic acid in erythrocyte survival, *Blood* **45**:11.

Emery, T., 1982, Iron metabolism in humans and plants, *Am. Sci.* **70**:626.

Fillet, G., 1977, *Le fer dans l'organisme: Metabolisme et reutilization*, Masson, Paris.

Fillet, G., Cook, J. D., and Finch, C. A., 1974, Storage iron kinetics. VII. A biologic model for reticuloendothelial iron transport, *J. Clin. Invest.* **53**:1527.

Finch, C. A., Cook, J. D., Eschbach, J. W., Harker, L. A., Funk, D. D., Marsaglia, G., Hillman, R. S., Slichter, S., Adamson, J. W., Ganzoni, A., and Giblett, E. R., 1970, Ferrokinetics in man, *Medicine* **49**:17.

Fineberg, R. A., and Greenberg, D. M., 1955, Ferritin biosynthesis. II. Acceleration of synthesis by the administration of iron, *J. Biol. Chem.* **214**:97.

Freireich, E. J., Miller, A., Emerson, C. P., and Ross, J. F., 1957, The effect of inflammation on the utilization of erythrocyte and transferrin bound radioiron for red cell production, *Blood* **12**:972.

Ganzoni, A. M., Oakes, R., and Hillman, R. S., 1971, Red cell aging *in vivo*, *J. Clin. Invest.* **50**:1373.

Garby, L., and Noyes, W. D., 1959, Studies of hemoglobin metabolism. I. The kinetic properties of the plasma hemoglobin pool in normal man, *J. Clin. Invest.* **38**:1479.

Gattegno, L., Fabia, F., Bladier, D., and Cornillot, P., 1979, Physiological aging of red blood cells and changes in membrane carbohydrates, *Biomedicine* **30**:194.

Gayer, G., Linss, W., and Schaaf, P., 1972, The distribution pattern of anionic sites at the human erythrocyte surface as revealed by the colloidal iron method, *Acta Histochem.* **42**:138.

Gordon, S., and Bearn, A. G., 1966, Hemoglobin binding capacity of isolated haptoglobin polypeptide chains, *Proc. Soc. Exp. Biol. Med.* **121**:846.

Granick, S., 1946, Ferritin, its properties and significance for iron metabolism, *Chem. Rev.* **38**:379.

Green, S., and Mazur, J., 1957, Relation of uric acid metabolism to release of iron from hepatic ferritin, *J. Biol. Chem.* **227**:653.

Greenwalt, T. J., and Lau, F. O., 1978, Evaluation of toluidine blue for measuring erythrocyte membrane loss during *in vivo* aging, *Br. J. Haematol.* **39**:545.

Grieger, T. A., and Kluger, M. J., 1978, Fever and survival: The role of serum iron, *J. Physiol.* (*London*) **279**:187.

Gutnisky, A., and Van Dyke, D., 1963, Normal response to erythropoietin or hypoxia in rats made anemic with turpentine abscess, *Proc. Soc. Exp. Biol. Med.* **112**:75.

Haidas, S., Labie, D., and Kaplan, J. C., 1971, 2,3-Diphosphoglycerate content and oxygen affinity as a function of red cell age in normal individuals, *Blood* **38**:463.

Hanstein, A., and Muller-Eberhard, U., 1968, Concentration of serum hemopexin in healthy children and adults and in those with a variety of hematological disorders, *J. Lab. Clin. Med.* **71**:232.

Harrison, P. M., Hoare, R. J., Hoy, T. G., and Macara, I. G., 1974, Ferritin and haemosiderin: Structure and function, in: *Iron in Biochemistry and Medicine* (A. Jacobs and M. Woorwood, eds.), pp. 73–114, Academic Press, New York.

Haurani, F. I., and Green, D., 1967, Primary defective iron reutilization: Response to testosterone therapy, *Am. J. Med.* **42**:151.

Haurani, F. I., and O'Brien, R., 1972, A model system for the release of iron from the reticuloendothelial system, *J. Reticuloendothelial Soc.* **12**:29.

Haurani, F. I., and O'Brien, R., 1973, The erythropoietic effect of a reticuloendothelial blocking agent, *J. Reticuloendothelial Soc.* **13**:126.

Haurani, F. I., Young, K., and Tocantins, L. M., 1963, Reutilization of iron in anemia complicating malignant neoplasms, *Blood* **22**:73.

Haurani, F. I., Burke, W., and Martinez, E. J., 1965a, Defective reutilization of iron in the anemia of inflammation, *J. Lab. Clin. Med.* **65**:560.

Haurani, F. I., Green, D., and Young, K., 1965b, Iron absorption in hypoferremia, *Am. J. Med. Sci.* **249**:537.

Haurani, F. I., Meyer, A., and O'Brien, R., 1973, Production of transferrin by the macrophage, *J. Reticuloendothelial Soc.* **14**:309.

Herman, S. P., Golde, D. W., and Cline, M. J., 1978, Neutrophil products that inhibit cell proliferation—Relation to granulocytic chalone, *Blood* **51**:207.

Hershko, C., Cook, J. D., and Finch, C. A., 1974, Storage iron kinetics: Effects of inflammation on iron exchange in rat, *Br. J. Haematol.* **28**:67.

Hillman, R. S., and Henderson, P. A., 1969, Control of marrow production by the level of iron supply, *J. Clin. Invest.* **48**:454.

Hume, R., Dagg, J. H., and Goldberg, A., 1973, Refractory anemia with dysproteinemia: Long-term therapy with low-dose corticosteroids, *Blood* **41**:27.

Jacobs, A., and Worwood, M., 1975, The biochemistry of ferritin and its clinical implications, *Prog. Hematol.* **9**:1.

Jacobs, A., Miller, F., Worwood, M., Beamish, M. R., and Wardrop, C. A., 1972, Ferritin in the serum of normal subjects and patients with iron deficiency and iron overload, *Br. Med. J.* **4**:206.

Jandl, J. H., Greenberg, M. S., Yonemoto, R. H., and Cartle, W. B., 1956, Clinical determination of the sites of red cell sequestration in hemolytic anemias, *J. Clin. Invest.* **35**:842.

Kadlubowski, M., 1978, The effect of *in vivo* aging of the human erythrocyte on the protein of the plasma membrane: A characterization, *Int. J. Biochem.* **9**:67.

Kadlubowski, M., and Agutter, P. S., 1977, Changes in the activities of some membrane-associated enzymes during *in vivo* aging of the normal human erythrocyte, *Br. J. Haematol.* **37**:111.

Kadlubowski, M., and Harris, J. R., 1974, The appearance of a protein in the human erythrocyte membrane during aging, *FEBS Lett.* **47**:252.

Kampschmidt, R. F., and Arredondo, M. I., 1963, Some effects of endotoxin upon plasma iron turnover in the rat, *Proc. Soc. Exp. Biol. Med.* **113**:142.

Kanakakorn, K., Cavill, I., and Jacobs, A., 1973, The metabolism of intravenously administered iron-dextran, *Br. J. Haematol.* **25**:637.

Katz, J. H., 1961, Iron and protein kinetics studied by means of doubly labeled human crystalline transferrin, *J. Clin. Invest.* **40**:2143.

Kay, M. M., 1978, Role of physiologic autoantibody in the removal of senescent human red cells, *J. Supramol. Struct.* **9**:555.

Keene, W. R., and Jandl, J. H., 1965, The sites of hemoglobin catabolism, *Blood* **26**:705.

Keitt, A. S., Smith, T. W., and Jandl, J. H., 1966, Red cell "pseudmosaicism" in congenital methemoglobinemia, *N. Engl. J. Med.* **275**:397.

Kirpatrick, F. H., Mushs, A. G., Kostok, R. K., and Gabel, C. W., 1979, Dense (aged) circulating red cells contain normal concentrations of adenosine triphosphate (ATP), *Blood* **54**:946.

Kochan, I., 1973, The role of iron in bacterial infections with special consideration of host–tubercle bacillus interaction, *Curr. Top. Microbiol. Immunol.* **60**:1.

Kumar, R., Arora, B. B., Singh, U., and Mehrotra, G. C., 1978a, Storage iron in acute and chronic infection, *Indian J. Med. Res.* **68**:503.

Kumar, R., Singh, U., and Mehrotra, G. C., 1978b, Mechanisms of hypoferraemia in acute and chronic infection, *Indian J. Med. Res.* **68**:508.

LaCelle, P. L., and Arkin, B., 1970, Acquired rigidity: A possible determinant of normal RBC life span, *Blood* **36**:837.

Leonhardt, H., Grigoleit, H. G., and Reinhardt, I., 1978, Erythrocyte deformability in a red cell aging model, *Ric. Clin. Lab.* **8**:65.

Lichtman, M. A., and Murphy, M. S., 1975, Red cell adenosine triphosphate in hypoproliferative anemia with and without chronic renal disease: Relationship to hemoglobin deficit and plasma inorganic phosphate, *Blood Cells* **1**:467.

Lichtman, M. A., and Weed, R. I., 1972, Divalent cation content of a normal and ATP depleted erythrocyte membrane, *Nouv. Rev. Fr. Hematol.* **12**:799.

Lipschitz, D. A., Cook, J. D., and Finch, C. A., 1974, A clinical evaluation of serum ferritin as an index of iron stores, *N. Engl. J. Med.* **290**:1213.

Lorand, L., Weismann, L. B., Epel, D. L., and Bruner-Lorand, J., 1976, Role of the intrinsic transglutaminase in the Ca^{2+}-mediated crosslinking of erythrocyte proteins, *Proc. Natl. Acad. Sci. USA* **73**:4479.

Lukens, J. N., 1973, Control of Erythropoiesis in rats with adjuvant-induced chronic inflammation, *Blood* **41**:37.

Luner, S. J., Szklarek, D., Knox, R. J., Seaman, G. V., and Josefowicz, J. Y., 1977, Red cell charge is not a function of cell age, *Nature* (*London*) **269**:719.

Lutz, H. U., and Fehr, J., 1979, Total sialic acid content of glycophorins during senescence of human red blood cells, *J. Biol. Chem.* **254**:11177.

Lutz, H. U., Liu, S. C., and Polek, J., 1977, Release of spectrin free vesicles from human erythrocytes during ATP depletion, *J. Cell Biol.* **73**:548.

Masson, P., 1970, in: *La Lactoferrin*, Editions Arscia, SA, Brussels.

Muller-Eberhard, U., 1976, Hemopexin, *N. Engl. J. Med.* **283**:1090.

Noyes, W. D., and Garby, L., 1967, Rate of haptoglobin synthesis in normal man: Determinations by the return to normal levels following hemoglobin infusion, *Scand. J. Clin. Lab. Invest.* **20**:33.

Noyes, W. D., Bothwell, T. H., and Finch, G. A., 1960, The role of the reticuloendothelial cell in iron metabolism, *Br. J. Haematol.* **6**:43.

Nyman, M., 1959, Serum haptoglobin methodological and chemical studies, *Scand. J. Clin. Lab Invest.* **39**:1.

Pekarek, R. S., Bostian, K., Bartelloni, P. J., Calia, F. M., and Beisel, W. R., 1969, The effects of *Francisella tularensis* infection on iron metabolism in man, *Am. J. Med. Sci.* **258**:14.

Pekarek, R. S., Wamemacher, R. W., Jr., and Beisel, W. R., 1972, The effect of leukocytic endogenous mediator (LEM) on the tissue distribution of zinc and iron, *Proc. Soc. Exp. Biol. Med.* **140**:685.

Peschle, C., Marone, G., Genovere, A., Rappaport, I. A., and Condorelli, M., 1976, Increased erythropoietin production in anephric rats with hyperplasia of the reticuloendothelial system induced by colloidal carbon or zymosan, *Blood* **47**:325.

Pessina, G. P., Paulesu, L., and Bocci, V., 1980, Studies of factors regulating the aging of human erythrocytes. II. Metabolic depletion of erythrocytes is not accompanied by a decrease of their sialic acid content during blood bank storage, *Vox Sang.* **37**:338.

Peterson, R. E., 1953, Plasma radioactive iron turnover in acute viral hepatitis, *Proc. Soc. Exp. Biol. Med.* **84**:47.

Phillips, M. E., and Thorbecke, G. J., 1966, Studies on the serum proteins of chimeras. I. Identification and study of the site of origin of donor type serum proteins in adult rat into mouse chimers, *Int. Arch. Allergy Appl. Immunol.* **29**:553.

Powell, L. W., Alpert, E., Isselbacher, K. J., and Drysdale, J. W., 1975, Human isoferritins: Organ specific iron and apoferritin distribution, *Br. J. Haematol.* **30**:47.

Reeves, W. B., and Haurani, F. I., 1980, Clinical applicability and usefullness of ferritin measurements, *Ann. Clin. Lab. Sci.* **10**:529.

Reeves, W. B., Fairman, R. M., and Haurani, F. I., 1981, Influence of hormones on the release of iron by macrophages, *J. Reticuloendothelial Soc.* **29**:173.

Rich, I. N., and Kubanek, B., 1982, Release of erythropoietin from macrophages mediated by phagocytosis of crystalline silica, *J. Reticuloendothelial Soc.* **31**:17.

Richter, G. W., 1959, The cellular transformation of injected colloidal iron complexes into ferritin and hemosiderin in experimental animals, *J. Exp. Med.* **109**:197.

Roeser, H. P., Lee, G. R., Nacht, S., and Cartwright, G. E., 1970, The role of ceruloplasmin in iron metabolism *J. Clin. Invest.* **49**:2408.

Ross, J. D., and Muller-Eberhard, U., 1970, Pharmacologic induction of serum hemopexin by 3-methylcholanthrene and allylisopropylacetamide, *J. Lab. Clin. Med.* **75**:694.

Sears, D. A., Friedman, J., and White, D. R., 1975, Binding of intracellular protein to the erythrocyte membrane during incubation, *J. Lab. Clin. Med.* **86**:722.

Shattil, S. J., and Cooper, R. A., 1971, Maturation of macroreticulocyte membranes *in vivo, Blood* **38**:806.

Shiga, T., Maeda, N., Suda, T., Kon, K., and Sekiya, M., 1979, The decreased membrane fluidity of *in vivo* aged human erythrocytes: A spin label study, *Biochim. Biophys. Acta* **553**:84.

Shoden, A., Gabrio, B. W., and Finch, C. A., 1953, The relationship between ferritin and hemosiderin in rabbits and man, *J. Biol. Chem.* **204**:823.

Song, S. H., and Groom, A. C., 1972, Sequestration and possible maturation of reticulocytes in the normal spleen, *Can. J. Physiol. Pharmacol.* **50**:400.

Sturgeon, P., and Shoden, A., 1964, Mechanisms of iron storage, in: *Iron Metabolism: An International Symposium* (F. Gross, ed.), pp. 121–146, Springer-Verlag, Berlin.

Tenhunen, R., Marver, H. S., and Schmid, R., 1968, The enzymatic conversion of heme to bilirubin by microsomal heme oxygenase, *Proc. Natl. Acad. Sci. USA* **61**:748.

Thomas, D. B., and Briscoe, C. V., 1973, Effects of intravenous injections of endotoxin on distribution of cells between murine blood and bone marrow, *J. Anat.* **114**:407.

Udupa, K. B., and Lipschitz, D. A., 1982, Endotoxin-induced suppression of erythropoiesis: The role of erythropoietin and a heme synthesis stimulating factor, *Blood* **59**:1267.

Vannotti, A., 1957, The role of the reticuloendothelial system in iron metabolism, in: *Physiopathology of the Reticuloendothelial System* (B. N. Halpern, B. Benacerraf, and J. F. Delafresnaye, eds.), pp. 172–187, Thomas, Springfield, Ill.

van Snick, J. L., Markowitz, B., and Masson, P. L., 1977, The ingestion and digestion of human lactoferrin by mouse peritoneal macrophages and the transfer of its iron into ferritin, *J. Exp. Med.* **146**:817.

Ward, H. P., Kurnich, J. E., and Pisarczyk, M. J., 1971, Serum levels of erythropoietin in anemia associated with chronic infection, malignancy and primary hematopoietic disease, *J. Clin. Invest.* **50**:332.

Weed, R. I., and Reed, C., 1966, Membrane alterations and red cell destruction, *Am. J. Med.* **41**:681.

Weinberg, E. D., 1974, Iron and susceptibility to infectious disease, *Science* **184**:952.

Williams, A. R., and Morris, D. R., 1980, The internal viscosity of the human erythrocyte may determine its life span *in vivo, Scand. J. Haematol.* **24**:57.

Wintrobe, M. M., Grinstein, M., Dubach, J. J., Humphreys, S. R., Ashenbrucker, H., and Worth, W., 1947, The anemia of infection. VI. The influence of cobalt on the anemia associated with inflammation, *Blood* **2**:323.

Yamada, H., 1968, Clinical studies on iron kinetics. II. Iron-kinetics studies in patients with malignant neoplasms with special references to ferrokinetics and ^{59}Fe labeled iron-dextran studies, *Nagoya J. Med. Sci.* **30**:491.

Zucker, S., Lysik, R. M., and Di Stefan, J. F., 1980, Cancer cell inhibition of erythropoiesis, *J. Lab. Clin. Med.* **96**:770.

16

Lead and Cadmium
Effect on Host Defense Mechanisms and Toxic Interactions with Bacterial Endotoxin

JAMES A. COOK, W. J. DOUGHERTY, W. C. WISE, and P. V. HALUSHKA

There is growing concern over the presence of significant quantities of nonbiodegradable pollutants, such as lead and cadmium, in the environment (Wessel and Dominski, 1977; Goyer and Rhyne, 1973; Webb, 1979). Widespread environmental contamination with lead and cadmium occurs from many sources ranging from industrial or automobile emissions (Haley, 1968; Beliles, 1975; Cox, 1974) to consumable items and appliances in the home (Chisolm, 1973; Cox, 1974). Understanding of the interaction of these toxic trace metals with biologic systems is thus becoming increasingly important. Such interactions may result from accidental overt intoxication or potential toxic effects of subclinical body burdens of these heavy metals manifested in the presence of costressors. The impact on host defense mechanisms is an important variable in toxicity assessment of these environmental contaminants. Selye *et al.* (1966) reported that a single injection of lead acetate increased the sensitivity of rats to various endotoxins from gram-negative bacteria by approximately 100,000-fold. This initial observation of an extreme toxic synergism between lead and endotoxin has inspired a number of studies of environmental significance as well as an understanding of the fundamental mechanisms of endotoxic shock. This chapter reviews some toxicologic effects of lead and cadmium on phagocyte function

---

JAMES A. COOK and W. C. WISE • Department of Physiology, Medical University of South Carolina, Charleston, South Carolina 29425. W. J. DOUGHERTY • Department of Anatomy, Medical University of South Carolina, Charleston, South Carolina 29425. P. V. HALUSHKA • Departments of Pharmacology and Medicine, Medical University of South Carolina, Charleston, South Carolina 29425.

and host resistance to infection and examines potential mechanisms by which these heavy metals render animals hypersensitive to bacterial endotoxin.

1. RELATIVE TOXICITY OF LEAD AND CADMIUM ON PHAGOCYTES

Several studies have demonstrated *in vitro* cytotoxic properties of cadmium salts (Waters *et al.*, 1975; Ward *et al.*, 1975; Bell *et al.*, 1979; Loose *et al.*, 1978) and of lead salts (Ruiter *et al.*, 1977; Kaminski *et al.*, 1977; Aranyi *et al.*, 1979) on alveolar or peritoneal macrophages as reflected by loss of viability or by morphologic alterations. Soluble cadmium salts are particularly potent and relatively more cytotoxic *in vitro* than a variety of other heavy metals e.g., VO^-, Ni^{2+}, Mn^{2+}, Cr^{3+} (Waters *et al.*, 1975). The cytotoxic effects of lead or cadmium on alveolar phagocytic cells are of particular interest in view of the proximity of these cells to the content of ambient air and the observations that toxic trace metals preferentially concentrate in respirable particles (Natusch *et al.*, 1973). The airborne lead pollutant Pb_2O_3 produces significant changes in pulmonary phagocytic cell populations. Prolonged exposure, i.e., several months, of rats to vapors of Pb_2O_3 at concentrations of 10 to 150 $\mu g/m^3$ markedly reduced numbers of alveolar macrophages recoverable by lavage (Bingham *et al.*, 1968). To evaluate the specificity of this response, Bingham *et al.* (1972) subsequently compared the toxicity of soluble and insoluble aerosols of Pb_2O_3, $PbCl_2$, NiO, and $NiCl_2$. Although numbers of cells obtained by lavage from the lungs of rats exposed to Pb_2O_3 fumes were significantly reduced 2 days after exposure, NiO vapors produced a proliferation of alveolar macrophages which increased with duration of the exposure. The chloride salt of both of these metals failed to elicit any response. These quantitative cellular responses following Pb_2O_3 exposure, however, differ from the observations of Kaminski *et al.* (1977). The latter investigators reported significant increases in recoverable alveolar macrophages in rats following intratracheal instillation of Pb_2O_3 (0.25–1.0 mg/rat). These divergent observations may reflect differences in the method of exposure to Pb_2O_3 (i.e., airborne particulate versus intratracheal instillation) as well as in the duration of exposure. Furthermore, studies by Aranyi *et al.* (1979) have demonstrated that the cytotoxicity of surface-PbO-coated fly ash on alveolar macrophages is dependent on particle size as well as dosage, and that the greater toxicity of smaller particles (< 2 μm) is apparently due to increased surface area.

In contrast to lead, exposure to cadmium aerosols typically produces a cellular proliferative response. Gardner *et al.* (1977) exposed mice to an aerosol of $CdCl_2$ (1500 $\mu g/m^3$) for 2 hr and observed an initial decrease both in total number of macrophages recovered from the lungs and in cellular viability. This response, however, was associated with a concomitant influx of PMN (approximately 100-fold over controls). In essential agreement with these findings, Bouley *et al.* (1977) reported an initial but more delayed (48 hr) decrease in relative number of recoverable alveolar macrophages as a result of the appearance of numerous neutrophils subsequent to a 15-min exposure of rats to microparticles of cadmium oxide (10 ± 2 mg/m^3). Over a 24-day period, however, there was a significant increase in absolute numbers of alveolar macrophages. Such cellular proliferation following acute pulmonary exposure to cadmium aerosols is consistent with observations of other investigators (Palmer *et al.*, 1975; Strauss *et al.*, 1976) and may play a role in the etiology of respiratory dysfunction observed in man (e.g., fibrosis and emphysema) associated with inhalation of cadmium fumes.

2. EFFECT OF LEAD AND CADMIUM ON PHAGOCYTIC CAPACITY

2.1. *IN VITRO* STUDIES

Although exposure of macrophages to toxic trace metals can result in loss of cell viability and changes in inflammatory cell populations, functional defects at concentrations below those which affect viability may be as detrimental.

One vital functional parameter affected by heavy metals is phagocytosis. Graham *et al.* (1975) assessed the phagocytic capacity of rabbit alveolar macrophages incubated *in vitro* with several toxic trace metals at concentrations which did not markedly reduce viability. Cadmium (2.2×10^{-5} M) significantly depressed the phagocytic index of these cells as did several other trace metals (Ni^{2+}, Cr^{3+}, and Mn^{2+}). Loose *et al.* (1978) evaluated relative cytotoxic effects of cadmium salts on phagocytic and microbicidal capacities of mouse peritoneal macrophages, alveolar macrophages, and PMN. Phagocytic capacity of all three cell types was uniformly suppressed in a dose-associated manner when incubated with $CdCl_2$ (8.0×10^{-3} to 8.0×10^{-1} meq/liter). Intracellular microbicidal activity, however, was selectively impaired in cadmium-exposed alveolar macrophages but not in cadmium-exposed peritoneal macrophages or PMN. This increased sensitivity of alveolar macrophages to cadmium may reflect certain unique metabolic characteristics of these cells (Allen and Loose, 1976).

Macrophages can phagocytize by nonspecific as well as specific (immune) recognition. The latter mechanism of phagocytosis involves the attachment of the Fc portion of immunoglobulin (IgG) with specific Fc receptors of the surface of the macrophage (Gaafer and Doyle, 1971). These cytophilic antibodies therefore enhance attachment and engulfment of antibody-coated bacteria. One mechanism whereby cadmium may interfere with this type of recognition is by altering the Fc receptor site on the macrophage (Hadley *et al.*, 1977). Both cadmium chloride and nickel chloride *in vitro* were potent inhibitors of antibody-mediated rosette formation by rabbit alveolar macrophages. Although both metals inhibited rosette formation in a dose-dependent manner, cadmium was approximately five times more potent than nickel (Hadley *et al.*, 1977).

2.2. RESPIRATORY EXPOSURE

These toxic effects of cadmium may contribute to changes in pulmonary clearance capacity that occur in animals following exposure to cadmium aerosols or microparticulates. Gardner *et al.* (1977) demonstrated impaired pulmonary clearance of streptococci in mice following a 2-hr exposure to an aerosol of $CdCl_2$ (325 μg/m³). On the fourth day postinfection, the log titer of streptococci in cadmium-treated groups was approximately five times greater than that of control groups. Similar impairment in pulmonary bacterial clearance was reported by Bouley *et al.* (1977) following exposure of rats to microparticles of cadmium oxide. In these studies, the clearance kinetics of inhaled aerosols of *Salmonella enteritidis* was significantly depressed between 1 and 6 hr following infection.

2.3. PARENTERAL ADMINISTRATION

Parenterally administered lead salts are effective phagocytic inhibitors. Trejo *et al.* (1972) demonstrated impaired intravascular clearance of colloidal carbon in rats administered lead acetate in (5 mg/iv). A depressed clearance half-time was evident within 6 hr following lead injection, and this phagocytic depression persisted for 72 hr. Similar suppressive actions of intravenous lead salts measured by colloidal carbon clearance in rats or mice (Filkins and Buchanan, 1973;

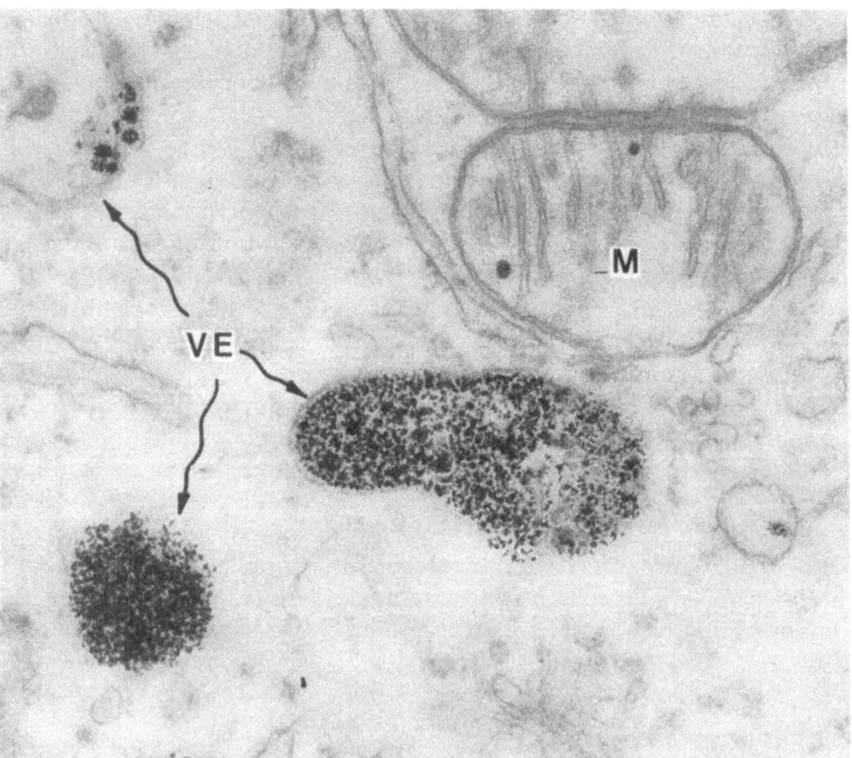

FIGURE 1. High-magnification electron micrograph illustrating the natural electron density of the precipitates within vesicles (VE) in Kupffer cell cytoplasm in livers of animals treated with lead only. M, mitochondria. Section was not stained with uranyl acetate or lead citrate. 50,000 ×.

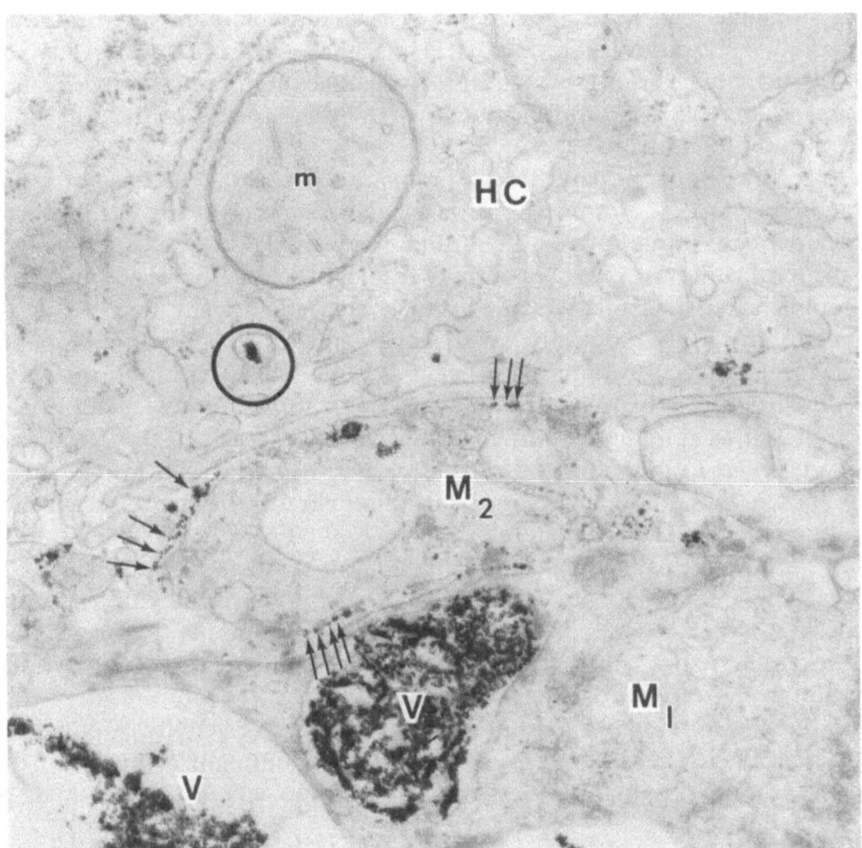

FIGURE 2. Portions of two sinusoidal macrophages (M_1, M_2) and a hepatic cell (HC) in the liver of an animal treated with lead and endotoxin. One macrophage (M_1) contains vesicles (V) with coarse precipitates, while the surface of the other (M_2) is covered with small, fine, electron-dense particles (arrows). One electron-dense particle (circle) appears to be confined within a vesicle that either has just invaginated into the hepatic cell cytoplasm or is just about to release its dense content into the space of Disse. M, mitochondria. 29,000 ×.

Seyberth *et al.*, 1972) and with the ^{131}I RE test lipid emulsion (Trejo *et al.*, 1972; Cook *et al.*, 1974) have been reported. This lead-induced phagocytic depression appears to be a cellular defect rather than an opsonic deficiency since prior opsonization of colloids with normal serum did not alter the clearance kinetics (Trejo *et al.*, 1972). The depressed phagocytic activity following lead is associated with progressive ultrastructural changes in hepatic Kupffer cells (Hoffmann *et al.*, 1974, 1975; Cook *et al.*, 1975a). Granular electron-dense aggregates are seen in cytoplasmic vesicles of approximately 30% of Kupffer cells 5 hr after lead acetate injection (20 mg/kg) (Fig. 1). This material is predominately associated with large vacuoles and smaller lysosome-like granules. Less frequently the deposits are observed in the sinusoidal space and in membrane vesicles of hepatic parenchymal cells (Fig. 2). It is probable that this electron-dense material is nondiffusible lead ligands since our studies have demonstrated a tissue distribution of ^{203}Pb administered with a lead acetate carrier (20 mg/kg) characteristic of particulate clearance by the RES (Table 1). The rapid clearance of isotope was associated with an approximate 70% hepatic uptake of the total injected dose within 5 hr (Table 1).

In contrast to lead, parenteral administration of an acute dose of cadmium acetate (6 mg/kg) did not impair intravascular clearance of the ^{131}I RE test lipid emulsion in rats 3 hr following the cadmium injection (Cook *et al.*, 1974). Indeed, hepatic localization of the labeled lipid emulsion was significantly increased in cadmium-treated rats relative to controls. Such effects may be dose and time dependent but are not entirely surprising in view of the diverse metabolic and biochemical effects of cadmium (Vallee and Ulmer, 1972). Furthermore, Kupffer cell ultrastructure 8 hr following the same dose of cadmium acetate was minimally altered as compared to more extensive cytoplasmic degradation and necrosis observed in parenchymal cells (Hoffmann *et al.*, 1975). Previous studies have shown that following parenteral administration, hepatic localization of cadmium is rapid (Decker *et al.*, 1957; Perry *et al.*, 1970), but unlike parenterally administered lead salts, most of this cadmium appears to be sequestered by nonphagocytic cells. Hepatic cell separation techniques following cadmium injection in rats demonstrated that the cation was predominantly associated with hepatic parenchymal cells, perhaps as a result of higher concentrations of metallothionin in these cells (Cain and Skilleter, 1980).

Paradoxically, however, agents which influence RE activity can modify cadmium toxicity. Pretreatment of rats with the RE stimulant zymosan or the RE depressant methyl palmitate significantly protected rats from acute toxic doses

TABLE 1. DISTRIBUTION OF ^{203}Pb FOLLOWING LEAD ACETATE INJECTION WITH AND WITHOUT CONCURRENT ENDOTOXIN ADMINISTRATION[a]

	Tissue distribution 1 hr % ID/TO[b]		Tissue distribution 5 hr % ID/TO	
	Pb	Pb + LPS	Pb	Pb + LPS
Liver	55.6 ± 4.6	59.13 ± 8.04	69.45 ± 8.07	74.42 ± 4.22
Spleen	0.98 ± 0.10	1.10 ± 0.20	1.21 ± 0.2	1.35 ± 0.4
Kidney	0.85 ± 0.04	0.57* ± 0.07	1.68 ± 0.29	1.64 ± 0.14
Lung	0.58 ± 0.06	0.67 ± 0.06	0.39 ± 0.06	1.08 ± 0.32
Heart	0.25 ± 0.02	0.21 ± 0.02	0.073 ± 0.010	0.10 ± 0.01
Brain	0.038 ± 0.008	0.03 ± 0.006	0.05 ± 0.004	0.05 ± 0.006

[a] ^{203}Pb (0.5 μCi/100 g body wt) was added to carrier lead acetate and injected i.v. 2 mg/100 g body wt with or without *S. enteritidis* endotoxin (LPS) 10 μg/100 g body wt ($N = 8$).
[b] Tissue distribution expressed as percent injected dose per total organ (% ID/TO).
* $p < 0.05$ compared to Pb group.

of cadmium chloride and reduced the severity of hepatic lesions (Lazar *et al.*, 1974). The mechanism of protection afforded by these two agents which have divergent actions on RE function was not delineated but both compounds accelerated plasma clearance and hepatic uptake of cadmium. One possibility suggested by Lazar *et al.* (1974) was that zymosan and methyl palmitate may affect hepatic metallothionin levels.

2.4. ORAL ADMINISTRATION

It remains to be firmly established that chronic low-level exposure to lead or cadmium is deleterious to macrophage phagocytic function. The results of Koller and Roan (1977) suggest that phagocytic function is enhanced. In these studies, mice were maintained on deionized water containing cadmium chloride (3 to 200 ppm) and lead acetate (13 to 1300 ppm) for 70 days. Percent *in vitro* phagocytosis of erythrocytes by peritoneal macrophages was subsequently evaluated along with cellular concentrations of acid phosphatase. Phagocytic capacity was uniformly increased in those mice receiving the higher doses of cadmium and lead, and macrophage acid phosphatase activity was increased at median toxic doses of these heavy metals. The duration and level of exposure, however, may be critical determinants of the toxicity of these pollutants since more prolonged exposure (18 months) of mice to lead acetate in drinking water (13 to 1300 ppm) suppressed phagocytic activity of peritoneal macrophages (Koller *et al.*, 1977). Likewise, Knutson *et al.* (198O) observed depressed intravascular clearance of stable heat aggregates of [^{125}I]-IgG in mice given cadmium (10 to 300 ppm) in drinking water for 7 months. This appeared to be a specific defect since clearance of aggregated human serum albumin and colloidal carbon was normal in cadmium-treated mice. The investigators suggested that cadmium may alter the Fc or complement receptors of Kupffer cells. Such a mechanism of action of cadmium therefore would be consistent with inhibitory actions of this heavy metal on antibody-mediated rosette formation by rabbit alveolar macrophages (Hadley *et al.*, 1977).

3. METABOLIC EFFECTS OF LEAD AND CADMIUM ON PHAGOCYTES

Mustafa *et al.* (1971a,b) demonstrated that endogenous respiratory activity of intact sheep alveolar macrophages was depressed approximately 60% when incubated *in vitro* with 50 μM Cd^{2+}. Studies with isolated macrophage mitochondria also demonstrated that cadmium at the same concentration disrupted the electron transport system. It was suggested that since flavin-linked and pyridine nucleotide-linked substrates reduced cytochromes in the presence of Cd^{2+}, flavoproteins or other dehydrogenases rather than cytochromes were targets for cadmium (Mustafa *et al.*, 1971a,b). The increased dependency of alveolar macrophages on aerobic metabolism compared to other phagocytes (Karnovsky *et al.*, 1970) may make these cells particularly vulnerable to cadmium toxicity.

Since alveolar macrophages also are exposed directly to the contents of ambient air, pollutants may be deleterious to outer membrane-associated transporting enzymes. Mustafa and Cross (1971) observed that a number of divalent cations including Cd^{2+} and Pb^{2+} inhibited membrane ATPase activity of alveolar macrophages and that cadmium inhibited the ATPase of isolated mitochondria as well. Both Na^+-K^+-dependent and Na^+-K^+-independent ATPase were inhibited in the presence of 2 mM Cd^{2+}. These effects of cadmium on membrane-bound enzyme systems thus could affect cellular ionic gradients and energy utilization (Mustafa and Cross, 1971).

Phagocytosis is associated with concomitant metabolic events including an increase in O_2 consumption, lactate production, and oxidative pathways of glucose metabolism (Sbarra and Karnovsky, 1959). Loose *et al.* (1977) assessed the effects of cadmium salts on O_2 consumption of murine phagocytic cells during rest and phagocytosis. No significant alterations in resting O_2 consumption could be demonstrated when alveolar macrophages, peritoneal macrophages, or PMN were incubated in media containing either cadmium chloride or cadmium acetate (3.6×10^{-1} to 3.6×10^3 meq/liter). However, when opsonized *Pseudomonas aeruginosa* was added as a phagocytic stimulus, marked depression in respiratory burst activity occurred in the presence of cadmium. The investigators postulated that cadmium inhibition of myosin ATPase, essential in the phagocytic event, may contribute to this impairment.

In addition to inhibition of respiratory activity, heavy metals may alter other metabolic events necessary for intracellular killing. Significant impairment of granulocyte myeloperoxidase (MPO) activity and iodination ability occurred following chronic lead poisoning in dogs (Caldwell *et al.*, 1979). The decrease in MPO activity and iodination capacity correlated with cumulative lead toxicity in these animals and these toxic effects were associated with reduced intracellular killing of *Staphylococcus albus* by granulocytes. Aside from intracellular bactericidal mechanisms, macrophages and certain other cells actively secrete lysozyme, a polypeptide capable of lysing some microorganisms (McLelland and van Furth, 1975). The activity of this enzyme has been shown to be progressively reduced in the spleen and serum of dogs following prolonged administration of lead salts (deBruin, 1971).

Heavy metal-induced impairment of macrophage metabolism may be exacerbated by an affinity of these cells for incorporating these soluble cations. Kinetic studies by Hart (1978) suggest that macrophages may take up soluble cadmium by a facilitated transport process since the rate of cadmium incorporation was concentration and temperature dependent, and exhibited saturation kinetics. These cells apparently were capable of transporting only free cadmium ions as the presence of protein and particularly low-molecular-weight sulfhydryl-containing ligands reduced the uptake. In contrast, zinc, which has previously been shown to reduce cadmium toxicity (Freiberg *et al.*, 1971), potentiated the rate of incorporation and blocked the release of cadmium from macrophages.

Cox and Waters (1978) isolated a soluble cadmium-binding protein, from rabbit alveolar macrophages, with properties similar to metallothionin. Such a protein also may contribute to the affinity of macrophages for incorporating cadmium and may provide a limited cellular protective mechanism which diminishes the toxicity of cadmium. This protein has an approximate molecular weight of 11,000 but differs from hepatic metallothionin by an apparent lack of zinc in protein. A similar low-molecular-weight macromolecule with characteristics of metallothionin has been isolated from blood lymphocytes (Hildebrand and Cram, 1979). Human blood lymphocytes incubated with low levels of cadmium chloride were found to have a thousandfold elevation in cellular cadmium concentrations relative to the media. This affinity of lymphocytes and macrophages for incorporating the toxic trace metal may contribute to the immunosuppressive actions of cadmium following prolonged low-level exposure (Koller *et al.*, 1975).

4. EFFECT OF LEAD AND CADMIUM ON HUMORAL AND CELLULAR IMMUNITY

4.1. HUMORAL IMMUNITY

Effective immune surveillance involves a close coordination between macrophages and both B and T lymphocytes. Considerable evidence now exists that toxic trace metals and other environmental contaminants are immunosuppressive. For a comprehensive treatise of this subject the reader is referred to the previous review by Koller (1979). Acute lead or cadmium exposure produces variable effects on humoral immunity depending on the time and route of administration. Jones *et al.* (1971) reported that a single intravenous injection of cadmium 14 days before antigen inoculation (human gamma globulin) enhanced antibody synthesis, whereas administration of cadmium 7 days before the antigen suppressed antibody formation. Similar variable responses following acute cadmium intoxication can be produced by different routes of exposure (Koller *et al.*, 1976). A single intraperitoneal dose of cadmium (0.15 mg) in mice produced an increase in IgM and IgG antibody response to SRBC. On the other hand, oral administration of the same dose produced a slight decrease in IgM response and an even more suppressive effect on IgG response. Depressed primary immunity following cadmium also can be elicited following a respiratory route of exposure. Graham *et al.* (1978) reported significant reductions of specific antibody-producing spleen cells following a 2-hr exposure to an aerosol of cadmium chloride (190 μg Cd/m^3). Impaired primary immunity was not evident when mice were administered an intramuscular injection of $CdCl_2$ even though the doses employed far exceeded effective aerosol concentrations. Similar discrepancies can be seen following acute lead intoxication. Administration of 4 mg of lead acetate via either intraperitoneal or oral routes in mice produced suppression of both IgM and IgG response to SRBC (Koller *et al.*, 1976). Studies by Trejo *et al.* (1972), however, demonstrated that intravenous administration of lead acetate to rats (5 mg/rat) failed to alter primary or secondary immune response to the same antigen. The diverse effects of different routes of acute exposure may thus reflect differences in clearance and detoxification mechanisms of target tissues (Graham *et al.*, 1978).

Prolonged exposure to these toxic trace metals, however, appears to consistently suppress humoral immunity. Koller (1973) reported that rabbits given cadmium chloride (300 ppm) or lead acetate (2500 ppm) for 70 days in drinking water and then challenged with pseudorabies virus produced significantly lower neutralizing antibody titers than did controls. Similar immunosuppressive actions of cadmium were observed in mice maintained on cadmium chloride (3 or 300 ppm) in drinking water for 10 weeks. When these animals were administered antigen (SRBC) 6 weeks after discontinuance of cadmium, a highly significant decrease in antibody formation occurred, particularly the secondary IgG response. The most remarkable observation was that this suppression occurred even with the lower exposure level of cadmium (3 ppm) (Koller *et al.*, 1975). Chronic low-level exposure of mice to lead in drinking water (13.75 to 1375 ppm) also produced impaired primary and secondary immune responses (Koller and Kovacic, 1974). In subsequent studies, direct inhibitory effects of cadmium or lead on bone marrow-derived B lymphocytes were demonstrated (Koller and Brauner, 1977). B lymphocytes from mice exposed 70 days to cadmium (130 ppm) or lead (130 to 1300 ppm) were significantly inhibited from forming rosettes in an antibody-complement assay.

4.2. LYMPHOCYTE BLAST CELL TRANSFORMATION

Lymphocyte blast cell transformation is an *in vitro* correlate of immune responsiveness. Splenic lymphocytes from mice maintained on drinking water containing $CdCl_2$ (160 ppm) or lead acetate (2000 ppm) for 30 days exhibited significant reductions in response to the mitogens phytohemagglutinin (PHA) and pokeweed mitogen (PWM) (Gaworski and Sharma, 1978). Both of the heavy metals depressed lymphocyte membrane ATPase activity and the investigators suggested that such inhibition may represent an early stage by which heavy metals alter the immune process. Koller *et al.* (1979) also evaluated lymphocyte mitogen response of mice exposed for 10 weeks to drinking water containing lead or cadmium. Under these conditions, lymphocytes from lead-treated mice generally exhibited an impaired mitogenic response to endotoxin, a specific B-cell mitogen, and to purified protein derivative (PPD) after sensitization with BCG. In contrast, cadmium exposure tended to potentiate the mitogenic response of lymphocytes at higher doses (30 and 300 ppm) but reduced the lymphocyte response to PPD and endotoxin at the 3-ppm dose. Neilan *et al.* (1980) reported that lymphocytes from mice maintained on drinking water containing 1300 ppm of lead acetate displayed a suppressed mitogenic responsiveness to both PHA and Con A but exhibited no impairment in the mitogenic action of endotoxin. They concluded, therefore, that lead predominantly suppressed cellular immunity.

4.3. CELLULAR IMMUNITY

Altered cellular immunity induced by lead can be demonstrated by other immune parameters. Daily intraperitoneal injections of lead acetate in mice (0.25

to 0.025 mg) for 30 days produced a dose-dependent inhibition of delayed-type hypersensitivity (DTH) response elicited by foot pad inoculation with RBC (Muller *et al.*, 1977). This DTH inhibition was evident in both the primary and the secondary response. Faith *et al.* (1979) reported that pre- and postnatal exposure of rats to lead altered delayed hypersensitivity responsiveness and other cell-mediated immune functions. In these studies rats were exposed prenatally to 25 to 50 ppm of lead in drinking water and subsequently maintained on this dose up to 35 to 45 days after being weaned. The DTH response in the lead-treated groups, measured with PPD following challenge with Freund's adjuvant, was reduced relative to controls. Thymic weights in lead-treated groups also were significantly decreased and splenic lymphocyte responsiveness to PHA and Con A were reduced. Of particular interest is that this lead exposure produced blood levels comparable to those of children living in urban environments (Caprio *et al.*, 1974).

5. EFFECT OF LEAD AND CADMIUM ON SUSCEPTIBILITY TO INFECTIONS

5.1. BACTERIAL INFECTIONS

One sequela of altered immunocompetence produced by lead or cadmium is a compromised capacity of the host to tolerate bacterial infections. Hemphill *et al.* (1971) exposed mice to daily intraperitoneal injections of lead nitrate for 30 days. The mice were subsequently challenged with a dose of *Salmonella typhimurium* and demonstrated to be 10 times more susceptible to lethal *Salmonella* infections than control groups. Acute intravenous doses of either lead acetate or cadmium acetate also markedly sensitized rats to *E. coli* septicemia (Cook *et al.*, 1975b). Although normal rats displayed significant tolerance to systemic infection with *E. coli* (5×10^9/100 g body), much lower doses of *E. coli* (5×10^6/100 g body wt) resulted in high mortality in lead- or cadmium-treated rats. Bouley *et al.* (1977) reported increased lethal infections with aerosols of *Salmonella enteritidis* and *Pasturella multocida* in rats and mice, respectively, subsequent to acute pulmonary exposure of rats to microparticles of cadmium oxide. In similar studies, Gardner *et al.* (1977) observed enhanced mortality of mice to infections with *Streptococcus pyrogenes* following acute exposure to cadmium chloride aerosol at concentrations which exceeded 100 μg Cd/m^3. The effect of chronic low-dose cadmium administration on susceptibility of mice to infection with BCG, a microorganism involving primarily cell-mediated immunity, was investigated by Bozelka and Burkholder (1979). Mice were administered daily intraperitoneal injections of cadmium chloride (2 mg Cd^{2+}/kg) for 4 weeks before challenge with viable BCG organisms. This infection produced a higher mortality in cadmium-treated mice (60%) compared to control mortality (2%). Cultured spleen cells from cadmium-treated mice after infection with BCG, however, produced larger amounts of the macrophage inhibitory factor, with or without inclusion of PPD, than did control spleen cells. Despite the enhanced lymphokine response, suggestive of improved T lymphocyte function, cadmium-treated mice exhibited persistent severe infections and subsequent high mortality. It was suggested that cadmium may be deleterious to other effector cells, e.g., macrophages mediating cellular immunity to BCG.

5.2. VIRAL INFECTIONS

In addition to altering host defense to bacterial infections, lead and cadmium also enhance susceptibility to viral infections. White *et al.* (1975) reported altered responses of lead-poisoned dogs to experimental infection with distemper virus. The responses which indicated an earlier response to lead poisoning included vulnerability to parenchymal liver cell damage, changes in hemoglobin synthesis, and histopathological alterations of the thymus. Gainer (1973) investigated the susceptibility of mice to Rauscher leukemia virus after they were maintained on drinking water containing lead acetate or cadmium chloride for a period of 6 weeks. Spleens from the heavy metal-treated group were enlarged and were found to have higher titers of virus than those of control mice. Additionally, mortality was enhanced in lead-treated mice exposed to lytic viral infections with encephalomyocarditis virus (EMCV) or pseudorabies virus (Gainer, 1974). Similar enhancement of EMCV-induced mortality was observed following pretreatment of mice with cadmium (Gainer, 1977a). Since lead suppressed the EMCV antiviral action of polynucleotide interferon inducers, it was suggested that lead may aggravate viral infections through a suppression of interferon synthesis (Gainer, 1974). Lead given orally also inhibited the protective activity of Newcastle disease virus against EMCV infections but effects of lead on synthesis of interferon by brain tissue of EMCV-infected mice could not be demonstrated (Gainer, 1977b). Neither did lead depress the *in vitro* actions of interferon against vesicular stomatitis virus, although the effectiveness of interferon was suppressed when cadmium chloride was added at near-toxic concentrations (Gainer, 1977b). Heavy metal-induced impairment of interferon metabolism thus has not been fully substantiated. Exon *et al.* (1979) evaluated the effect on EMCV-induced mortality of exposure of mice to lead and cadmium simultaneously. In this study, mice were maintained for 10 weeks on drinking water containing either cadmium chloride (3 to 600 ppm) and/or lead acetate (13 to 2600 ppm). Viral-induced mortality was significantly enhanced in lead-treated mice but reduced in mice treated with cadmium. Cadmium also reduced mortality in the lead-treated group. The observed protective action of cadmium thus differs from the observations of Gainer (1977a) who reported an increased susceptibility to EMCV infection. The investigators postulated that cadmium may offset the lead effect by inhibiting viral replication or certain catabolic enzymes that normally degrade interferon or antibodies (Exon *et al.*, 1979).

5.3. PARASITIC INFECTIONS

Heavy metals may interact with other microbial pathogens producing synergistic lethality. Exon *et al.* (1975) observed that mice maintained chronically on drinking water containing cadmium (300 ppm) exhibited an unexpected 26% mortality. The animals did not present cadmium lesions and the causative agent of death based on clinical signs and histopathology was established as the protozoan *Hexamita muris*. They suggested that mortality may be due to synergism between *H. muris* and cadmium.

6. TOXIC INTERACTIONS OF LEAD AND CADMIUM WITH BACTERIAL ENDOTOXIN

Selye *et al.* (1966), while investigating the pharmacologic action of a number of heavy metals, observed that intravenous administration of well-tolerated doses of lead acetate (5 mg) to rats markedly sensitized these animals (100,000-fold) to the lethal effects of bacterial endotoxin. This increased susceptibility of lead-sensitized rats to endotoxin has subsequently been reported by several groups of investigators (Trejo and DiLuzio, 1971; Trejo *et al.*, 1972; Filkins, 1973; Filkins and Buchanan, 1973; Cook and DiLuzio, 1973; Cook *et al.*, 1974, 1975a; Shumer and Erve, 1973; Jones *et al.*, 1974). The potentiating effect of lead acetate on endotoxin-induced lethality is not species specific and can be demonstrated in a variety of other animals including mice (DeClerq and Merigan, 1969; Seyberth *et al.*, 1972), baboons (Holper *et al.*, 1973), and chicks (Truscott, 1970).

Since the RES plays an essential role in sequestering intravascular endotoxin thus preventing systemic toxicity (Braude, 1964), Selye *et al.* (1966) initially postulated that lead acetate may induce a blockade of the RES. Although intravascular clearance of certain colloidal agents is depressed by lead (Trejo *et al.*, 1972), it is questionable whether phagocytic alterations per se contribute to the enhanced susceptibility of these animals to endotoxin. Increased sensitivity of lead-treated rats to endotoxic shock did not correlate well with periods of maximal phagocytic depression (Trejo *et al.*, 1972). Likewise, an actual depression of intravascular clearance of endotoxin following lead administration was not apparent (Filkins and Buchanan, 1973) or was quantitatively minor (Seyberth *et al.*, 1972). Studies conducted in our laboratory also have assessed the possibility that endotoxin may alter the tissue distribution and thus toxicity of lead when these two agents are given conjointly (Cook *et al.*, 1981). Tissue distribution of ^{203}Pb injected with a lead acetate carrier (20 mg/kg) was measured in rats with and without concurrent dose of *S. enteritidis* endotoxin (Table 1). However, with the exception of a lower renal localization of the isotope seen at 1 hr postinjection in the lead–endotoxin group, endotoxin did not significantly affect tissue distribution of the heavy metal. Therefore, it is unlikely that the toxic synergism between lead and endotoxin can be attributed to any unique clearance of these agents when conjointly administered.

The high hepatic clearance of this intravenous dose of lead, i.e., 60–70% of the injected doses (Table 1), however, provides a basis for the hepatotoxic effects induced by heavy metal interaction with endotoxin. The hepatic dysfunction was denoted by marked elevations in serum transaminase and ornithine carbamyl transferase activity or impaired bromsulphalein (BSP) clearance (Cook and DiLuzio, 1973; Holper *et al.*, 1973; Cook *et al.*, 1974). Ultrastructural observations also reveal extensive disruption of sinusoidal integrity involving endothelial and Kupffer cells in rats 5 hr after lead–endotoxin administration (Figs. 2 and 3). Kupffer cells frequently contained pleomorphic vesicles with fine electron-dense particles (Fig. 2) or coarse precipitates and autophagic vacuoles (Fig. 3). White blood cells, platelets, and fibrin deposits were also observed in the sinusoidal space. These hepatic alterations appear to be progressive since sinusoidal microthrombosis and marked degenerative changes in hepatic parenchymal cells can be observed 24 hr after lead–endotoxin interaction in the baboon (Hoffmann *et al.*, 1974).

One sequela of hepatic dysfunction secondary to the toxic synergism between lead and endotoxin is severe hypoglycemia. Indeed, the mechanism of lead sensitization to endotoxin has been attributed to lead-induced impairment of hepatic gluconeogenesis (Filkins, 1973; Cornell and Filkins, 1974; Rippe and Berry, 1973). Studies by Hadbavny *et al.* (1978) also have shown that lead acetate administration sensitized rats to lethal insulin-induced hypoglycemia, increased glucose tolerance, and enhanced whole body oxidation of glucose. Based on these observations it was postulated that the RES plays a modulating role in insulin metabolism which is disrupted by lead poisoning.

Several agents have been demonstrated to protect against lead sensitization to endotoxin and reduce the severity of hepatic dysfunction and associated

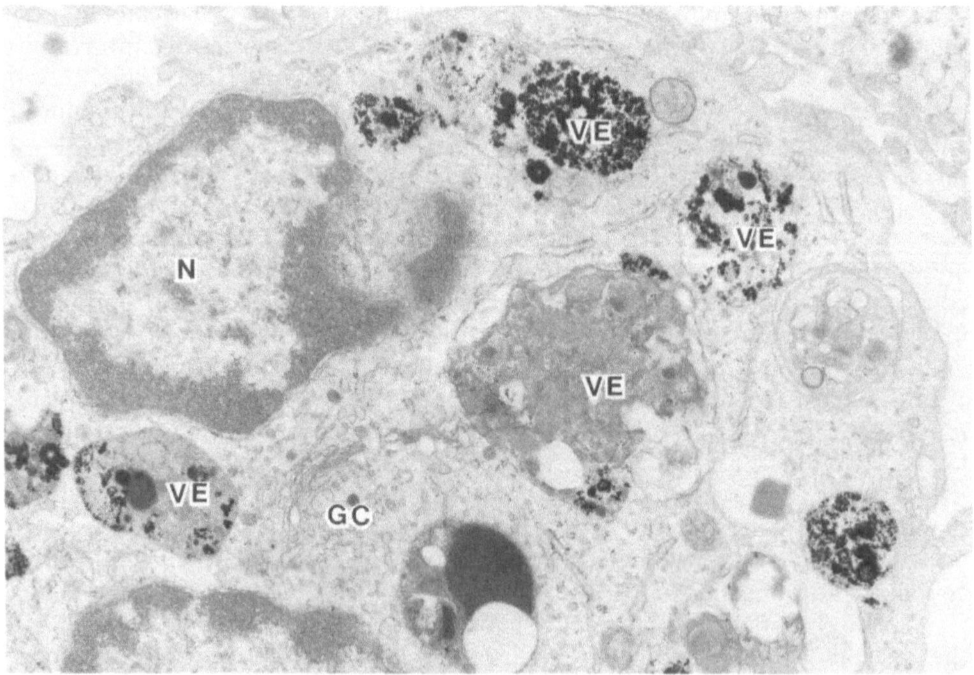

FIGURE 3. Portion of a sinusoidal macrophage in the liver of an animal treated with lead and endotoxin. The macrophage contains enlarged vesicles (VE) with dense, coarse precipitates. N, nucleus; GC, Golgi complex; M, mitochondria. 16,500 ×.

sequelae. These include diverse pharmacologic agents such as methylprednisolone (Cook and DiLuzio, 1973), cysteine (Bertók, 1968; Cook *et al.*, 1981), and the RE phagocytic depressant methyl palmitate (Cook *et al.*, 1981). When the protective agent cysteine is administered concurrently with lead acetate in the rat, there is a significant change in organ distribution of the heavy metal characterized by a severalfold increase in renal concentration and a concomitant 80% reduction in hepatic clearance of lead (Cook *et al.*, 1981). This chelator, therefore, may prevent lead interaction with hepatic sulfhydryl-containing enzymes (Bertok, 1968) or other metabolic factors essential in host resistance to endotoxin. Neither methylprednisolone nor methyl palmitate, however, significantly reduced hepatic uptake of lead, and it is likely that these compounds have other salutary mechanisms, e.g., induction of gluconeogenic enzymes (Shuler *et al.*, 1976) or stabilization of membranes (Galvin *et al.*, 1978).

Changes in arachidonic acid metabolism also may play a role in lead sensitization to endotoxin. The arachidonic acid metabolite thromboxane (Tx) A_2, a potent platelet aggregator and vasoconstrictor, has been shown to be markedly elevated in early endotoxic shock in the rat (Cook *et al.*, 1980). This vasoactive factor therefore may play a significant role in the pathophysiology of endotoxemia since TxA_2 has been implicated in the disseminated intravascular coagulation (DIC) (Cook *et al.*, 1980; Wise *et al.*, 1980) and pulmonary hypersensitive sequelae of endotoxemia (Harris *et al.*, 1980). On the other hand, the arachidonic acid metabolite prostacyclin (PGI_2), a vasodilator and platelet antiaggregator, increases in the delayed phase of endotoxin shock and may contribute to the terminal hypotensive events (Bult *et al.*, 1978; Harris *et al.*, 1980).

These observations, coupled with the previous report by Jones and Kiesow (1974) that lead acetate potentiates endotoxin-induced consumptive coagulopathies, prompted us to evaluate plasma levels of TxA_2 (Burch *et al.*, 1979) and PGI_2 (Wise *et al.*, 1980) by radioimmunoassay of their stable immunoreactive (i) metabolites $iTxB_2$ and i6-keto-$PGF_{1\alpha}$, respectively. In these studies, plasma from

TABLE 2. PLASMA TxB_2 LEVELS DURING LEAD-POTENTIATED ENDOTOXIC SHOCK[a]

	Plasma $iTxB_2$ (pg/ml)	
Group	30 min postinjection	4 hr postinjection
Control	ND[b] ($N = 5$)	ND ($N = 5$)
Pb	183 ± 59 ($N = 5$)	496 ± 157 ($N = 5$)
LPS	345 ± 44 ($N = 8$)	528 ± 99 ($N = 9$)
Pb + LPS	625 ± 120 ($N = 5$)	1282 ± 122* ($N = 8$)

[a]Lead acetate was administered (20 mg/kg) i.v. with or without a concurrent dose of i.v. *S. enteritidis* LPS (100 μg/kg).
[b]ND, nondetectable (< 150 pg/ml).
*$p < 0.01$ compared to Pb or LPS controls.

TABLE 3. PLATELET COUNTS AND FIBRIN DEGRADATION PRODUCTS (FDP) FOLLOWING LEAD-POTENTIATED ENDOTOXIC SHOCK[a]

Group	Platelet count (x 10^3/mm^3)	FDP (μg/ml)
Pb (N = 5)	520 ± 18	0.3 ± 0.1
LPS (N = 5)	420 ± 35	5.3 ± 3.7
Pb + LPS (N = 5)	130 ± 51*	51.2 ± 14.0*

[a]Studies were conducted at 4 hr after lead acetate administration (20 mg/kg) with or without a concurrent dose of *S. enteritidis* LPS (100 μg/kg).
*$p < 0.001$ compared to Pb or LPS control groups.

rats was collected at 30 min or 4 hr following an intravenous dose of lead acetate (20 mg/kg), with or without *S. enteritidis* endotoxin (100 μg/kg). Four hours after administration of either agent alone, plasma levels of $iTxB_2$ rose from nondetectable levels (< 200 pg/ml) to 496 ± 157 and 528 ± 99 pg/ml, respectively, in lead- or endotoxin-treated control groups (Table 2). When these agents were conjointly administered, the plasma $iTxB_2$ levels rose to 1282 ± 112 pg/ml. This level of $iTxB_2$ may have, in part, contributed to systemic coagulopathies since there is a marked synergism between lead and endotoxin with respect to plasma levels of fibrin split products, an index of the severity of DIC (Table 3). The marked elevation in fibrin split products in lead–endotoxin-treated rats (51.2 ± 14.0 μg/ml) above the level of those treated with lead (0.33 ± 0.07 μg/ml) or endotoxin alone (5.33 ± 3.7 μg/ml) was associated with a severe thrombocytopenia at 4 hr following injection (Table 3). There was a synergistic increase in plasma levels of 6-keto-$PGF_{1\alpha}$ in lead–endotoxin-treated rats (2350 pg/ml) above lead-treated (417 ± 160 pg/ml) or endotoxin-treated (408 ± 112 pg/ml) groups (Table 4). The elevated levels of PGI_2, however, did not influence the severe thrombocytopenia and coagulopathies following administration of lead and endotoxin.

Several inhibitors of arachidonic acid metabolism have been demonstrated to attenuate the severity of endotoxic shock (Cook *et al.*, 1980; Wise *et al.*, 1980). Furthermore, essential fatty acid-deficient (EFAD) rats which are depleted of the thromboxane and prostacyclin precursor, arachidonic acid, are more resistant to the lethal effects of endotoxin (Cook *et al.*, 1980). Studies were undertaken to determine whether either the fatty acid cyclooxygenase inhibitor indomethacin or essential fatty acid deficiency would prevent lead sensitization to endotoxic shock. Rats were pretreated intravenously with indomethacin (10 mg/kg) 30 min prior to administration of lead and endotoxin. Both the indomethacin and EFA deficiency prolonged survival time ($p < 0.05$) as indicated by respective mortalities of 40 and 18% compared to a 100% mortality in the shocked control group at 12 hr (Table 5). In the case of indomethacin, however, the overall 24-hr mortality was not changed. This finding concurs with the previous report of Reichgott and Engelman (1975) that oral administration of indomethacin did not prevent the lethal lead–endotoxin interaction in the rat. Nevertheless, survival time was significantly improved and such an observation suggests that arach-

TABLE 4. PLASMA 6-KETO-PGF$_{1\alpha}$ LEVELS IN RATS FOLLOWING LEAD-POTENTIATED ENDOTOXIC SHOCK[a]

Group	6-Keto-PGF$_{1\alpha}$ (pg/ml) 30 min postinjection	4 hr postinjection
Control	ND[d] (N = 5)	ND (N = 5)
Pb	ND (N = 5)	417 ± 160 (N = 5)
LPS	ND (N = 5)	408 ± 112 (N = 9)
Pb + LPS	ND (N = 5)	2350 ± 496* (N = 8)

[a]Lead acetate was administered (20 mg/kg) i.v. with or without a concurrent dose of i.v. *S. enteritidis* LPS (100 μg/kg).
[b]ND, nondetectable (< 200 pg/ml).
*$p < 0.01$ compared to Pb or LPS controls.

idonic acid metabolites (i.e., TxA_2 and PGI_2) may play an early pathogenic role in the toxic synergism between lead and endotoxin. It is probable, however, that additional factors independent of arachidonic acid metabolites, e.g., lead inhibition of gluconeogenic function, may be the determining lethal factor in sensitization to endotoxin.

Acute administration of cadmium, like lead, also sensitizes rats to lethal endotoxemia and produces extensive alterations of hepatic function (Cook *et al.*,

TABLE 5. EFFECT OF INDOMETHACIN PRETREATMENT AND ESSENTIAL FATTY ACID DEFICIENCY (EFAD) ON SURVIVAL DURING LEAD-POTENTIATED ENDOTOXIC SHOCK[a]

Group[a]	Time (hr) 6	12	18	24
	Dead/total			
Pb	0/5	0/5	0/5	0/5
LPS	0/5	0/5	0/5	0/5
Pb + LPS	3/10	10/10	10/10	10/10
Pb + LPS (indomethacin-treated)	0/10*	4/10*	8/10	10/10
Pb + LPS (EFAD)	0/11†	2/11†	6/11	6/11

[a]Lead acetate was administered i.v. 20 mg/kg with or without concurrent *S. enteritidis* LPS 100 μg/kg. Indomethacin was injected 10 mg/kg 30 min prior to lead–endotoxin administration.
*$p < 0.05$ compared to Pb + LPS group.
†$p < 0.05$ compared to Pb + LPS group.

1974). There are, however, several features which distinguish this toxic interaction from the lethal synergism between lead and endotoxin. Parenteral administration of cadmium alone is comparatively less well tolerated and more hepatotoxic than lead. These acute doses of cadmium also did not suppress RE phagocytic function and produced distinctly different hepatic lesions consisting predominantly of parenchymal cell alterations rather than sinusoidal changes (Cook *et al.*, 1975a). The potentiation between cadmium and endotoxin therefore may differ from the lead sensitization mechanism.

7. CONCLUSION

The extreme toxic synergism between lead and bacterial endotoxin is an intriguing toxicologic phenomenon. Since lead also potentiates shock resulting from trauma, ischemia, and anaphylaxis (Filkins and Buchanan, 1973), studies in this area may provide important information with respect to the fundamental mechanisms of shock. It is probable that heavy metal-induced alterations in macrophage function will prove to be a common denominator in such lethal synergisms. The comparatively acute doses of lead or cadmium employed to alter shock susceptibility in animals may be of questionable environmental relevance. Greengard (1966), however, reported that shock occurs frequently in children with symptomatic plumbism and is an ominous complication. These observations, therefore, may be the clinical equivalent of experimental studies.

It has been estimated that 400,000 children in the United States exhibit significant body burdens of lead, the implications of which are unknown (Wessel and Dominski, 1977). Unfortunately, even less is known of the distribution and consumption of the toxic trace metal cadmium. Potentially deleterious effects resulting from agent interaction with these pollutants following chronic low-level exposure have yet to be fully assessed. Sufficient data are at hand to suggest that the adverse health effects of toxic trace metals may be related not only to the intrinsic toxicity of the metal, but, just as importantly, to toxic manifestations in the presence of costressors, e.g., infection. Under these conditions, agents which alone produce moderate or minimal toxicity may produce severe functional impairment when present together. In view of the essential role of the macrophage in mediating immunity, the impact of environmental pollutants on such host defense mechanisms is of significant health concern and an important area for further investigation.

ACKNOWLEDGMENTS. We appreciate the excellent technical assistance of Ms. Mary Dougherty, Barbara Bayley, Kathy Anderegg, and Marsha Black. Supported in part by NIH, GM-27673, and the American Heart Association.

REFERENCES

Allen, R. C., and Loose, L. D., 1976, Phagocytic activation of luminal-dependent chemiluminescence in rabbit alveolar and peritoneal macrophages, *Biochem. Biophys. Res. Commun.* **69:**245

Aranyi, C., Miller, F. J., Andres, S., Ehrlich, R., Fenters, J., Gardner, D.E., and Waters M. D., 1979, Cytotoxicity to alveolar macrophages of trace metals absorbed on fly ash, *Environ. Res.* **20:**14.

Beliles, R. P., 1975, Lead and cadmium: Effect on host defense mechanisms and toxic interactions with bacterial endotoxin, in: *Toxicology, the Basic Science of Poisons* (L. J. Cassarett and J. Doull, eds.), p. 454, Macmillan Co., New York.

Bell, S. W., Masters, S. K., Ingram, P., Waters, M., and Shelburne, J. D., 1979, Ultrastructure and X-ray microanalysis of macrophages exposed to cadmium chloride, *Scan. Electron Microsc.* **3:** 111.

Bertok, L., 1968, Effect of sulfhydryl compound on the lead acetate-induced endotoxin hypersensitivity of rats, *J. Bacteriol.* **95:**1974.

Bingham, E., Pfitzer, E. A., Barkley, W., and Radford, E., 1968, Alveolar macrophages: Reduced number in rats after prolonged inhalation of lead sesquioxide, *Science* **162:**1297.

Bingham, E., Barkley, W., Zerwas, M., Stemmer, K., and Taylor, P., 1972, Responses of alveolar macrophages to metals. I. Inhalation of lead and nickel, *Arch. Environ. Health* **25:**406.

Bouley, G., Dubreuil, A., Despaux, N., and Boudene, C., 1977, Toxic effects of cadmium microparticles on the respiratory system, *Scand. J. Work Environ. Health* **3:**116.

Bozelka, B. E., and Burkholder, P. M., 1979, Increased mortality of cadmium-intoxicated mice with the BCG strain of *Mycobacterium bovis*, *J. Reticuloendothelial Soc.* **26:**229.

Braude, A. L., 1964, Absorption distribution and elimination of endotoxins and their derivatives, in: *Bacterial Endotoxins* (M. Landy and W. Braun, eds.), Rutgers University Press, New Brunswick, N.J.

Bult, H., Beetens, J., Vercrysse, P., and Herman, A. G., 1978, Blood levels of 6-keto-$PGF_{1\alpha}$, the stable metabolite of prostacyclin, during endotoxin induced hypotension, *Arch. Int. Pharmacodyn. Ther.* **236:**285.

Burch, R. M., Knapp, D. R., and Halushka, P. V., 1979, Vasopressin stimulates thromboxane synthesis in the toad urinary bladder: Effects of imidazole, *J. Pharmacol. Exp. Ther.* **210:**344.

Cain, K., and Skilleter, D., 1980, Selective uptake of cadmium by parenchymal cells of liver, *Biochem. J.* **188:**285.

Caldwell, K. C., Taddeini, L., Woodburn, R. L., Anderson, G. L., and Lobell, M., 1979, Induction of myeloperoxidase deficiency in granulocytes in lead-intoxicated dogs, *Blood* **53:**588.

Caprio, R. J., Margulis, H. L., and Joselow, M. M., 1974, Lead absorption in children and its relationship to urban traffic densities, *Arch. Environ. Health* **28:**195.

Chisolm, J., Jr., 1973, Management of increased lead absorption and lead poisoning in children, *N. Engl. J. Med.* **289:**1016.

Cook, J. A., and DiLuzio, N. R., 1973, Protective effect of cysteine and methylprednisolone in lead-endotoxin induced shock, *Exp. Mol. Pathol.* **19:**127.

Cook, J. A., Marconi, E. A., and DiLuzio, N. R., 1974, Lead, cadmium, endotoxin interaction: Effect on mortality and hepatic function, *Toxicol. Appl. Pharmacol.* **28:**292.

Cook, J. A., Hoffmann, E. O., and DiLuzio, N. R., 1975a, Factors modifying susceptibility to bacterial endotoxin: The effect of lead and cadmium, *Critical Reviews in Toxicology*, pp. 201–229, CRC Press, Cleveland, January.

Cook, J. A., Hoffmann, E. O., and DiLuzio, N. R., 1975b, Influence of lead and cadmium on the susceptibility of rats to bacterial challenge, *Proc. Soc. Exp. Biol. Med.* **150:**741.

Cook, J. A., Wise, W. C., and Halushka, P. V., 1980, Elevated thromboxane levels in the rat during shock: Protective effects of imidazole 13-azaprostanoic acid, or essential fatty acid deficiency, *J. Clin. Invest.* **65:**227.

Cook, J. A. Dougherty, W. J., and Holt, T., 1981, Distribution of ^{203}Pb during lead potentiated endotoxic shock, *Exp. Mol. Path.* **34:**253.

Cornell, R. P., and Filkins, J. P., 1974, Depression of hepatic gluconeogenesis by acute lead administration. *Proc. Soc. Exp. Biol. Med.* **147:**371.

Cox, C. C., and Waters, M. D., 1978, Isolation of a soluble cadmium binding protein from pulmonary macrophages, *Toxicol. Appl. Pharmacol.* **46:**385.

Cox, D. B., 1974, Cadmium; trace element of concern in mining and manufacturing, *J. Environ. Health* **36:**361.

deBruin, A., 1971, Certain biological effects of lead upon the animal organism, *Arch. Environ. Health* **23:**249.

Decker, C. G., Byerrum, R. U., and Hoppert, C. A., 1957, A study of the distribution and retention of cadmium-115 in the albino rat, *Arch. Biochem. Biophys.* **66:**140.

DeClerq, E., and Merigan, T. C., 1969, An active interferon inducer obtained from *Hemophilus influenza* type B, *J. Immunol.* **103:**889.

Exon, J. H., Patton, N. M., and Koller, L. D., 1975, Hexamitias in cadmium exposed mice, *Arch. Environ. Health* **30:**463.

Exon, J. H., Koller, L. D., and Kerkvliet, N. I., 1979, Lead–cadmium interactions: Effects on viral-induced mortality and tissue residue in mice, *Arch. Environ. Health* **34:**469.

Faith, R. E., Luster, M. I., and Kimmel, C. A., 1979, Effect of chronic developmental lead exposure on cell mediated immune functions, *Exp. Immunol.* **35:**413.

Filkins, J. P., 1973, Hypoglycemia and depressed hepatic gluconeogenesis during endotoxicosis in lead sensitized rats, *Proc. Soc. Exp. Biol. Med.* **142:**915.

Filkins, J. P., and Buchanan, B. J., 1973, Effects of lead acetate on sensitivity to shock, intravascular carbon and endotoxin clearances, and hepatic endotoxin detoxification, *Proc. Soc. Exp. Biol. Med.* **142:**471.

Freiberg, R., Piscator, M., and Nordberg, M. B., 1971, *Cadmium in the Environment*, Chemical Rubber Co., Cleveland.

Gaafer, H. A., and Doyle, J., 1971, Specificity of macrophage receptors, *Proc. Soc. Exp. Biol.Med.* **136:**121.

Gainer, J. H., 1973, Activation of the Rauscher leukemia virus by metals, *J. Natl. Cancer Inst.* **55:**609.

Gainer, J. H., 1974, Lead aggravates viral diseases and represses the antiviral activity of interferon inducers, *Environ. Health Perspect.* **7:**113.

Gainer, J. H., 1977a, Effects of heavy metals and deficiency of zinc on mortality rates in mice infected with encephalomyocarditis virus, *Am. J. Vet. Res.* **38:**87.

Gainer, J. H., 1977b, Effects of interferon on heavy metals excess and zinc deficiency, *Am. J. Vet. Res.* **38:**863.

Galvin, M. J., Shupe, K., and Lefer, A. M., 1978, Antiendotoxin actions of methylprednisolone in isolated perfused cat liver, *Pharmacology* **17:**181.

Gardner, D. E., Miller, F. J., Wing, J. W., and Kirtz, J. M., 1977, Alterations in bacterial defense mechanisms of lung induced by inhalation of cadmium, *Bull. Eur. Physiopathol. Respir.* **13:**157.

Gaworski, C. L., and Sharma, R. P., 1978, The effects of heavy metals on [^{3}H] thymidine uptake in lymphocytes, *Toxicol. Appl. Pharmacol.* **46:**305.

Goyer, R. A., and Rhyne, B. C., 1973, Pathological effects of lead, *Int. Rev. Exp. Pathol.* **12:**2.

Graham, J. A., Gardner, D. E., Waters, M. D., and Coffine, D. L., 1975, Effect of trace metals on phagocytosis by alveolar macrophages, *Infect. Immun.* **11:**1278.

Graham, J. A., Miller, F. J., Daniels, M. J., Payne, E. A., and Gardner, D. E., 1978, Influence of cadmium, nickel, and chromium on primary immunity in mice, *Environ. Res.* **16:**77.

Greengard, J., 1966, Lead poisoning in childhood: Signs, symptoms, current therapy, clinical expressions, *Clin. Pediatr. (Philadelphia)* **5:**268.

Hadbavny, A. M., Buchanan, B. J., and Filkins, J. P., 1978, Insulin and regulatory alterations of RES depression. *J. Reticuloendothelial Soc.* **24:**57.

Hadley, J. G., Gardner, D. E., Coffin, D. L., and Menzel, D. B., 1977, Inhibition of antibody-mediated rosette formation by alveolar macrophages: A sensitive assay for metal toxicity, *J. Reticuloendothelial Soc.* **22:**417.

Haley, T. J., 1968, A review of the toxicology of lead, *Air Quality Monogr.* **69**(7)**:**53.

Harris, R. H., Zmudka, M., Maddox, Y., Ramwell, P. W., and Fletcher, J. R., 1980, Relationship of TxB_2 and 6-keto-PGF_1 to the hemodynamic changes during baboon endotoxic shock, in: *Advances in Prostaglandin and Thromboxane Research* (B. Samuelson, P. W. Ramwell, and R. Pavoletti, eds.), Vol. 7, pp. 843–849, Raven Press, New York.

Hart, B. A., 1978, Transport of cadmium by alveolar macrophages, *J. Reticuloendothelial Soc.* **24:**363.

Hemphill, F. E., Kaeberle, M. L., and Buck, W. B., 1971, Lead suppression of mouse resistance to *Salmonella typhimurium*, *Science* **172:**1031.

Hildebrand, C . E., and Cram, L. S., 1979, Distribution of cadmium in human blood cultured in low levels of CdCl: Accumulation of Cd in lymphocytes and preferential binding to metallothionin, *Proc. Soc. Exp. Biol. Med.* **161:**438.

Hoffmann, E. O., DiLuzio, N. R., Holper, K., Brettschneider, L., and Coover, J., 1974, Ultrastructural changes in the liver of baboons following lead and endotoxin administration, *Lab. Invest.* **30:**311.

Hoffmann, E. O., Cook, J. A., DiLuzio, N. R., and Coover, J., 1975, The effects of acute cadmium administration on the liver and kidney of the rat, *Lab. Invest.* **32**:655.

Holper, K., Trejo, R. A., Brettschneider, L., and DiLuzio, N. R., 1973, Enhancement of endotoxic shock in the lead sensitized subhuman primates, *Surg. Gynecol. Obstet.* **136**:595.

Jones, R. B., and Kiesow, L. A., 1974, Potentiation of endotoxin-induced consumptive coagulopathy by lead acetate administration, *Infect. Immun.* **10**:1343.

Jones, R. J., Williams, R. L., and Jones, A. M., 1971, Effect of heavy metals on the immune response and preliminary findings for cadmium in rats, *Proc. Soc. Exp. Biol. Med.* **137**:1231.

Jones, R. B., Wise, J. L., and Kiesow, L. A., 1974, Failure of methylprednisolone to protect lead sensitized rats against endotoxin, *Infect. Immun.* **8**:683.

Kaminski, K. J., Fischer, C. A., Kennedy, G. L., Jr., and Calandra, J. C., 1977, Response of pulmonary macrophages to lead, *J. Exp. Pathol.* **58**:9.

Karnovsky, M. L., Simmons, S., Glass, E. A., Gaafar, A. W., and Hart, P. D., 1970, *Mononuclear Phagocytes* (R. van Furth, ed.), p. 145, Davis, Philadelphia.

Knutson, D. W., Vredevoe, D. L., Aoki, K. R., Esther, J., and Levy, L., 1980, Cadmium and the reticuloendothelial system (RES), a specific defect in blood clearance of soluble aggregates of IgG by the liver in mice given cadmium, *Immunology* **40**:17.

Koller, L. D., 1973, Immunosuppression produced by lead, cadmium and mercury, *Am. J. Vet. Res.* **34**:1457.

Koller, L. D., 1979, Effects of environmental contaminants on the immune system, *Adv. Vet. Sci. Comp. Med.* **23**:267.

Koller, L. D., and Brauner, J. A., 1977, Decreased B-lymphocyte response after exposure to lead and cadmium, *Toxicol. Appl. Pharmacol.* **42**:621.

Koller, L. D., and Kovacic, S., 1974, Decreased antibody formation in mice exposed to lead, *Nature (London)* **250**:148.

Koller, L. D., and Roan, J. G., 1977, Effects of lead and cadmium on mouse peritoneal macrophages, *J. Reticuloendothelial Soc.* **21**:7.

Koller, L. D., Exon, J. H., and Roan, J. G., 1975, Antibody suppression by cadmium, *Arch. Environ. Health* **30**:598.

Koller, L. D., Roan, J. G., and Exon, J. H., 1976, Humoral antibody response in mice after single dose exposure to lead or cadmium, *Proc. Soc. Exp. Biol. Med.* **151**:339.

Koller, L. D., Roan, J. G., Brauner, J. A., and Exon, J. H., 1977, Immune response in aged mice exposed to lead, *J. Toxicol. Environ. Health* **3**:535.

Koller, L. D., Roan, J. G., and Kerkvliet, N. I., 1979, Mitogen stimulation of lymphocytes in CBA mice exposed to lead and cadmium, *Environ. Res.* **19**:177.

Lazar, G., Serra, D., and Tuchweber, B., 1974, Effect on cadmium toxicity of substances influencing reticuloendothelial activity, *Toxicol. Appl. Pharmacol.* **29**:367.

Loose, L. D., Silkworth, J. B., and Warrington, D., 1977, Cadmium-induced depression of respiratory burst in mouse pulmonary alveolar macrophages, peritoneal macrophages and polymorphonuclear neutrophils, *Biochem. Biophys. Res. Commun.* **79**:326.

Loose, L. D., Silkworth, J. B., and Warrington, D., 1978, Cadmium induced phagocyte cytotoxicity, *Bull. Environ. Contam. Toxicol.* **20**:582.

McLelland, D. B. L., and van Furth, R., 1975, *In vitro* synthesis of lysozyme by human and mouse tissues and leukocytes, *Immunology* **28**:1099.

Muller, T. E., Gilbert, K. E., Krause, C. H., Gross, U., Age-Stehr, J., and Diamantstein, T., 1977, Suppression of delayed type hypersensitivity of mice by lead, *Experientia* **33**:667.

Mustafa, M. G., and Cross, C. E., 1971, Pulmonary alveolar macrophage-oxidative metabolism of isolated cells and mitochondria and effect of cadmium ion on electron and energy transfer reactions, *Biochemistry* **10**:4176.

Mustafa, M. G., Cross, C. E., and Tyler, W. S., 1971a, Interfence of cadmium ion with oxidative metabolism of alveolar macrophages, *Arch. Intern. Med.* **127**:1050.

Mustafa, M. G., Cross, C. E., Munn, R. J., and Hardie, J. A., 1971b, Effects of divalent metal cations on alveolar macrophage membrane adenosine triphosphatase activity, *J. Lab. Clin. Med.* **77**:563.

Natusch, D. F. S., Wallace, J. R., and Evans, C. N., 1973, Toxic trace elements; preferential concentration in respirable particles, *Science*, **183**:202.

Neilan, B. A., Taddeini, L., McJilton, C. E., and Handwerger, B. S., 1980, Decreased T cell function in mice exposed to chronic low levels of lead, *Clin. Exp. Immunol.* **39**:746.

Palmer, K. C., Snider, G., and Hayes, J. A., 1975, Cellular proliferation induced in the lung by cadmium aerosol, *Am. Rev. Respir. Dis.* **112**:173.

Perry, H. M., Erlanger, M., Unice, A., Schoepfle, E., and Perry, E. F., 1970, Hypertension and tissue metals following intravenous cadmium, mercury and zinc, *Am. J. Physiol.* **219**:755.

Reichgott, M. J., and Engelman, K., 1975, Indomethacin: Lack of effect on lethality of endotoxin in rats, *Circ. Shock* **2**:215.

Rippe, D. F., and Berry, J. L., 1973, Metabolic manifestations of lead acetate sensitization to endotoxin in mice, *J. Reticuloendothelial Soc.* **13**:527.

Ruiter, N., Seemayer, N., and Manglovic, G., 1977, Einfluz von zindk-ionen auf die toxische wirkung von bleichlord (PbCl) untersucht an mausemakrophagen *in vitro*, *Zentralbl. Bakteriol. Parasitenkd. Infektionskr. Hyg. Abt. 1 Orig. Reihe B* **164**:90.

Sbarra, A. J., and Karnovsky, M. L., 1959, The biochemical basis of phagocytosis. I. Metabolic changes during the ingestion of particles by polymorphonuclear leucocytes, *J. Biol. Chem.* **234**:1355.

Selye, H., Tuckweber, B., and Bertok, L., 1966, Effect of lead acetate on susceptibility of rats to bacterial endotoxins, *J. Bacteriol.* **91**:884.

Seyberth, H. W., Schmidt-Gayk, H., and Hackenthal, E., 1972, Toxicity clearance and distribution of endotoxin in mice as influenced by actinomycin D, cycloheximide, α-amanitin and lead acetate, *Toxicology* **10**:491.

Shuler, J. J., Erve, P. R., and Schumer, W., 1976, Glucocorticosteroid effect on hepatic carbohydrate metabolism in the endotoxin-shocked monkey, *Ann. Surg.* **183**:345.

Shumer, W., and Erve, P. R., 1973, Endotoxin sensitivity of adrenalectomized rats treated with lead acetate, *J. Reticuloendothelial Soc.* **13**:122.

Strauss, R. H., Palmer, K. C., and Hayes, J. A., 1976, Acetate lung injury induced by cadmium aerosol, *Am. J. Pathol.* **84**:561.

Trejo, R. A., and DiLuzio, N. R., 1971, Impaired detoxification as a mechanism of lead acetate-induced hypersensitivity to endotoxin, *Proc. Soc. Exp. Biol. Med.* **136**:889.

Trejo, R. A., DiLuzio, N. R., and Hoffmann, E. O., 1972, Reticuloendothelial and hepatic function alterations following lead acetate administration, *Exp. Mol. Pathol.* **17**:145.

Truscott, R. B., 1970, Endotoxin studies in chicks: Effect of lead acetate, *Can. J. Comp. Med.* **34**:135.

Vallee, B. L., and Ulmer, P. D., 1972, Biochemical effects of mercury, cadmium and lead, *Annu. Rev. Biochem.* **41**:91.

Ward, P. A., Goldschmidt, P.,Norbert, D., and Greene, J., 1975, Suppressive effects of metal salts on leukocyte and fibroblastic function, *J. Reticuloendothelial Soc.* **18**:313.

Waters, M. D., Gardner, D. W., Aranyi, C., and Coffin, D. L., 1975, Metal toxicity for rabbit alveolar macrophages *in vitro*, *Environ. Res.* **9**:32.

Webb, M., 1979, *Topics in Environmental Health*, Vol. 2, *The Chemistry and Biochemistry of Cadmium*, Elsevier/North-Holland, Amsterdam.

Wessel, M., and Dominski, A., 1977, Our children's daily lead, *Am. Sci.* **65**:294.

White, D. J., Marshall, A. J., and McLeod, S., 1975, The influence of experimental distemper infarction on the distribution of lead in dogs previously subacutely intoxicated with lead carbonate, *Br. J. Exp. Pathol.* **56**:544.

Wise, W. C., Cook, J. A., Halushka, P. V., and Knapp, D. R., 1980, Protective effects of thromboxane synthetase inhibitors in rats with endotoxic shock, *Circ. Res.* **46**:854.

Index

Actinomycin D, 117
Adenylate cyclase, 120
Adrenocorticotrophic hormone, 177
Albumin
  adsorption of, to bacteria, 11–12
  affinity of, for phagocyte surface, 81, 82
  in hemoglobin and heme clearance, 353
Alcoholism, 123, 126
Alveolar macrophages
  affinity of Fc receptor for IgG on, 83
  in interferon production, 148
  lead and cadmium effects on, 380–381
  in quantitation of macrophage phagocytosis, 59
Aminolevulinic acid, 199–200
Ampicillin, 23
Anemia
  of chronic disease, 368–370
  hemolytic, 208, 247
  iron deficiency, 203–204
  sideroblastic, 205, 209
Angiotensin, 129
Antibiotics
  bacterial surfaces affected by, 22–23
  in choline deficiency, 125
Antibodies
  to erythropoietin, 154
  lead and cadmium effects on, 387
  as opsonins, 91
  response of, to endotoxin, 132
Antigelatin factor, 17
Antigens
  fibroblast surface, 17
  gut-derived, transfer of, across gut wall, 122
Antimicrobial immunity, 280–281
Antineoplastic drugs, 149
Antiproteases, 224–225
α_1-Antitrypsin, 224, 332–334
Apoferritin, 365
Aqueous media
  opsonization with IgG in, 14
  phagocytic particle adhesion and engulfment in, 4, 7–10
Aqueous media (*cont.*)
  phagocytosis in, in vitro, 28–29
  protein adsorption in, 11, 12
Asbestos, 61, 63
Atherosclerosis, 325–326
Autoimmune thrombocytopenic purpura, 246

B lymphocytes
  in immune response to endotoxin, 132–133
  lead and cadmium effects on, 387
  reticuloendothelial cell interaction with, 279–280
Bacteria
  agents affecting surface of, 22–23
  as coagulation activator, 223
  complement affecting phagocytosis of, 17
  contact angles of, 26
  detachment of IgG from, 14
  IgA affecting phagocytosis of, 17–18
  IgG affecting phagocytosis of, 14–15
  IgM affecting phagocytosis of, 17
  lead and cadmium effects on infection from, 389–390
  opsonization of, 14–15
  phagocyte engulfment of, 6, 7–10
  phagocytic adhesion of, 4–6
  protein adsorption to, 11–13
  recognition of, for phagocytosis, 30
  vascular clearance of, 104–106
Bile
  bilirubin excretion via, 199
  endotoxin absorption affected by, 123
Bilirubin
  from biliverdin, 203
  drugs affecting biosynthesis of, 209
  heme catabolism to, 199
  metabolism of, 210–212
    hepatic pathways, 210–211
    intestinal pathway, 211–212
Biliverdin, 201, 203
Blind gut loop syndrome, 125
Blood
  mononuclear phagocytes from, 38

Blood (*cont.*)
  opsonization in, 18, 19
  particle compatibility with, for vascular clearance, 85
  phagocytic engulfment in, 7
Blood cells
  electrokinetic surface potential of, 23
  endotoxin effect on, 130
Blood lipids
  macrophage clearance of, free, 307
  macrophage clearance of, tissue, 314–315
Bone marrow
  in erythroclasia, 195, 196–197
  erythrophagocytosis in, 148
  heme oxygenase of, 204
  mononuclear phagocytes from, 38–40
    genetics of colony formation by, 51
    proliferative capacity of, 43–45
  reticular cells of, 148–149
Bradykinin, 221

C-reactive proteins, 336–338
Cadmium, *see* Lead and cadmium
Calcium
  with aging of erythrocyte, 192
  in coagulation system, 221
  in platelets, 238
Carbon, 61, 63
  in delayed-onset RES paralysis, 93
  in erythropoiesis stimulation, 163
  platelet phagocytosis of, 250
  stability of, in studies of vascular clearance by phagocytes, 84
Catecholamines, 128
Cell-attachment factor, 17
Cell-surface protein, 17
Cell-spreading factor, 17
Cetrimide, 41
Chalones, 173
Cholesterol
  dietary, clearance of, 314
  hepatic removal of, 318
  injected, clearance of, 315
  macrophage clearance of, free, 307
Cholesterolesterase, acid, 311
Choline deficiency, 125
Chromium phosphate, 84, 85
Coagulation system
  activators of, 221, 222–224
  in hemostasis, 221, 222
  reticuloendothelial system affected by, 231
Collagen
  as coagulation activator, 221, 223
  fibronectin receptor for, on platelets, 249
Colony inhibitory factor, 50
Colony-stimulating factor, 148, 283
Complement
  endotoxin effect on, 120
  in immune response to endotoxin, 133
  as opsonins, 91
  in phagocytic ingestion of bacteria, 17
  in platelet–macrophage interactions, 255
  in recognition of bacteria in vivo, 30
  tissue factor affected by, 229
Con A, 22
Contact angle measurements, 25–26
Coproporphyrinogen, 200
Cysteine, 393

Dendritic cells of RES, 273–275
Dextran, 106, 308
Disseminated intravascular coagulation, 129

Endocarditis, 105
Endotoxemia, 115–136
Endotoxin
  absorption of, in disease, 123–125
  in alcoholic liver disease pathogenesis, 123, 126
  biological properties of, 119
  in catecholamine release, 128
  cellular level effects of, 1i9–120
  chemical components of, 115–116
  in cirrhosis of liver, 124
  clotting time affected by, 131
  as coagulation activator, 223
  definition of, 115
  detection of, 116–118
  disseminated intravascular coagulation from, 129
  in glucose homeostasis, 292–294
  hepatotoxin potentiation by enteric, 125–126
  histamine protection from, 129
  host immune responses to, 132–136
  host responses induced by, 128–132
  hypotension from, 128
  Kupffer cell processing of enteric, 127–128
  lead and cadmium interactions with, 391–396
  leukocytosis from, 128
  platelets affected by, 130
  pyogenic response to, 117, 128, 132
  in shock, 124
  sources of infection with, 118–119
  species variation in susceptibility to, 116
  tolerance to, 116
  transfer of, across normal gut wall, 120–123
  tumor-necrotizing effect of, 117
  vascular clearance of, 116

Enzymes
of aging erythrocyte, 190–191, 355, 356
in bilirubin metabolism, 210–211
in heme degradation, 203–204
in heme metabolism, 200, 201, 202
lipolytic, 307
Erythroclasia, 189–212
bilirubin metabolism and, 210–212
erythrocyte changes with aging and, 190–194
erythrocyte life span and, 189–190
heme degradation with, 203–210
hemoglobin and heme metabolism and, 199–203
mechanisms of, 195–199
sites of, 194–195
Erythrocytes
aging, changes in, 190–194
cytoplasmic, 355–357
with enzymes and metabolic pathways, 190–191
with membranes, 192, 357–359
with physical properties, 192–194
deformability of, 193–194
destruction of, 189–199, 359
in bone marrow, 195, 196–197
via erythrophagocytosis, 198
in liver, 195, 197–198
mechanisms of, 195–199
via osmotic cell lysis, 198
via progressive fragmentation, 198
sites of, 194–195
in spleen, 194, 195–196
electrokinetic surface potential of, 23
endotoxin effect on, 120
in erythrophagocytosis, 147
growth-promoting activity of, for MNP, 49–50
as iron carrier, 355–359
iron utilization and reutilization by, 366–367
iron-deficient, 204–205
life span of, 189–190
metabolism of, 190
production of, *see* Erythropoiesis
as test particles for phagocytosis in vitro, 61, 63–64
Erythroid burst-forming units, 160
Erythropoiesis
extramedullary, 149–150
fetal or neonatal, 162
hormones and chemical agents in, 160–178
inhibitors of, 163
macrophage in, 148–178
medullary, 148–149
reticular cells of bone marrow in, 148–149
reticular cells of liver and spleen in, 149–160
Erythropoiesis (*cont.*)
reticuloendothelial system and, 147–178
stimulating agents for, 163
thorium dioxide in, 372
in vitro, 160–162
Erythropoietin
action of, 162
in hepatic regeneration, 173–174
hypoxia affecting, 162
from kidney, 162
Kupffer cell origin of, 148, 171–172
from liver, 162
prostaglandins and, 160–161
in renal insufficiency, 175
from reticular cells of bone marrow, 149
reticuloendothelial system and, 162–172
stimulation of levels of, 163–164
Escherichia coli
adherence of, to aortic tissue, 106
complement and IgM in phagocytosis of, 17
phagocytic ingestion of, 9
protein adsorption onto, 12, 13
Estrogens, 177–178
Exotoxins, 115
Extramedullary erythropoiesis, 149–160

Ferritin, 364–365
Ferrochelatase, 204
Fibrin, 91
clearance of, 225, 226, 228
endotoxin effect on, 131
fibrinogen–fibronectin interaction with, 226–227
in lead-potentiated endotoxic shock, 394
Fibrinogen
derivatives of, clearance of, 225–226
fibrin–fibronectin interaction with, 226–227
Fibrinolysis, 229–230
Fibroblast surface antigen, 17
Fibroblasts
lipoprotein interaction with, 307
proliferative capacity of, 43
protein metabolism and, 331–344
Fibronectin
in acute respiratory distress syndrome, 227
cells secreting, 341
in fibrin clearance, 225
fibrin–fibrinogen interaction with, 226–227
forms of, 341
functions of, 226, 341–342
in gelatin binding, 94
as opsonizing agent, 17
in platelet recognition by RES during thrombosis, 249–259

Fibronectin (*cont.*)
in shock syndromes, 227
vascular clearance rates affected by, 91
Fucoidin, 308
Fungi, 106

Galactoprotein-a, 17
Gelatin, 84, 85, 94
Genetics
of accessory cell–T cell interactions, 272
of antimicrobial immunity, 281
of mononuclear phagocyte colony formation, 51–52
Gentamycin, 23
Globulin, cold-insoluble, 17, 341–342
Glucagon, 173
Glucan
in erythropoiesis stimulation, 163, 177
in RES activation, 157, 177
Glucocorticoid-antagonizing factor, 296–297
Glucose
in endotoxicosis, 292–294
changes in input and output, 292–293
insulin and, 293–294
phases of responses, 292
in erythrocyte vs. reticulocyte metabolism, 190
oxidation of, in measurements of phagocytosis, 67
in phagocytic activity of PMNL, 24
reticuloendothelial system in regulation of, 291–300
endocytic functions, 298
glucocorticoid-antagonizing factor in, 296–297
during hypoglycemia, 298
leukocytic endogenous mediator in, 296
macrophage insulin-like activity in, 297
macrophage insulin-releasing activity in, 297
monokines in, 296–297
after reticuloendothelial system perturbations, 294–296
changes in input and output, 294–296
insulin and, 295–296
Glycoprotein, opsonic alpha$_2$ surface-binding, 17
Gold, 205–206
Growth hormone, 177

Hageman factor, 22
Hanks balanced salt solution
phagocytic adhesion and engulfment in, 4, 7
protein adsorption in, 11, 12
Haptoglobin, 353–355

Heinz bodies, 191
in hemolytic anemia, 201
removal of, 196
Hematopoiesis
prostaglandins in, 161
reticuloendothelial system in, 160–178
Heme degradation, 203–210
alternate pathways for, 210
factors influencing, 208–209
hematological disorders and, 209–210
heme oxygenase in, 203–204
iron metabolism and heme oxidation in, 204–207
in sideroblastic anemias, 205
Heme metabolism, 199–203
Heme oxygenase
gold effect on, 205–206
heavy metals affecting, 208
hematological disorders and, 209–210
in heme metabolism, 203–204
hemoglobin effect on, 208
hormones affecting, 208
prostaglandins affecting, 208
Hemoglobin
of aging erythrocytes, 192
in erythroclasia, 198
heme degradation affected by, 208
iron metabolism affecting, 204
metabolism of, 199–203
structure of, 199
Hemolytic anemia
autoimmune nature of, 247
heme oxygenase in, 208
Hemopexin, 353, 355
Hemosiderin, 365–366
Hemostasis, 221–232
coagulation system activator clearance in, 222–223
coagulation system affecting RES in, 231
fibrinogen derivative clearance in, 225–228
plasma protein coagulation system in, 221
platelet aggregation in, 221
procoagulants and antiprotease clearance in, 224–225
production of active substances in, 228–230
coagulation system moieties, 228–229
fibrinolytic moieties, 229–230
regulation of, 230–231
vascular contraction in, 221
Heparin, 22
Hepatic erythropoietic factor, 174, 175–177
Hepatocytes, 109
Hepatotoxins, 125–126
Histamine, 129
Histiocytic metabolic apparatus, 291

Hormones
  in erythropoiesis, 160–178
  heme degradation affected by, 208
Hypersensitivity responses, delayed-type, 281, 389
Hypersplenism, 239–243
Hypoglycemia
  with lead poisoning, 392
  reticuloendothelial system and, 298
Hypotension, 128

IgA
  in gut lumen, 122
  in opsonization, 17–18
  in recognition of bacteria *in vivo*, 30
IgG
  adsorption of, in phagocytosis, 10–13
  affinity of, for phagocyte surface, 82, 83
  of aging erythrocytes, 358
  lead and cadmium effects on, 387
  opsonizing role of, 14–15
  in platelet–macrophage interactions, 254
  in recognition of bacteria *in vivo*, 30
IgM
  lead and cadmium effects on, 387
  in phagocytic ingestion of bacteria, 17
Immune complexes
  as coagulation activator, 223
  phagocytic ingestion of, 15–17
    with antigen vs. antibody excess, 85
  in platelet activation, 253
  vascular clearance of, 109
Indomethacin, 177
Inflammatory bowel disease, 123
Insulin
  endotoxicosis causing changes in, 293–294
  in hepatic regeneration, 173
  reticuloendothelial system perturbations affecting, 295–296
Interferon, 148, 282–283
Intestinal bypass surgery, 124
Iron
  absorption of, 367
  carriers of, 353–359
    albumin, 353
    erythrocyte, 355–359
    haptoglobin, 353–355
    hemopexin, 353, 355
  ferritin and, 364–365
  ferrous, 204
  hemosiderin and, 365–366
  labile pool of, 360–364
  lactoferrin and, 364
  metabolism of, 353–372
    heme oxidation and, 204–207
Iron (*cont.*)
  plasma turnover of, 366
  processing and release of, 360–366
    factors decreasing, 367–371
    factors increasing, 371–372
  stable pool of, 364–366
Iron deficiency anemia, 203–204
Isoferritins, 365

Kallikrein, 224
Kidney
  diseases of, erythropoiesis affected by, 175–177
  erythropoietin of, 162, 163
  hepatic erythropoietic factor affected by, 174
Kininogen, 221
Kupffer cells
  in colony-stimulating factor production, 148
  in enteric endotoxin processing, 127–128
  erythropoietin from, 148, 171–172
  in extramedullary erythropoiesis, 150, 153
  glucocorticoid-antagonizing factor from, 296
  lead and cadmium effects on, 392
  liposomes of, 313
  lymphocyte interactions with, 268
  numbers of, affecting vascular clearance of particles, 90–91
  phenylhydrazine effect on, 163
  plasma protein synthesis in, 332
  in prostaglandin production, 148
  in quantitation of macrophage phagocytosis, 60
  receptors on surface of, 109
  in thrombin clearance, 224

Lactoferrin, 364
Langerhans cell, 270
Large external transformation-sensitive protein, 17
Latex beads, 62
Lead acetate, 118
Lead and cadmium, 379–396
  in cellular immunity, 388–389
  in host defense mechanisms, 379–391
  in humoral immunity, 387–388
  in lymphocyte blast cell transformation, 388
  metabolic effects of, on phagocytes, 385–387
  phagocytic capacity affected by, 381–385
    with oral administration, 385
    with parenteral administration, 382–385
    with respiratory exposure, 381–382
    in vitro studies of, 381
  relative toxicity of, on phagocytes, 380–381
  in susceptibility to infections, 389–391
    bacterial, 389–390

Lead and cadmium (*cont.*)
in susceptibility to infections (*cont.*)
parasitic, 390–391
viral, 390
toxic interactions of, with endotoxin, 391–396
Lectins, 22
Leukocytes, polymorphonuclear
agents affecting surface of, 22
electrokinetic surface potential of, 23
leukocytic endogenous mediator of, 296
phagocytic ingestion of bacteria by
complement and IgM in, 17
glucose effect on, 24
IgA in, 18
IgG in, 14, 15
immune complexes in, 16
media effect on, 8–10
opsonization in, 14–185
particle adhesion in, 4, 5
quantitation of phagocytosis by, *in vitro*, 57, 58
reticuloendothelial system interaction with, 267–283
lymphoid, 267–270, 271–280
nonlymphoid, 270–271
in vascular clearance of bacteria, 105
Leukocytic endogenous mediator
as glucoregulatory monokine, 296–297
lymphocyte activation factor and, 332
Leukocytosis, 128
Leukotrienes, 256
Levamisole, 22
Ligandin, 210
Limulus amebocyte lysate test, 117
Lipase, acid, 311
Lipid metabolism
free macrophages in, 306–309
tissue macrophages in, 309–325
Lipid test emulsion, 309–310
Lipopolysaccharides, 115
affinity of, for biological membranes, 119
appearance of, microscopic, 119
immune response to, 133
Lipoproteins
chylomicron, 339
free macrophages in metabolism of, 306–309
high-density, 340
low-density, 339–340
tissue macrophages in metabolism of, 309–325
types of, 339
Liposomes
as enzyme or drug carriers, 312
location of, 313
Liposomes (*cont.*)
reticuloendothelial cell clearance of, 306, 312–314
Lipoxygenase, 256
Listeria monocytogenes
antibiotics affecting phagocytosis of, 23
opsonizing role of IgG with, 14, 15
phagocytic ingestion of, 9, 10
protein adsorption onto, 12, 13
Liver
in antitrypsin synthesis, 333
in bilirubin metabolism, 210
in cholesterol clearance, 318
in erythroclasia, 195, 197–198
erythrophagocytosis in, 148
erythropoietin of, 162, 163
heme oxygenase of, 204
plasma protein synthesis in, 331–332
regeneration of, 160, 172–175
reticular cells of, 149–160
Lymphocytes
B, *see* B lymphocytes
electrokinetic surface potential of, 23
lead and cadmium effects on, 388
macrophages in transformation of, 160
protein metabolism in relation to, 331–344
in recognition of bacteria in vivo, 30
in reticuloendothelial cell activation, 280–283
in vitro, 281–283
in vivo, 280–281
reticuloendothelial cell interaction with, 267–270
T, *see* T lymphocytes
Lymphokines, 282
Lysostaphin, 6

α_2-Macroglobulin, 224, 334–336
Macrophage growth factor
macrophage consumption of, 47–48
in mononuclear phagocyte growth regulation, 51
for murine MNP colony formation, 38
Macrophage insulin-like activity, 297
Macrophage insulin-releasing activity, 297
Macrophages
activation of, 21–22
alveolar, 59, 83, 148, 380–381
in atherosclerosis, 325–326
cell suspensions of, 60
in cholesterol and triglyceride clearance, 307
endotoxin-induced activation of, 126
in erythropoiesis, 148, 160–162
free, lipids interacting with, 306–309
blood lipids, 307
lipoproteins, 307–309

Macrophages (*cont.*)
liposomes, 306
test substances, 306
glucose effect on, 24
heparin effect on, 22
immune complex attachment by, 16
in immune response to endotoxin, 134
inhibition of proliferation of, 50–51
Kupffer cells, 60
lead and cadmium effects on, 381, 387
lectins affecting surface of, 22
lipoprotein recognition site on, 307–308
in liposomal clearance, 306
lymphocyte interaction with, 267–268, 281–283
in lymphocyte transformation, 160
macrophage growth factor consumption by, 47–48
monocytes, 58–59
monolayer cultures of, 61
in myeloid differentiation, 160
peritoneal, 43, 59, 83
platelet interactions with, 254–257
immune, 254–255
leukotrienes in, 256
platelet-activating factor in, 256–257
prostaglandins in, 255–256
prostaglandin effect on, 161, 256
quantitation of phagocytosis *in vitro* by, 57–68
morphological and bacterial counting methods for, 64–66
radioactive and chemical methods for, 66–67
sources of macrophages for, 58–61
test particles used in, 61–64
receptors for IgG on, 17
sources of populations of, 58–61
surface markers for differentiating, 45–46
tissue, lipids interacting with, 309–325
blood lipids, 314–315
lipoproteins, 315–325
liposomes, 312–314
test substances, 309–312
Mechloretha-mine, 149
Media
for murine MNP colony formation, 38
opsonization affected by, 14, 15, 18, 19
particle adhesion and engulfment affected by, 7–10
protein adsorption affected by, 10–13
thioglycollate, 40
Medullary erythropoiesis, 148–149
Mesenteric artery occlusion, 124
Mnalmitate, 393
Methylprednisolone, 393
Migration inhibition factor, 21, 282
Monocytes
in colony-stimulating factor production, 148
leukocyte interactions with, 271
in prostaglandin production, 148
in quantitation of macrophage phagocytosis, 58–59
as stem cells, 166
Monokines, glucoregulatory, 296–297
glucocorticoid-antagonizing factor, 296–297
leukocyte endogenous mediator, 296
macrophage insulin-like activity, 297
macrophage insulin-releasing activity, 297
Mononuclear phagocyte system, 147
Mononuclear phagocytes
bone marrow, 43–45
colony formation by, 38–41
appearance of, 39, 40
genetics of, 51–52
erythrocyte effect on growth of, 49–50
functions of, 37
inhibition of proliferation of, 50–51
local proliferation of, 52–53
peritoneal, 40–43
regulation of proliferation of, 37–53
surface markers for differentiating, 45–46
Mumps virus, 104

Neuraminidase, 245
Neutrophils, polymorphonuclear, 271

Opsonins
definition of, 91
depletion of, 93–94
endothelial, 106
vascular clearance rates affected by, 91
Opsonization
complement in, 17
fibronectin in, 17
IgA in, 17–18
IgG in, specific and aspecific, 14–15
IgM in, 17
immune complex size in, 15–17
in immune response to endotoxin, 132
in phagocytic adhesion and engulfment, 7–10
protein adsorption in phagocytosis and, 10
in vivo, 18

Paraffin oil, 61, 62, 306
Parasitic infections, 390–391
Particle adhesion
in liquid media of lower surfacetension than saline water, 7–10
thermodynamics of, in phagocytosis, 4–6

Particle engulfment
  in liquid media of lower surfacetension than saline water, 7–10
  thermodynamics of, in phagocytosis, 6–7
Peritoneal exudate cells
  bone marrow cells vs., 40
  genetics of colony formation by, 50
  macrophage growth factor and, 48
  proliferative capacity of, 41–43
  serum dialyzable activity regulating growth of, 48–49
Peyer's patches, 121
Phagocyte
  activation of, 21–22
  affinity of particles for surface of, 81–84
  agents affecting surface of, 22, 23
  cell shape of, for phagocytosis, 23, 24
  contact angles of, 26
  detachment of IgG from, 14
  electrokinetic surface potential of, 23
  in immune response to endotoxin, 134–135
  kinetics of vascular clearance by, 73–97
  lead and cadmium effects on, 380–387
    metabolic, 385–387
    with oral administration, 385
    with parenteral administration, 382–385
    in phagocytic capacity, 381–385
    relative toxicity, 380–381
    with respiratory exposure, 381–382
  mononuclear, *see* Mononuclear phagocytes
  pathological, 18–21
  particle adhesion to, thermodynamics of, 4
  particle engulfment by, 6–7
    opsonization affecting, 7–10
  in recognition of bacteria *in vivo*, 30
  regulation of proliferation of, 37–53
Phagocytosis
  assay conditions for, in vitro, 60–61
  definition of, 57
  of erythrocytes, 147–148
  in immune response to endotoxin, 134
  kinetics of, 73–97
    association constant in, 81–84
    with competitive inhibition between particles, 94–96
    with delayed-onset RES paralysis, 93
    host factors modulating, 88–91
    with immediate-onset RES paralysis, 92–93
    with inhibition by particle injection, 91–96
    Kupffer cell number affecting, 90–91
    liver blood flow effect on, 88–89
    maximum phagocytic velocity in, 78–80
    with opsonin depletion, 93–94
    opsonins affecting, 91
    particle membrane constant in, 78–80
Phagocytosis (*cont.*)
  kinetics of (*cont.*)
    particle selection for studies on, 84–87
    species effect on, 89–90
  lead and cadmium effects on, 386
  metabolic concomitants of, 67
  by platelets, 250–252
  of platelets, 239–246
    in hypersplenic and thrombocytopenic states, 239–243
    in normal and thrombotic states, 243–246
  quantitation of in vitro, by macrophages, 57–68
    morphological and bacterial counting methods for, 64–66
    radioactive and chemical methods for, 66–67
    sources of macrophages for, 58–61
    test particles for, 61–64
  surface forces in, 3–30
    adhesion and adsorption methods for measurement of, 26–27
    agents affecting, 22–23
    in biological fluids, 29–30
    cell shape affecting, 23–24
    contact angle methods for measurement of, 25–26
    droplet sedimentation method for measurement of, 28
    electrokinetic surface potential affecting, 23–24
    with macrophage and phagocyte activation, 21–22
    measurement of, in liquids of different surface tensions, 28
    opsonization, 14–18
    particle adhesion, 4–6, 7–10
    particle engulfment, 6–10
    partition methods for measurement of, 24
    with pathological phagocytes, 18–21
    protein adsorption, 10–14
    solidification front method for measurement of, 27–28
    in vitro in aqueous media, 28–29
    in vivo, with recognition, 30
Phenobarbital, 209
Phenylhydrazine, 149, 157, 163
Phospholipase A, 306
Phytohemagglutinin, 22
Plasma proteins
  degradation of, 332
  in host defenses, 331
  lymphocyte, fibroblast, and RES interaction with, 342–343
  synthesis of, 331–332

Plasmin
  action of, 230
  in fibrinolysis, 229
  macroglobulin effect on, 224
  in platelet clearance, 245–246, 250
Plasminogen, 229, 230
Plasminogen activator, 229, 230
Platelet-activating factor, 256–257
Platelets
  activators of, RES uptake of, 253–254
  aging of, 247–248
  anatomy and physiology of, 238–239
  cell shape of, in cellular contact, 24
  as coagulation activator, 223
  conglutination or congregation function of, 252
  EDTA, 243–244
  electrokinetic surface potential of, 23
  endotoxin effect on, 130
  in hemostasis, 221
  in lead-potentiated endotoxic shock, 394
  macrophage interactions with, 254–257
    immune, 254–255
    leukotrienes in, 256
    prostaglandins in, 255–256
  particle interaction with, 250–253
    in colloid clearance 252–253
    in phagocytosis, 250–252
  phagocytosis of, 239–243
    in hypersplenic and thrombocytopenic states, 239–243
    in normal and thrombotic states, 243–246
  reticuloendothelial system interaction with, 237
  reticuloendothelial system recognition of, 246–250
    immune, 246–247
    senescent, 247–248
    during thrombosis, 248–250
Pneumococci, 106
Polymyxin B, 135–136
Polystyrene latex, 62, 250–251
Porphobilinogen, 200
Porphyrin biosynthesis, 199
Potassium of aging erythrocyte, 192
Prekallikrein, 221
Procoagulants, 224–225, 228–229
Pronase cetrimide procedure, 41
Properdin, 107
Prostacyclin, 238
Prostaglandins
  cells producing, 148
  erythropoietic effect of, 160
  erythropoietin and, 160–161
  hematopoietic effects of, 161
Prostaglandins (*cont.*)
  heme degradation affected by, 208
  in hepatic regeneration, 173
  in lead-potentiated endotoxic shock, 393–394
  macrophages affected by, 161
  in mononuclear phagocyte inhibition, 50
  in platelet-macrophage interactions, 255–256
Protein
  adsorption of, in phagocytosis, 10–13
    and opsonization, 10
    thermodynamics of, 10–13
  α_1 antitrypsin, 332–334
  c-reactive, 336–338
  cadmium-binding, 386
  cell surface, 17
  contact angles of, 26
  fibroblast function and, 331–344
  fibronectin, 341–342
  galactoprotein-a, 17
  haptoglobin, 353–355
  in hemoglobin and heme clearance, 353–355
  hemopexin, 353, 355
  large external transformation-sensitive, 17
  lipoproteins, 339–341
  lymphocyte interaction with, 331–344
  α_2 macroglobulin, 334–336
  macrophage metabolism affected by, 60
  microfibrillar, 17
  opsonic, 17
  plasma, 331–332
  reticuloendothelial system interaction with, 331–344
  transferrin, 338–339
Prothrombin, 221
Protoporphyrin, 200, 204
Pseudomonas aeruginosa, 106
Pyelonephritis, 105

Radiogold, 84, 85
Red blood cell, *see* Erythrocytes
Renal inhibitory factor, 175–177
Respiratory distress syndrome, acute, 227
Reticular cells
  of bone marrow, 148–149
  of liver and spleen, 149–160
Reticulocytes
  membrane changes in, with aging, 192
  metabolism of, 190
Reticuloendothelial system
  blood lipid and lipoprotein metabolism interaction with, 305–326
  in coagulation system activator clearance, 222–223
  coagulation system effect on, 231
  in endotoxemia, 115–136

Reticuloendothelial system (*cont.*)
in erythroclasia, 194
in erythropoiesis, 147–178
erythropoietin and, 162–175
in erythropoiesis, 162–172
in hepatic regeneration, 172–175
in fibrinogen derivative clearance, 225–228
fungal clearance by, 106
glucan stimulation of, 157
in glucose regulation, 291–300
endocytic and exocytic functions, 298–299
in endotoxicosis, 292–294
glucocorticoid-antagonizing factor, 296–297
during hypoglycemia, 298
macrophage insulin-like activity, 297
macrophage insulin-releasing activity, 297
monokines, 296–297
after perturbations, 294–296
in hematopoiesis, 160–178
hormonal effects on, 177–178
in iron metabolism, 353–372
iron carrier uptake, 353–359
iron processing and release, 360–372
methods of study of, 366–367
lead and cadmium effects on, 379–396
leukocyte interactions with, 267–284
lymphoid, 267–270, 271–278
nonlymphoid, 270–271
lymphocyte activation of, 280–283
in vitro, 281–283
in vivo, 280–281
lymphocyte interactions with, 267–270
B, 279–280
T, 271–279
paralysis of, 92–93
in platelet activator uptake, 253–254
in platelet recognition, 246–250
immune, 246–247
sialic acid and, 247–248
during thrombosis, 248–250
in platelet regulation, 237
protein metabolism in relation to, 331–344
viral clearance by, 104, 105
Reticulohistiocytic system, 147
Rhopheocytosis, 148

Salmonella typhimurium, 107, 108
Serotonin, 129, 238
Serum
in assay conditions for phagocytosis in vitro, 60–61
in endotoxin inactivation, 135
mononuclear phagocyte colony formation affected by, 49
pathological phagocytes affected by, 18, 20
protein adsorption in, 11, 13, 18
Serum dialyzable activity
for mononuclear phagocyte growth regulation, 51
peritoneal exudate cell growth regulation by, 48–49
Shock, 124
fibronectin blood levels in, 227
lead-potentiated endotoxic, 391–396
Schwartzman reaction, 115, 129, 130
Sialic acid
erythrocyte, 247–248
erythrocyte, 358
Sideroblastic anemias, 205, 209
Sinusoidal cells, hepatic, 109, 110
Sodium of aging erythrocyte, 192
Spermine, 163
Spherocytes, 193
Spleen
in erythroclasia, 194, 195–196
erythrophagocytosis in, 147–148
mononuclear phagocytes from, 38
reticular cells of, 149–160
Staphylococcus aureus
opsonizing role of IgG with, 14, 15, 19
phagocytic ingestion of, 9
protein adsorption onto, 12, 13
Staphylococcus epidermidis
opsonizing role of IgG with, 14, 15, 18, 19
phagocytic engulfment of, 7, 8, 9, 10
protein adsorption onto, 12, 13
vascular clearance of, 105
Streptococcus faecalis, 105
Streptococcus viridans, 105
Surfactants, 23

T lymphocytes
helper, 268, 273
in immune response to endotoxin, 132
lead and cadmium effects on, 387
reticuloendothelial cell interaction with, 271–279
dendritic cells, 273–275
in development of T repertoire, 275–276
mechanism of, 276–279
proliferative responses, 272–273
in tissue factor increases, 228–229
Testosterone, 177, 372
Thioglycollate medium, 40–41
Thrombi and thromboemboli, 238
Thrombin
action of, 221–222
clearance of, 224
macroglobulin effect on, 224
in platelet activation, 253
reticuloendothelial system regulation of, 253
Thrombocytopenia, 239–243, 246

Thromboplastin, blood, 222
as coagulation activator, 223
Thromboplastin, tissue, 223
Thromboxane A2, 238, 393
Tissue factor, 228–229
Togaviruses, 104
Transferrin, 338–339
Triglycerides
dietary, clearance of, 314
injected, clearance of, 315
macrophage clearance of, free, 307

Urobilinogen, 199, 211
Uroporphyrinogen, 200

Vascular clearance
of microorganisms, 103–111
bacteria, 104–106
fungi, 106
future directions on, 109–110
liver perfusion studies of, 106–109
viruses, 103–104
in vivo studies of, 103–106
of particles, kinetics of, 73–97
affinity of particles for pagocyte surface in, 81–84
with competitive inhibition between particles, 94–96
with delayed-onset RES paralysis, 93
Vascular clearance (*cont.*)
of particles, kinetics of (*cont.*)
host factors modulating, 88–91
with immediate-onset RES paralysis, 92–93
with inhibition by particle injection, 91–96
Kupffer cell number effect on, 90–91
liver blood flow effect on, 88–89
maximum phagocytic velocity in, 78–80
monodispersity in, 87
with opsonin depletion, 93–94
opsonins affecting, 91
particle charge in, 86
particle membrane constant in, 78–80
particle stability in, 84–86
species effect on, 89–90
surface-saturation, 73
zero- or first-order rate equations in, 73
Venezualian equine encephalitis viruses, 104
Vincristin, 149
Viruses
cell shape of, in cellular contact, 24
IgA in prevention of cellular penetration by, 18
lead and cadmium effects on infection from, 390
vascular clearance of, 103–104

Yeast, 106

Zeta, 17